MHD ENERGY CONVERSION
PHYSIOTECHNICAL PROBLEMS

Edited by V.A. Kirillin and A.E. Sheyndlin

Volume 101
PROGRESS IN
ASTRONAUTICS AND AERONAUTICS

Martin Summerfield, Series Editor-in-Chief
Princeton Combustion Research Laboratories, Inc.
Monmouth Junction, New Jersey

Published by the American Institute of Aeronautics and Astronautics, Inc.
1633 Broadway, New York, N.Y. 10019

American Institute of Aeronautics and Astronautics, Inc.

Library of Congress Cataloging in Publication Data

Magnito-girdrodinamicheskoe preobrazovanie energii.
English
MHD energy conversion
physiotechnical problems

(Progress in astronautics and aeronautics; v. 101)
Translation of: Magnito-girdrodinamicheskoe preobrazovanie energii.
Includes index.
1. Magnetohydrodynamic generators. I. Kirillin, V. A. (Vladimir Alekseevich) II. Sheyndlin, Aleksandr Efimovich. III. Title. IV. Series.
TL507.P75 vol. 101 629.1 s [621.31'245] 86-10787
[TK2970]
ISBN 0-930403-05-3

Table of Contents

Chapter 1 Engineering Technical Problems of Developing MHD Power Plants

Chapter 2 Calculation of MHD Channel Flows

Chapter 3 Characteristics of a Nonideal MHD Generator

Chapter 4 Plasma Diagnostics in MHD Generators

Chapter 5 Phenomena in the Near-Electrode Region of a Constricted Discharge

Foreword

Magnetohydrodynamics (MHD) – particularly the technology of converting the thermal energy in a hot flowing gas stream into electrical energy by means of an electromagnetic field – has been the subject of research and development for many decades. Despite the scientific attention that has been given over the years to the processes of combustion, fluid mechanics, heat transfer and cooling, ionization of the hot (sometimes seeded) gases, etc., the problems have only been partially solved. As a result, if a practical operating plant were built today, the efficiency of conversion would be less than competitive with other forms of energy conversion and the materials problems would still tend to limit the operating lifetime between servicings.

In the U.S., funding authorities have hesitated to apply the still imperfect design technologies to the development and construction of a costly experimental operating plant, where the real-world problems would be confronted and, hopefully, might be solved. But such caution may be too conservative and may be tending to slow the rate of progress. Meanwhile, in the Soviet Union a bolder decision has been made, and a commercial MHD plant is now in construction, one that will generate 500 Mwe and 1,000 Mwt, the so-called U-500. This plant is to be commissioned in two phases: the downstream portion, a conventional gas-fired steam plant, is to become operational first, in the late 1980s, and the MHD topping plant is to be added several years later. Research scientists and development engineers around the world have much to learn from a detailed description of the technical data underlying the design of the plant and its intricate processes, described for Western readers comprehensively for the first time in the present volume.

The author-editors are authorities in the field. Academician A.E. Scheyndlin, a key figure in the MHD program of the Soviet Union, currently heads the USSR's Institute for High Temperature, where the MHD program is centered. Academician V.A. Kirillin is the former chairman of the State Committee for Science and Technology; in this capacity he fostered and backed the MHD development work that led to the U-500 plant. Together, they have produced a book that provides a broad insight into the state of development of MHD power in the Soviet Union and a detailed analysis of the data base of the U-500 system. The book was first published in the Soviet Union. A decision was made by the U.S. Dept. of Energy (DOE) to have it translated into English and published in the U.S. in order to make the subject more widely available to Western specialists.

The Russian text was translated under DOE auspices by Charles Shiskevich, an expert in the Soviet MHD literature. We, the American editors, caution the reader, however, that *MHD Energy Conversion* has been edited in only a limited sense. The technical accuracy of the Russian text could not be verified. Inconsistencies that prevailed in the text could not be resolved for the translation. Although an attempt has been made to minimize the identified inconsistencies, some were allowed to stand as they were in the original text, and some may not even have been identified. Translation problems notwithstanding, *MHD Energy Conversion* is a book that will surely provide English-speaking readers with valuable information on the USSR program in this important technology area.

Our gratitude is due to Michael Petrick, Director, Fossil Energy Programs, Argonne National Laboratory, in whose division the decision was made to make the book available in English, and to Betty Cheever, of that same division, for her careful work in preparing the translation for printing and for her diligence in spotting many inconsistencies and resolving them. We wish to add our appreciation as well for the supportive efforts of AIAA's editorial staff.

Martin Summerfield
Editor-in-Chief
Progress in Astronautics and Aeronautics
and
President, Princeton Combustion Research Laboratories, Inc.
June 1986

Introduction

The present rapid growth of energy production and demand has been worldwide in scope. Analysis of energy production trends shows that, when planning the required growth of electric power production utilizing natural deposits of organic fuels, it is necessary to take into account the increase of energy costs due to depletion of available energy resources and increasing costs of mining and exploration. The most important aspects of present-day power engineering involve development of new types of energy (nuclear, solar, geothermal, thermonuclear, etc.) as well as attempts to increase the efficiency of using energy resources.

A promising method of solving the second problem is the magnetohydrodynamic (MHD) method of energy conversion, in which both internal and kinetic energy of combustion products are converted directly into electric power at a considerably higher initial temperature. Initial experimental and pilot facilities that have clearly demonstrated the feasibility of the MHD energy conversion method for electric power generation have been developed in both the Soviet Union and in other countries.

At the present time, work on the development of MHD energy conversion has reached the stage where it is feasible to design and construct high-power, commercial-scale electric power plants with MHD generators. The gas at the outlet from the MHD channel in such electric power plants is still at a high temperature and is, therefore, fed into the steam generator. Thus, thermal energy conversion into electric power is a two-stage process, first in an MHD channel (direct conversion) and then in a conventional thermal power plant. The efficiency may reach 50-60%, i.e., it may be considerably higher than in conventional thermal power plants (30-40%).

In addition to high efficiency, MHD energy conversion is characterized by other important advantages, for example, a sharp decrease in the required amount of cooling water; considerable decrease in the harmful effects on the environment, due to a decrease in heat and toxic oxide exhausts; feasibility of generating high power in a single plant, and variable-load operation throughout broad ranges of loads. These advantages demonstrate the desirability of utilizing this form of energy conversion.

Principal technoeconomic justifications of the possibility and feasibility of constructing magnetohydrodynamic power plants have been demonstrated. Nevertheless, a number of physiotechnical problems of MHD energy conversion methods are of considerable interest, and their investigation is very important from the viewpoint of improving MHD facilities.

This monograph, written by a team of authors at the Institute of High Temperatures of the Soviet Academy of Sciences, summarizes the results of such investigations, representing many years of scientific research in the area under consideration. The book consists of five chapters devoted both to general analysis of characteristics of an MHD generator and to various aspects of magnetohydrodynamic, plasma, and near wall phenomena occurring in an MHD channel. The material in each chapter is so arranged as to be independent of the other chapters.

Considerable attention is devoted to analysis of the results of investigations conducted on the large operating U-02 and U-25 MHD facilities. The principal results obtained on the U-25B facility with a superconducting magnet system, forming a bypass loop of the U-25 facility, is described. The bypass loop was designed and constructed in accordance with a program of U.S.A.-U.S.S.R. scientific-technical cooperation in research in the area of MHD energy conversion.

The problems associated with the design and construction of a first commercial-scale system of an MHD power plant are considered. The analysis performed of thermal loops of an MHD power plant demonstrates techniques of increasing their thermal efficiency considerably in comparison with conventional thermal power plants. Regenerative heating of the oxidizer, thermochemical processing of fuel, amd various methods of utilizing low-potential heat in an MHD facility have been investigated. A typical energy balance of a commerical-scale MHD facility is analyzed. The major requirements imposed on such important systems and individual types of equipment of an MHD power plant: combustion chambers, MHD generator channels, magnetic systems, steam generators, electrotechnical equipment, etc., are discussed and formulated. These requirements are formulated on the basis of theoretical investigations, as well as the experience gained in operating MHD facilities.

At the present stage of development of MHD energy conversion, modeling problems of flows in channels, and mathematical simulation aimed at optimizing the design and operating modes of an MHD generator become of primary importance in the area of theoretical analysis. Channel design and loading schemes require solution of spatial electrodynamic problems. This book describes certain new results obtained during investigation of this class of problems. Various problems of electrodynamics have been formulated and solutions have been obtained that provide data on the optimal loading and optimal design of diagonal channels. A number of axial electrodynamic problems, including nonuniform loading of an operating section of a segmented MHD channel, have been solved. Analysis of the latter aspect made it possible to indicate the most radical method of regulating a multicomponent load. A method of analyzing characteristics of frame channels with a nonuniform distribution of flow parameters over the cross section has been developed, and calculations that led to a determination of the recognized advantages of frame channels over a Faraday channel have been performed. Numerical investigations of supersonic and subsonic, inviscid, two-dimensional flows are of considerable interest. Characteristics and the influence of two-dimensional fields of gasdynamic variables in MHD channels on the output characteristics of various types of MHD generators have been analyzed. A comparison of two-dimensional calculations with quasi-one-dimensional calculations has been made.

An important practical application is the development of engineering methods of calculating flows in MHD channels. This monograph discusses a method of comparing the results of calculations with the experimental data acquired on the U-25 facility, and the improvement of electrical elements of the hydraulic flow model on the basis of such a comparison. Problems associated with the influence of nonideal behavior of an MHD channel on local characteristics of an MHD generator have been investigated. Nonideal behavior refers to all the phenomena neglected in idealized flow models, characteristics of real MHD channels that depend on specific features of design and construction. These effects include nonuniformity of the flow of parameter distribution over the channel cross section (primarily nonuniformity of the distribution of electrical conductivity of plasma in the boundary layers at electrode and insulation walls of the channel), nonideal electrical isolation properties of the channel structure, nonideal electrode segmentation, near-electrode potential drop, current losses from the plasma flow to the side (insulation) channel walls. The influence of these factors, separately and jointly, were considered. Analytical results are compared with the experimental data throughout a broad range of operating conditions of an MHD channel.

Results of these investigations were used to develop a physical model of electrodynamic processes in a nonideal MHD channel, and an engineering method of calculating local characteristics of a nonideal MHD generator on the basis of that model, taking into account all the aforementioned effects. Although relatively simple, this methodology provides adequate accuracy for engineering calculations, and can be recommended for cases that require a great deal of computation, e.g., optimizing local conditions. The method suggested was used to estimate possible influence of the nonideal behavior of an MHD channel on the external characteristics and the internal relative efficiency. Such an analysis makes it possible to formulate technical requirements for the design of an MHD channel.

Successful MHD energy conversion depends considerably on the characteristics of the working fluid. The working fluid of an equilibrium MHD generator is adequately characterized by the following parameters: gas temperature, concentration of alkali seed atoms, electron concentration, and collision rate of electrons with neighboring particles. These must be measured because the electrical conductivity of a working fluid and, consequently, also the power output of an MHD generator are dependent upon these parameters. Knowledge of each of them makes it possible to determine and control optimum operating modes of important subsystems and components of an MHD generator facility.

In the book, considerable attention is devoted to the methods of diagnosing (primarily spectroscopic, laser and probe) the working fluids of an MHD generator.

Spectroscopic methods make it possible to measure temperature of the gas and atom concentration of seed. An electric probe is used to determine local temperature of the gas. Submillimeter interferometry is designed to determine the concentration and collision rate of electrons.

Finally, one should note the problem associated with phenomena in a constricted discharge. Such discharge, named by analogy with the classical "arc" discharge, always occurs on electrodes of an MHD generator channel in a large-scale facility. The principal problems that have to be solved in the analysis of such discharges involve determination of conditions under which their occurrence will cause minimal destruction of the electrode systems of the channel.

Complexity of this problem in both experiments and theory can be attributed to the multitude of processes that occur in a fairly narrow near-electrode zone.

The monograph deals in detail with the available experimental and theoretical data. It is shown that the burning characteristics of an arc discharge typical of an MHD generator can be analyzed on a basis of the mechanisms for arc burning in vapors of the electrode material. Theoretical models are suggested that make it possible to explain both the behavior of discharges of this type and the erosion mechanism of metal electrodes.

Thus, the book describes large amounts of data dealing with important problems of the MHD method of energy conversion. The multi-sided review of problems of modern power engineering makes the book useful for both researchers investigating the subject, and for specialists working in associated areas of science and technology. The monograph can be useful to students in these specialties.

In preparing this book, an especially important contribution was made by laureate of the state prize, an honored member of science and technology, doctor of technical sciences B. Ya. Shumyatskiy.

The work was divided in the following manner. Chapter 1 was written by G. M. Karyagin, G. N. Morozov, A. E. Sheyndlin, B. Ya. Shumyatskiy; Chapter 2 by V. A. Bityurin, G. A. Lyubimov, S. A. Medin; Chapter 3 by V. V. Kirillov; Chapter 4, by I. A. Vasil'eva; and Chapter 5, by I. I. Beylis.

Chapter 1

Engineering Technical Problems of Developing MHD Power Plants

1.1. Introduction

At the present time, constantly increasing attention is devoted to both purely theoretical aspects of MHD energy conversion and to technoeconomic and engineering problems that, to a considerable extent, will be the determining factor in whether or not MHD power plants find practical application in industry. Reference [1] to some extent generalizes the engineering aspects of the MHD energy conversion, and complements and extends this monograph in regard to engineering applications.

Technical progress in all branches of industry is closely related to development of power generation. This explains the extremely rapid increase in power consumption in all industrially developed countries during the last ten years. The high demand leads to numerous problems associated with depletion of the cheapest and most accessible energy sources. All of this is responsible for the attention concentrated on all aspects of energy problems during the past two years. The efforts of specialists in different branches of science and technology are directed both toward a rational use of energy and to finding means of increasing the efficiency of all aspects of power generation, first of all, electrical power generation, which is the most widely used and convenient source of energy.

Modern thermal power stations generate approximately 90 % of all electrical power. For more than 80 years, such stations have incorporated electric generators driven by steam turbines. Considerable success in increasing technoeconomic indices of thermal power plants has not been achieved as a result of major changes in the present method of generating electrical power, but by improving the construction of electrical power machinery. Today, conventional thermal power stations have practically reached the limit of their thermal efficiency (mean annual efficiency of present-day stations reaches 37-39 %). Further increase in efficiency requires increasing the initial temperature of the working fluid. However, this is hindered by limitations imposed by the strength and cost of steam generator and steam turbine materials. The efficiency of present-day nuclear power plants with steam turbine generators is considerably lower.

According to the second law of thermodynamics, further increase in efficiency of the heat cycle requires new techniques of generating electric power with considerably higher initial temperature of the working fluid.

High efficiency of power generating facilities with a single working fluid requires high pressure and high temperature in the same unit. This can be avoided in facilities with two working fluids (binary cycles). Naturally, this simplifies solution of a whole number of technical problems associated with development of highly economical power generating facilities. For example, incorporating a turbine operating on potassium vapor in the topping (high-temperature) stage of the binary cycle will result in the pressure not exceeding 0.5 MPa. However, to attain adequate efficiency, initial temperature of potassium vapors has to be about 1100-1300^{0}C. Relatively low pressure simplifies development of a steam generator operating at such a high temperature. However, the problem of choosing the metal for heat exchangers in the potassium cycle cannot be considered solved, from the point of view of either durability or cost.

The use of a thermionic converter (TEC) jointly with the steam generating cycle, in the temperature range of 1500-500^{0}C as a high temperature addition, is being discussed. In this case, the efficiency of the binary facility has been estimated to be 48-50 %. However, at present, the available experimental data on TEC were acquired under the conditions differing substantially from those in an electrical power plant burning organic fuel. Therefore, a prediction of the future use of thermionic converters in electric power generation is difficult.

The problem is much easier to solve when the high-temperature stage of the binary cycle operates directly on the combustion products of fuel. Then, depending on the thermal loop and fuel, either a high-temperature heat exchanger is not required, or it is sufficient to use a heat exchanger with a heat-retaining ceramic lining, such as blast furnace-type cowpers, to preheat the oxidizer to 1400-1500^{0}C (MHD facility). It is exactly for this reason that the Brayton-Rankine binary cycle attracts the greatest attention. Binary facilities with gas turbines and with an MHD generator operate on such a cycle.

Success achieved in developing cold vanes of gas turbines made it possible to develop gas turbines operating at an initial temperature of 1200°C with efficiency of approximately 33 %. The temperature of the exhaust gases of such a turbine is 600-650°C. Because fuel combustion in a gas turbine occurs at high degree of oxidizer excess (α = 2.5-2.6) and because air is supplied to the combustion products to cool the blades, exhaust gases contain a fairly high concentration of oxygen (13-14 %).

There are two methods of using gas turbines as high-temperature additions to the steam generating cycle. In one, exhaust gases are routed into the steam generator and the oxygen they contain is used for combustion of additional fuel, approximately 50 % more than the amount that is consumed in the turbine's combustion chamber. In this case, the flue gases are at a temperature close to that in standard steam generators, and the heat of these gases may be used to produce steam with standard parameters (17-25 MPa, 540-545°C). Then, only 40 % of the heat of the fuel burned in the facility participates in the topping cycle and the remaining heat (approximately 60 %) is used to convert energy only in the bottoming, low-temperature Rankine cycle, decreasing the efficiency of the facility as a whole. The efficiency of binary facilities with such gas turbines may be approximately 44-45 % [2].

If no fuel is combusted in the steam generator, all of the heat of the working fluid will pass through the binary cycle. However, high oxidizer stoichiometry results in considerable heat losses to the exhaust gases (approximately 30 %). As a result of the low temperature of the combustion products leaving the gas turbine, it is necessary to use a steam turbine with fairly low steam parameters (9 MPa, 530°C). Then, the efficiency turns out to be approximately the same as in the first case.

If the MHD generator is used as the topping cycle, all heat passes through the high-temperature stage. The temperature of the exhaust gases from the MHD generator is 2000-2200°C, high enough for production of steam with high parameters. Therefore, the advantages of a binary cycle are completely utilized in MHD facilities.

1.2. Present Stage of Investigations and Future Developments

Electric generators convert mechanical energy into electrical energy by using the rotational motion of a solid conductor in a magnetic field. However, a fluid or gaseous conductor moving in a magnetic field can be used

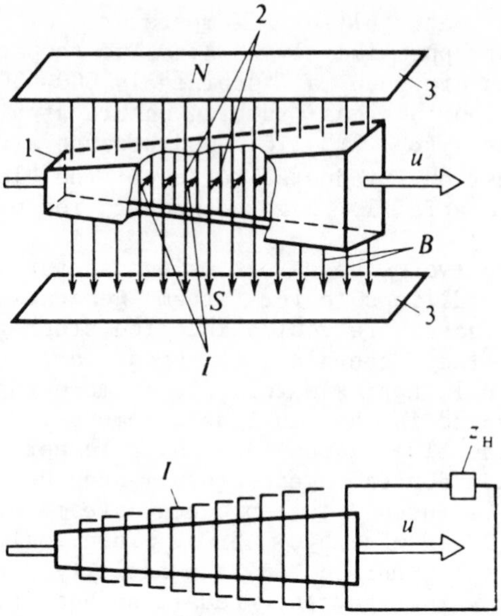

Fig. 1.1 Schematic diagram of a linear MHD generator, 1--channel, 2--electrodes, 3--magnetic system

for this purpose [3]. The MHD generator is based on the same principle (with the thermal-kinetic-electrical energy conversion cycle) in which the moving, electrically conducting fluid performs the same function as the rotor in the conventional electric generator (with thermal-kinetic-mechanical-electric cycle). In the simplest linear MHD generator (Fig. 1.1), the current induced by the emf flows between the channel walls, parallel to the magnetic field, confining the fluid conductor flow. Electrodes are located on these walls. If the circuit consisting of the electrodes and the working fluid also includes a load, z, a current will flow in the circuit in the same way as in a conventional electric generator. Part of the energy of the fluid (gas) can be converted directly into electric energy. The flow of fluid (flow rate u) performs work against the stagnation forces resulting from interaction of current, I, with the magnetic field induction, B.

An MHD generator performs the functions of two components of a conventional thermal power station: steam turbine and electric generator. An MHD generator, being in principle a simpler device than the conventional turbogenerator, consists of a channel with a set of electrodes, electrically insulating elements, and a magnetic system. The three possible variants of MHD energy converters are: open-cycle MHD generators with plasma of fossil fuel combustion products, closed-cycle MHD generators with nonequilibrium plasma of inert gas, and liquid-metal MHD generators. In this monograph, we will deal with the most promising variant for the near future, i.e., open-cycle MHD generators.

<u>1.2.1. Future Application of MHD Conversion in Power Generation</u>. One of the most promising arguments in favor of developing MHD energy conversion is the better protection of the biosphere offered by this method, which is expected to become more important in the near future. Another argument in its favor is the problem of fuel-energy resources in a country that depends not only on its own known geological resources, but also on the vagaries of the international market, economic interrelationships, etc. These nontechnical factors exert substantial influence on the development of the MHD energy conversion in each country. It is interesting to trace the approach to the development of MHD energy conversion in various countries during the past 15 years.

In Great Britain, serious investigations of the MHD energy conversion were conducted between 1959 and 1968. However, increased emphasis on nuclear energy production and a decrease in coal mining resulted in discontinuation of large-scale funding of MHD studies [4]. Other countries also discontinued their MHD energy conversion research, at somewhat later dates. MHD energy conversion research and development is being most actively pursued in U.S.S.R., U.S.A., and Japan, which support national MHD programs. It should also be noted that, beginning in 1976, MHD energy conversion research has been initiated in Bulgaria, Hungary, Poland, Rumania, and Czechoslovakia within the framework of a program for MHD energy conversion conducted by the Council for Economic Aid.

From the technical point of view, MHD systems are very attractive for numerous reasons. An MHD generator differs from a turbine by not being equipped with high-stress rotating parts operating at high temperatures and having components, such as turbine vanes, inaccessible to cooling. Therefore, it can operate at temperatures of the working fluid that are above maximum operating temperatures of turbines. Consequently, in accordance with the second law of thermodynamics, MHD power plants are more efficient, i.e., they operate at higher efficiency than conventional steam- or gas-turbine electric power plants.

It has been estimated that the efficiency of an MHD power plant will be 47-48 % during the first stage of its operation, saving 20 % of fuel over the present-day steam generating plants. Further increase of efficiency of MHD power plants, to 55-60 %, will result in a 30 to 33 % saving. No other known method of improving electric power generation from thermal energy can result in such a substantial increase in power plant efficiency.

An MHD power plant is a two-stage facility with an MHD generator operating in the first, high-temperature stage and another thermal unit, usually steam turbine, in the lower stage. Gradual improvement of thermal schemes is expected to lead to the following efficiencies (%) of energy systems with an MHD generator, using the indicated oxidizer preheating techniques:

Separate, up to 1700°C	47-49
Regenerative, up to 1700°C	50-52
with chemical regeneration. . . .	53-56
Regenerative, up to 2000°C (with chemical regeneration at B = 6-8 T).	57-58

An open-cycle MHD generator can operate in the temperature range of 2200 to 2700°C, inaccessible to any of the known thermal facilities. Therefore, one of the advantages of MHD energy conversion is that the MHD generator can, in principle, be added to any other thermal cycle. What's more, the more effective the lower stage of the combined MHD facility, the greater the thermal efficiency of an MHD power plant as a whole.

When estimating the future of an electric facility, the development of which requires restructuring various branches of industry and additional capital expenditures, it is necessary to take into account not only the direct economic effects gained as a result of higher efficiency, but also the operating lifetime of new facilities. In connection with this, it should be noted that an MHD generator can be used in nuclear power generation together with high-temperature gas reactors operating in inert gas (argon, helium) or uranium hexafluoride. Interesting proposals have been put forward, such as the use of MHD generators with thermonuclear facilities. Hence, MHD energy conversion appears to be a promising method for the future. The MHD generator differs from other thermal systems by very intense harmful surface phenomena (frictional losses, thermal losses, current losses to the walls, near-electrode losses). Consequently, as the power output and dimensions of an MHD generator increase, the ratio of losses with the total power output decreases and,

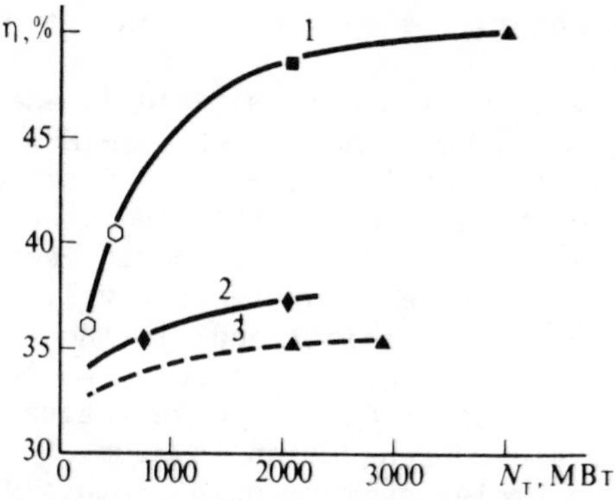

Fig. 1.2 Variation of thermal efficiency of steam-turbine and MHD power systems with thermal power output, 1--MHD power system, 2,3--steam power systems operating on coal

therefore, the efficiency of the MHD generator increases. Figure 1.2 shows a plot of efficiency of an MHD power plant, η, as a function of its power output N_T, based on the data of U.S. investigations [5].

An MHD generator is characterized by a high power output, estimated to reach 2 x 10^6 kW. This follows the overall trend of higher specific power output of individual units to satisfy the ever-increasing demand for electricity.

The relative simplicity of the MHD energy conversion method in comparison with the conventional steam turbine method was already mentioned. However, because MHD power generation is a new technology (new materials and new component and assemblies), the capital expenditures for construction of an MHD power plant will probably be higher, initially, than for construction of conventional thermal power plants. However, in principle, it appears that their cost will eventually approach that of conventional power plants.

As already noted, the ever-increasing concern for the environment, involving pollution of the biosphere and exhaustion of natural resources, also has revived interest in MHD facilities. Thermal pollution and the emission of harmful gases and particulates with flue gases per unit power output will decrease substantially at higher efficiency of the MHD power system. An especially large drop is expected in the amount of cooling water required. As the efficiency of a thermal power plant increases from 37 to 47 %, the cooling water rate decreases approximately 1.5 times. Increasing efficiency to 58 % results in a 2.25-time decrease of cooling water rate. It should be noted that the problem of cooling water is one of the most important problems encountered in electric power generation.

An extremely unpleasant effect is the exhaust of sulfur oxides into the atmosphere. In MHD power plants, potassium seed injected in the form of potassium carbonate, K_2CO_3, can completely bind the sulfur in all known varieties of oil and coal. During cooling, sulfur, being the most active substance, will combine with potassium and will be precipitated in the form of a solid sulphate, K_2SO_4. However, this requires regeneration of seed in the form of K_2CO_3. Nevertheless, it has been shown that a seed regeneration system is several times cheaper than a special sulfur oxide removal system. This resolves the problem of MHD power plants polluting the atmosphere with sulfur oxides. In MHD facilities, removal of nitrogen oxides from gases exhausted into the atmosphere is somewhat complicated by the fact that the combustion process takes place at fairly high temperatures. The most promising near-future solution is to reduce emission of nitrogen oxides by decomposition. This requires incomplete combustion of the fuel in the MHD generator channel, at a stoichiometric ratio of 0.85-0.95, accompanied by formation of strong reducing agents, hydrogen and carbon oxides.

At temperatures of 2270-1600 K, these gases will remove oxygen from NO_x compounds and will restore nitrogen. Incompletely oxidized fuel remaining in the combustion products after decomposition of nitrogen oxides has to be burned at a temperature of 1500-1600 K. Practical realization of reduction of nitrogen oxides obviously encounters certain difficulties. However, it has been shown that it is possible to attain nitrogen oxide exhaust levels that are even lower than those of conventional thermal power plants [6].

A second approach to solving the problem of nitrogen oxides is to use them for the production of nitric acid and nitrogen fertilizers. This will reduce substantially the cost of electric power generation given in the foregoing discussion.

The following variants were considered [7] in comparing the cost of electric power generation by MHD power plants operating on coal with sulfur content 3.2 %, equipped with scrubbing systems for removal of sulfur oxides, with the cost of electric power generated by conventional thermal electric power plants operating on coal and not equipped with sulfur oxide scrubbing systems [7].

1. A 3000 MW steam turbine power plant with 500 MW units and the following initial steam parameters: pressure of 24 MPa, temperature of 540/540°C. The power plant is equipped with a flue gas treatment system to decrease emission of SO_x.

2. A 2820 MW MHD power plant with 940 MW units incorporating 500 MW turbine assemblies. The air for the MHD facility is preheated to 1900°C in separate ceramic preheaters. Coal is the main fuel for the principal loop of the MHD facility and natural gas is the fuel for cowpers. In order to decrease the emission of nitrogen oxides, they are decomposed by burning coal in two stages. Sulfur oxides are trapped by seed.

3. A 2820 MW MHD power plant with 940 MW units with the remaining components as in 2 above, but with removal of nitrogen and sulfur oxides from the flue gas at the outlet from the facility. Seed is not regenerated and is introduced into the MHD combustion chamber in the form of sulfates.

Figure 1.3 shows variation of the cost of electric power ζ for the variants considered (1-3) of thermal power plants (with respect to the cost of electric power of the standard variant (0) as a function of the cost of fuel c_T). A comparison of the curves shows that prevention of emission of SO_x in an MHD power plant is considerably cheaper than for steam turbine plants. This can be attributed to the MHD power plant being equipped

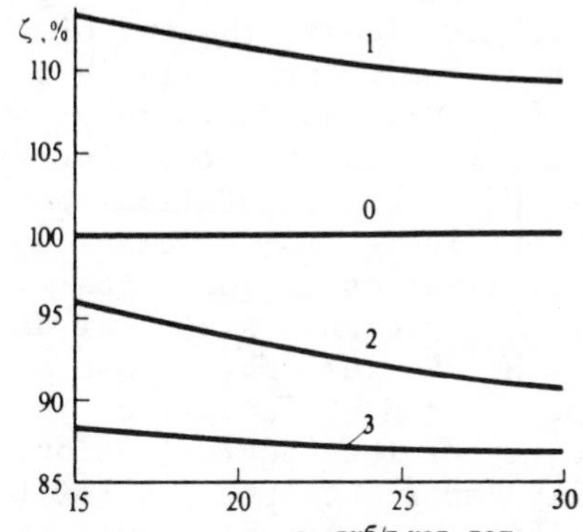

Fig. 1.3 Variation of the relative cost of electric power generated at various power plants with the cost of fuel, 0,1--steam turbine power plant operating on coal without emission controls and with flue gas scrubbing of SO_2, respectively, 2,3--MHD power plants with decomposition of NO_x and with removal of SO_2 and NO_x and their utilization for production of sulfuric and nitric acids

with seed injection and removal systems. The purge of flue gas of SO_x is actually reduced to seed regeneration.

At the same emission levels of harmful compounds into the atmosphere, limited by the emission standards (curves 1-3), MHD power plants with removal of SO_x and NO_x from flue gases, accompanied with production of sulfuric and nitric acids, are characterized by the lowest cost.

An advantage of an MHD power plant is its ability to change rapidly the loads of operation. Investigations, as well as operation of the U-25 facility, have shown that, if the high-temperature ceramic air preheater is maintained in high state (such air preheaters usually used in blast furnaces are always maintained in hot states), an MHD generator can reach its rated power output in 10-15 s after its start-up. Shutdown of an MHD generator does not encounter any difficulties. Thus, highly variable load operation of an MHD system can be attributed to the feasibility of rapid start-up and shutoff of the MHD generator. During this process, the power output of the MHD power plant drops about 100 %. When the MHD generator is shut down, the steam generator will operate on independent burners with subsequent control of the steam generating system. Assuming that the steam power portion will use a variable load steam turbine operating at subcritical parameters (e.g., K-500-130), the total power output of the system may be controlled between the limits of 100 % and 20 % of nominal.

Even though the present worldwide trend is to electric power generation by nuclear fuel, all predictions indicate that a considerable portion of electric power requirements will be met, until at least the end of the 20th century, by burning organic fuel, primarily coal. According to published data [8, 9], approximately one-half of world coal resources are located in the Soviet Union and one-fourth in the U.S.A.. Utilization of MHD energy conversion may contribute to a considerable degree to a more effective utilization of world coal resources.

Analysis of the present state of development of MHD facilities in the Soviet Union makes it possible to consider incorporating MHD facilities into electric power generation industry of the Soviet Union. MHD facilities become especially important because of the peculiarities of fuel availability, power requirements (fuel-power balance) of the country during the past few years. In particular, the fuel-power imbalance of the European part of the U.S.S.R. and the Urals, because of a shortage of natural deposits of fuel in these regions, the transportation of fuel from distant eastern parts of the country, the continuously varying power requirements during the day and consequent necessity of developing variable-load electric power plants, are the most important aspects of the problem [10, 11]. The last factor is especially important because Soviet industry does not fabricate special variable-load power units for operating during periods of sharp, sudden increase of power demand (load increase). It should be pointed out that no variable-load nuclear power plants exist at the present time and the feasibility of their development is not yet clear.

The peculiarities of MHD power plant operation make it possible to consider these facilities, operating on all types of organic fuel, as most expedient for incorporation into the power-producing capacity of the U.S.S.R., particularly its European parts. Future MHD power plants should operate on coal. However, at the present time, only MHD power plants operating on gaseous fuels are ready for production. Active development of MHD power plants operating on coal will apparently require an additional five to seven years. In order to expedite the development of MHD power production and to prepare the machine building industry for mass production of equipment for MHD facilities, it is expedient to construct MHD assemblies operating on gas.

Since MHD power plants operating on gaseous fuel and MHD power plants operating on coal do not differ significantly, i.e., little more than coal-combusting thermal power plants differ from gas-combusting power plants (such components as the magnetic system, converter systems, seed injection system actually do not differ, nor the MHD channel; combustion chamber, high-temperature air preheater, steam generator, and seed trapping system are more complicated for coal burning facilities, but still maintain their principal features), the development of industry to fabricate MHD systems operating on gaseous fuel will actually make it possible to organize and produce MHD assemblies operating on coal.

At the present pace of development, it can be expected that the first commercial-scale, pilot MHD system operating on coal will be developed in the 1990s. However, construction of such a facility has to be preceded by solution of a number of problems, the most important of which are as follows.

- Attainment of high temperatures (2800-2850 K) of coal combustion products. This can be achieved in two ways. The first involves development of air preheaters, such as cowpers, heating air to 1800-2000 K, operating under complex conditions when the heating medium is contaminated with ash and seed. The second approach requires development of an oxygen-enrichment facility, increasing oxygen content of air up to 40-50 %, at relatively low power consumption (0.2-0.3 kW·h/m^3 of oxygen under normal conditions, when, as in the case in standard air preheating facilities, approximately 0.4 kW·h is spent for production of the same quantity of oxygen). The second technique is to be preferred (it should be noted that the Federal Republic of Germany and U.S.S.R. have developed such oxygen-enrichment facilities [13]).

- Development of a combustion chamber with liquid removal of slag (up to 80 %) and low heat losses (not exceeding 3 %).
- Development of an MHD generator with slag films on electrode and insulation walls.
- Finding effective methods of separating seed from fly ash, with no more than a 2 % seed loss.

Apparently, most of these problems can be solved by renovating the existing U-02 and U-25 facilities and the facility in Kokhtla-Yarva and on special stands.

From the economic point of view, it appears to be most expedient to construct MHD power systems in the European part of the U.S.S.R., where the cost of Donetsk and Kuznetsk coal or enriched Kausko-Achinsky coal is not too high. These coals are characterized by a fairly high heat content, required for successful use in MHD power plants. It can be assumed that a 1000-1500 MW commercial-scale pilot MHD system operating on coal will be constructed in the middle 1990s. This might be followed by a 3000-5000 MW annual increase of electric power generation by MHD power plants.

1.2.2. Basic Results of Investigation of the Development of the First Commercial-Scale, Coal-Combusting, MHD Energy System in the U.S.S.R. Investigations on the MHD energy conversion in the U.S.S.R. began in the early 1960s. A large number of organizations from various ministries and departments participated during the first, initial stages of investigations. Further investigations were concentrated primarily at the Kurchatov Institute of Atomic Energy, Institute of High Temperatures of the Soviet Academy of Sciences, and the Krzhizhanovskiy Power Institute.

Experimental investigations conducted on facilities of the Institute of High Temperatures were especially important to the development of MHD power plants.

1.2.2.1. The U-02 Facility

Investigations performed on the U-02 facility, which

was designed and constructed at an accelerated pace and became operational in April of 1964, occupy an important place in the overall MHD energy conversion program. A special characteristic of the U-02 facility compared with other experimental MHD facilities of that period is its compactness. The U-02 facility incorporates all components of an MHD power plant. The principal directions of scientific research programs conducted on the U-02 facility are investigation of the principal physical processes in the MHD channel and other components of MHD power facilities and the development of basic principles of designing various assemblies of the MHD facility with concurrent experimental verification and improvement of design solutions, using models involving long-duration tests under conditions close to those in commercial-scale MHD power plants.

An important portion of the research conducted on the U-02 facility was devoted to problems involving design, construction, and completion of the U-25 facility. The initial program for experimental investigations on the U-02 facility, operating on gaseous fuels, was practically completed between 1973-1975 [14].

Let us briefly discuss the principal results of investigations conducted on the U-02 facility.

Basic concepts of operating characteristics of a nonideal MHD generator, determined on this facility, enabled the testing of methods of engineering calculation. Different seed injection and removal systems were investigated and methods of removing deposits from heat transfer surfaces were developed. First loading of an MHD generator by an inverter was accomplished.

A high-temperature air preheater with a fixed pebble bed, capable of stable operation with air heated to 1700-2000^{0}C, was developed and investigated. The pilot unit of this type of industrial air preheater was put into operation at the Cherepovets Metallurgical Combine. The use of such air preheaters in metallurgy, oil processing, and other branches of industry is being investigated.

Long-duration tests of electrode walls of an MHD generator were conducted on the U-02 facility [15, 16].

The U-02 facility was renovated and modified in 1974 for investigation of the operation of an MHD facility operating on solid coal. A method was developed that made it possible to model operation of an MHD channel combusting coal by introducing ash into the combustion chamber. By 1982, the U-02 facility equipped with ash injection and direct coal combustion had operated for a total of approximately 750 hours. Data were obtained on the behavior of the slag film and its influence on the characteristics of the principal components of the gasdynamic duct, on the interaction of slag with the seed, on the methods of supplying pulverized coal, and on operation of a cyclone burner with liquid slag removal; removal of 85-90 % of slag was achieved. These data established a number of characteristics that have to be taken into account in further investigations [17, 18].

1.2.2.2. The U-25 Experimental, Commercial-Scale Facility

As in the case of the U-02, the U-25 facility is a complex power facility that incorporates all of the principal components of future commercial-scale MHD power plants [19]. One of the important problems in developing the U-25 facility was the determination of its dimensions and particularly of its power output.

It was natural to attempt to maximize the operating conditions of the facility, making it approach commercial-scale MHD power plants, while minimizing its construction cost. Preliminary analysis has shown that, in order for the electric current generation of the MHD generator to be an order of magnitude above the harmful near-wall effects, the combustion products flow has to be several dozen kilograms per second. This constraint dictated the size of the facility.

Studies based on the data on electrical conductivity of combustion products flow, available in 1963-1965, have shown that, at maximum attainable parameters of the MHD generator using reliable standard auxiliary equipment with maximum parameters (high-temperature air preheater, cowper to heat the air to 1200^{0}C, magnetic system with magnetic field induction of 2 T at an acceptable length of the working sector of the magnet, on the order of 5 m, near-sonic velocity of the combustion products flow in the MHD channel), it is possible to provide combustion products flow rate of 50-60 kg/s and a 25-MW maximum power output. In accordance with this estimate of the power output, the first commercial-scale MHD facility was named U-25. In designing the U-25 facility, it was decided to make maximum use of standard equipment. This was necessary in order to accelerate construction of the facility and to decrease its cost.

Construction of the U-25 facility was begun in 1966 and it became operational in March 1971 [20]. A photograph of the room showing the MHD generator, steam generator, seed removal system, and smoke stacks is shown in Fig. 1.4. The rated electric power output of the MHD generator in a normal operating mode of the U-25 facility

Fig. 1.4 MHD generator room of the U-25 facility

utilizing potassium seed, operating at a flow rate of the working fluid of 50 kg/s, is 10 MW.

The main problem during the first stage of operation of the U-25 facility was verification of the operational capability of all components of the facility and improvement of its components so as to bring its characteristics to their projected values. The behavior of the MHD channel at high electric fields and considerable load currents was investigated. The current flow, near electrode losses, heat fluxes, electrical insulation, etc. were investigated [21]. Experience was gained in operating important assemblies of the MHD facility (steam generator [22], combustion chamber [23, 24], seed injection and removal system, inverter subsystem, air-preheaters and oxygen station) [25-27].

The aim of the second stage of operation of the U-25 facility was attainment of the maximum power output of the MHD generator. In order to accomplish that end, it was necessary to improve many components of the facility, including the MHD channel. The use of seed containing both potassium and cesium salts and an increase in oxygen enrichment of air from 40 % (planned value) to 43 %

(maximum allowable value based on the operating conditions of a compressor) made it possible to increase by approximately 50 % the conductivity of the plasma of products of combustion. In addition, increasing the flow rate of products of combustion was also analyzed.

All of these factors, as well as careful optimization of channel construction and loading of the U-25 facility conducted at the existing level of understanding of processes in an MHD channel, made it possible to solve the problem of obtaining the maximum power output of 20.4 MW. The 1D Faraday channel was designed to achieve this level of power output. The U-25 facility with the 1D channel reached the projected power output of 20 MW in December, 1975. This was repeated in April, 1976 [28]. The principal parameters of the U-25 facility are given in Table 1.

The aim of the third stage of operation of the U-25 facility is to investigate the operation of the MHD generator at the power output level close to the nominal (10 MW), in relatively long-duration, continuous operating modes. According to calculations, this power output should occur in normal operating modes of the U-25 facility. The R and RM channels were developed for the third stage. The R channel, rated at a power output below the nominal power output of the U-25 facility, operated for 100 hours, generating a continuous power output of up to 4 MW [29]. Experience gained with the R channel made it possible to develop the RM frame channel, rated at a 10 MW power output and designed to operate continuously for 200-250 hours. The RM channel was tested in 1976, reaching its projected power output level. In April 1977, the RM channel operated continuously during a period of 250 hours at a power output of 2-4 MW, reaching 11.5 MW at the end of the run [30].

In addition to this, another modification of No. 1 channel (1V channel) was developed. The 1V channel has shown satisfactory characteristics at a power output of 8-9 MW [31].

MHD Generator. One of the most important factors in developing the U-25 facility is the investigation and testing of the MHD generator. At first, to achieve reliability, the electrode and insulation walls of the MHD generator channel were made of modular water-cooled metal elements. This decreased the effective plasma conductivity and increased the thermal losses in the MHD generator channel. Raising the temperature of electrode walls up to 1500-1700 K in subsequent modifications of the

Table 1

	Theoretical	Experimental
Maximum power output (more precise value) of the MHD generator, MW	20.4	20.1
Nominal power output of an MHD generator (at parameters given below), MW	10	13.8
Oxidizer	oxygen-enriched air (up to 40%)	oxygen-enriched air (up to 40-43%)
Preheat temperature of the senriched air, °C	1200	1200-1500
Initial temperature of combustion products, °C	2600	2600-2650
Seed	up to 1% K	up to 1% K or K+Cs
Combustion products flow rate kg/s	50	50-60
Stagnation pressure at the inlet into the MHD generator, kPa	270	275-315
Velocity at the inlet into the MHD generator, m/s	800	800-900
Pressure at the outlet of the MHD generator, kPa	105	98-108
Power consumption of the magnetic system, MW	2.6	2.4

1A and 1B channels, and structural changes made it possible to increase the power output of the MHD generator up to 6.5 MW (at the rated 9-MW power output of the 1B channel). Incorporation of insulating methods developed earlier on the U-02 facility [32] increased considerably the electrical strength of insulation and the reliability of channel operation. Further investigations of the MHD generators on the U-25 facility followed two directions. One of these is improvement of the Faraday MHD generator utilizing the existing No. 1 channel construction, and the other, development of an experimental MHD generator of radically new design.

Faraday Channel. A series of channels consisting of modifications of the No. 1 channel was characterized by principally the same type of design scheme with a modular wall. The 1, 1A, 1B, and 1D channels underwent thorough, longer duration tests. Detailed analysis of the experimental data on runs of the 1A and 1B channels [21] had demonstrated the feasibility of increasing the power output of the MHD generator. The principal problem of designing the 1D channel [28] was to eliminate or decrease as much as possible the influence of factors decreasing the power output of the MHD generator, while in principle maintaining the existing design of the Faraday channel. An increase of the flow through cross section of the channel made it possible to increase the flow rate of the working fluid 20 %. The active part of the channel was elongated by utilizing the tapered zones of the magnetic field induction at the ends of the magnet and by utilizing curvilinear geometry for electrode walls. The temperature of the walls in the 1D channel was increased considerably: electrode walls, to 2100-2250 K; insulation walls, to 1800 K. The degree of segmentation of electrodes was also increased. In the 1D channel, the segmentation was chosen so that the axial potential drop between adjacent electrodes in a nominal mode does not exceed 30-40 V. To achieve this, the number of electrode pairs was increased to 107.

The inner joints in the channel along the electrode and insulation walls were hermetically sealed, and the resistance of the external components of the loop was decreased. The frame of the 1D channel is insulated from the ground, considerably simplifying operating conditions of channel components at high potential. Particular attention was paid to the distribution of electrical load along the length of the channel.

Diagonal Channel. A logical continuation of work on MHD generators of the U-25 facility was the development of a diagonal-type MHD channel (R channel) [29]. The potential advantages of such a channel include feasibility of a maximum degree of segmentation and a minimum number of components at different potentials, simplicity of the electrical loading scheme, and improved flow gasdynamics achieved by means of fairly free variation of the shape of the cross section of the channel, high electrical efficiency due to the flow of current to the side walls of the channel, simplicity of construction technology, high reliability, etc.

For simplicity, the parameters of the diagonal R channel were chosen below the nominal parameters of the U-25 facility. However, duration of operation was quite long. The calculated electric power output of the R channel is 3 MW. The angles of modules of the diagonal wall with the longitudinal channel axis were chosen in accordance with the calculated potential distribution (along the working sector of the channel the angle is 21^0). The degree of segmentation corresponds to the maximum potential difference between adjacent modules (30 V). The experimental data acquired verified the usefulness of the basic design of the diagonal channel, the choice of materials (in particular, for long-duration electrodes), flow calculation methods, electrical loading, etc. for future applications on the U-25 facility.

The experience gained in designing, fabricating, and operating the diagonal R channel, as well as the large amount of data acquired during investigations of several Faraday-type channels were utilized in designing the RM channel.

The RM channel is of frame construction, with hermetically sealed housing made of electrically insulating material. The angle between the frames and the axis along the active sector of the channel is constant. To provide high electrical efficiency, the angle along the active sector was also fairly small (35^0 with the axis of the channel). Bent frames and equipotential current takeoff components were used at the end sectors. This design of the end sectors of the channel, which underwent successful testing on the R channel, made it possible to attain effective transition from the active sector of the channel to the nonsegmented components of the gasdynamic duct (nozzle and diffuser) at minimum length of the end sectors. The distance between the segments of the RM channel was 45 mm along the axis of the channel. The potential difference between the frames did not exceed 15 V.

To decrease both hydraulic and electrical losses, the gasdynamic duct of the channel was made with rounded angles. This also resulted in a more uniform distribution of current density along the channel cross section. The channel diffuser located in the fringe magnetic field was segmented.

The RM channel underwent a cycle of tests in the U-25 facility. Stable operation of the channel was achieved during the long-duration run. It was not accompanied by a drop of electrical characteristics, even though the electrodes along the initial sector underwent considerable erosion. Channel insulation was maintained at a high level during the whole run. No current leakage was observed along the structure [30]. Approximately 20 runs with the RM channel were carried out before 1982, involving testing of electrodes, loading schemes, investigations of fluctuations and instabilities, electrical strength of the channel, and all of the auxiliary systems.

Principal Equipment of the Thermal Loop. The attainment of rated parameters of the U-25 facility in accordance with the principal stages of operation of the MHD generator required certain changes of the components of thermal loops and, in certain cases, development of principally new equipment. In particular, investigations of the combustion chamber are of great interest.

Combustion Chamber [23, 24]. An MHD power system plays the role of a generator and determines to a considerable degree the operating capacity and indices of the facility as a whole. The major peculiarities of the combustion chamber for an MHD power system are the use of high-temperature (1500-2000 K) oxidizer, the presence of seed and high electrical potential (for a large MHD power plant, 20-35 kV). Two types of flow through combustion chambers were developed and tested under long-duration operating conditions. The combustion chambers with "cold walls" provided a high degree of combustion and acoustic stability of combustion. At the same time, higher heat losses to the walls and insufficiently effective injection and ionization of seed during its residence time of up to 7 ms in the combustion chamber failed to provide the required plasma ionization. Consequently, an improved circular-cross section combustion chamber with warm walls was designed and fabricated. This combustion chamber was used in all major runs of the U-25 facility. Its principal characteristics are as follows.

- The presence of two combustion zones formed by the jets of oxidizer and gas flowing from the nozzles of the end and peripheral burners and injection of seed in the forechamber zone, increasing residence time of seed at high temperatures up to 15 ms, i.e., twice as long as in the first type combustion chambers.
- Use of high (140-180 m/s) flow rate of premixed oxidizer and natural gas from the nozzles of the burner assembly, as well as development of recirculation zones of combustion products, resulting in intense fuel combustion and seed vaporization and ionization.

- Insulation of walls by lining the surfaces of the combustion chamber with high-current clay concrete on the oxidizer side and a rammed mass, based on zirconium dioxide, on the fire size, making it possible to decrease thermal losses to 3-4 % as against 6-7 % losses in the first type of combustion chamber.

The principal results of recent work on the investigation and improvement of the combustion chamber with insulated walls include stable, long-duration operation at nominal parameters (the plasma temperature at the combustion chamber outlet is 2850 K, the pressure is 315 kPa); stabilization of combustion in the combustion chamber, attained as a result of increasing stability of ignition by supplying part of the fuel (approximately 30 %) into the peripheral zones of gas-air jets, and an increase in the uniformity of fuel distribution in the oxidizer along the nozzles of the burners.

Experimental investigations made it necessary to devote considerable attention to problems associated with thermal fatigue strength of thermally stressed components of the combustion chamber, and to search for methods to decrease thermal stresses in thermally stressed cooled components. Such methods include attempts to decrease heat flux to the wall by utilizing thermally protected coatings capable of operating at high temperatures, or provide internal cooling of the walls by oxidizer purge and by choosing high-strength materials of high thermal conductivity for thermally stressed components. In particular, good results were obtained in the combustion chamber with insulated walls, by replacing radiant sector pipes made of austenite steel with pipes of rectangular cross section, made of copper, and placing them so as to protect the welded seams of the main burner assembly nozzles.

Two important directions of investigations dealing with combustion chambers have been outlined and are being implemented on the U-25 facility. The first is the development of insulated combustion chambers of the flow-through type that are preferable for use in gas-combusting MHD power plants. The second direction is combustion chambers for coal. The research program on methods of utilizing coal in MHD power plants is also being conducted in the Krzhizhanovskiy Power Institute facility at Kokhtla-Yarva [32], on the U-02 facility [17, 18], and also on the modernized facility of the U-25G bypass loop [33].

Seed Injection and Removal System [25]. A closed seed recirculation cycle is achieved on the U-25 facility. The seed is injected in the form of a 50 % aqueous solution of K_2CO_3 (solutions of other potassium and, also, cesium compounds were used during experimental runs of the facility). Vaporization and ionization occurs either in the combustion chamber or in the hot air duct. Seed is removed in the form of melt from the steam generator, and in the form of an aqueous solution from the hydromechanical scrubbing system. Regeneration and preparation of seed is carried out prior to the injection cycle.

One of the important problems is attainment of thermodynamic equilibrium at the inlet of the MHD generator, which controls the level of electrical conductivity of plasma achieved in the channel. Pneumatic nozzles of special design were developed for the purpose of increasing the electrical conductivity of plasma due to vaporization and ionization of seed. Tests performed on a stand have shown that they make it possible to obtain a range of drops with median dimensions of 75-100 μm from a 50 % solution of K_2CO_3. The use of these nozzles with an insulated combustion chamber made it possible for the electrical conductivity to approach its theoretical value. Special investigations were performed on selecting the seed injection site. Different methods of seed injection were considered: into the hot air ducts upstream of the combustion chamber, directly into the chamber, and into both.

Calculations and the experimental data have shown that injection of seed at the oxidizer temperature of 1200°C into the air duct provides practically no gain. What's more, a lot of seed is lost on walls of the air duct and at the inlet into the combustion chamber. Therefore, in later investigations, the only technique used was seed injection directly into the combustion chamber.

However, injection of seed into the combustion chamber in the form of an aqueous solution exerts a negative influence on operation of the MHD facility. The ballast water (solution) decreases plasma temperature by almost 70°C, resulting in the substantial drop of electrical conductivity and a 1.5 to 2 % decrease of the power output of an MHD generator. Injection of potash (K_2CO_3) in dry powder form is hindered by its tendency to stick together, resulting in compaction of lumps, leading to clogging of the seed storage bunker, transport pipes, and nozzles. Means of improving friability of potash are being researched on the U-25 facility. A promising technique is the use of aerosol, i.e., highly dispersed silica powder with dimensions of particles on the order of 100 Å that, mixed with potash, covers the potash particles with a film. Experiments have shown that only 0.5 to 1.0 %, by mass, of the aerosol is required to achieve the necessary friability [34].

Considerable attention was devoted to investigating behavior of the ionized seed in the gas duct of the facility and, particularly, to removal of seed in a molten form from the radiant section of the steam generator. It was determined that seed condensation (in the form of KOH vapors) on the cold surfaces of radiant section pipes during the first two to three hours leads to formation of solid deposit, consisting primarily of potassium carbonate formed when KOH interacts with carbon dioxide. After reaching a certain thickness, the surface of deposits melts and the seed being condensed flows down in the form of a mixture of K_2CO_3-KOH, with a KOH content of up to 20 %.

Approximately 40 % of the injected seed is removed in the molten state. The melt flows into a collection tank with a circulating solution that is gradually enriched with seed. Obtaining a 50 % solution of seed is not difficult. An investigation of the efficiency of seed removal from the flue gases by a system of scrubbing devices including Venturi scrubbers has shown that this system can remove up to 99 % of the seed reaching it. The total hydraulic resistance (together with three-shelf foaming device) does not exceed 15 kPa. Trapping of seed by means of electrostatic precipitators appears to be more promising for commercial-scale MHD power plants. Tests of these filters on the U-25 facility have shown that this method also provides trapping of up to 99 % of the seed.

Steam Generator. A flow through steam generator [22] with standard parameters of the steam (540°C/100 atm), and a steam flow rate of 270 t/h, is used in the U-25 facility. Operation of the steam generator was investigated at nominal thermal loads in a force-through operating mode. The boundaries of circulation in the evaporation tubes of the radiant portion of the steam generator were calculated throughout a wide range of heat loads. The principal characteristic of operation of the steam generator as part of the MHD facility is the presence of seed deposited on the fire surface. Difficulties of bringing the generator to operate at its rated parameters and its otherwise unnecessarily complex construction is associated with this phenomenon.

The seed deposits in the radiant and convective sectors of the steam generator behave differently. The seed in the radiant sector, in the form of a fluid, flows into a special tank and is regenerated. Investigations of the behavior of seed in the convective sector of the steam generator show that the condensed seed cooled below the melting temperature (900°C) is in the form of solid, submicrometer particles of K_2CO_3 deposited on the heat transfer surface of pipes. These porous deposits cannot be dislodged and removed by the gas flow, even when it is accelerated by the narrowing of the channels. This lowers the heat transfer and leads to a more substantial increase of hydraulic resistance.

Attempts were made to develop effective means of cleaning convective fire surfaces of the steam generator. The use of periodic shot cleaning did not provide the required results. Shot cleaning was much more effective when used continuously; but, even this did not achieve the desired results. It became necessary to extend investigations of optimizing the cleaning of convective surfaces of the steam generator from deposits. Ultrasonic cleaning methods were also utilized in the U-25 facility. These techniques are based on effective interaction of acoustic waves in a gas flow with seed deposits. Impulse, nonstationary combustion chambers were used to generate acoustic waves. These chambers are characterized by simple design and are capable of providing conditions for breaking up and removing mechanically weak impurities. The impulse chamber operates on an air-gas mixture. The acoustic device, consisting of a pipe acoustically closed at one end and equipped with a pulse guide at the other end, is mounted near the steam generator. High-impulse power makes it possible to generate pressure up to 800 kPa. The system operating periodically, with each operating cycle in the stationary operating mode initiated at 15-minute intervals, decreases considerably the resistance of the convective shaft. The average level of resistance was 1-2.5 kPa. The method of cleaning by flushing with steam, widely used in boiler practice, was also tested on the U-25 facility. At the present time, a combination of all three methods of cleaning complementing each other is optimum for commercial-scale MHD power plants.

Inverter System. The technical approach used in developing the inverter system [26] for the MHD generator of the U-25 facility incorporated all of the best features known at that time from experience gained in dc current transmission. Specific characteristics of inverters operating with an MHD generator in the U-02 facility were also taken into account.

However, a limited number of inverter bridges (24 sets) and, as a result, low degree of segmentation of the MHD generator decreased efficiency and duration of operation. To increase the degree of segmentation of the MHD generator, ignitrons were replaced with thyristors in half of the inverter sets. After this change, the loading circuit of the MHD generator could be decoupled from the Hall potential at three levels. First decoupling was by means of six transformers, each of which is equipped with separate windings, providing 24 independent electrode groups (one each for each inverter set). Second decoupling consists of a capacitor bank that makes it possible to connect two thyristor bridges to each winding of the transformer. Third decoupling is in the form of ballast impedances and, for Faraday loading of the MHD generator, makes possible connection of from two to five electrodes to a single inverter bridge. A control system making possible constant-voltage and constant-current operation modes of electrodes was developed, to expand the scope of experimental investigations with an MHD generator. The control system thus maintains the voltage at the station near zero and the current at the station 0.07 A/B. Experimental investigations of the control system have demonstrated its dynamic stability, sufficiently rapid operation (50-100 ms), and a small degree of over-control (5-10 %) in response to a jump-wise disturbing effect.

Investigations have shown that current control is advantageous in terms of stability of the joint operation of the generator-inverter system. Current control also lowers, by approximately 2 to 3 times, the short-circuit current of individual electrode pairs and leads to a uniform current distribution along the channel length. However, it should be noted that the load factor in current control mode depends on the plasma conductivity. Group control of the facility is also feasible. A more modern control (individual and group) of the inverter modes by means of a computer is being used at the present time.

High-Potential Decoupling. The high Hall potential generated during operation of an MHD channel imposes serious requirements on the insulation of all components of the MHD system. Investigation of the shape and magnitude of the voltage acting on insulation shows that a variable component is superimposed on the constant component. The maximum frequency of the spectrum reaches 1.5 kHz. The fundamental frequency of the ac component is 300 Hz and its amplitude is 10-15 % of the constant component.

The decoupling with respect to Hall potential on the U-25 facility is provided by electrically isolating assemblies adjacent to the inlet into the MHD channel, as well as isolation of the latter by inclusion of insulation gaps in the water cooling system, insulation of inverters from the ground, diagnostic equipment, etc. [35]. The most complex problem turned out to be electrical isolation of the combustion chamber at a maximum Hall potential up to 5-6 kW from the grounded hot oxidizer feed pipes, and decoupling from seed, gas injection systems, and the combustion chamber cooling system. These problems were resolved by careful design of electrical insulating joints. An insert with above normal breakdown voltage and additional potential break that decreases chances of a breakdown was installed between the combustor and the air duct. The oxidizer parameters are carefully regulated in order to avoid moisture condensation that may lead to an accidental breakdown. Preheating of the loop during start-up of the facility is designed to preclude the settling of moisture from the gas flow on the electrical insulation of the facility.

Control and testing of insulation, in accordance with methods developed during assembly of individual components and completely assembled MHD generator loop, increase reliability of the thermal and electrotechnical equipment of the MHD system of the facility. A substantial effect in verifying that the MHD generator is ready for operation is attained by performing preliminary high-voltage tests with the assembled loop of the U-25 facility. These tests are performed twice, in a cold loop and in a loop preheated by a flow of oxidizer heated to 1200°C. A special technique, developed to conduct high-voltage tests, provides close to the real potential distribution over all components of the non-standard equipment assembly. The voltage during tests exceeds by 50 % the operating voltages of the components of the non-standard equipment assembly.

Control System. The function of recording and processing the primary data, as well as control of an experiment on the U-25 facility, is performed by an automatic recording and processing system (ARPS) that incorporates, as its major element, components of computers NR-2116V and NR-21005 computers and auxiliary equipment with a distributed network of peripheral devices. During the past four years, the ARPS system was continuously improved, and its functions were expanded. This system performs the following operations.

- Sorting and storage of data;
- Data processing, including computation of primary physical parameters and preparation for their visual display on a monitor, curve plotting, etc.;
- Control of computer functions during construction of curves, command output to the printer, etc.;
- Control functions, primarily of the electric mode of operation of the U-25 MHD facility, with the aid of inverters, determination of current-voltage characteristics, etc.

The amount of experimental data acquired on the U-25 facility is on the average 10^6 bits. A bank utilizing an NR-3000 computer is utilized for storage, processing, and data output. The data are stored on magnetic tapes and floppy disks. Standard and recently developed equipment make it possible to sample, sort, and graphically represent the required experimental data. The data bank is used for analysis of processes occurring in U-25 and U-25B facilities, development and improvement of the mathematical model of the overall facility and of its major subsystems, as well as prediction of its characteristics and preparation of an algorithm for conducting a run of the facility.

By mid-1982, the U-25 facility had completed practically the entire series of planned investigations for which it was designed. The principal conclusion of the investigations is that it is feasible to fabricate a commercial-scale MHD power plant operating on gaseous fuel. All of the major systems of the MHD facility have now progressed to a point that it is possible to design and fabricate these systems on commercial scale. Initially, these investigations made it possible to complete technoeconomic justification of such a power plant and then to design the 500-MW MHD system being constructed in Ryazan. Unfortunately, many years of operation since 1971, with a large number of experiments conducted at extreme conditions (a total of 110 runs and shutdowns), resulted in a tremendous amount of wear of some components of the U-25 facility, such as the steam generators, hindering long-duration operation of the facility.

A major renovation of the facility is planned for 1982-1983. After renovation, the U-25 MHD facility will operate in long-duration mode, with a steam turbine. Its operating mode will be similar to that of base thermal power plants, i.e., it will operate for up to 5000-6500 hours per year. The aim of this operation is to obtain sufficient experience prior to bringing on line the first commercial-scale MHD power system.

1.2.2.3. U-25B Facility

In accordance with the U.S.-U.S.S.R. cooperative program on MHD energy conversion, conducted during 1973-1979, a special bypass loop [33] was designed and fabricated at the Institute of High Temperatures of the Soviet Academy of Sciences for the U-25 facility by mid-1977. The bypass loop is intended to operate with a superconducting magnet system developed at the Argonne National Laboratory in the United States. The facility was designed to operate at a flow rate of 5 kg/s of combustion gases. It can operate with the existing high-temperature air preheaters (cowpers) with additional supply of gaseous oxygen from a liquid oxygen supply located at the combustion chamber.

A combustion chamber, diagonal frame MHD channel [36], diffuser, and other components and systems were designed and fabricated for the bypass loop. The overall facility has been named the U-25B. It includes a unique superconducting magnetic system with the following parameters: warm bore volume inlet diameter, 400 mm; outlet diameter, 670 mm; length of the active sector, about 2500 mm. The system generates a magnetic field of up to 5 T, which is close to that required for commercial-scale MHD facilities.

The U-25B facility is equipped with apparatus that make it possible to conduct an entire series of measurements of the principal gasdynamic and electrical parameters. A data acquisition system is used when conducting experiments. The system makes possible automatic high-resultant recording, processing, and storage of all necessary parameters.

The U-25B facility operates under conditions that, in terms of some of the most important parameters, are close to those of commercial-scale facilities. It is unique with respect to its characteristics, i.e., until the present time, neither the Soviet Union nor foreign countries had developed long-duration facilities characterized by a high degree of MHD interaction. Therefore, investigations of the MHD generator of the U-25B facility under conditions of high electrical field intensity and magnetic field inductance are of considerable theoretical and practical interest.

The first (physical) run of the U-25B facility was performed in December, 1977. Six runs were conducted during 1978-1979 [37]. The MHD channel operated successfully for approximately 100 hours, including approximately 50 hours at the rated magnetic field induction of 5 T. One should note the stable operation of the principal assemblies and components of the U-25B facility. A large volume of unique data was acquired on the operation of the principal components in strong electric and magnetic fields. Oxygen enrichment of the oxidizer reached 60 %. The initial temperature of combustion products was 2600-2900 K. Record high electrical field intensity (exceeding 2.5 kV/m), current density (up to 1 A/m^2), heat flux (3 MW/m^2), specific power output (10 MW/m^3) were reached. The Hall potential reached 5 kV.

The maximum power output during several hours of operation was 1.5 MW. The magnetic field induction and the electrical field strength obtained are nearly the values required for commercial-scale MHD generators.

Another bypass loop, the U-25G operating on coal, is being assembled next to the U-25B bypass. Its completion will make it possible to perform a unique experiment with an MHD generator operating on solid fuel under conditions of strong electrical and magnetic fields [38].

1.2.3. Investigations of Certain MHD Facilities Intended for Other Purposes. Experimental investigations on MHD energy conversion performed in the Soviet Union were conducted on a broad scale and, in addition to the Institute of High Temperatures of the Soviet Academy of Sciences, were also carried out by a number of organizations. Although no direct attempt was planned to determine optimal conditions for developing a commercial-scale MHD system and somewhat other goals were being pursued, the results obtained are unquestionably useful to the development of a stationary MHD power plant. Let us briefly describe the most important investigations.

The ENIN-3 Facility. For many years, investigations have been conducted on the experimental ENIN-3 facility of the Krzhizhanovskiy Power Institute and Estonenergo, at a thermal power plant at Kokhtla-Yarva [32, 39]. This facility operates on coal and is equipped with a channel with an applied electrical field.

The thermal power output of the facility is about 10 MW, the pressure in the combustor is 150 kPa, and the combustion products flow rate is 2 kg/s. The coal is combusted in a 1000-mm vertical cyclone chamber of 650 mm diameter, equipped with a jet-coal powder supply system operating under pressure, injection system of an aqueous solution of seed (potash), a preheater heating air up to 1000 K, enriched with oxygen up to 40-50 % by mass, and a slag removal system from the combustor, also operating under pressure.

The channel is made of water-cooled copper electrodes with different electrical insulation. The inner diameter of the channel is 150 mm, the length is 1800 mm, and the velocity of products of combustion is up to 1000 m/s. The channel is installed vertically and is equipped with an electric power supply system of inverted electric current at a voltage up to 3 kV and a current up to 300 A. Investigations performed on this facility include combustion of Kuznets, Ekibastuz, near Moscow, and Kansko-Achinskiy coals. The temperature of combustion products reached 3000 K. The specific electrical conductivity of gases at the inlet was 15-17 S/m. The duration of continuous operation of the facility was 5-6 h and was limited only by the oxygen supply. Stable operation of the channel was achieved with an electric field in the presence of liquid slag flow along the electrodes. Overall, the channel was operated with an electric field for approximately 100 hours without exhibiting any traces of electrode erosion. Experiments have shown that the channel can operate in a diffuse discharge mode at a current up to 60 A. At a higher current, it operates in an arc discharge mode. The Estonenergo MHD facility is being overhauled, in order to increase considerably its power output and other indices.

"Pamir," "Khibin," "Ural" Facilities. The "Pamir," "Khibin," and "Ural" facilities developed and tested by the Kurachatov Institute of Atomic Energy are of considerable value to the development of MHD power plants. These facilities generate a power output of 50-100 MW during a short period of time (up to 10 s) [40]. Even though pulsed facilities are intended for deep electromagnetic sounding of the earth's surface, in terms of a number of indices (specific power output of 500 MW/m^3, specific power takeoff of 0.6 MJ/kg, velocity M = 2.4), they exceed commercial-scale MHD generators planned for production of electric power at large thermal electric plants. In addition to phenomena specific to this type of facility, other problems of importance for MHD generators of stationary MHD power plants were investigated. This includes electrode and insulation materials, current flow phenomena in the boundary layers and in the near-electrode region, boundary layer separation and its interaction with density gradients, boundary layer slant and its influence on the near-electrode losses and on the boundary layer separation, etc. It should be noted that, in terms of MHD energy conversion, Kurchatov Institute of Atomic Energy conducted a wide range of fundamental investigations of both practical and theoretical value. In particular, these investigations have shown that overheating and acoustic and convective instabilities may occur in low-temperature plasma. Ionization instability may also occur in nonequilibrium plasma.

ENIN-2 Facility. The Energiya Scientific Industrial Trust of Minenergo SSSR incorporated the ENIN-2 facility in developing an MHD generator with a power output of 10 MW, with continuous operation of 6 min. A large number of runs indicate stability of operation, stability of characteristics, as well as excellent overall stability of the structure in short-duration operating modes. When using cesium seed, the power output of the MHD generator reached 20 MW [41].

K-1 Facility. Important investigations of MHD energy conversion in accordance with a joint program with the Institute of High Temperatures are performed on a facility developed at the Institute of Electrodynamics of the Academy of Sciences of the U.S.S.R. [42]. All components of the facility, i.e., combustor, fuel supply system, oxidizer, and MHD generator, are being continuously upgraded. A series of experiments was conducted on the MHD generator equipped with segmented electrode modules with copper electrodes and rammed masses based on zirconium dioxide.

Supersonic operating modes with Mach number up to 1.25 along practically the whole length of the generator (3 m) were achieved in a non-load run at a magnetic field induction of 2 T and along a length of 0.8 m in a short-circuit mode. Different MHD generator designs were investigated, as well as phenomena in the end zones, the influence of various parameters on the nature and development of the interelectrode breakdown.

Various structural solutions of electrode and insulation walls of an MHD channel utilizing small-module design were investigated on the K-1 facility. This design is considered an alternative for commercial-scale MHD power plants.

1.2.4. Development of the First Commercial-Scale MHD Power Plant. The results achieved in theoretical and experimental investigations in the area of MHD energy conversion made it possible to initiate development of the first pilot, commercial-scale MHD power system now under construction at the Ryazan thermal power plant.

The following assumptions were made in formulating the design of the commercial-scale MHD power plant [43].

1. Specific fuel rate of the pilot MHD power plant has to be considerably lower than that of known alternative electric-power-generating facilities.

2. Technical solutions accepted in this project have to be verified experimentally. In particular, this requirement imposed limitations on the type of fuel to be used, i.e., at the present time, the only experience gained was that of MHD facilities operating on natural gas.

3. The technical level of new equipment has to be such that most of it can be fabricated by existing industrial facilities, or should require minimum modification, using minimum amounts of new materials. This condition was the limiting factor in choosing specific power output of assemblies and parameters of the working fluid. In this connection, the project includes the following compromise solutions. The power output of the facility is on the order of 500 MW (steam turbine, 300 MW, and MHD generator, approximately 200 MW). This power output is 2 to 4 times lower than optimal. At the same time, it is the minimum power output that demonstrates principal advantages of the MHD power plant. The oxidizer is to be preheated to the maximum temperature of 1700^{0}C in separate high-temperature cowpers, because greater oxidizer preheating or replacement of separate preheaters with regenerative units would require new materials, the fabrication of which would necessitate re-equipping refractory industry facilities.

4. Except for the steam generator, the steam turbine portion of the steam turbine has to be standard. When changing from the pilot system to small-lot production, it is possible to replace standard components of the steam turbine portion with components having characteristics optimum for MHD power plants.

The MHD generator of the commercial-scale MHD system is so calculated that enthalpy of its exhaust gases is adequate to provide the necessary quantity of steam required by the K-300-240 turbine. According to the plan, atmospheric air will be used as the oxidizer (flow rate of 230 kg/s) and, if preheating is insufficient, air enriched with up to 27 % oxygen (flow rate of 190 kg/s) will be used.

The same equipment is used in both cases. The variation involving operation with atmospheric air is somewhat more advantageous economically. However, utilization of enriched air makes the MHD generator less sensitive to possible deviations of other equipment from the calculated characteristics. This can be attributed to the higher initial enthalpy of the combustion products at the inlet into the MHD channel. The MHD power plant is equipped with a 2-cylinder axial compressor with an intermediate cooling unit. Each cylinder is driven by a 40-MW synchronous motor. Installation of a compressor driven by the turbine receiving gas after intermediate preheating is more economical. However, in this case, it would be necessary to develop a new steam turbine for the generator drive (i.e., steam prior to and after intermediate preheating would be substantially different). This solution is unacceptable for a pilot system. However, it will be reviewed when work is begun on nonpilot MHD power plants, and it is expected that a change will be made to a steam compressor drive.

Rational utilization of low-potential heat in the MHD cycle is very important. An MHD power plant differs from conventional steam turbine stations in that cooling exhaust gases at temperature below 300°C cannot be attained by means of an oxidizer, because it leaves the compressor at a high temperature. Part of the heat of the combusted fuel (7-9 %) is removed from the cooling system of the combustion chamber and the MHD generator at a temperature not exceeding 250-260°C. Utilization of this potential heat requires partial replacement of steam regeneration, as well as installation of additional heat exchangers, in which the heat of exhaust gases is used to preheat the fuel as well as air for high-temperature preheaters and for other purposes. Preliminary designs have shown that replacement of steam-turbine electric power plants by MHD power plants will make it possible to save up to 20 % of fuel.

1.3. Thermal Loops and Layout of MHD Power Systems

As with large steam-turbine electric power plants, power plants with MHD generators must consist of individual assemblies. The MHD power system includes the MHD facility (MHD generator with auxiliary devices and components), and steam generating equipment that is basically the same as that used in conventional thermal power plants.

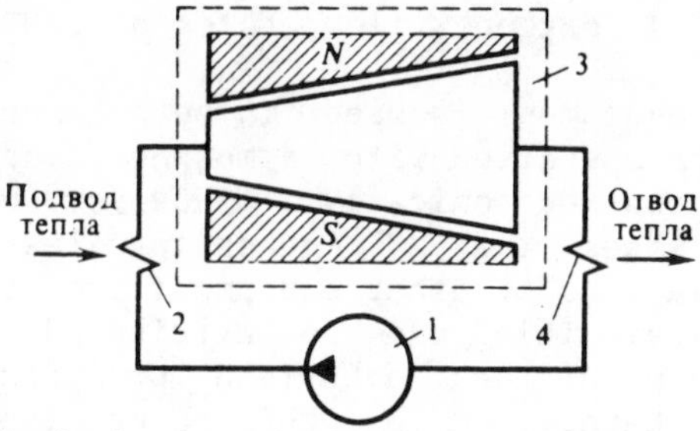

Fig. 1.5 Principal thermal loop of the simplest possible MHD facility, 1--compressor, 2,4--heat exchangers, 3--MHD generator

1.3.1. Thermodynamic and Basic Thermal Schemes of an MHD Facility. An MHD facility with gaseous working fluid, including combustion products of fuel, operates according to the Brayton thermal cycle (Fig. 1.5). Compressor 1 compresses the working fluid adiabatically, after which it is heated isobarically in heat exchanger 2. Downstream, in MHD generator 3, working fluid expands adiabatically and thermal energy is converted into electric energy. The cycle is closed by the heat transfer at atmospheric pressure utilizing heat exchanger 4.

In terms of thermodynamics, the cycles of gas-turbine and MHD facilities are identical. The difference is that thermal energy in MHD facilities is converted directly into electric rather than into mechanical energy.

A necessary condition of converting energy in an MHD generator is the capacity of the working fluid expanding in the MHD generator to conduct electric current. The higher the electrical conductivity of the working fluid, the more effective is energy conversion. According to calculations, minimum electrical conductivity of the working fluid at which energy conversion in an MHD facility is economically justified, is 2-4 S/m. In open-cycle MHD facilities in which the working fluid consists of combustion products of organic fuel, this conductivity can be reached only by introducing easily-ionizable seed into the combustion products. The seed consists of potassium, rubidium, or cesium compounds (usually potash, K_2CO_3). The static temperature of the flow cannot be below 2200-2300 K. Obviously, the static temperature of the working fluid from the MHD generator has to be at this level. This corresponds to stagnation temperature of 2450-2550 K. Naturally, the initial temperature of the working fluid must be even greater (preferably 2900-3000 K), presenting a fairly complex technical problem, because the combustion temperature of high-energy-content types of

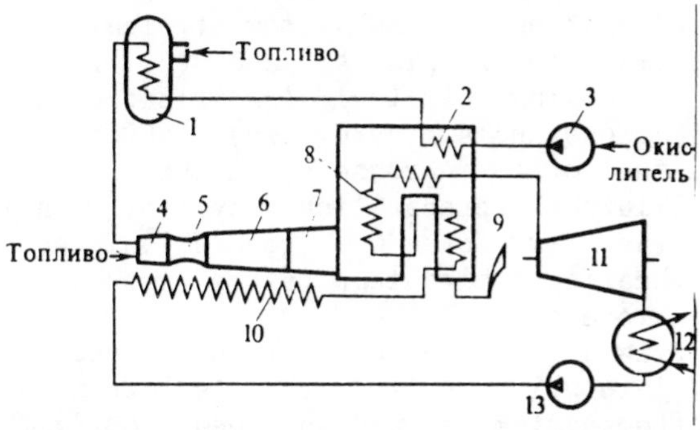

Fig. 1.6 Simplified thermal scheme of an open-cycle MHD power system, 1--separate oxidizer preheater, 2--oxidizer preheater utilizing flue gases from the MHD generator, 3--compressor, 4--combustor, 5--nozzle surfaces, 6--MHD generator, 7--diffuser, 8--steam generating surfaces, 9--exhaust, 10--cooling system of the combustor, MHD channel nozzle and diffuser, 11-turbine, 12--condenser, 13--pump

fuel in atmospheric air, even without heat conduction from the flame, does not exceed 2200-2250 K. Rational utilization of the heat of high-temperature gases leaving the MHD generator channel is just as important for the economical operation of the MHD facility. Devices utilizing this heat are part of the MHD power system and will be considered in detail below. Here, we want to become familiar with the simplified thermal cycle of an MHD power system (Fig. 1.6) that incorporates both conditions for effective operation. High temperature is attained by heating the air with initially exhausted gases (regenerative preheating) and then in a separate preheater (external heating). The heat of the exhaust gases is used not only to heat the air, but also to generate steam for the steam turbine. This basic scheme is being developed differently, according to specific goals and conditions. For example, one can include an air separator with the MHD facility. Oxygen produced by this facility is used to enrich air, increasing the combustion temperature.

1.3.2. Systems Plan. The simplified MHD power system under consideration includes only components through which the working fluid flows. A more complete representation is given by the schematic diagram of the MHD power system (Fig. 1.7) that includes the following subsystems and components.

I. Fuel preparation system, in which two subsystems may be isolated: (1) for the primary preparation of fuel (receiving, storing, pulverization, grinding, partial drying, and heating), a subsystem typical of all conventional thermal power plants, and (2) for the thermochemical processing of fuel (gasification or conversion and, for gaseous fuels, pyrolysis), which may be used to clean the fuel or to increase efficiency.

II. Oxidizer preparation system, including an optional facility for O_2 generation, equipment for oxidizer heating (high-temperature air preheater), and oxidizer compressor.

III. Plasma generator-combustion chamber for burning the fuel, into which is fed the fuel from system I, the oxidizer from system II, and the seed from system VI.

IV. MHD generator, the most important elements of which are the superconducting magnet and the channel.

V. Bottoming systems to utilize the heat of the MHD generator: (1) steam generation; (2) oxidizer heating, which can also be considered part of the oxidizer preparation system, and (3) some elements of the primary and thermochemical fuel processing subsystem, utilizing heat of the exhaust gases.

VI. Seed system, incorporating devices for receiving and storing, injecting the seed into the combustion chamber, recovering seed by removing it from the combustion products, regenerating the seed, and transporting the seed between different components.

VII. Electric system, incorporating the inverter, the high-voltage substation with transformers, and auxiliary subsystems.

VIII. Pollutant removal system.

IX. Plant control system, which includes the necessary plasma diagnostics instrumentation (connection between this system and other systems and components of the MHD power system is shown in Fig. 1.7 by dashed lines).

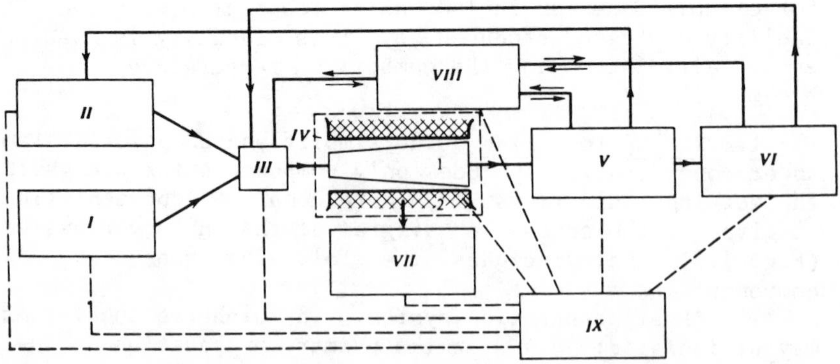

Fig. 1.7 MHD plant systems and components

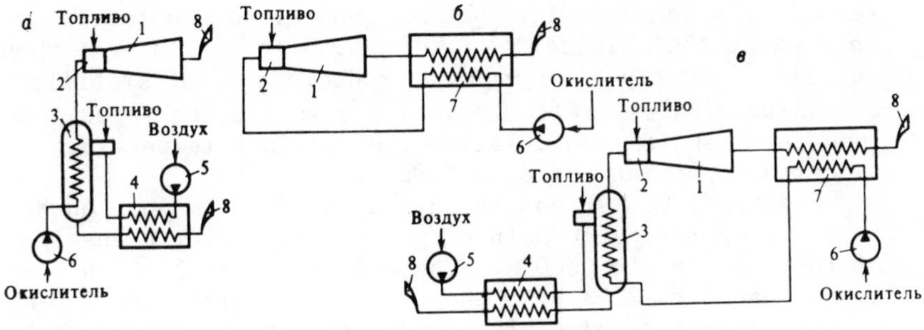

Fig. 1.8 Block diagram illustrating oxidizer preheating methods for simple MHD facilities: a--with separately fired oxidizer preheater, b--with directly fired oxidizer preheater, c--with both directly and indirectly fired preheaters in combination, 1--MHD generator, 2--combustor, 3--separately fired oxidizer preheater, 4--air preheater for separately fired preheater, 5--fan, 6--compressor, 7--directly fired oxidizer preheater, 8--exhaust

In addition, the MHD power plant must provide, as does a conventional power plant, for adequate water supply and for maintenance, and must perform auxiliary functions, such as slag and ash removal.

Open-cycle MHD power plants are classified in accordance with (1) the type of fuel used (coal, gas, oil), (2) the function (production of power only or combined production of electrical power and cogeneration of thermal energy), and (3) the operating mode (base load, intermediate load, or peak load). The characteristics and parameters of the components and equipment and the type of thermal loop are chosen in accordance with application of the MHD power plant. The thermal scheme depends on two most important factors: the method of obtaining high-temperature combustion products and the method of utilizing heat of the exhaust gases from an MHD generator.

1.3.3. Methods of Obtaining High Initial Temperature. Three different methods of obtaining high temperature combustion products are recognized. Two of these require preparation of the oxidizer (preheating and oxygen enrichment), and the third is connected with fuel preparation (heating, drying and thermal chemical processing) [44].

The oxidizer may be heated by the fuel combusted in the separate preheater (separate or indirect heating) (Fig. 1.8a), or by the combustion products leaving the MHD generator (regenerative or direct heating) (Fig. 1.8b).

A high-temperature ceramic blast-furnace preheater is utilized for separate preheating of the oxidizer. For gas combustion in this type of preheater, it appears to be technically feasible for first generation MHD facilities to preheat the oxidizer to 2000 K. In the future, when domestic industry will produce products made of stabilized zirconium dioxide refractories and sufficient experience in using them is accumulated, the oxidizer temperature is expected to be increased to 2300 K.

Figure 1.9 is an energy flow diagram of an MHD facility operating on natural gas [1] with direct preheating of the oxidizer to 2000 K. The efficiency is 19 % and can be increased to 21 % by preheating the oxidizer to 2300 K.

Regenerative (directly-fired) oxidizer preheating by heat of the exhaust gases (see Fig. 1.8) eliminates burning fuel in a separate preheater. Facilities with directly-fired oxidizer preheater temperature up to 2000-2300 K save 37-40 % of fuel, and their efficiency increases to 30-35 %, respectively. Regenerative heating of the oxidizer encounters a number of difficulties. The seed solidifies during combustion of coal and heavy oil. This also happens to slag during cooling of the combustion products, and may lead to slagging over gas ducts in the lining. In order to prevent slagging, combustion products must exit the preheater at a temperature at which the mixture is fluid, i.e., 1400-1600 K, which leads to considerable complexity in designing the preheater. Additional difficulties appear in connection with the fact that lowering the temperature of the gas down to 1500 K leads to seed condensation and to penetration of the pores of ceramic refractories by the liquid phase, destroying them by physiochemical processes. Apparently, this phenomenon can be eliminated by using high-density ceramics made of fairly pure magnesium dioxide that, according to current investigations, adequately resists the harmful effect of seed up to 2050 K (preheat temperature of the oxidizer, up to 1900 K).

Recuperative-type preheaters may be used to heat the oxidizer. However, in this case, tubes for heating surfaces must be sufficiently heat-resistant to withstand internal pressure and high thermal compensation stresses. In addition, the material of the tubes must be easily weldable. Present day materials (metals) having such properties [45, 46] may be used to construct a preheater to heat the oxidizer to 1100 K. Tube-type preheaters to such temperature may be used in facilities in which the oxidizer consists of air enriched with up to 50 % of oxygen and, also, in combined oxidizer preheating (Fig. 1.8c). In the latter case, a tube-type preheater is a

[1] Here, and in subsequent analysis, it will be assumed that the gas consists only of methane, CH_4.

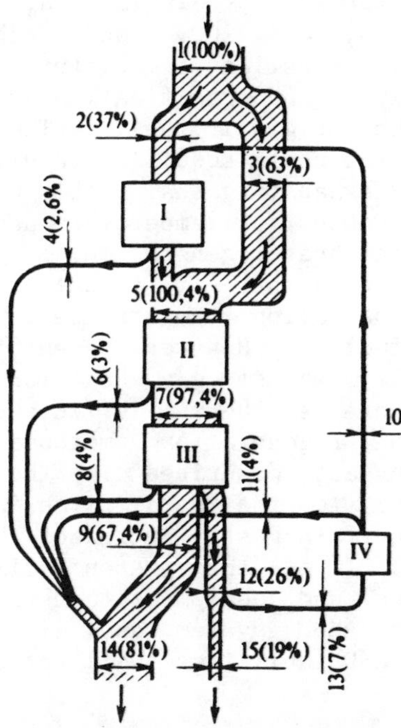

Fig. 1.9 Energy flow diagram of an MHD facility with separate oxidizer preheating to 2000 K, I--oxidizer preheater, II--combustor, III--MHD generator, IV--compressor and auxiliary machinery; 1--incoming fuel, 2--fuel into the oxidizer preheater, 3--fuel into the combustor, 4--heat losses in the oxidizer preheater, 5--oxidizer and fuel into combustor, 6--heat losses in combustor, 7--combustion products upstream of MHD generator, 8--heat losses in MHD generator, 9--losses with exhaust gases, 10--oxidizer heat, 11--losses in the compressor and auxiliary mechanisms, 12--resulting electrical power, 13--electrical power to compressor (6 %) and auxiliary mechanisms and losses in inverter (1 %), 14--total losses, 15--net output

Here, and in subsequent analysis, it will be assumed that the gas consists only of methane, CH_4.

first stage of heating the air by combustion products leaving the MHD generator (regeneration), with subsequent heating to a higher temperature in a separate, blast-furnace-type preheater. As a result of the partial regenerative preheating of the oxidizer of the MHD facility, the efficiency of the combined preheating facility is approximately 2 % (absolute) higher than for a nonregenerative facility (see Fig. 1.8a), i.e., approximately 4 % of fuel is saved.

As has been noted, the combustion temperature can also be increased by preheating, drying, and thermochemically processing the fuel. Examples of thermochemical fuel processing are methane pyrolysis, $CH_4 = C+2H_2$, coal gasification, $C+CO_2 = 2CO$, and other endothermal reactions. The combustion temperature of the products obtained by pyrolysis and gasification is higher than for combustion of the original fuel. These reactions are endothermic, and the heat required should be supplied by the MHD generator exhaust gases. This makes it possible to increase the combustion temperature and simultaneously to achieve chemical heat regeneration.

It should be noted that it is not always possible to obtain higher combustion temperature by thermochemical processing of fuel. However, even in such cases, utilization of the exhaust gas heat for the endothermic reactions can increase thermal efficiency considerably. Numerous chemical regeneration methods of increasing thermal efficiency are described in the literature [44, 47-50]. According to the data given in [44], chemical regeneration in certain cases increases the efficiency by 4-6 % (absolute). At the present time, however, the necessary methods and equipment are not sufficiently developed for the MHD system and, in all probability, first-generation MHD facilities will not utilize chemical regeneration.

1.3.4. Using Generator Exhaust Gas Heat in Low-Temperature Cycles. The preceding discussions show that the efficiency of an MHD facility utilizing regenerative oxidizer preheating, even up to 2300 K, is only 35 %. Forty percent of the heat of fuel is exhausted into the atmosphere through the stack. The temperature of the exhaust gases is approximately 1500 K. Adding chemical regeneration increases the efficiency to 40 %. However, even in this case, approximately 30 % of the heat of the combusted fuel would be lost with the exhaust gases.

In first-generation MHD facilities, the heat of the exhaust gases will most likely be used for limited preheating of the oxidizer in the metal oxidizer preheater. The efficiency of such a facility is estimated to be 21-23 %, and the heat losses through exhaust gases would exceed 60 %. Therefore, using exhaust gases solely for regenerative heating cannot make effective use of their heat content. A number of binary schemes have been suggested for utilizing exhaust gas heat in low-temperature cycles [51-53].

As shown in the schematic diagram, Fig. 1.10a, the air turbine (6) operates on the MHD generator (1) oxidizer. The oxidizer temperature upstream of the turbine is limited by heating conditions in the metal oxidizer preheater (4) and does not exceed 1000 K. Two-stage expansion of the working fluid in the turbine and in the

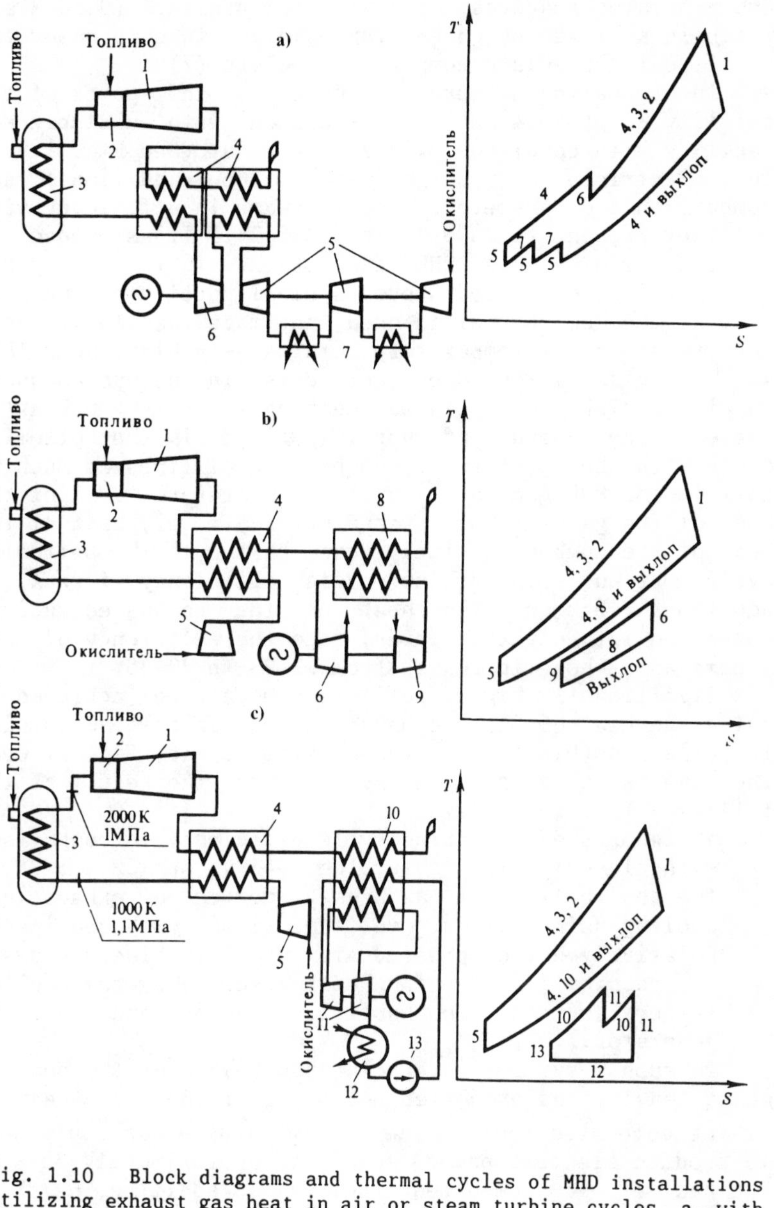

Fig. 1.10 Block diagrams and thermal cycles of MHD installations utilizing exhaust gas heat in air or steam turbine cycles, a--with first-stage air turbine, b--with turbine, c--with steam turbine, 1--MHD generator, 2--combustor, 3--separately fired air preheater, 4--directly fired air preheater, 5--compressor, 6--air turbine, 7--intermediate compressor-cooler, 8--air preheater, 9--turbine compressor, 10--steam generator, 11--steam turbine, 12--condenser, 13--feed pump. Numbers in the T,S diagrams correspond to the elements in the block diagram in which the processes occur

MHD generator requires high initial pressure (9-10 MPa), which is achieved by combusting oxidizer in compressor (5) with two intermediate compressor-coolers (7).

Incorporating a turbine increases the power of the facility 12 to 14 % in comparison with a simple MHD facility with combined oxidizer preheat (see Fig. 1.8c). This electric power is generated without additional fuel combustion and, therefore, the increase in efficiency with oxidizer preheat to 2000 K is 23.5-25.5 %, as opposed to 21-23 % for a simple MHD facility. In the binary MHD configuration described above, a small portion of the heat leaving the MHD channel is used, no more than 15 %. As a result, the stack temperature remains very high, near 1800 K. This disadvantage does not exist in the system shown in Fig. 1.10b which is a combination of two Brayton cycles. The amount of working fluid in the bottoming cycle with the air turbine can be several times as much as that in the MHD generator cycle. Therefore, 90 % of the heat of the gases from preheater (4), i.e., 55 % of the heat of the combusted fuel, can be fed to the bottoming cycle. At an air turbine cycle efficiency of 22 %, an additional 12 % of the heat of the combusted fuel is converted into electric power, and the efficiency of this scheme as a whole increases from 21-23 to 33-35 %.

Significantly higher efficiency can be achieved by using the heat of the exhaust gases in a steam turbine installation (binary Brayton-Rankine cycle, Fig. 1.10c). The advantages of such a system are the result of the following factors:

- a higher steam turbine-cycle efficiency in comparison with closed-cycle gas turbines (43 versus 22 %);
- the possibility of a higher degree of exhaust gas cooling using water, in comparison with cooling by relatively hot compressed air in air turbine systems;
- the possibility of using heat absorbed by the cooling systems of the combustion chamber and the MHD generator.

In such a system, water can absorb 95 % of the heat of gases leaving the preheater (4). This is 60 % of the heat of the combusted fuel. The steam turbine uses this heat to produce electric power equal to approximately 25 % of the heat of the combusted fuel. In addition, 6-7 % of combusted heat transferred by the combustion chamber and the generator cooling systems can be used at an efficiency of approximately 25-30 %. The additional electric power produced will be approximately 2 % of the heat released. The efficiency of the system with a steam turbine will be 47-48 %, compared with 33-35 % for the system with a gas turbine.

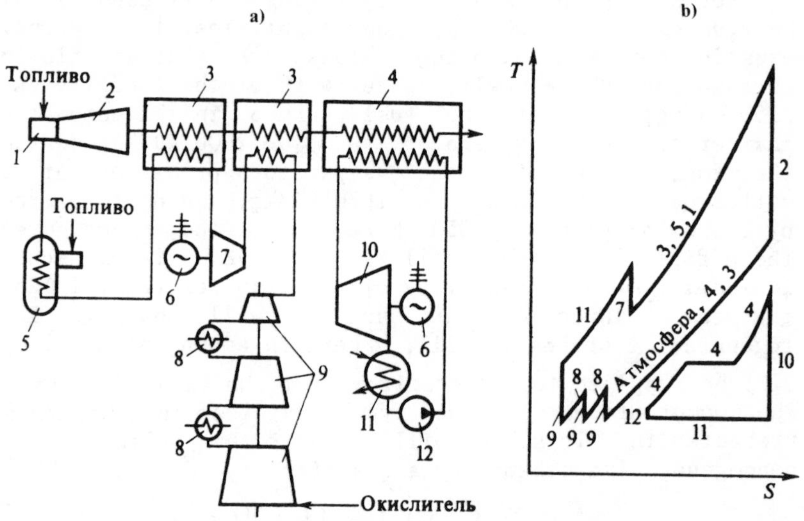

Fig. 1.11 Block diagram (a) and thermal cycle (b) of an MHD energy system with a combined steam gas-turbine system, 1--combustor, 2--MHD generator, 3--regenerative oxidizer preheater, 4--steam generator, 5--separate oxidizer preheater, 6--turbogenerator, 7--air turbine, 8--intermediate cooler, 9--compressor, 10--steam turbine, 11--condenser, 12--pump

The preceding analysis clearly demonstrates a substantial difference in thermal efficiency of facilities with different turbines in the low-temperature (bottoming) stage of the binary cycle (see Fig. 1.10). A facility with a steam turbine is the most efficient.

In addition to a steam turbine (10), the block diagram in Fig. 1.11 also includes an air turbine (7) operating at a pressure gradient between 4.7 MPa and 1 MPa at an air turbine inlet temperature of 1000 K. This scheme results in a gain of 0.25-0.35 % of fuel in comparison with the previous scheme (the temperature of the separate air preheat in schemes shown in Figs. 1.10 and 1.11 is 2000 K). Such a small increase of thermal efficiency does not justify additional facility complexity. Hence, the most rational scheme for first-generation MHD power plants is that utilizing the heat of the exhaust gases in the steam generating cycle.

1.3.5. Preheating and Oxygen Enrichment of Oxidizer in Binary MHD Power Systems. The efficiency of a binary MHD power system with a steam turbine and a separate preheater (Fig. 1.12a) may be represented in the form

$$\eta = L\,(q_{c.ch.} + q_{sep}) = L\,(q_{c.ch.} + q_{ox}/\eta_h)\ .$$

For MHD power systems operating on atmospheric air heated to T_3 = 2000 K, the quantities in the above equation have the following values per kg of gas flowing through the MHD channel: useful work of the facility, L = 2150 kJ/kg; heat of the fuel burned in the combustion chamber of the MHD generator, $q_{c.ch}$ = 2800 kJ/kg; heat of the fuel burned in the separator preheater at an efficiency η = 0.93, q_{sep} = 1670 kJ/kg, and heat absorbed by the oxidizer q_{ox} = 1550 kJ/kg. For the values given, the efficiency of the facility is 0.48. If this scheme includes combined preheating of the oxidizer (Fig. 1.12b), the heat balance of the facility will change. In a regenerative preheater (11), a certain amount of heat, q_p, will be supplied to the oxidizer from the exhaust gases. The amount of heat absorbed by the oxidizer in a separate preheater in this scheme will be lower than in the nondegenerative scheme, by a quantity

$$q'_{ox} - q_p = q_{ox} \times (1 - \mu) ,$$

where $\mu = q_p/q_{ox}$ is the regenerative coefficient.

The fuel combusted in the separate preheater in a facility with combined oxidizer preheating is given by the expression

$$q'_{sep} = (q_{ox} - \mu q_{ox})/\eta'_h = q_{ox}(1 - \mu)/\eta_h ,$$

where η'_h is the efficiency of the separate preheater for a combined oxygen preheating scheme.

Owing to preheating of the oxidizer by exhaust gases, the heat transferred from the exhaust gases to water and vapor will be q_p times smaller than in the nondegenerative scheme.

As a result, the power output of the turbine and of the overall facility decreases by an amount

$$\Delta L = \eta_{s.t.} q_p = \eta_{s.t.} \mu q_{ox}$$

($\eta_{s.t.}$ = 0.43 is the efficiency of the steam turbine) and the efficiency of the overall facility with combined oxidizer preheating is given by the following expression:

$$\eta' = \frac{(L - \Delta L)}{(a_{c.ch.} + q'_{sep})} = \frac{(L - \eta_{s.t.} \mu q_{ox})}{(q_{c.ch.} + q_{ox}(1 - \mu)/\eta'_h)} .$$

Assuming the same data as above and setting $\eta'_h = \eta_h$, the regeneration coefficient from μ = 0 (nondegenerative

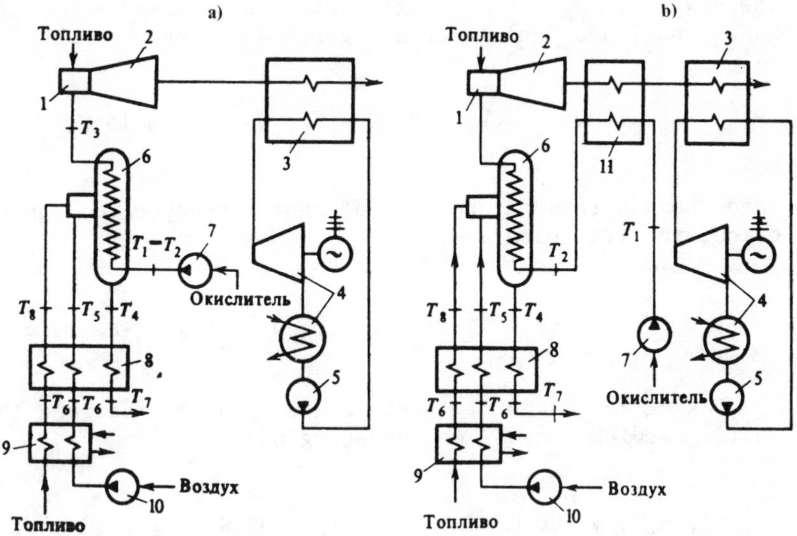

Fig. 1.12 Block diagram of binary MHD facilities with separate (a) and combined (b) oxidizer preheating, 1--combustor, 2--MHD generator, 3--steam generator, 4--steam turbine with a condenser, 5--pump, 6--separate preheater, 7--compressor, 8--preheater of air to be combusted and fuel for separate oxidizer preheater, 9--preheater, 10--fan, 11--oxidizer preheater operating on exhaust gases of the MHD generator

scheme) to $\mu = 0.3$ (regenerative heating to approximately 1000 K) leads to an almost linear growth of efficiency from 48 % to 49.2 %. This corresponds to a decrease of the specific fuel consumption by approximately 2.5 %.

Assumption of equality of efficiencies of separate oxidizer preheaters for different preheating schemes ($\eta_h = \eta_h'$) is not always fulfilled. This can be attributed to an increase of losses with gases from separate preheaters that may occur with increasing intermediate (regenerative) oxidizer preheat temperature T_2. Therefore, in a number of cases $\eta_h' < \eta_h$. This can be attributed to the following characteristics of the scheme (see Fig. 1.12b). Flue gases exiting preheater (6) at temperature T_4 heat the air and fuel gas to be combusted, in heat exchanger (8) to temperature T_5 and T_8, respectively, and then exit the heat exchanger at temperature T_7. To eliminate moisture formation in the heat exchanger (8), the air and fuel gas are heated in preheater (9) to a temperature $T_6 \simeq 350$ K. For the scheme under consideration, heat balance of heat exchanger (8) can be written in the form

$$W_{s.g.}(T_4 - T_7) = W_a(T_5 - T_6) + W_{n.g.}(T_8 - T_6) ,$$

where $W_{s.g.}$, W_a, $W_{n.g.}$ are water equivalents of flue gases, air, and fuel (natural) gas. Assuming

$$W_{s.g.} = W_a + W_{n.g.} \ (W_a \simeq 0.85\ W_{s.g.}\ W_{n.g.} \simeq 0.15\ W_{s.g.}) \ ,$$

then the gas temperature drop at heat exchanger (8) can be calculated from equation

$$T_4 - T_7 = 0.85\ (T_5 - T_6) + 0.15\ (T_8 - T_6) \ .$$

Taking into account temperature gradient between the heating medium and the medium being heated

$$(T_4 = T_2 + 100\ \mathrm{K},\ T_5 = T_8 = T_4 - 50\ \mathrm{K} = T_2 + 50\ \mathrm{K}) \ ,$$

we obtain

$$T_7 = T_6 + 50 \ .$$

Because, to prevent appearance of moisture, T_6 is assumed to be 350 K, from previous equality it follows that

$$T_7 = 400\ \mathrm{K}\ (127°\mathrm{C}) \ , \qquad (1.1)$$

i.e., temperature of the exhaust gases under conditions assumed is independent of the regenerative oxidizer preheat temperature, T_2. This is a consequence of equality of the water equivalent of the heating medium being heated. Improved calculations show that, in practice, water equivalents are somewhat different. However, the error does not exceed 10-12 K.

In deriving this relation, it was assumed that $T_5 = T_8 = T_2 + 50$ K. This condition cannot always be maintained, i.e., when $T_2 > 700$ K the preheat temperature of fuel gas $T_8 > 750$ K, which is not allowable, because natural gas decomposes at temperature exceeding $T_8^{max} = 750$ K, forming pyrographite. For this reason, in the case when $T_2 > 700$ K, heat balance has the form

$$W_{s.g.}(T_4 - T_7) = W_a(T_5 - T_6) + W_{n.g.}(T_8^{max} - T_6) \ ,$$

or

$$T_2 - 100 - T_7 = 0.85(T_2 + 50 - 350) + 0.15\ (750-350)\ .$$

From this, when $T_2 > 700$ K we obtain

$$T_7 = 0.15\ T_2 = 0.15\ T_2 + 295\ . \tag{1.2}$$

When $T_2 \leq 700$, Equation 1.1 remains valid.

From Equations 1.1 and 1.2, it follows that, as T_2 is increased from 700 K to 1000 K, the temperature of exhaust gases increases by 45 K. This results in a noticeable drop in the efficiency of the separate preheater. At the same time, when the temperature of the oxidizer at the inlet of the separate preheater (T_2) changes from its value downstream of the compressor (T_1) to 700 K, the temperature of the exhaust gases does not increase. For this reason, it is expedient to limit oxidizer preheat to $T_2 \simeq 700$ K. This is very important, because at this temperature, it is not necessary to use alloyed steels to fabricate heat exchangers (8) and (11).

In developing this scheme, it should be kept in mind that even if regenerative oxidizer preheating is excluded, i.e., assuming

$$T_2 = T_1 \simeq 570\ K,\ T_5 = T_8 = T_2 + 50 = 630\ K\ ,$$

at this preheat temperature of fuel and air, the flame temperature in the combustor of preheater 6 approaches 2400 K. Such high flame temperature may result in local heating of lining to 2250-2300 K. Modern refractories can withstand heating to temperatures up to 2000-2050 K. Therefore, the flame temperature cannot exceed 2100-2150 K.

The flame temperature is lowered by recirculating part of the flue gases from the separate oxidizer preheater (6 in Fig. 1.13) into the combustion zone. Two ways of connecting recirculation smoke pumps are possible: directly downstream of the separate oxidizer preheater (6 in Fig. 1.13a) and downstream of the air and fuel preheater (8) in Fig. 1.13b. The use of the first variant

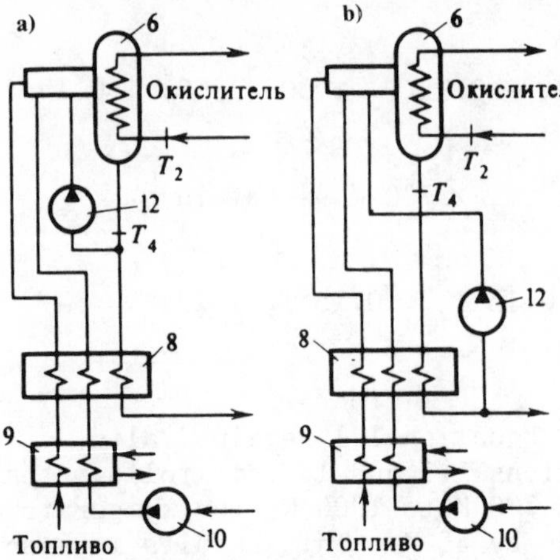

Fig. 1.13 Recirculation scheme with partial gas rerouting downstream from the separate oxidizer preheater (a) and combined (b) oxidizer preheating, 1--combustor, 2--MHD generator, 3--steam generator, 4--steam turbine with a condenser, 5--pump, 6--separate preheater, 7--compressor, 8--preheater of air to be combusted and fuel for separate oxidizer preheater, 9--preheater, 10--fan, 11--oxidizer preheater operating on exhaust gases of the MHD generator, 12--recirculation induced-type (ID) fan

is limited by the fact that, at the high temperature of the flue gases downstream of preheater 6 ($T_4 = T_2 - \Delta T$, where ΔT is the temperature gradient), it is not always possible to find the required smoke pump (high-power smoke pumps normally operate at a temperature of 600-620 K). Temperature T_4 can be decreased by decreasing ΔT. However, this will result in an increase of the size of the separate preheater and of its cost. A drop of T_4 in a scheme that does not include a regenerative oxidizer preheater (see Fig. 1.12a) may be attained by decreasing T_2, using a compressor with an intermediate cooler. This may lead to a drop in facility efficiency. Temperature T_2 may be decreased by using regenerative preheating (see Fig. 1.12b). This, however, may lead to a decrease of T_4; in which case the efficiency of the assembly will also drop. Finally, it is possible to use a scheme in which the recirculation smoke pump is connected downstream of the air and fuel preheater (8), where the temperature of flue gases is sufficiently low (see Fig. 1.13b). An increased amount of flue gases passing through the heat exchanger (8) is cooled by the same amount of air and fuel as in the first variant. This apparently leads to a drop in the temperature of exhaust gases and a drop of efficiency.

The best solution would be to increase the operating temperature of the smoke pump, with direct installation downstream of the oxidizer preheater (see Fig. 1.13a) at optimal values of T_2 and ΔT. Otherwise, it is necessary to compare the technoeconomic variants, taking into account the factors listed above that decrease the efficiency and increase capital expenditures. In blast-furnace-type oxidizer preheaters, the fuel is compressed at a pressure close to the atmospheric pressure, while the pressure of the oxidizer being heated is 1-1.2 MPa. Such a pressure difference complicates construction of the preheater. In this connection, schemes have been considered in which the pressure of the combustion products and of the oxidizer is the same (Fig. 1.14). This is achieved by supplying compressed air that also serves as the working fluid of the air turbine for the combustion gas. In such schemes, transition from seed preheating to its cooling is simplified considerably. As a result of this, the optimum duration of a preheater operation cycle decreases from 1-1.5 h to 10-15 min. This allows the use of highly effective pebble bed lining, which decreases the size and cost of the separate preheater. In addition, the design of gate valves is simplified and air losses from poor sealing are decreased. In the high-pressure oxidizer preheater scheme, the

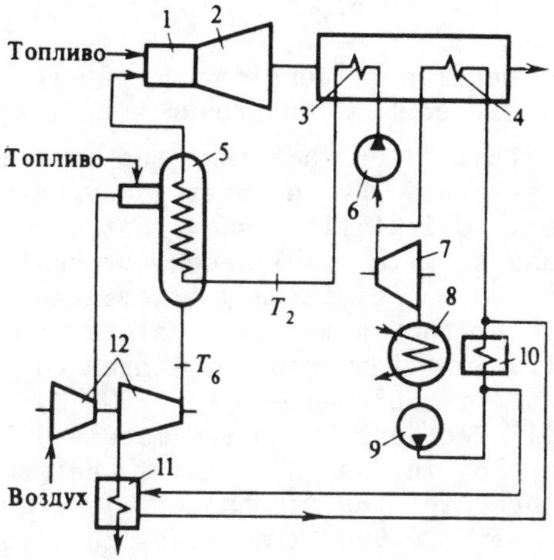

Fig. 1.14 Block diagram of an MHD facility with high pressure oxidizer heating, 1--combustor, 2--MHD generator, 3--oxidizer preheater using exhaust gases of the MHD generator, 4--steam boiler, 5--separate oxidizer preheater, 6--oxidizer compressor, 7--steam turbine, 8--condenser, 9--feed pump, 10--steam regeneration system, 11--turbine condensate preheater utilizing the exhausted gases, 12--gas turbine

temperature of flue gases upstream of the turbine, T_6, coincides with the temperature downstream of the separate preheater that, similar to the scheme in Fig. 1.13, is determined by the temperature of the regenerative oxidizer preheater, T_2, and the temperature gradient, $\Delta T = T_6 - T_2$, at the "cold" end of the preheater. At initial temperature typical of a gas turbine $T_6 = 1023$ K (750°C), and $\Delta T = 100$ K, the temperature of the regenerative oxidizer preheater, T_2, is 923 K (650°C). Technically, this is quite feasible. The principal thermodynamic deficiency of the scheme with high-pressure oxidizer preheat is that cooling of exhaust gases from the turbine is by means of the steam turbine condensate of that part of the facility, which leads to a decrease of deficiency of the latter because of partial replacement of steam regeneration. However, even with this factor taken into account, the additional production of electric power by the gas turbine is at a high efficiency, i.e., efficiency of 48-50 %. Because the efficiency of an MHD facility is at the same level, connection of a gas turbine in accordance with this scheme exerts practically no influence on the overall efficiency. Even though some complication is introduced, because of the presence of three types of energy converters (MHD generator, steam and gas turbines), as was expected, this scheme will make it possible to decrease specific capital expenditures, primarily as a result of the lower cost of a separate preheater.

In the schemes considered above, increasing temperature of the combustion products upstream of the MHD generator was attained by heating the oxidizer. Similar effect can be attained by increasing oxygen concentration in the oxidizer. Calculations show that oxygen enrichment of up to 30 mass % with simultaneous preheating to 2000 K does not increase efficiency compared with using atmospheric air at the same temperature as the oxidizer. If the concentration of oxygen is increased to 50 mass %, and the oxidizer is preheated to 2000 K, specific heat consumption will decrease by approximately 3 %.

Supersonic flow in the MHD channel and pressure on the order of 4.0 MPa in the combustor is required in this case, instead of 800-900 kPa with fuel compression in atmospheric air. Development of such equipment encounters certain technological difficulties because of the high pressure, increased danger of fire, etc.

The danger of fire decreases considerably when operating with up to 50 % oxygen-enriched oxidizer, but with relatively low regenerative preheating (1000-1050 K) in a metal heat exchanger only (separate high-temperature preheater is not installed, see Fig. 1.8b). In this case, specific heat consumption is 3-3.5 % higher than when air preheated to 2000 K in a separate preheater is used as the oxidizer. At the same time, the absence of a costly high-temperature preheater in this scheme can, in some cases, compensate both the drop in heat efficiency and additional capital expenditures for oxygen generating facilities. In particular, similar schemes may be rational for power systems intended to meet half-peak load requirements, because:

- productivity and cost of the oxygen equipment may be decreased considerably because oxygen may be produced and stored during the time the power system is not operational;
- electric energy consumed to produce oxygen during the night is calculated only from the heat component;
- additional losses associated with the necessity of maintaining operating mode temperature when other parts of the facility are not operational become necessary in schemes utilizing regenerative-type preheaters.

<u>1.3.6. Compression of the Oxidizer.</u> It is known that the minimum amount of compression work required in a compressor with a single intermediate cooling unit (Fig. 1.15) will be performed when

$$\pi_{LP} = \pi_c^{1/2}(T_1/T_3)^{-\sigma/2}, \quad \pi_{HP} = \pi_c^{1/2}(T_1/T_3)^{\sigma/2}, \tag{1.3}$$

where $\pi_{LP} = p_1/p_2$ is the increase in pressure of the low pressure cylinder, $\pi_{LP} = p_4/p_3 \approx p_4/p_2$ is the increase in pressure of the high pressure cylinder, $\pi_c = \pi_{LP}\pi_{HP}$ is the increase in the total pressure of the compressor,

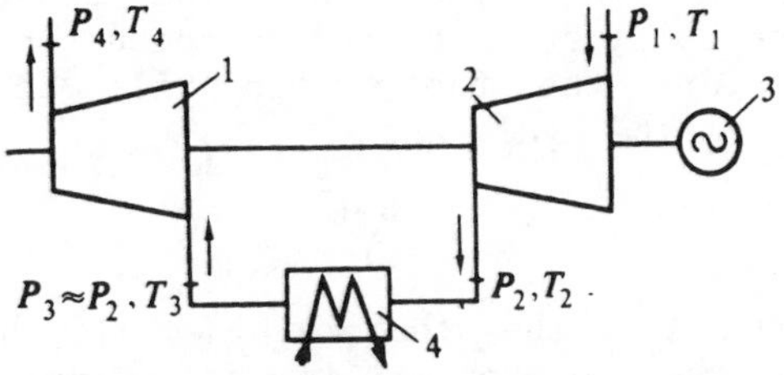

Fig. 1.15 Oxidizer compressor, 1--single intermediate cooling unit high-pressure cylinder, 2--low-pressure cylinder, 3--compressor drive, 4--intermediate cooler

$\sigma = (k - 1)/k\eta$, k is the adiabatic factor and η is the polytropic efficiency of the expansion/compression processes.

In the absence of intermediate cooling and when π_c = 10, T_1 = 300 K, the work performed to compress the air is 345 kJ/kg, whereas the work performed for intermediate cooling to T_3 = 300 K requires 290 kJ/kg, i.e., is lower by ΔL = 55 kJ/kg. At the same time, in the first case, the air temperature downstream of the compressor is 620 K, whereas it is 420 K in the second case. For this reason, heating of the oxidizer in the second case will require 200 kJ/kg more of heat, q. Such an energy balance is justified only when the heat expended to preheat the oxidizer is used in the cycle of a station with efficiency $\Delta L/q$ = 0.274. In real life, in the absence of intermediate cooling, the heat, q, is used in a steam turbine with an efficiency of 42-43 %. This is more advantageous in terms of the energy. In this connection, it is recommended that a compressor without intermediate cooling should be used in the facilities [54]. However, such a solution is not always optimal. It can be shown that the highest overall efficiency of the MHD generator with a single intermediate cooling unit as the compressor is achieved when the pressure increase in the low pressure combustor is

$$\pi_{LP}^{oHT} = \pi_c^{1/2}(T_1/T_3(1 - \eta_Q\eta_M\ \phi))^{-1/2\sigma}, \tag{1.4}$$

where η_M is the mechanical efficiency of the compressor, $\eta_Q = \Delta L/q$ is the efficiency at which the heat of the oxidizer is used to preheat the oxidizer downstream from the compressor, T_4, to the temperature of the oxidizer at the compressor outlet without intermediate cooling, i.e., to temperature $T_1\pi_c^{\sigma}$. ϕ is the factor that takes into account the type of drive being used.

If, before being fed into the separately-fired preheater, the oxidizer is heated by combustion gases from the MHD generator, the fuel saved will be used in the steam portion of the cycle and

$$\eta_Q = \eta_{St}^{net},$$

the efficiency net of the steam turbine. If, however, the oxidizer from the compressor is fed directly into the separately-fired preheater, real fuel economy will be achieved and η_Q will be equal to the efficiency of the overall facility, i.e., for an MHD plant, η_0 = 0.48-0.50.

For an electric drive, $\phi = \eta_{el}$ is the efficiency of the electric drive, including losses in the transformer and the distribution network. For a turbine drive, $\phi = \eta_{aux}/\eta_{st}$, the ratio of efficiencies at the shaft of the drive turbine to that of the main turbine.

For a steam-operated drive turbine, where the steam is diverted from the main turbine, with sufficient accuracy $\phi = 1$. For typical values, π_c = 10–12, we obtain from Equation 1.4

$$\pi_{LP}^{opt} = 1.5\text{–}1.6 ,$$

and the temperature of air downstream from the low pressure compressor will be on the order of 360 K. If the water used for cooling is at a temperature close to T_2, the air will be cooled only to a small degree. In this case, one has either to increase compression of the low pressure compressor, but not by more than

$$1.5\ \pi_{LP}^{opt},$$

or eliminate an intermediate cooler (i.e.,

$$\pi_{LP} > 1.5\ \pi_{LP}^{opt}$$

practically does not increase efficiency of the facility). Increasing π_{LP} above

$$\pi_{LP}^{opt}$$

is necessary, even when the oxidizer pressure in the cooler is lower than the cooling water pressure. This occurs because, in this case, it is possible to have unwanted water leakages into the oxidizer through seals in the tube-type cooling system. Intermediate cooling may be required to drop the temperature of smoke gases downstream of the separate oxidizer preheater to a level that allows switching on the recirculation smoke fan downstream of the oxidizer preheater (see Fig. 1.13a). For MHD facilities operating on high-temperature oxidizer with a high oxygen concentration, it is necessary to have high initial pressure of the working fluid. In this case, it is possible to install a compressor with several intermediate stages.

For a number of cooling units, z, (corresponding to z + 1 cylinders), the maximum efficiency of an MHD facility is attained for the following compression in each cylinder:

for a cylinder at the oxidizer inlet

$$\pi_1 = \pi_c^{1/(z+1)}(T_1/T_3(1 - \eta_Q\eta_M\phi))^{-1/(z+1)\sigma} ;$$

for all intermediate cylinders (except the first and the last)

$$\pi_i = \pi_1(T_1/T_3)^{1/\sigma} ;$$

for a cylinder at the oxidizer outlet

$$\pi_{z+1} = \pi_c/\pi_1\pi_i^{z-2} .$$

1.3.7. Methods of Utilizing Low-Potential Heat. One of the principal problems encountered in selecting the system configuration for MHD generator facilities is the development of methods for utilizing the heat of heat-transfer media whose temperatures are below 570-590 K. When separately-fired preheaters are used and the oxidizer is air at a temperature of 2000 K, such low-potential heat per kilogram of combustion products flowing through the MHD channel is estimated to be (kJ/kg):

In the combustion products
(cooled from 570-590 K to 390-410 K)
of the MHD generator 215
separate air preheater 125

In the heat transfer medium
of the cooling system
of the MHD channel (electrode wall temperature of 2000-2100 K, insulation wall temperature of 1500-1600 K) 200
combustion chamber (wall temperature of 2000 K) 140
Total 680

For several reasons, the heat transferred from the cooling systems of the combustion chamber and the channel is low-potential heat, at least for the first generation of MHD power plants. Cooling the combustion chamber and the MHD generator channel is a complex technical problem, because of large specific-heat fluxes and the necessity of insulating the cooled combustion chamber and the channel components from the Hall field. (They are connected by means of ducts for circulation of the heat transfer medium.) The higher the temperature of the heat transfer medium, the more effectively the heat can be used, but the more complex and less reliable the cooling system. For this reason, the estimates given below assume that the maximum temperature of the heat-transfer medium of the channel and combustor cooling system is 520-540 K. Steam at a pressure of 5 MPa can be generated at this temperature.

As the technology matures, an MHD channel and the combustor might be cooled by feedwater circulated through all the conventional seed-water regenerative heaters, in which case its heat content will no longer be considered low potential. Low-potential heat can be used to heat "cold" flows to a temperature range of 520-540 K; the heat requirements of such flows, recalculated per 1 kg of combustion products flow through the channel, are:

Air	
for separately fired preheater.	135
to complete combustion in the steam generator[2]	25
Fuel gas	
for MHD generator	35
for separately fired preheater.	20
Total	215

Based on the information presented above, one can make a conclusion that the amount of the low potential heat available is more by 465 kJ/kg, or 10 % of the fuel combusted in the system, than can be used for heating the "cold" flows. In a coal-based power plant, the low potential heat would be required to dry coal, but it would not change significantly the overall heat balance.

2 In order to suppress nitrogen oxides, the fuel upstream of the MHD generator must be combusted at 5-10 % oxygen deficiency. The incomplete products of combustion are burned in the steam generator; however, this requires an amount of air equal to 10-15 mass % of the combustion products.

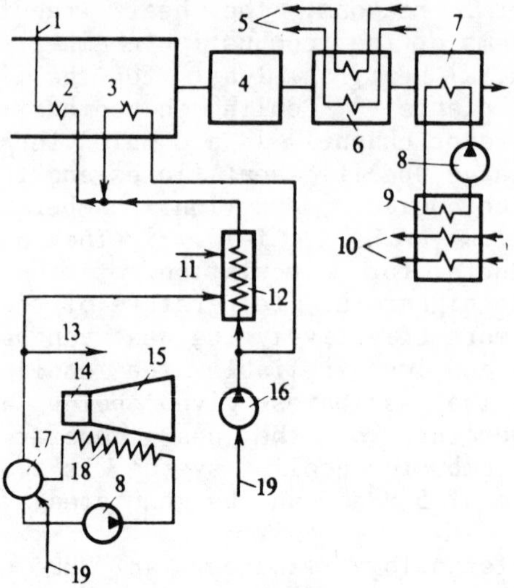

Fig. 1.16 A scheme for utilizing low-potential heat, 1--feedwater into the radiant heat surfaces, 2,3--main and first stage economizers, 4--electric filters, 5--air to complete combustion and fuel for the combustor, 6--fuel and air preheaters, 7--water preheater for 9, 8--blower, 9--air preheater, 10--fuel and air for separate oxidizer preheaters, 11--steam from the turbine, 12--high-pressure preheaters, 13--steam into the intermediate superheater, 14--combustor, 15--MHD generator, 16--feed pump, 17--separator, 18--cooling elements of the combustor and the MHD channel, 19--feedwater from the deaerator

In designing the specific thermal components, it is not always possible to preheat cold flows to the temperature of 520-540 K by using the low-potential heat and, hence, the excess may increase to 500 kJ/kg. This excess heat may be utilized to heat the feedwater, e.g., in turbines, to supercritical steam pressure (Fig. 1.16). In this case, the feedwater flow, G, at a temperature of 170^{0}C is split downstream of the pump into two flows. Flow G′ is fed into the first-stage economizer, cools the exhaust gases from 590 K to 490 K, and itself is heated to 540 K. To achieve this cooling of gases and heating of water, it is necessary to maintain an approximate equality of water equivalents of the heating medium and medium being heated. This is attained when $G' \simeq 0.3\ G$. The other flow, $G'' = G-G'$, fed through a group of high-pressure preheaters receiving steam both from the cooling system and from the turbine, is at a temperature 540 K and is mixed with the flow from the first-stage economizer. After this, the total flow enters the main economizer.

If more steam than is required to preheat flow G′′ is generated in the cooling system, the excess can be directed toward intermediate steam superheater and then into the medium-pressure cylinder of the turbine.

Seed is removed from the products of combustion leaving the economizer, before further cooling. The temperature at which the seed is removed (490 K) is chosen by taking into account the fact that the seed has a tendency to adhere to the heating surface at lower temperatures. Further cooling of the exhaust gases from the natural gas-fired MHD generator is achieved by introducing additional air supplied for completing combustion of incompletely burned combustion products from the MHD generator, and flue gas for the MHD generator. Air and flue gas heated to 370-400 K can cool the exhaust gases only to 440 K. Therefore, provisions are made for additional water cooling of exhaust gases in the heat exchanger. The water is then utilized to preheat the air and fuel for the separate oxidizer preheater, to 350-360 K. If cooling is inadequate, the exhaust gases can be cooled by using them to heat a portion of the condensate from the turbine, eliminating low-pressure steam regeneration. For example, to lower the exhaust gas temperature from 430 K to 400 K, it would be necessary to divert approximately one third of the turbine condensate from the low-pressure steam generating system and heat it from 370 K to 420 K by the exhaust gases. This will lower the efficiency of the turbine by approximately 0.35 %. Nevertheless, fuel consumption of the assembly would still decrease by 0.6-0.7 % in comparison with the case where the temperature of the exhaust gases is 430 K. It should be noted that viability of the scheme based on the use of a condensate is questionable. Considering all factors, it may be preferable to accept lower thermal efficiency rather than increase the complexities of the thermal portion of the installation.

1.3.8. System and Layout of the MHD Power System. Figure 1.17 is a schematic block diagram of the first commercial-scale MHD power system under construction in the Soviet Union [43]. The MHD power system is intended to operate on gas. The scientific and experimental results of operation of this system will be used in developing solid-fuel MHD facilities. The power system is planned to operate as part of a condensation thermal power plant. Because the MHD generator becomes effective at a power output of not less than 250-350 MW, which corresponds to the total power output of an MHD power system of 550-750 MW, the steam-turbine section of the facility must produce the balance of the necessary power output. This requirement is met by the K-300-240

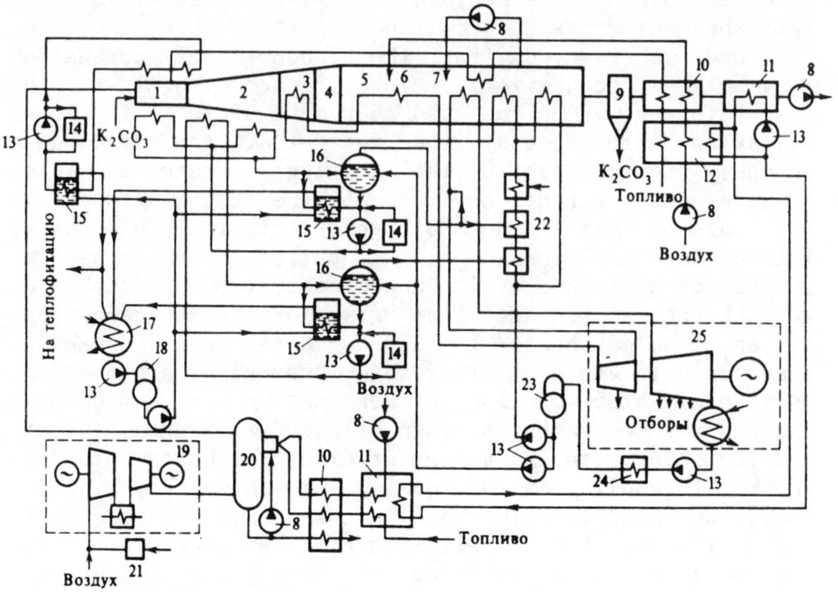

Fig. 1.17 A schematic block diagram of a pilot MHD power system, 1--combustor, 2--MHD generator, 3,4--water-cooled and ceramic diffusers, 5--steam generator, 6--air injection to complete combustion, 7--input for recirculation, 8--blowers, ventilators, 9--electric filters, 10--fuel-gas and air preheaters, 11--water heaters for preheater, 12--preheater, 13--pump, 14--salt- and iron-extractor (from water), 15--evaporator, 16--separator, 17--condenser, 18--deaerator, 0.12 MPa, 19--two-shaft compressor with intermediate cooling, 20--separate oxidizer preheater, 21--air separator, 22--high-pressure preheater, 23--deaerator, 0.7 MPa, 24--low-pressure preheaters, 25--steam turbine

condensation turbine at nominal power output of 300 MW, operating on steam with parameters of 23.5 MPa, 540/540°C. The oxidizer preparation system includes the following components:

- A compressor rated at 207 kg/s and pressure of 1.07 MPa (electric drive),
- Separate blast-furnace-type preheaters to heat the oxidizer to 1973 K,
- A reserve air separator to enrich oxygen up to a concentration of 30 mass %.

The oxidizer preheated in a separate preheater, fuel-gas heated by the exhaust gases of the MHD generator, and seed enter the flow-through combustor. In order to decrease concentration of nitrogen oxides in stack gases, combustion in the combustion chamber takes place at an oxygen deficiency, $\alpha = 0.9$. A cooling system is incorporated to protect the lining and components from excessively high temperatures. Plasma with stagnation parameters of 2950 K, 1.0 MPa leaving the combustor enters a diagonal MHD generator channel with the following parameters:

Magnetic field induction at the inlet/outlet . . . . 6/4 T
Electric power output . . . . 250 MW
Combustion products flow. . . . . 220 kg/s

The gases from the MHD generator exit at a velocity of 850 m/s and stagnate in the diffuser. The initial section of the diffuser is actually an extension of the channel and shares the cooling system with it. The velocity of the flow along this section decreases to 300 m/s. Farther on, the gases enter the section of the diffuser formed by the water-cooled pipes connected to the boiler circulation loop, which actually are part of the boiler. The temperature of gases at the outlet from the water-cooled section of the diffuser is approximately 2000 K. The flow past the uncooled section of the diffuser, made of refractory materials, then enters the radiation chamber of the boiler, analogous to blast furnace chambers of conventional boilers. In this chamber, the gases are cooled to 1670 K. The total residence time of gas at a temperature above 2670 K is approximately 2 s, including approximately 1.5 s in the radiation chamber.

This time is adequate for the reduction of nitrogen oxides by incompletely combusted products of combustion (carbon oxide and hydrogen in the exhaust gases of the MHD generator, produced during combustion at an oxidizer deficiency $\alpha = 0.9$). Combustion of residual carbon oxide and hydrogen is completed downstream of the cooling chamber. A specific characteristic of operation of the steam generator of the MHD facility is that seed deposited in liquid form on the fire surface solidifies, forming a fairly dense layer, i.e., a process analogous to slagging during coal combustion takes place. This problem is solved by recirculating gases exiting from the economizer, into the temperature zone of 1350 K. The amount of recirculated gas is chosen so that the flow temperature after mixing drops to 1200 K. Thus, the slagging temperature range is excluded from the heat exchange.

Cooling of the section of the combustor and the channel under the highest thermal stresses (burners and accelerator), as well as certain auxiliary components, is by means of water at a temperature of 130^{0}C, which, after being heated in the combustor in the MHD channel, passes through the evaporator. The heat from the water is transferred to the secondary steam that flows into a condenser. The remaining components of the gas- dynamic duct are cooled by boiling water at a pressure of 5 MPa (combustor in outlet section of the MHD channel) and a pressure of 2 MPa (inlet section of the MHD channel). From components being cooled, the steam-water mixture enters separators, the steam from which is diverted to high-pressure preheaters and into intermediate superheaters. During runs, evaporators transfer heat from the circulating cooling water, the steam from which flows into a condenser. All cooling loops are closed. Salt and products of corrosion in the cooling water are extracted in each of the loops separately, in specially constructed condensate desalinization devices.

Feeding the steam generator with water, cooling of gases leaving the steam generator and the high-temperature oxidizer preheater takes place in accordance with schemes shown in Figs. 1.13a and 1.16. Technical solutions for the pilot system are not necessarily optimal, but have been accepted on the basis of reliability and simplicity considerations, in order to simplify operation of new, unique equipment. In particular, oxygen enrichment of the oxidizer makes it possible to attain rated plasma parameters without relying on the development of an air preheater heated with exhaust gases. A compressor with an electric drive used in the MHD power system (see Fig. 1.18) is less economical than a compressor with a turbine drive. However, because bleeding a large amount of steam to the drive turbine of the compressor cannot be attained in a conventional turbine, a new turbine for the generator drive would have to be designed and constructed. In addition, frequent start-ups and shutdowns of the compressor, required during the first stage of the

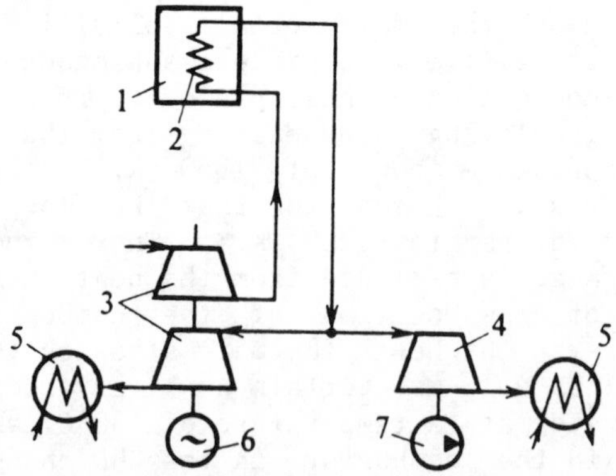

Fig. 1.18 Block diagram of the compressor drive, 1--steam generator, 2--intermediate steam superheater, 3--main turbine, 4--compressor turbine, 5--condenser, 6--turbogenerator, 7--compressor

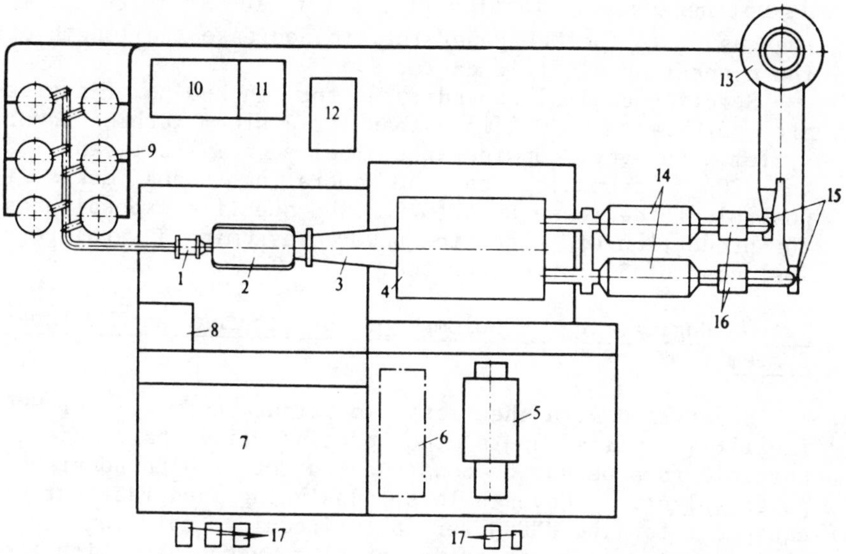

Fig. 1.19 Layout of the MHD power system (planned), 1--combustor, 2--MHD generator, 3--diffuser, 4--steam generator, 5--steam turbine, 6--condenser, 7--inverter substation, 8--seed injection unit, 9--high-temperature oxidizer preheaters, 10--compressor compartment of the air separator, 11--cryogenic facility, 12--oxidizer compressor, 13--smokestack, 14--electric filters, 15--pumps, 16--heat exchangers of the exhaust gases cooling system, 17--transformers

shakedown tests, are easier with an electric drive. This simplification will not be needed in the future. What's more, separate oxidizer preheating can probably be replaced with regenerative preheating. Adding chemical regenerative preheating and making other improvements will make it possible to increase the efficiency from 47-49 % at the present level, to 54-56 %.

An example of the layout of an MHD power system is shown in Fig. 1.19. In order to provide optimum conditions for the exhaust gases at the inlet of the boiler, the front of the latter is turned toward the side of the MHD room, rather than the turbine, as in the case of conventional thermal power plants. High-temperature oxidizer preheaters are located so that a straight sector of the oxidizer duct, approximately 20 times its I.D. in length, extends in front of the combustor. This is necessary for aerodynamic flow stabilization. Combustor, MHD generator, and diffuser are located farther on along the same axis. The diffuser configuration is chosen to provide effective (low-loss) stagnation of the flow. Increasing the length of the oxidizer duct or the diffuser is not necessary. The inverter substation is moved closer than usual to the MHD generator, to decrease the length of the connecting electric cable.

Specific capital expenditures for an MHD power system are approximately 20 % more than for a steam turbine power system. However, considerable fuel economy (between 17.5 and 20 % depending on the operating mode) makes it possible to decrease by 6 to 9 % the specific expenditures for production of electric power, assuming an arbitrary fuel cost of 30 rubles per ton.

1.4. Engineering Problems of Developing an MHD Power System

Experience with the most important subsystems of MHD facilities, gained up to the present time, have made it possible to proceed to construction of a high power MHD power plant. Nevertheless, designing and fabricating equipment for the MHD power plant requires solutions to a whole complex of engineering problems associated with the peculiarities of operation as part of an MHD power plant.

Let us consider the major requirements imposed on the most important subsystems of the MHD power system and individual structural solutions.

1.4.1. Combustor.

Major requirements imposed on the combustor of an MHD power plant. The combustor of an open-cycle MHD power plant is intended to produce ionized plasma at specified parameters (temperature, electrical conductivity, pressure) as the working fluid in an MHD generator. Accordingly, construction of the combustor must meet the following requirements:

- temperature of 2600-3000 K,
- high degree of ionization of alkaline seed and calculated electrical conductivity of the combustion products at the inlet of the MHD generator channel,
- uniformity of the flow and constant, time-independent flow parameters at the inlet of the channel,
- limitation of heat transfer from plasma to the combustor walls,
- suppression of formation of toxic nitrogen oxides in the combustor and provisions for proper conditions for their decomposition in the gas duct of the MHD power plant.

Effective operation of the combustor requires considerable length (in perspective, many thousands of hours) of operation. This imposes a qualitatively new level of requirements on the reliability of high-temperature combustors, and must be taken into account in choosing materials, design, fabrication, and operating conditions of combustor walls, mixers, and their cooling systems.

Heat transferred from the walls of the combustor must be used with the greatest efficiency in the thermal loop of the electric power plant. A heat carrier with high energy parameters, e.g., high pressure water or oxidizer, can be used for this purpose to extract the high heat fluxes from the plasma to the walls.

Axial Hall potential is induced in MHD generator channels. It can reach 20-30 kV in commercial-scale MHD generators. The steam generator adjacent to the channel is grounded. Therefore, the combustor floats at a high electrical Hall potential. It thus becomes necessary to isolate the combustor electrically with respect to all pipes and ducts (fuel, oxidizer, seed, cooling water, trapped slag), support structures, and communication lines with the control and measuring equipment.

Cooling of the combustor. The construction and material of the combustor walls of an MHD power plant must provide reliable, long-duration operation at high heat flux densities and cooling of the combustor. When coal is used, this problem is complicated by the presence of liquid slag in the combustion products.

Various investigations have demonstrated that a considerable drop in the density of the heat flux to the walls may be attained by utilizing the following:

- refractory coating of water-cooled surfaces and in between surfaces,
- slag coating of water-cooled surfaces,
- porous ceramic walls of the combustor, through which part of the oxidizer is injected.

Porous cooling of the walls can apparently be used only in gas and residual oil combustors. Studies have shown that partial separation of the plasma flow from the walls makes it possible to decrease convective heat transfer and to decrease heat flux density in the cylindrical portion of the combustor to 1 MW/m^2. According to investigations performed at the Institute of High Temperatures, a promising technique is to construct the fire walls of the combustor of porous ceramic cooled by injection of some of the oxidizer through the wall. The combustion chamber must be designed to utilize this oxidizer for combustion. Development and experimental tests of porous ceramic capable of long-duration operation under those conditions is required.

The influence of seed on the construction of the combustor. Analysis of data of theoretical investigation of the evaporation time and ionization of seed injected into the combustor in the form of a dry powder and spray droplets of a 50 % aqueous solution of K_2CO_3 shows that, in the case of natural gas and heavy oil, the combustor residence time required for complete combustion of the combustion products is shorter than the vaporization and seed ionization time. Therefore, when choosing the size of the chamber for combustion of gas and residual oil, it is necessary to start from the following seed vaporization times given below.

Dry injection of K_2CO_3 . . . . 30-50 ms
50 % aqueous solution of K_2CO_3 . . . . . 25-35 ms

When coal power is combusted, the residence time of combustion products in the chamber, assuming complete combustion, exceeds the vaporization time of dry seed. Therefore, the size of the chamber for coal combustion should be chosen in accordance with the time required to combust the powder (on the order of 80 ms).

Choice of design guaranteeing plasma uniformity. The provision for providing the most uniform flow parameter distributions along the cross section at the inlet into the MHD generator channel is best met for pure ash-free fuel (either gas or residual oil) by utilizing flow-through design of the combustor. High thermal stresses and short residence time of combustion products make it difficult to equalize flow parameters over the cross sections of large combustors. Therefore, the principal means of equalizing and stabilizing processes of obtaining a uniform plasma within the chamber is highly uniform distribution of oxidizer, fuel, and seed in mixers at the inlet end wall of the combustor.

When choosing a gasdynamic scheme of the combustor and analyzing combustion, it is necessary to determine conditions that prevent or decrease acoustic oscillations. In addition, fuel self-ignition in the high-temperature oxidizer requires external combustion stabilization, achieved by generating small local recirculation zones at the outlet from the mixer. It is also necessary to design the profile of the channel so as to eliminate unwanted compression and vibration of jets and to provide quiet entrance of the hot mixture into the combustor at a moderate velocity.

The experience gained in investigating the operation of combustors in the U-02 and U-25 facilities [23, 24] shows that combustion stability, with pressure pulsations not exceeding 5 % of the mean pressure in the chamber, can be attained by changing the nature of flow in slits (choice of proper geometry) and by controlling combustion in neighboring jets.

Single-chamber, flow-through combustors are now used for gas combustion. Effective combustion of fuel and production of uniform ionized plasma at the inlet to the MHD channel are achieved by proper mixing, self-combustion and stabilization of combustion, and proper seed ionization and seed distribution along the flow cross section. Single-chamber flow-through combustors are also used for residual oil combustion.

Figure 1.20 shows one possible design of a combustor for a 1000-MW thermal power plant. According to the layout of the combustor and the MHD generator, the axis of the combustor is horizontal. Structurally, the combustor consists of a frontal burner, 1; cylindrical sector, 2; and acceleration nozzle, 3. Seed, 4, is injected some distance from the frontal device, responsible for uniform injection of natural gas and oxidizer. Combustion and seed ionization are completed in the cylindrical portion of the chamber. The relationship between the diameter and length of the chamber is determined from conditions of minimum heat losses.

The internal surface of the combustor will be made of ceramic based on Al_2O_3 rammed between water pipes cooled by boiling water. According to the plan, cold water will be used to cool the frontal device and nozzle sector where heat fluxes reach 3 MW/m^2. Section A-A along the collector of the frontal device shows the holes for outlet of the natural gas and the cooling system pipes. All communication cables of the combustor and the support structure are electrically isolated.

The use of pure fuels, natural gas, and residual oil makes it possible to simplify the fuel preparation system of the MHD power plant, i.e., combustor and fuel injection system. Combustion of such fuels eliminates contamination of the fire surface of the steam generator by ash deposits, and simplifies the seed regeneration scheme.

Characteristics of coal combustors. The advantages of coal-fired MHD power plants, i.e., protection of the walls of high-temperature components of the system by slag coating, which lowers stability requirements of high-refractory materials, as well as higher electrical conductivity of the combustion products in comparison with those from the gas and residual oil at the same temperature, may simplify the problem of high-temperature oxidizer preheating.

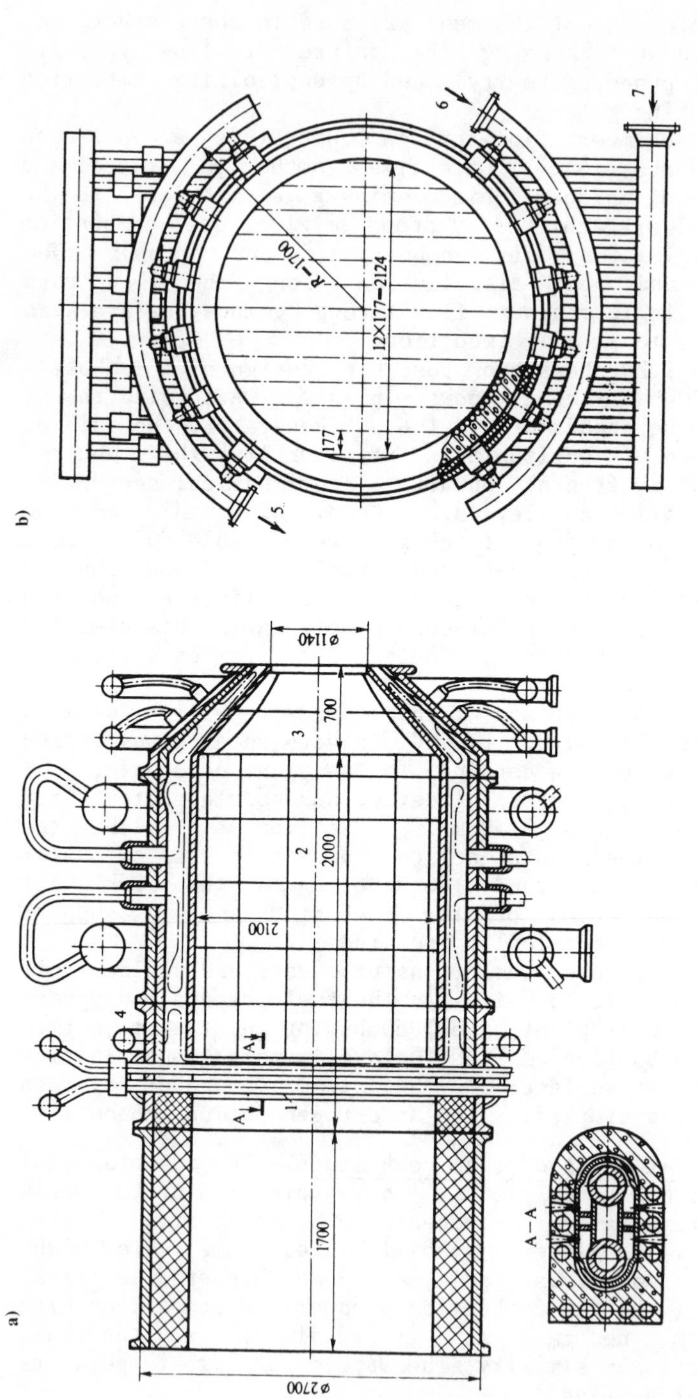

Fig. 1.20 Longitudinal section (a) and cross section (b) of the combustion chamber of the MHD power plant, 1--frontal burner, 2--cylindrical section, 3--nozzle, 4--inlet seed collector, 5,6--cooling water input and output, 7--natural gas supply

Development of an MHD power plant operating on coal requires solution of many complex, specific problems, such as feeding of pulverized coal into the combustion chamber under pressure, extraction of the trapped slag from the chamber, prevention of seed losses to slag, electrical isolation of the combustion chamber from coal supply and trapped slag removal lines, cleaning of fire surfaces from ash deposits with seed, seed regeneration in the presence of ash, and additional complexity of high-temperature oxidizer preheating.

The following aspects are being considered at the present in the development of coal combustors.

- Fuel combustion without slag entrapment when all of the fuel ash is exhausted with combustion products from the combustor, passing through the MHD channel and the steam generator.
- Fuel combustion in the combustor equipped so that most of the fuel ash is entrapped in the form of liquid slag, and its removal outside the facility. The untrapped mineral part of the fuel, in the form of fluid and gas, is exhausted from the chamber with the combustion products, passing through the MHD channel and steam generator.
- Combustion of fuel in the facility equipped with a device for intermediate fuel gasification and complete separation of gasification products and ash. The pure gas obtained is combusted in the combustor.

The choice of the combustion method and the design of the combustor for solid fuel depend on the degree to which fuel or its combustion products are purified of ash (slag) and the methods used to achieve this goal, as well as techniques to prevent loss of seed with slag. A flow-through aerodynamic flow, possibly with a swirl to provide stability of the slag coating of the walls, is used when developing one-chamber combustors without slag entrapment.

Some of the molten mineral particles formed during combustion of pulverized fuel in a cyclone combustor are separated from the gas flow and thrown onto the combustor walls by centrifugal forces. Particles of different chemical composition react, forming an almost homogeneous mineral melt, i.e., liquid slag. Proper tapping of liquid slag on combustor walls results in the walls being covered by a flowing, continuously replenished liquid slag. To maintain the stable operating mode of the combustor, a tapping system continuously removes slag melt flowing down the walls. Investigations performed on small-size combustors [55] (diameter, 250-450 mm) have shown that, as a result of interaction of seed with slag deposited on the walls, the latter may include up to 20 % of the potassium injected into the chamber. Such losses cannot be allowed in commercial-scale operation.

Promising new ways of limiting the amount of seed in slag are being developed. One is the use of aerodynamic means to decrease the incidence of particles of seed on the combustor walls by proper injection of pulverized coal and seed. Another technique is the use of thermal factors to prevent the reaction of seed and slag when they come into contact. To prevent seed losses with slag, the temperature of the slag film must exceed the transition temperature of seed compounds into condensed phase.

Protection from emission of nitrogen oxides. At present, it is planned that two-stage, high-temperature fuel combustion will be used to decrease emission of nitrogen oxides. Incomplete combustion of a fuel-rich mixture occurs during the first stage, at a deficiency of the high-temperature oxidizer (α = 0.9–0.95). This provides plasma with temperature and electrical conductivity close to the maximum values. The second stage (complete combustion of incompletely combusted products of combustion) takes place in the steam generator at a temperature of approximately 1600 K, at which no significant amounts of nitrogen oxides are produced.

The most important aspects of research and development of combustion chambers for commercial-scale MHD power plants are: increasing efficiency of mixing, injection, and seed ionization systems, increasing stability and thermal conductivity of refractory linings, experimental tests of cooling the walls by injecting part of the oxidizer through porous walls, determination of the optimum degree of removal of fuel ash from products of combustion, elimination or sharp reduction of seed losses to slag, improvement of injection systems of pulverized coal under pressure, development and investigation of systems of removing trapped slag, electrical isolation of the combustion chamber with respect to all external connections.

1.4.2. MHD Generator Channel. The MHD generator includes a channel, magnet system, and current converter system. All of this equipment is new to a power station, and the design of the MHD generator (high enthalpy extraction) to a considerable degree will determine the success of the MHD energy conversion method. Special requirements are imposed on construction of the MHD generator, because direct conversion of thermal and kinetic energy into electric energy takes place in the channel.

Experimental results obtained on various MHD facilities, particularly, stable operation of the MHD generator of the U-25 facility [56, 57] verify the feasibility of developing real MHD generators with predetermined parameters.

The flow-through section of the MHD generator includes the nozzle, the channel, and the diffuser. The shape of subsonic and supersonic nozzles is determined by using well-known gasdynamic calculation methods. Because of the high heat fluxes, cooling is the most difficult aspect of nozzle design.

Design and development of diffusers with effective restoration of plasma pressure at the outlet of the MHD generator is apparently a complex problem. This can be attributed to the fact that the data available on aerodynamic diffusers do not correspond to conditions existing at the inlet into the diffuser of an MHD generator, where electromagnetic interaction in the channel leads to significant distortion of boundary layer profiles.

The design of an MHD channel takes into account calculated Faraday and Hall electric fields, pressure distribution and heat fluxes to the walls. In addition, the MHD generator must have mechanical integrity, must be gasproof, must make effective use of the magnet system, and must effectively transfer heat fluxes of 1-5 MW/m^2.

Electrical connections. Hall field makes it possible to load the channel in several different ways, resulting in a Faraday, diagonal, and Hall MHD generator channels.

In a Faraday generator, the duct is axially segmented, and each pair of transverse electrodes is connected to a separate load. Induced emf induces a current transverse to the flow into the opposing electrodes. The channel consists of two segmented electrode and two segmented insulation walls. Because the current in each cross section can be controlled independently by its own inverter, the Faraday channel provides flexible and reliable loading of the generator. Theoretical investigations and calculations have shown that a Faraday generator is highly efficient in nominal mode and somewhat less efficient in part load mode.

However, a Faraday generator is quite complex structurally, because of the fine segmentation of electrode and side insulation walls, which requires electrical isolation of a multitude of individual pegs from each other and from the channel frame. Furthermore, it is necessary to provide reliable cooling of a multiplicity of pegs and electric isolation along the feed lines of cooling water. Operation of such a complex structure may result in damage from water leakage.

In a Hall generator, opposing electrodes are short-circuited and the transverse current works through its Lorentz force against the expanding gas. However, the Hall field drives a current to a pair of power electrodes located at the inlet and outlet of the generator, where power is extracted. It is a single-load generator of relatively simple electromechanical design (each transverse section of the channel is an equipotential). However, at Hall numbers below five, feasible in commercial-scale MHD power plants, the internal efficiency of a Hall generator is lower than acceptable for base-type MHD power plants.

A Faraday generator can be easily converted into a diagonal generator by connecting an anode with a cathode at the same potential. This connection can be made externally with a cable, or internally by using diagonal structural bars. The slant of the equipotentials can vary from the inlet to the outlet. Single or multiple loads can be used. In the nominal mode, a diagonal generator operates similarly to a Faraday generator in the absence of transverse Hall current. Hall current is always present when the channel deviates from the nominal operating mode and the efficiency of the generator drops.

Principal types of construction. Structurally, MHD generator channels consist of a support housing made of electrically insulating and thermally stable material (sometimes steel), inside of which are located modules or frames cooled by water, electrically isolated from each other and forming the flow-through section of the channel. The cross section may be rectangular, circular or oval. External cooling manifolding is used. The MHD generator is connected with the combustion chamber and the steam generator by decoupling flanges. Three principal types of MHD channels are known at the present time. In the first type, two segmented electrode and two finely segmented insulation walls form a channel of rectangular cross section. The load-bearing panels are joined and sealed, forming an impervious vessel. Because the insulation walls are finely segmented, any type of electrical load (Faraday, diagonal, Hall, mixed) can be used in such a channel. The load-bearing panels are made of glass reinforced plastic or steel. In the latter case, it is necessary to isolate each module of the channel from the housing carefully [34]. This type of channel is used in generators at the U-25 facility, as well as facilities of the Krzhizhanovskiy Power Institute and the Avco-Everett and Westinghouse firms.

In the second type of channel construction, insulation walls are made of full or segmented diagonal members and the electrode walls are segmented. The load-bearing wall panels are made of glass reinforced plastic. The cross section of such a channel is rectangular. Diagonal modules and electrodes may have external connections. Such channels are intended for operation with diagonal loads. However, they can also be used with a Faraday load. This mode of construction was used for the Mark VI generator and for the U-25 facility.

In the third type of construction method, the channel is made of cooled window frames, oriented primarily along the assumed ideal equipotentials. Window-frame channels should be used with a Hall or a diagonal load. These channels use external coolant manifolding with good integrity of the cooling system. This is the type of design to be used for the planned MHD generator of a prototype power plant.

Figure 1.21 shows a longitudinal section of the channel of a pilot MHD power plant (conceptual design). The channel is placed within the warm bore of a superconducting magnet system. The flow-through part of the channel diverges nonlinearly along the length with the cross section having the shape of a regular octagon. Water-cooled frames forming the flow-through portion of the channel are at an angle with the longitudinal axis and are insulated from each other with high-temperature magnesium oxide-based ceramic material. The plasma facing surface of the frame is covered with thermally stable,

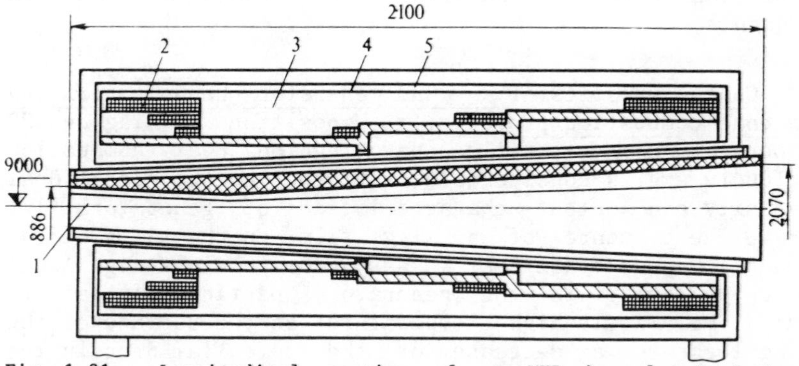

Fig. 1.21 Longitudinal section of an MHD channel and of the superconducting magnetic system, 1--active volume of the MHD generator channel, 2--superconducting coil, 3--helium cavity of the cryostat, 4--radiation shield, 5--vacuum cavity

electrically conducting zirconium oxide-based mass. Several frames are fed water from a single manifold, electrically insulated from another manifold and from the ground. The cooling water inlet and outlet connecting pipes also serve as current leadouts.

One of the most important problems is that of developing electrodes and electrode walls capable of long-duration operation. Numerous attempts to develop electrodes dealt with fabrication, and investigation of special electrode materials and fabrication of electrodes, providing optimum conditions of their operation and long lifetimes. The advantages of metal electrodes are easy fabrication; long lifetime at very intense cooling; extremely high thermal conductivity; electrical conductivity and thermal stability, and excellent strength characteristics. However, arcs are formed on metal electrodes, with Hall breakdown easily occurring between the arcs. Such electrodes are also characterized by high erosion.

At surface temperature above 2000 K, high-temperature electrodes made of graphite and ceramic based on zirconium dioxide and chromites can provide diffusion current flow from plasma to the electrode. This decreases the probability of interelectrode breakdown and electrical erosion of material. An important property of sprayed or rammed-mass electrodes based on zirconium dioxide is the fairly simple restoration by repeated spraying or ramming concrete mass without dismounting the channel.

The presence of strong induced and Hall fields and, consequently, the probability of breakdown both along the channel and transverse to it necessitates good electric isolation between individual electrodes and between segments.

Characteristics of an MHD generator channel operating on coal combustion products. Deposition of slag on MHD generator channel walls operating on coal combustion products exerts considerable influence on the operational capacity and output characteristics of generators [17, 18]. The presence of a slag film on the walls of the channel may decrease heat losses and maintain a high level of conductivity in the generator, particularly at its exit. Relatively high temperature on the surface of the slag leads to a decrease of the potential drop in the boundary layer. However, current losses in transverse and axial directions in highly-conductive slag degrade parameters of the generator.

The greatest difficulties encountered in developing a commercial-scale MHD power plant [4] is the Hall field, the intensity of which cannot exceed 4 kV/m for linear segmented generators (at the present time, it is not likely that segmentation can be increased to increase this value). The second limitation is the maximum voltage between adjacent electrodes, which, based on safety considerations, must be on the order 20-40 V. The third limitation is the current density level required to prevent arc formation on electrodes (1 A/cm^2 maximum).

One of the most important problems facing the MHD generator is that of materials, i.e., prevention of electrical erosion and chemical degradation of materials of the structure. An additional requirement imposed on the materials of the channel is that they must be nonmagnetic, to prevent a disturbing influence on the magnetic field in the channel volume.

1.4.3. Superconducting Magnet System. In order to obtain high efficiency in a facility with an MHD generator, it is necessary to generate a magnetic field of up to 6 T in the channel. This is quite a bit higher than the 2 T that can be reached with water-cooled conventional iron magnets at a moderate power. In a case of a 6 T magnet with a water-cooled conductor, the required power supply may exceed the power generated by the MHD generator.

Superconducting magnets that do not require high power supplies are promising for MHD power plants. However, additional equipment is required (a large cryogenic facility, special power supplies to charge the magnet, and protection equipment).

Development of a cryogenic system on a scale required by a 300-500-MW MHD generator is within the realm of possibility of present-day technology. In a stable mode, the superconducting magnet requires practically no energy. However, it is necessary to charge the magnet (magnetization), e.g., 1 MW low-voltage power source would charge a magnet during several hours and the energy of the magnetic field stored may reach 10^{10} J.

Basic problems. Development of large superconducting magnet systems requires solutions of many principally new problems including:

- development of technology of fabricating the conductor stabilizing its superconducting properties,
- control and prevention of malfunctions, and
- development of the support structure.

Present-day technology makes it possible to fabricate superconducting materials based on niobium-titanium alloys, stabilized with copper, capable of operating in a magnetic field of up to 8 T, while providing magnetic field induction of 6 T in the MHD generator channel. The current in experimental magnets reaches only 1 kA, whereas it is necessary to fabricate composite conductors operating at high currents of 10 to 50 kA.

The total mass of the coil of a superconducting magnet of an MHD generator for an MHD power plant is 200-500 tons. The coil will require a large number of conductor joints. The technology of making these joints without destroying superconductivity and while providing the required mechanical and electrical properties, presents a problem. The electrical resistance of each joint should not exceed 10^{-11} Ω. This requires that the superconducting coil be made in a form of multistrand cables, with joints made on individual strands rated for low current.

Cooling the superconductor in large magnets may be accomplished by means of two techniques: pool boiling or forced flow of liquid helium. The open-bath cooled conductors are more widely used and have a simple cryogenic system. The forced flow conductors require a pumped flow of coolant, but their internally cooled character allows integral insulation. Furthermore, the inventory of liquid helium in a hollow conductor winding is less than that in an open bath, which may be an important consideration. Analytical work indicates that forced-flow conductors can achieve overall current densities equal to or higher than pool-boiling conductors [59].

A serious problem facing superconducting magnets of an MHD power plant is the potential possibility of a malfunction. Incorrect operation (too high a rate of current increase, excess current, some disturbance from the MHD channel, or unsatisfactory operation of the charging source) should not lead to an emergency condition. Under such conditions, it should be possible to return rapidly to normal operation, thus limiting helium loss to perhaps 10 % of inventory.

The electrical faults may involve shorts between turns or to ground, destructive arcs, and conductor burnout. Destructive arcs can occur when a coil is discharged rapidly during an emergency dump. Gaseous helium is not a good insulator and arcs can easily start at points where the metal cryostat or coil support structure is not perfectly insulated. Requirements of reliable interturn isolation excludes shorts, but conflict with the requirement that the conductor must generally be cooled over much of its surface. The winding must also be protected against damage by any disturbance that drives the conductor into normal state. All materials have very low heat capacity at liquid helium temperature. Therefore, small heat inputs cause temperature excursions sufficient to drive regions of a superconducting magnet normal [58]. Coil design must reduce these disturbances or effects associated with them to a minimum. Construction of coil in a form of individual modules may eliminate conductor motion and, thus, further reduce the possibility of strains.

The support structure of a superconducting magnet system is required to withstand forces acting on the conductor during the flow of high currents and to prevent relative motion within the winding structure, thus avoiding release of heat, which could quench the magnet if the cooling is insufficient to enable the superconductor to recover.

Previous studies of linear MHD magnets indicate that preference should be given to saddle-type windings [60], because of their low ratio of peak field-to-duct field with sufficiently economical use of materials. Various types of windings can be fabricated. In general, windings are either layer or pancake wound. In the layer-wound technique, adjacent turns are laid down sequentially, generally parallel to the winding axis. In the pancake configuration, turns are generally transverse to the winding axes. Most MHD magnets require that the field at the inlet be higher than the field at the outlet end. It appears that the simplest way to achieve such a tapered field is to spread out the windings from the inlet to the outlet of the magnet [61]. In many instances, however, it is impossible to give the desired field taper by this technique. In that case, there can be large longitudinal forces between windings that will be absorbed by the support structure.

Examples of superconducting magnetic systems. The design of magnets constructed at Argonne National Laboratory for the bypass loop of the U-25 facility utilizes layer-wound circular saddles. Each winding layer is tightly secured by stainless steel banding. Outward forces are transmitted to the bore-tube through the bands and the winding fillers. Longitudinal forces on the windings are taken by steps and flanges of the bore tube.

The magnet winding is made up of saddle coils in ten layers. It is a modular design, each module consisting of two saddle coils supported in an aluminum half shell. The ten modules, five for each half, are bolted together around the bore tube to make the complete winding. Each saddle layer consists of two coil sections (small and large) with aluminum filler pieces between, all on the same cylindrical surface.

The structural support for the magnet is a thick bore tube with interlayer tension bands. This is opposed to the external girder approach and yields a lighter, more compact system allowing the magnet to be more readily transported and cooled down. The cryostat is relatively conventional. Pool-boiling helium at 4.5 K is contained in an inner chamber. Cooldown is by circulating cold gaseous helium.

A conceptual design of a superconducting magnet system with saddle coil windings on a cylindrical core, intended for the pilot unit of the MHD power plant, is shown in Fig. 1.21. It must generate a specified magnetic field in the MHD generator channel, 1. Variation of the magnetic field induction along the length of the channel between 6 T and 4 T is achieved by using a three-section winding. Current-carrying coils, 2, made of composite wire based on NT-50 superconductor, stabilized with copper, are placed within the cryostat, 3, maintained at a temperature of 4.5 K. Thermal insulation of helium vessel is by means of radiative shield, 4, and vacuum cavity, 5. The MHD generator channel is installed inside the inner tube of the cavity.

Biological effect of the magnetic field and its influence on the structures. The strong magnetic field around the MHD generator imposes special requirements on protection of building and structures of metal construction, placement of equipment, and safety considerations.

The biological limits for magnetic field environments have not been established. However, biological effects of a strong magnetic field on humans, even though weak, do exist. Various preventive measures, such as shielding the magnet or control panel are possible. The magnet should be closed during charge and discharge; it should be remotely controlled during operation. It is desirable to decrease the scattered magnetic field. This can be attained by dividing a plant into two parallel half-power magnet-channel units that operate with opposite polarity. The safe magnetic field induction level during prolonged exposure of humans for the pilot MHD power system is set below 0.01 T.

1.4.4. Inverter System. Most MHD generators, being developed at the present time, generate dc current. In general, connecting an MHD generator to an ac current transmission network requires interposition of a power conditioner inverter that also performs the functions of regulation and control. The Institute of High Temperatures of the Soviet Academy of Sciences investigated operation of an MHD generator interfaced with a commercial power grid by means of an inverter substation. It demonstrated the feasibility of stable power transfer from an MHD generator to the power grid across static converters [27].

Source interface. Different source interfaces and the number of necessary inverters correspond to the various, different methods of arranging MHD generator electrodes. The most common ones are the Faraday, Hall, and diagonal configurations.

A Faraday segmented generator exhibits certain advantages over other type generators in terms of energy indices. However, in view of the large number of individual loads operating at relatively low voltage, this generator requires the most complex inverter system. The Hall emf, which may be several times higher than the Faraday emf, increases requirements imposed on electrical isolation of individual electrode pairs, and also necessitates isolation of the inverter substation components rated for full Hall emf, from the ground. An advantage of a diagonal generator over a Faraday generator is the smaller number of individual loads and higher voltage. A deficiency of this generator is that it operates efficiently only in the nominal mode. An advantage of a Hall MHD generator is the single load; however, its efficiency is prohibitively low.

Characteristics of Inverter Systems for MHD Power Plants. With respect to many parameters, inverter systems for MHD generators are analogous to the conventional inverter systems. However, MHD power converters display certain specific characteristics: relatively high internal impedance of MHD generators, necessity of summing currents of multi-electrode units, high isolation of transformers, necessity of protecting electrodes from faults stemming from inner electrode breakdown, and necessity for stable operation of various electrode pairs in the common load system of MHD generators. When selecting configuration, control, and protection of the inverter substation, it is necessary to take into account the characteristics of its joint operation with the MHD generator.

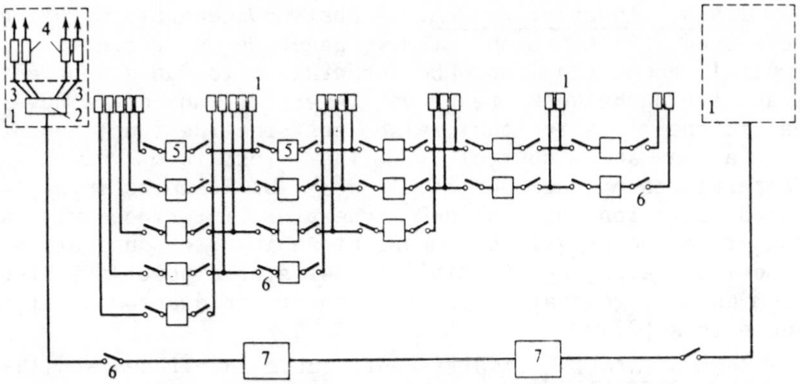

Fig. 1.22 A block diagram of an inverter system of a 250-MW diagonal MHD generator, 1--electrode-inverter interface, 2--current summator, 3--MHD generator electrodes, 4--frame current regulators, 5--8 kV inverters, 6--quick response switches, 7--20 kV inverters

The principal components of an inverter subsystem are smoothing inductors, control rectifiers (thyristors), transformers, harmonic filters, reactive power regulators, commutation equipment, and rectifier control systems. Smoothing inductors are required to smooth the electrode current and to provide stable operation of the inverter. Commutating devices serve to improve the power factor of the inverter system and to decrease losses. Transformers transmit electric power from inverters to the commercial grid.

As already noted, the diagonal generator allows a single load, connected to the two outer frames. However, optimization of power generation in an MHD channel shows that the load current has to vary along the channel length in accordance with a specified law, i.e., rapid current increase at the inlet, relatively slow drop along the length, and rapid drop at the exit. Loading of the channel has to be in accordance with the optimum current profile along the length of the MHD generator.

Loading and inverting schemes of a diagonal MHD generator with a power output of 250 MW at a nominal Hall field of up to 40 kV is shown in Fig. 1.22. Standardized inverters 5 and 7 at two nominal voltages (8 and 20 kV) are used in this scheme. The optimum current and load voltage distributions at the points of intermediate power takeoff and at the ends of the channel are attained by combination of series and parallel switching of inverters. The voltage of a single inverter is chosen such that the total voltage of the generator for the scheme selected is an even multiple of a single unit and the maximum current is an even multiple of the current of a single inverter.

Junction 1 between inverters and electrodes includes a device, 2, for summing currents of several electrode pairs, 3, and also a device to regulate current of frames, 4. The current summing device decouples electrodes with respect to the Hall potential and divides current among the electrodes. The regulating device limits the maximum current of the frame to a given level. The output of an inverter is connected with the windings of a low-voltage intermediate grid transformer, with a single high-voltage coil, and two low-voltage windings. Inverter substation-commercial grid interface is described elsewhere [1].

Inverters can be regulated by using one of the following modes of operation: constant current, constant voltage, constant power, maximum power, or constant load factor.

The choice must be made on the grounds of efficiency and performance under fault conditions. In a constant-current system, the FR losses are constant at their full-load value, even at reduced outputs. In a constant-voltage (C.V.) system, the losses are reduced in proportion to the square of the power produced. Therefore, considerations of efficiency favor the constant-voltage system. Such a system would be endangered, however, by very large fault currents. It follows that it is desirable to operate in the constant-voltage mode up to rated value of current, and in the constant-current mode beyond that point.

Current dividers and protective devices. MHD generators differ from most of the existing energy conversion facilities by generating hundreds of megawatts of electrical power at a relatively low inlet voltage and high current. Specific characteristics of an MHD generator make it difficult to design and develop a power semiconductor unit. In order to decrease the cost of inverter substations, it is necessary to decrease the number of takeoff points where inverters are connected with electrodes, by combining electrode pairs into groups (such summation of power is possible when the Hall emf between electrode pairs is compensated). Current distribution along the branches, especially at overloads and in fault modes when currents may exceed the nominal current three-to-sixfold, creates a problem. The need for protection of a thyristor inverter, i.e., fast-acting protective equipment for malfunction situations, is associated with this requirement.

The design of transformers for MHD facilities is determined by special requirements:

- Because transformers are the basic insulation element for the Hall potential, the class of insulation used for their windings must be adequate to withstand the maximum possible electrode-to-ground potential (several dozen kV).
- The leakage reactance of the transformer must be kept low, because it hinders the commutation of the inverter elements; however, the low-leakage reactance results in high fault currents, so that special bracing of the windings is needed.
- The situation is particularly unsatisfactory in the case of multi-electrode Faraday generators that require a large number of coils.
- The presence of harmonics and, possibly, dc components of current dictates operation at relatively low flux density.
- The transformer rating must allow for substantial flow of VARs.

For the foregoing reasons, MHD plant transformers are considerably larger and more expensive than standard utility transformers of equivalent rating. Because it may be impractical to go from the natural inverter line voltage (a few kilovolts) to the transmission line voltage (hundreds of kilovolts) in one transformation, two transformations in cascade may be required. Such a case imposes no special requirements on the high-voltage transformers.

Each part of the current inverter subsystem has to be designed to operate under the worst conditions. In order to decrease the cost of the inverter substation, it is necessary to attempt to increase the power of individual converters and to decrease the number of inverters in each power inversion system.

1.4.5. Oxidizer Preparation System. The oxidizer preparation system includes an air compressor and a high-temperature preheater and, in certain cases, a device for enriching air with oxygen. Direct preheating of the oxidizer to 2000 K by combustion products from the MHD generator is thermodynamically most efficient, but, at present, cannot be attained using existing equipment.

1.4.5.1. High-Temperature Air Preheaters

Principal types of air preheaters differ with respect to the structural design and heat transfer technique.

As a result of their mechanical strength, metal recuperative heat exchangers are utilized to preheat the air to a temperature not exceeding 1200-1250 K. In the case of austenitic chromium-nickel steels now used in the power industry, the upper temperature limit is 970 K (oxidizer temperature of 900 K). An additional limitation imposed on preheat temperature is imposed by metal corrosion in combustion products containing compounds of seed and ash.

Regenerative heat exchangers with refractory linings can preheat the oxidizer to the required temperatures. Operation in the range of direct high-temperature oxidizer preheating by the combustion products from the MHD generator are confined at present to investigations of corrosion of refractory materials by seed and ash. Papers on this subject were presented at the International Symposia on MHD Energy Conversion in Warsaw (1968), Munich (1971), and Washington (1975 and 1980).

The present level of technological development is adequate to fabricate regenerative, high-temperature oxidizer preheaters with independent heating, operating on pure combustion products. The thermal loop of the MHD power plant may include two stages of oxidizer heating, first by the combustion products from the MHD generator, in a metal recuperator to temperature of 700-850 K, then by high-temperature heating in a regenerative heat exchanger with a separate combustion chamber operating on pure fuel. This is the arrangement for oxidizer heating in plans for the pilot assembly of an MHD power plant under development in the U.S.S.R.. Various aspects of the choice of oxidizer preheat temperature in a recuperator are considered in Section 1.3.

An example of an air preheater. Figure 1.23 shows a cross section of a high-temperature regenerative oxidizer preheater with a vertical matrix (one of the designs for the pilot system of the MHD power plant). Four such units will be installed on a single, continuous reinforced-concrete basement for each subsystem of the MHD power plant. Blast furnace-type air preheaters, widely used in metallurgy in all industrially developed countries, will be used as the basic units.

The high-temperature air preheater intended to heat the oxidizer (air) to 1973 K consists of a matrix chamber, 7, and combustion chamber, 6, completely interconnected by a cavity in a form of a half-torus. The principal element of the air preheater is the matrix chamber that determines its thermal and hydraulic characteristics. The matrix consists of core elements with core diameter of 25 mm. The temperature stability of the matrix is provided by utilizing combined refractory components. The overall height of the matrix layer is 30 m, including a 13.5-m

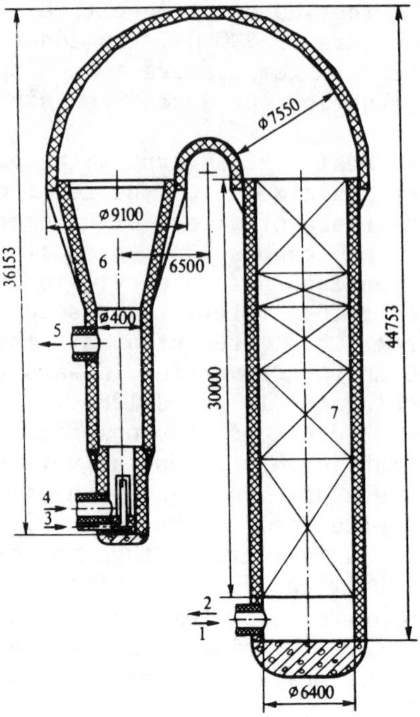

Fig. 1.23 Cross section of a high-temperature air preheater, 1--oxygen supply, 2--combustion products outlet, 3--fuel supply, 4--air supply and smoke gas recirculation, 5--heated oxidizer outlet, 6--combustion chamber, 7--heat absorbing lining

corundum section, a 6.5-m mullite-corundum section, and a 10-m shamot-caolinite section. In order to provide the required temperature conditions, maintaining strength of the refractory lining and preventing the outer housing from overheating, various sections of the combustion chamber and the matrix chamber, defined by the temperature zones, are lined by corundum, mullite-corundum, and shamot-caolinite refractory compounds.

It can be seen from Fig. 1.23 that the preheater is of complex configuration. Therefore, a simpler preheater is being considered, in which the combustion chamber is smaller in diameter and is located in the space under the dome. However, this design will necessitate a larger number of units.

Characteristics of Air Preheaters for an MHD Power Plant. Because the temperature of the oxidizer entering the separately-fired preheater from the recuperator or compressor is considerably higher than that commonly used in metallurgical air preheaters, it became necessary to design and develop a new support structure for the air preheater, one intended to distribute the flow along the matrix uniformly and to support the weight of the overall structure.

Another characteristic of the air preheater for the MHD power plant is the high oxidizer pressure. The housing of the air preheater is a vessel that operates under pressure of approximately 1 MPa. This fact determines the requirements imposed on its shape, thickness of steel members, materials, and quality of fabrication.

The air preheater matrix is alternately exposed to the combustion products and oxidizer being heated. To switch over from the matrix heating mode to oxidizer heating mode and back, water-cooled gate valves are installed on the ducts. Unfortunately, no ceramic gate valves for regenerative air preheaters are available at the present time.

Because the products of combustion from the separate air preheater are exhausted into the atmosphere, they must satisfy environmental pollution restrictions on their nitrogen oxide, carbon oxide, and hydrocarbon content. In order to reduce the exhaust of nitrogen oxides from the air preheater of the pilot system of the MHD power plant, combustion in the combustion chamber of the air preheater takes place at an air deficiency ($\alpha = 0.98$). In addition, the exhaust gases pass through a special catalytic converter that reduces nitrogen oxides and oxidizes carbon dioxide and hydrogen. The combustion temperature in the combustion chamber of the separate air preheater must be limited to 2100 K, because of the use of refractory materials. The products of combustion leaving the air preheater are recirculated in the combustion chamber, to decrease the combustion temperature without increasing the excess air. The temperature of combustion products leaving the air preheater is 700-800 K. Before the exhaust into the atmosphere, combustion products are used to preheat the fuel and the combustion air in special recuperators.

<u>Increasing Thermal Effectiveness</u>. The principal shortcoming of blast-furnace-type air preheaters is their low thermal effectiveness, resulting in large preheaters that require large amounts of refractory materials, and, hence, in high costs. Both the U.S.S.R. and other countries are now developing more effective regenerative heat exchangers. The principal method of increasing thermal effectiveness is to increase the heat transfer surface,

using other types of matrices, including pebble and core elements with inner channels and fixed or movable beds.

A fixed pebble bed regenerative preheater developed at the Institute of High Temperatures [62] has successfully operated at the U-02 facility for more than 15,000 hours. It provided stable air preheating to 1770-2270 K. The volume heat transfer constant of the pebble matrix is 30 to 50 times higher than of a ceramic element matrix. Low heat retention of the pebble matrix makes it possible to decrease the cycle time of the regenerator ten or more times and, thus, to decrease the amount of refractory materials used and, consequently, the cost of the regenerator.

1.4.5.2. Oxidizer Compression

MHD power plant compressors are intended to compress standard or oxygen-enriched air at the rate of 200-400 kg/s and feed the oxidizer at 1-1.2 MPa across the air preheater into the combustion chamber. In many respects, compressors of MHD power plants are analogous to gas-turbine compressors.

The principal requirement imposed on the MHD power plant compressor is to provide a wide range of variation in oxidizer mass flow and pressure. The operating mode of a compressor equipped with a turbine drive may be easily controlled by changing the rotation rate of the rotor. Synchronous electric motors can also drive the compressors. However, this will require fairly complex and expensive start-up devices. In particular, it is possible to utilize a start-up and rotation rate control of a synchronous electric drive by means of thyristor frequency converters.

The conceptual design of the pilot MHD power system included two electric motor-driven compressors operating in parallel. Their control is by means of variable pitch blades and by shutting off the compressors at a very low level of plant operation.

1.4.5.3. Oxygen Enrichment of Air.

At the present time, oxygen-enriched air for experimental MHD facilities and planned MHD power plants is obtained by diluting oxygen with air to the required concentration. Modern air separation devices utilize low-temperature rectification, based on the difference in boiling temperature of air components. Oxygen production requires considerable power (on the order of 0.4 kW h/m^3). Even when using oxidizer enriched with 30 % oxygen, an oxygen production facility will require approximately 3.5 % of the electric power output of an MHD power system.

Theoretically, incomplete separation of air requires less power. Several facilities using the low-pressure scheme to produce air enriched with 60 to 70 % oxygen have been developed in France, England, and the Federal Republic of Germany. Specific power required to produce oxidizer of such a concentration, using the oxygen production facility deployed at the metallurgical plant in Schwelgern [63] is 0.29 kW h/m^3 (recalculated for pure oxygen). The feasibility of using similar oxygen production facilities with MHD power plants should be considered.

1.4.6. Steam Generator. The heat of the combustion gases exiting the MHD generator must be used for high-temperature oxidizer preheating (heat regeneration in the MHD cycle) and to produce super-critical steam that serves as the working fluid in the bottoming steam cycle. Some experimental results, establishing and partially modeling operating conditions in a steam generator of an MHD power plant, were obtained on the U-02 and U-25 facilities [64].

Methods of high-temperature preheating of the oxidizer by combustion gases containing seed (and in the case of coal, containing ash), and materials for high-temperature preheaters operating on combustion products of an MHD generator have not yet been developed. Therefore, the separately-fired, high-temperature oxidizer preheater burning clean fuel appears to be the most promising approach to construction of an MHD power plant. In this case, combustion gases exiting the MHD generator at a stagnation temperature of 2300-2500 K will enter the steam generator directly.

Characteristics of a Steam Generator. The design of the steam generator and the layout of its components must reflect its operating characteristics as an element of an MHD power plant.

The temperature of combustion gases entering the steam generator (2300-2500 K) is not only considerably higher than the maximum flame temperature in steam generator furnaces of conventional power plants, but also higher than the theoretical combustion temperature of high-calorie fuel in air preheated to 473 K. The total enthalpy of 1 kg of combustion gases, taking into account the heat released by phase transitions of seed, is approximately 20 % higher than in conventional steam generators. In addition to the high temperature, the combustion gas flow downstream of the MHD generator is characterized by high velocity, requiring installation of a diffuser and connection of its cooling system into the overall circulation loop.

The heat transferred to the steam generator is used to produce superheated steam, to preheat the oxidizer entering the separately fired preheater, and to preheat the gaseous fuel for the main combustion chamber or heat and dry the solid fuel in the pulverizer.

The distribution of heat fluxes between the heat transfer surface of the steam generator of an MHD power plant has its own peculiarities. The allowable temperature levels of combustion gases at the outlet of the radiant section of the steam generator of an MHD power plant, and from the burner of conventional steam generators are practically independent of the initial enthalpy and are close to each other. Because the initial enthalpy of combustion gases in the steam generator of an MHD power plant is considerably higher, the fraction of heat transferred to the radiant surfaces is considerably greater and is as much as 80 % of the total heat transferred to the feed water-primary steam duct. As a result, the superheater operates at a relatively higher temperature of gases, necessitating a larger radiant section. The steam velocity must be increased correspondingly to maintain reliability of radiant surfaces. Of the surfaces that generate primary steam, the convective gas duct may include only the inlet sector of the economizer and outlet sector of the steam superheater. In addition, the secondary steam superheater and the oxidizer and fuel preheaters are also located in the convective gas duct.

In normal compressor operating modes, the oxidizer temperature upstream of the steam generator is approximately 500 K. This makes it difficult to cool the exiting gases by means of techniques used in conventional steam generators, and requires that additional measures be used to achieve this goal.

The effect of load changes of an MHD power plant on combustion products parameters at the inlet into the steam generator also differs considerably from those of a conventional thermal power plant. Effective regulation of the MHD electric power plant will occur when the electrical loading of the MHD plant is maintained within the range of 70-100 % of the full design power level. Under these conditions, change in steam generator load is subordinated to the operation mode of the MHD generator. Any reduction in MHD generator load results in a sharp reduction in the thermal efficiency of the plant. With reduction in MHD generator load, the enthalpy and temperature of the combustion gases at the entrance to the steam generator may increase. Consequently, the temperature of gas in the high-temperature section of the steam generator of an MHD power plant will depend considerably less on the load than in conventional steam generators. This leads to additional complications in regulating the load.

To increase the range of power levels over which an MHD steam electric power plant can operate will require a system design that permits operating the "steam portion" of the system separately and independently when the MHD generator is shut down. For such operation, the steam generator must be equipped with separate auxiliary combustors and separate supply of preheated air.

Increasing the range of operation, by making the system available for part-load operation with a steam generator that can be completely decoupled from the MHD generator, significantly increases the complexity of the system downstream of the MHD channel and will require protective cooling of auxiliary burners.

Influence of Seed. Seed in the combustion products exerts substantial influence on the steam generator of an MHD power plant. The amount of seed material contained in the combustion products is equivalent to 25 to 30 % of the mass flow rate of fuel burned.

In assessing the problems caused by the presence of seed within the steam generator, it is important to differentiate between the various hot gas temperature zones of the steam generator that correspond to differing physical and chemical states of the seed material. Seed deposits in each of these zones have unique physical and chemical characteristics.

In the hottest steam generator zone (above 1500 K), the seed is primarily in the form of atomic potassium and potassium hydroxide (KOH) which condense on relatively cold surfaces. The KOH condensate actively interacts with CO_2 and SO_2 to form potassium carbonates and sulfates. Because the rates of these reactions are somewhat slower than the condensation rate, mixtures of carbonates and sulfates with potassium hydroxide are formed. The melting temperature of those mixtures is below the condensation temperature of KOH. As a result, the solid deposits formed are coated with a flowing film of molten seed.

At a lower temperature of combustion gases (around 1500 K), KOH begins to interact with CO_2 and SO_x, forming K_2CO_3 and K_2SO_4. Because the vapor pressures of these compounds at the indicated temperature are low, supersaturation and condensation of seed droplets occur quickly. The resulting droplets settle out on the heat exchanger surfaces and, as they harden, form solid crustlike deposits with a molten outer layer.

Similarly to conventional-type steam generators operating on slag-forming fuels, semiradiant surfaces in steam generators of MHD power plants must be located in the 1400-1500 K temperature zone. These surfaces should be in the form of "U" shaped tubes with large pitch and must be equipped with an effective means of removing the seed.

When the melting temperature of potassium carbonate is reached (1174 K), finely dispersed potassium seed particles are formed in the combustion products and may contaminate convective surfaces in a way analogous to the ash of solid fuels. When operating on pure fuel, it is desirable to place radiant or semiradiant heating surfaces in the temperature zone about 1450 K and convective heating surfaces, below 1150 K. The most dangerous zone of 1450-1150 K may be excluded from heat transfer by recirculation of colder products of combustion into this zone. Sudden cooling will result in solidification of seed drops in the flow. Established boiler technology methods can be used to purify the porous seed deposits formed.

Particulate-capturing devices, such as electrostatic precipitators, should be located in the zone where the gas temperature is above 450 K, at which level seed does not absorb moisture and form the bicarbonate. After seed is removed, further decrease in the temperature of exhaust gases does not affect operation of the heating surfaces.

NOx Control. To minimize NO_x production in the combustion chamber of an MHD power plant, a fuel-rich mode of burning is required. In this case, incompletely combusted products of combustion with $\alpha = 0.9\text{-}0.95$ enter the steam generator. At a reduced cooling rate, NO_x compounds are decomposed in the reducing medium of the incompletely combusted products of combustion.

The required cooling rate of incompletely combusted combustion products may be attained by designing the initial portion of the steam generator as a radiant mode heat exchanger (gas holdup chamber) characterized by reduced heat transfer, which is achieved by reducing the velocity of the gas flow and by thermally insulating chamber walls.

A secondary molten slag recovery and removal stage can be incorporated in the gas holdup chamber when burning ash-laden fuel.

To exclude additional formation of NO_x, completion of combustion in the steam generator should take place at a temperature above 1600 K, achieving efficient combustion of large masses of products of incomplete combustion with very low calorie content (270-420 kJ/kg), which is a complex technological problem.

The steam generator of the MHD power plant will apparently be larger than conventional boiler systems, will require more metal and lining materials, and will be characterized by higher power consumption for its own needs. The service-free period of operation will depend on the service life of the lining in the delay chamber. As the fraction of regeneration heat in the MHD cycle increases and the method of decomposing NO_x and surface scrubbing techniques improve, the technical-economic indices of steam generators of an MHD power plant will improve considerably.

A conceptual design of a steam generator for the pilot system of the first commercial-size MHD power plant developed at the Institute of High Temperatures is shown in Fig. 1.24. The manufacturing plant proposed a new model of the steam generator, lacking a special shaft for the recirculating gas flow.

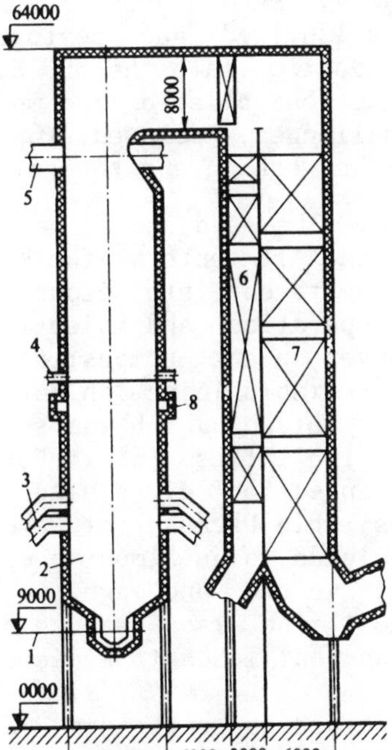

Fig. 1.24 Conceptual design of a steam generator of the 500-MW MHD power plant (cross section), 1--MHD generator axis, 2--radiant shaft, 3--air and fuel inlet for separate operation of the steam generator, 4--air inlet for secondary combustion, 5--recirculation gas inlet, 6--convective gas duct for the recirculating gas flow, 7--convective gas duct for the main gas flow, 8--insulated flange

1.4.7. Seed System. The seed system incorporates the injection, recovery, and regeneration subsystems. Both potassium carbonate and potassium sulfate appear to be promising seed material for industrial MHD power plants. The choice of material depends on the operating conditions of the plant and the type of fuel. When the facility operates on natural, sulfur-free gas, using K_2CO_3 as seed, potassium carbonate appears in the effluent. Potassium carbonate is reintroduced into the cycle with minimum losses for regeneration. It is desirable to introduce fresh potassium carbonate to compensate for seed losses. If coal or fuel oil containing sulfur is the MHD fuel, K_2SO_4, K_2CO_3, or a mixture of the two can be used as seed.

The presence of seed in the gas duct of an open-cycle MHD power plant creates a number of materials and maintenance problems: fouling of tubes, blockage of passages, attack on metal and refractory heat transfer surfaces, etc. A large proportion of the circulated seed must be captured and collected, to escape economic penalties associated with heavy seed losses. The spent MHD seed must also be desulfurized and restored to a condition suitable for introduction into the MHD combustion chamber. Undesirable interactions of seed and molten coal ash pose another set of challenges, related to both corrosion and seed recovery in the working environment.

Seed Injection. The seed should be ionized and evenly dispersed by the time it enters the MHD channel. These requirements must be taken into account when choosing the methods of seed preparation and injection and, also, when designing MHD power plant combustion chambers. The simplest and most reliable injection method is spraying of a water solution of potash, which was successfully used in the U-02 and U-25 facilities [65]. For commercial-scale MHD power plants, injection of a water solution of K_2CO_3 is extremely undesirable because of a drop in electrical conductivity that leads to an almost proportional increase of the MHD channel volume and magnetic system volume and greater capital cost, and also because of a sharp drop of enthalpy extraction that leads to a decrease of efficiency of the plant.

Injection of potassium carbonate into the combustion chamber in the form of a powder encounters technological difficulties. Even when well dried, potassium carbonate is not easily pulverized. Furthermore, the smaller the particles, the greater their tendency to stick together. Even the smallest amount of absorbed moisture causes formation of lumps and caking. Experiments in injection of dry seed were made at PERC (Pittsburgh Energy Research

Center) in a horizontal cyclone combustor burning 90 kg/h of coal at atmospheric pressure [66]. This system operated reliably so long as pulverized K_2CO_3 powder was maintained absolutely dry. Any absorption of moisture by the seed resulted in lumps and in caking of the powder, leading to clogging of feed pipes. It should be noted that such mixing of seed and coal in a single-stage combustor may result in increased loss of seed into the rejected slag stream.

The conceptual design of the pilot system of the MHD power plant includes a dry seed injection system from a hopper into the combustion chamber by means of a pump widely used to transport granulated materials. The seed is transported by air at a temperature of 473 K.

Seed Behavior in a Gas Flow. Physical and chemical transformations of the seed should be taken into account in formulating requirements imposed on the components and subsystems of an MHD power plant, particularly on the steam generator and the seed recovery subsystem. At a temperature above 1500 K, all seed in the plasma is in a gaseous phase, the bulk of seed in the form of KOH. As the temperature decreases, the amount of KOH increases. At approximately 1500 K, KOH interacts with CO_2 forming K_2CO_3, which has a low vapor pressure at this temperature. Supersaturation of gases with K_2CO_3 leads to volume condensation. When the temperature of the gas is below the melting temperature of K_2CO_3 (1174 K), the droplets solidify. In the low-temperature gas flows at a temperature below 438 K, carbonation of K_2CO_3 to $KHCO_3$ takes place.

Experiments on the U-02 facility [65] have shown that, in the high-temperature zone over 1500 K, the primary mechanism of deposit formation is condensation on the relatively cold surfaces. Having condensed on the heat exchange surfaces and reached its final temperature, KOH reacts with CO_2, forming K_2CO_3, which hardens at temperatures below 1174 K. The result is hard, crust-like deposits that are a mixture of 80-90 % potassium carbonate and 10-20 % KOH, with a melting temperature in the neighborhood of 1000 K, which is below the condensation temperature of KOH vapor. As a result, the thickness of deposits is stabilized, and the remainder of the condensing seed runs off. Experiments on the U-02 and U-25 facilities have shown that more than 40 % of seed can be recovered.

Long-term operation of the steam generator on the U-02 facility has demonstrated that, if the design of the semi-radiant surfaces is correct, the spaces between the pipes do not become clogged and do not thus limit service-free operation of the heat exchanger. Under such carefully controlled conditions, special removal of seed scale is not required.

The most unfavorable situation occurs in the 1450-1150 K temperature zone, close to the melting point of potassium. The seed in that zone consists of droplets of potassium carbonate, which will adhere to and solidify on the surface. The deposits thicken until they completely clog the passages between the pipes. Because of this problem, it is necessary to cool combustion products by recirculation, as was done in the U-25 facility and is planned in the conceptual design of the steam generator of the pilot system. The deposits formed at a temperature below 1150 K are porous and are easily removed or flushed with water.

Seed Behavior in a Coal-Fired Facility. The first problem that arises when coal is used is the physical and chemical interaction of seed with the mineral part of the solid fuel. Slag at a temperature of 1600-1800 K, which is close to the temperature of the molten state, can trap up to 50 % of the injected seed when all of the slag enters the MHD channel, and up to 10 % of the seed when 80 % of the slag is separated in the cyclone combustor [66]. To decrease losses, maximum slag removal from the combustor must be accomplished before the seed is introduced. Seed absorption by slag is determined by the kinetics, and needs further investigation. Preliminary analyses have, in principle, demonstrated the feasibility of separate removal of slag and seed from the combustion products, provided maximum slag removal from the combustor is accomplished before the seed is introduced [67].

The presence of ash in the combustion products changes the behavior of seed in the MHD power plant. Slag covers the cold surfaces of the steam generator pipes, and the temperature at the external surface of the deposits is raised to approximately 1500 K. This temperature precludes condensation of KOH vapors and practically eliminates settling of K_2CO_3 (or K_2SO_4). Therefore, all of the seed except that bound by slag passes through the radiant and the semi-radiant portions of the steam generator.

Unfortunately, the extensive deposits formed on heat transfer surfaces at gas temperatures slightly above the melting point of K_2SO_4 or a mixture of K_2SO_4 and K_2CO_3 apparently consist of fly ash cemented by molten seed. To prevent such deposits, it may be necessary to recirculate cooled combustion products, as is done in the U-25 facility. Dry, porous deposits apparently can be removed by methods used in conventional power plants, namely, shot and steam cleaning.

Seed Entrapment and Removal. The problem of maximum seed entrapment and re-use in the cycle is one of the most important aspects in the design of an MHD plant. Its solution is necessitated by both economic and environmental considerations. Up to 40 % of seed can be recovered as molten seed when it condenses on the surfaces on radiant and semi-radiant surfaces of the steam generator.

A significant portion of the seed condenses on the surfaces of the convective portion of the steam generator. This material is almost pure K_2CO_3 (or, in the presence of sulfur, K_2SO_4 or a mixture of K_2CO_3 and K_2SO_4). It is either flushed down or mechanically separated from the pipes, then taken from the loop and prepared for re-use. An important aspect of the overall seed removal process is the recovery of the seed from the exhaust gases. Particulate removal devices must be 98-99 % effective. Attaining this degree of gas purification is difficult, particularly in view of the submicrometer seed particle size. Three types of gas scrubbing devices were selected and tested in the U-02 facility: dry bag filters, Venturi scrubbers, and electrostatic precipitators.

Figure 1.25 is a block diagram of the conceptual design of the seed system. Seed is trapped by electrostatic precipitators (1) having a considerably lower hydraulic resistance than Venturi scrubbers. From the filters, the seed is fed into the hopper (4) by a pneumatic transport device consisting of compressor (2) and ejector (3). The transported air purified from seed in the cyclone (5) is exhausted with smoke gases. The seed is purified in the seed regeneration system (8) where it is solved and cleansed of mechanical impurities. The purified solution of seed, as well as solution of seed from the reserve capacity vat (12), is injected into the gas duct of the steam generator where the solution is evaporated.

Seed Regeneration in Coal-Fired Systems. Seed regeneration is a simple process in pure-fuel-fired systems. A central problem in seed regeneration in coal-fired plants, the problem that to a considerable extent determines seed losses, is separation of the seed from ash and slag. Recovery of the seed entrapped with fly ash seems readily achievable. Efficiency of 99.9 % has been achieved in the U-02 facility by means of bag filters.

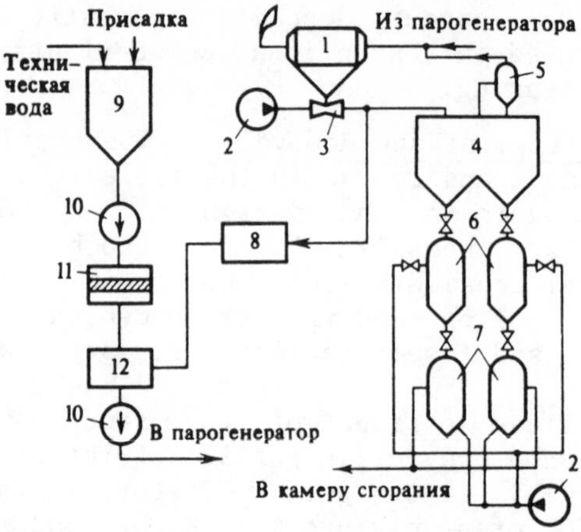

Fig. 1.25 Conceptual design of a seed system, 1--electrostatic precipitators, 2--compressor, 3--injector, 4--pneumatic transport device, 5--cyclone, 6--lock, 7--pump, 8--regeneration system, 9--seed solution storage tank, 10--centrifugal pump, 11--mechanical filter, 12--pure-seed solution storage vat

British, Japanese, and U.S. (Avco-Everett) researchers have successfully used electrostatic precipitators. It is generally believed that the presence of sulfur components in gases facilitates precipitator performance. Two methods of regeneration, which differ in the sequence of operation, are being investigated. In the first, the seed is separated from ash and, in pure form, is then regenerated. In the second technique, the seed is first regenerated with a mixture of ash and then is separated from the ash. It is not yet clear which method is preferable.

It has been theoretically and experimentally established that more than 99 % of the sulfur in fuel (coal, fuel oil) used in MHD power generation can be bound by potassium seed. Indications are that MHD power stations of the future will easily satisfy environmental restraints on SO_2 emissions. Several methods of seed desulfurization during regeneration are under investigation. Removal of sulfur oxides from gases will decrease the efficiency of the MHD power plant approximately in direct proportion to the amount of sulfur in the fuel.

1.4.8. Protection of the Biosphere. In the case of both conventional and MHD power plants, special measures must be taken to protect the environment from pollution by emission of sulfur and nitrogen oxides, particulates, discharge of hot water into the neighboring streams and lakes, and from heat discharge with flue gases.

Protection from sulfur oxides. Different systems to remove sulfur oxides and prevent their emission into the atmosphere have been developed. However, they require large capital expenditures and encounter technological difficulties. A high degree of sulfur dioxide removal from the combustion products has been achieved on open-cycle MHD power plants. This has been attained as a result of interaction of sulfur oxides produced during combustion of sulfur-laden fuel with potassium seed injected into the combustion products to increase electrical conductivity. This makes it possible to combust high sulfur-content coal in MHD power plants, the use of which in the power generating industry is difficult at present. Assuming the usual amount of seed injected, the concentration of sulfur oxides in the exhaust gases will meet the most stringent requirements for protection of the environment, because the seed in the MHD power plant can bind all of the sulfur introduced with fuel.

The sulfur-containing seed has to be regenerated (desulfurized) and returned into the cycle. Different seed regeneration methods [68] have been proposed that make it possible to produce sulfur as a byproduct [69].

Preliminary economic estimates of the cost of desulfurization of potassium seed, performed at PERC (Pittsburgh Energy Research Center), have shown that seed regeneration may be an insignificant part of the overall cost of an MHD power plant.

Protection from nitrogen oxides. Emission of nitrogen oxides exerting toxic influence on the biosphere must be limited in accordance with the clean air requirements imposed on the nitrogen dioxide in the atmosphere. What's more, the presence of sulfur oxides in the air results in cumulative harmful effects from the two kinds of oxides and in maximum permissible concentration of both sulfur oxides and nitrogen oxides. In MHD facilities, the nitrogen oxides are formed during combustion of fuel in the combustion chamber. When combustion products flowing through the MHD facility loop are cooled, nitrogen oxides are decomposed into elements. However, their decomposition rate decreases sharply as the temperature drops, so that the cooled combustion products may contain considerable amounts of nitrogen oxide, thus requiring special measures to lower their emission.

One of the possible measures for MHD facilities is staged combustion, i.e. fuel-rich (insufficient oxidizer air) combustion during the first stage, followed by fuel-lean combustion to complete burning during the second stage, at a lower temperature that precludes secondary formation of nitrogen oxides. In addition, methods of removing nitrogen oxides from the combustion products, thus making it possible to decrease emission of nitrogen oxides, are being considered. This involves removal of nitrogen oxides from the flue gas or their decomposition by means of catalysis.

Experimental investigations of the formation and decomposition of nitrogen oxides under conditions close to the actual conditions in MHD facilities, conducted in the U.S.A. [70], and kinetics calculations performed at the Institute of High Temperatures demonstrate the feasibility of lowering the concentration of nitrogen oxides in the exhaust gases of large-scale MHD facilities. This requires air combustion below the stoichiometric ratio (90-95 % of stoichiometric oxygen) during the first stage (in the main combustion chamber) and completion of combustion at a temperature on the order of 1600 K by injecting secondary air in sufficient quantity to provide air combustion at a certain air excess (105-110 % stoichiometric oxygen) during the second stage. Detailed kinetics calculations, including technical and economic optimization, determine the optimum cooling rate of a gas flow for various temperature ranges. It is recommended that minimum cooling rates occur in the range 2100-1900 K (in the delay chamber in the steam generator). At temperatures exceeding 2100 K and below 1900 K, the cooling rate decreases. These data were taken into account in the conceptual design of the first pilot, commercial-scale MHD energy system and, in particular, during development of the steam generator.

Extraction of nitrogen oxides from the cooled combustion products when the concentration is 0.003-0.004 % is a complex problem. Numerous schemes of scrubbing flue gas with high concentration of nitrogen oxides have been suggested (in this connection, it should be noted that conditions required for generation of a high concentration of NO_x in flue gases exist in MHD power plants), including production of a side product, nitric acid, a valuable product, useful in producing fertilizers. Technological schemes for simultaneous removal of sulfur and nitric oxides [71], still requiring experimental verification, have also been proposed.

Trapping of particulates. Electrostatic precipitators, fabric filters, and Venturi scrubbers can be used to remove particulates from the combustion products. Removal of particulates in an MHD facility will apparently differ from the methods used in conventional power stations. In a general case, particulates removed from the MHD power plants consist of the mineral part of the fuel and seed compounds that must be regenerated. The difference in the system of removing particulates is dictated by the distribution of particulate sizes, which differs from those in the combustion products of conventional power plants.

The size of the particles in MHD power plants. The size distribution of particles depends on the physical transformations of seed and ash in the gas duct [67]. Preliminary estimates show that the fly ash in an MHD power plant will be approximately an order of magnitude smaller in size than fly ash from conventional power stations. The choice of methods of purifying combustion products in an MHD power plant must be made on the basis of technoeconomic comparison of variants, taking into account the experimental data. In particular, in the conceptual design of the pilot system of the MHD power plant, preference is given to electrostatic precipitators characterized by lower hydraulic resistance than Venturi scrubbers.

Protection from thermal pollution. In view of the feasibility of an almost complete removal of sulfur oxides from the combustion products, substantial reduction of nitrogen oxide emissions, and the required degree of cleaning of flue gases from particulates, MHD energy conversion will also make it possible to decrease substantially the waste heat discharged with flue gases and with cooling water. This can be attributed to the considerably higher efficiency of the MHD power plants in comparison with conventional power plants. A reduction of waste heat discharged with flue gases per unit power output is proportional to the decrease of specific fuel consumption and, even in the case of the pilot MHD system, will be approximately 20 %. The amount of cooling water used by the MHD power plant is less by a factor of 1.5 than that of conventional power plants (efficiencies of 48 and 38 %, respectively) and by a factor of more than 2.0 when the efficiency of the MHD power plant is increased to 60 %.

REFERENCES

[1]Petrick, M., and Shumyatskiy, B. Ya. (eds.), Open-Cycle Magnetohydrodynamic Electrical Power Generation, a joint U.S.A./U.S.S.R. publication, 1978.

[2]Author not given, "Gas Turbines of 1980" (Gasovyye turbiny 80-kh godov), Teploenergetika, No. 19, 1976, p. 10-19.

[3]Faraday, M., "Experimental Research in Electricity," Philosophical Transactions, London, 1832, p. 125-194.

[4]Motulevich, V. P., Open-Cycle MHD Generators (Magnitogidrodinamicheskiye generatory otkrytogo tsikla), 836 p., translated from English, Publishing House Mir, Moscow, 1972.

[5]McCutchan, D. A., Lippert, T. E., and Retallick, F. D., "Economics of MHD Power Generation," Annual Meeting of the American Nuclear Society, Las Vegas, Nevada, 1980. Summaries, Transaction of the American Nuclear Society, 1980, No. 34.

[6]Bienstock, D., Bergman, P. D., Henry, J. M., et al., "Air Pollution Aspects of MHD Power Generation," Proceedings of the 13th Symposium on Engineering Aspects of Magnetohydrodynamics, Stanford, March 26-28, 1973, Stanford University, Stanford, California, 1973, Vol. 2, p. 1.1-1.10.

[7]Sheyndlin, A. E., Pishchikov, S. I., Shumyatskiy, B. Ya., et al., "Development of MHD Energy Conversion in the U.S.S.R.," Fourth U.S.-U.S.S.R. Colloquium on Magnetohydrodynamic Electrical Power Generation," Washington, D.C., October 5-6, 1978, Washington, D.C., 1978, p. 85-104.

[8]Olds, F. C., "The Availability of Fuels for Power Plants," Power Engineering, 1976, Vol. 80, No. 9, p. 42-49.

[9]Hubbert, M., "The Energy Resources of the Earth," Scientific American, Vol. 240, No. 3, 1971, p. 70-80.

[10]Nekrasov, A. M., and Pervakhin, M. G. (eds.), Power Production in the U.S.S.R. (Energetika v SSSR), Publishing House Energiya, Moscow, 1977, 287 p.

[11]Tikhonov, N. A., Principal Directions of Economic and Social Development of the U.S.S.R. During 1981-1985 and for the Period Until 1990 (Osnovnyye napravleniya ekonomichaeskogo sotsial'nogo razvitiya SSSR na 1981-1985 godyi na period do 90 goda), Publishing House Politizdat, Moscow, 1981, p. 24-25.

[12]Sheyndlin, A. E., Shumyatskiy, B. Ya., Shpil'rayn, E. E., et al., "Present State of Development of Electric Power Plants with an MHD Generator," Teploenergetika, No. 3, 1980, p. 2-5.

[13]Springmann, H., Greenberg, R., and Juhasz, A., "The Optimization of Air Separation Plants for Combined Cycle MHD Power Application," Seventh International Conference on Magnetohydrodynamics, Massachusetts, U.S.A., 1980, Linde AG Lotepro Corporation, NASA-Lewis, 1980, p. 403-525.

[14]Kirillin, V. A., Sheyndlin, A. E., Shumyatskiy, B. Ya., et al., "The U-02 Facility Under Long-Duration Operation Conditions," Teplofizika vysokikh temperatur, Vol. 9, No. 5, 1971, p. 1029-1046.

[15]Burenkov, D., Telegin, G., Sheyndlin, A., et al., "Joint Test of a U.S. Electrode System in the U.S.S.R. U-02 Facility," Proceedings of the 15th Symposium on Engineering Aspects of Magnetohydrodynamics, Philadelphia, May 24-26, 1976, University of Pennsylvania, 1976, p. 1-12.

[16]Telegin, G., Romanov, A., Peletskiy, V., et al., "Investigation of Thermophysical Properties of Refractory Materials used in MHD Generator Channels," High Temperatures-High Pressures, Vol. 8, No. 2, 1976, p. 199-209.

[17]Burenkov, D. K., Pishchikov, S. I., Shelkov, E. M., et al., "Preliminary Investigations of Operation of an Open-Cycle MHD Generator in the Presence of Ash in Combustion Products," Teplofizika vysokikh temperatur, Vol. 13, No. 6, 1975, p. 1136-1151.

[18]Burenkov, D. K., Dolinskiy, Yu. L., Zalkind, V. I., et al., "Certain Results of Investigations of the Effect of Slag Films on the Operation of Electrodes on the U-02 Facility," Sixth International Conference on MHD Electric Power Generation, June 9-13, 1975, Washington, D.C., Vol. 2, 1975, p. 321-334.

[19]Kirillin, V. A., Neporozhniy, P. S., and Sheyndlin, A. E., "Pilot Commercial Facility with an MHD Generator with a Power Output of 25,000 kW," Transactions of the 7th Worldwide Power Production Conference, Moscow, August 20-24, 1968, Section 4C, Paper No. 72, Publishing House Energiya, Moscow, 1968, 19 p.

[20]Kirillin, V. A., and Sheyndlin, A. E., "Certain Results of Investigations of the U-25 MHD Power Generating Facility," Teplofizika vysokikh temperatur, Vol. 12, No. 2, 1974, p. 372-389.

[21]Bityurin, V. A., Burakhanov, B. M., Dronov, Yu. A., et al., "Numerical Analysis of Characteristics of the MHD Generator of the U-25 Facility," Teplofizika vysokikh temperatur, Vol. 12, No. 2, 1974, p. 390-398.

[22]Miklashevkiy, L. P., Parnev, L. Kh., Pashkov, S. A., et al., "Characteristics of Operation of the Steam-Generator of the U-25 MHD Power Facility," Teplofizika vysokikh temperatur, Vol. 12, No. 2, 1974, p. 458-485.

[23]Glemba-Ovidskiy, O. A., Dreyzin, L. Z., Korotkevich, A. G., et al., "Development and Investigation of the Combustion Chamber with Hot Walls and Combustion in a System of Thin Jets," Teplofizika vysokikh temperatur, Vol. 12, No. 2, 1974, p. 431-439.

[24]Gronovskiy, E. A., Dreyzin, L. Z., Korolev, S. V., et al., "Certain Results of Tests of Once-Through Combustion Chambers with Cold Walls, Used in the U-25 Facility," Teplofizika vysokikh temperatur, No. 2, 1974, p. 423-430.

[25]Styrikovich, M. A., Zakharko, Yu. A., Mostinskiy, I. L., et al., "Problems Involving Injection and Exhaust of the Ionizable Seed on the U-25 Facility," Teplofizika vysokikh temperatur, No. 2, 1974, p. 440-445.

[26]Antonov, B. M., Barakayev, Kh. F., Iserov, A. D., et al., "Investigation of the Joint Operation of the MHD Generator and the Inverter Substation," Teplofizika vysokikh temperatur, Vol. 12, No. 2, 1974, p. 446-451.

[27]Bass, Z. R., Volovik, A. V., Gorlanov, A. V., et al., "Investigation of the Characteristics of the Oxidizer Preparation System for Oxidizer Injection System into the MHD Generator Loop of the U-25 Facility," Teplofizika vysokikh temperatur, Vol. 12, No. 2, 1974, p. 452-457.

[28]Sheyndlin, A. E., Buznikov, A. E., Iserov, A. D., et al., "Experimental Investigation of the MHD Generator of the U-25 Facility with the 1D Channel," Third US-U.S.S.R. Colloquium on Magnetohydrodynamic Energy Conversion, Moscow, October 20-21, 1976, Publishing House IVTAN, Moscow, 1978, p. 77-89.

[29]Sheyndlin, A. E., Barshak, A. E., Bityurin, V. A., et al., "Investigation of Diagonal-Conducting-Wall R-Channel of MHD Generator of U-25 Power Plant," 15th Symposium on Engineering Aspects of Magnetohydrodynamics, Philadelphia, May 25-26, 1976, University of Pennsylvania, 1976, p. IV.1.1-IV.1.7.

[30]Barshak, A. E., Bityurin, V. A., Buznikov, A. E., et al., "Diagonal Frame RM-Channel of the U-25 Power Plant," 17th Symposium on Engineering Aspects of Magnetohydrodynamics, Stanford, California, March 27-29, 1978, Stanford University, 1978, p. 2.1-2.9.

[31]Buznikov, A. E., Shelnin, V. A., Karpukhin, V. A., et al., "Studies of Diagonal MHD Channel of the U-25 Facility during Long-Duration Operation," Fourth U.S.-U.S.S.R. Colloquium on Magnetohydrodynamic Electrical Power Generation, Washington, D.C., October 5-6, 1978, p. 349-373.

[32]Kniga, A. A., Ziman, E. P., Garbuzov, V. N., et al., "Materials of Scientific-Technical Conference of Countries on Problem 1-4," Fuels, Methods of its Combustion, Development of Combustion Chambers (Toplivo, metody ego szhiganiya, razrabotka kamer sgoraniya), Publishing House IVTAN, Moscow, 1981, p. 28-37, 44-57, 80-98, 116-189.

[33]Kirillin, V. A., Sheyndlin, A. E., Maksimenko, V. I., et al., "U-25B MHD Facility to Conduct Investigations under Conditions of Strong Electric and Magnetic Fields," Teplofizika vysokikh temperatur, Vol. 16, No. 1, 1978, p. 148-159.

[34]Ochenko, V. M., and Bogomaz, V. M., "Investigation of Properties of Seed of K_2CO_3 used in MHD Facilities with Various Type Aerosol Additives," Scientific-Technical Report No. 173/80, Kiev, Institut fizicheskoy khimii AS U.S.S.R., 1981.

[35]Iserov, A. D., Kovbasyuk V. I., Kover'yanov, V. A., et al., "Certain Results of Investigations of the Dielectric Strength of Components of the U-25 MHD Generator," Teplofizika vysokikh temperatur, Vol. 12, No. 2, 1974, p. 412-416.

[36]Karpukhin, V. A., Maksimenko, V. I., Pashkov, S. A., et al., "No. 1 Channel of the U-25B Facility to Conduct Experiments in Strong Electric and Magnetic Fields," Teplofizika vysokikh temperatur, Vol. 16, No. 1, 1978, p. 160-168.

[37]Iserov, A. D., Maksimenko, V. I., Maslennikov, G. I., et al., "Study of the U-25B MHD Generator System in Strong Electric and Magnetic Fields," 18th Symposium on Engineering Aspects of Magnetohydrodynamics, Montana, June 18-20, 1979, Butte, MERDI, 1979, p. A.5.1-A.5.14.

[38]Kirillin, V. A., Sheyndlin, A. E., Shelkov, E. M., et al., "Principle Directions of Investigations and Development Works on Elaboration of a Coal-Fired MHD Power Plant in the U.S.S.R.," Specialists Meeting on Coal-Fired MHD Power Generation, Sydney, November 4-6, 1981, Sydney, 1981, p. 1.1.1-1.1.6.

[39]Golyashevich, S. L., Kalmaru, A. M., Nefedova, M. G., et al., "Investigations of Coal Combustion for MHD Energy Conversion," Sixth International Conference on MHD Electric Power Generation, Washington, D.C., June 9-13, 1975, Vol. 2, 1975, p. 321-334.

[40]Morokhov, I. D., Velikhov, E. P., and Volkov, Yu. M., "Pulsed MHD Generators and Deep Electromagnetic Sounding of the Earth's Crust," Atomnaya energiya, Vol. 44, No. 3, 1978, p. 213-219.

[41]Jimerin, D. G., Basilov, V. A., Makarov, Yu. V., et al., "Results of MHD Generator Testing," Proceedings of the 15th Symposium on Engineering Aspects of Magnetohydrodynamics, Philadelphia, May 24-26, 1976, University of Pennsylvania, 1976, p. 2.1-2.7.

[42]Shvets, I. T., Milyakh, A. N., Klimenko, E. P., et al., "Experimental MHD Generator of the Academy of Sciences and Ministry of Energetics of Ukrainian SSR," Fifth International Conference on MHD Electric Power Generation, Munich, April 19-23, 1971, Vol. 1, p. 171-186.

[43]Kirillin, V. A., Sheyndlin, A. E., Pishchikov, S. I., et al., "A Few Remarks Concerning the First Commercial-Scale System with an MHD Generator," Third U.S.-U.S.S.R. Colloquium on Magnetohydrodynamic Energy Conversion, Moscow, October 20-21, 1976, Publishing House IVTAN, Moscow, 1978, p. 12-31.

[44]Shumyatskiy, B. Ya., Sokol'skiy, A. G., Pereletov, I. I., et al., "The Influence of Methods of Obtaining High-Temperatures on Thermal Efficiency of Open-Cycle MHD Facilities," Teplofizika vysokikh temperatur, Vol. 7, No. 4, 1969, p. 754-756.

[45]Molchanov, I. A. (Ed.), Handbook of Boiler Maintenance (Sprayvochnik po ob'ektam kotlonadzora), Publishing House Energiya, Moscow, 1974, 439 p.

[46]Zhuravlev, V. N., and Nikolayeva, O. I., Machine Building Steels (Mashinostroitel'nyye stali), Publishing House Mashgiz, Sverdlovsk, 1962, 237 p.

[47]Poletavkin, P. G., Malyugin, Yu. S., and Golovkin, F. B., "Results of Experimental Investigations of a High-Temperature Regenerative Air Preheater with Stationary Pebble Bed," Teploenergetika, No. 7, 1972, p. 56-60.

[48]Karras, K., "Chemical Regeneration of Energy at an MHD Power Plant," Transactions of an International Symposium on Electric Power Generation with MHD Generators, Salzburg, Austria, July 7-8, 1966, Vol. 3, 1969, p. 155-156.

[49]Shumyatskiy, B. Ya., Sokol'skiy, A. G., Rossievskiy, G. I., and Koryagina, G. M., "Thermal Efficiency of Commercial-Scale MHD Plants in Terms of Different Methods of Obtaining High Temperature," Proceedings of the 10th Symposium on Engineering Aspects of Magnetohydrodynamics, March 26-28, 1969, Amherst, Mass., University of Massachusetts, 1969, p. 134-135.

[50]Rogochev, A. P., Koryagina, G. M., Ivanov, P. P., et al., "Technical-Economic Estimates of Optimized MHD Power Plants with Significant Heat Regeneration," Teploenergetika, No. 4, 1974, p. 26-30.

[51]Shumyatskiy, B. Ya., "Open-Cycle MHD Generators," Proceedings of a Symposium on MHD Electric Power Generation Organized by the International Atomic Energy and the European Nuclear Energy Agency, Salzburg, July 4-8, 1966, Vienna, International Atomic Energy Agency, Vol. 3, 1966, p. 742-743.

[52]Rossievkiy, G. I., Koryagina, G. M., Monastyrskaya, A. R., and Iglova, L. V., "Thermal Efficiency of Electric Power Plants with MHD Generators when Using Oxygen-Enriched Air Preheated by Means of Various Conventional Techniques," MHD Method of Generating Electric Power (Magnitogidrodinamicheskiy metod poluchenya elektroenergii), Publishing House Energiya, Moscow, 1968, p. 143-163.

[53]Hals, F. A., and Gannon, R. E., "Technical and Economic Aspects of MHD Power Systems and Component Development Work," First U.S.-U.S.S.R. Colloquium on MHD Energy Conversion, Moscow, February 25-27, 1974, Publishing House IVTAN, Moscow, 1974, p. 34-49.

[54]Gubarev, A. V., "Certain Characteristics of High-Temperature Combined Cycles," AN SSSR, Izvestiya, Energetika i transport, No. 5, 1969, p. 133-135.

[55]Shanklin, R. V., Crawford, L. W., Martin, J. F., et al., "The UTSI Coal Burning MHD Program," Proceedings of the 13th International Symposium on Engineering Aspects of Magnetohydrodynamics, Stanford, California, March, 1973, Stanford University, 1973, p. II.8.1-II.8.8.

[56]Barshak, A. E., Bityurin, V. A., Kovbasyuk, V. I., et al., "Investigations of the Diagonal MHD Channel on the U-25 Facility," Second U.S.-U.S.S.R. Colloquium on MHD Electric Power Generation, Washington, D.C., June 5-6, 1976, p. 111-151.

[57]Bityurin, V. A., Iserov, A. D., Kovbasyuk, V. I., et al., "Certain Results of Investigations of the Gasdynamics and Electrodynamics of the Faraday MHD Generator in the U-25 Facility," Sixth International Conference on Magnetohydrodynamic Electric Power Generation, Washington, D.C., June 9-13, 1975, p. 183-198.

[58]Al'tov, V. A., Zenkevich, V. B., Kremlev, M. G., et al., "Stabilization of Superconducting Magnetic Systems" (Stabilizatsiya sverkhprovosyashchikh magnitnykh system), Publishing House Energiya, Moscow, 1975, 270 p.

[59]Hoening, M. O., "Supercritical Helium-Cooled Cabled Superconducting Hollow Conductors for Large High-Field Magnets," Proceedings of the 5th International Cryogenics Engineering Conference, Grenoble, France, May, 1976, p. 216-228.

[60]Montgomery, D. B., Hatch, A. M., and Purcell, I. R., "Superconducting Magnets for Base Load MHD Generators," Sixth International Conference on MHD Electric Power Generation, Washington, D.C., June 9-13, 1975, Vol. 4, p. 115-130.

[61]Hatch, A. M., Becker, F. E., Marston, P. G., et al., "A Superconducting Magnet for Long-Duration Testing of MHD Channels," Proceedings of the 14th Symposium on Engineering Aspects of Magnetohydrodynamics, Tullahoma, Tennessee, April, 1974, University of Tennessee, 1974, p. II.4.1-II.4.6.

[62]Poletavkin, P. G., "Heating of Air and Gases up to 1500°C and Above in High-Temperature Regenerative Heat Exchangers with Stationary Pebble Bed," Heat Transfer in a High-Temperature Gas Flow (Teploobmen v vysoko-temperaturnom potoke gaza), Publishing House Mintis, Vil'nyus, 1972, p. 192-209.

[63]Schonpflug, E., "Hochofen-Sauerstoffanlage Schwelgern" (in German), Linde-Ber. Techn. und Wiss., 1973, No. 33, S.37-51.

[64]Morozov, G. N., Mostinskiy, I. L., and Rabkin, Yu. I., "Characteristics of a Steam Generator of a Power Plant with an MHD Generator," Teploenergetika, No. 9, 1976, p. 35-38.

[65]Styrikovich, M. A., Mostinskiy, I. L., Pinkhasik, D. S., et al., "Seed in U-02 and U-25 MHD Facilities," The First U.S.-U.S.S.R. Colloquium on MHD Energy Conversion (Perviy Sovetsko-Amerikanskiy kollokvium po MHD preobrazovaniyu energii)k, Moscow, 1974, p. 314-326.

[66]Bienstock, D., Demski, R. J., and Kurtzrock, R. D., "High-Temperature Combustion of Coal Seeded with Potassium Carbonate in the MHD Generation of Electric Power," U.S. Department of the Interior, Bureau of Mines, Washington, D.C., Report 7361, 1970.

[67]Dicks, J. B., Crawford, L. W., Muelhauser, J. W., et al., "The Direct Coal-Fired MHD Generator System," Proceedings of the 14th Symposium on Engineering Aspects of Magnetohydrodynamics, Tullahoma, Tennessee, April, 1974, University of Tennessee, 1974, p. 11.1.1-11.1.10.

[68]Bienstock, D., Demski, R. J., and Demeter, J. J., "Environmental Aspects of MHD Power Generation," Intersociety Energy Conversion Engineering Conference, Boston, August 3-5, 1971, p. 1210-1217.

[69]Joubert, J. I., Mossbauer, P. F., Ruppel, I. C., et al., "Kinetics of Regeneration of Spent Seed from MHD Power Generation System," Proceedings of the 15th Symposium on Engineering Aspects of MHD, Philadelphia, Pennsylvania, May 24-26, 1976, University of Pennsylvania, 1976, p. VII.4.1-VII.4.9.

[70]Hals, F. A., and Gannon, R. E., "Auxiliary Component Development Work at AVCO Everett Research Laboratory," Proceedings of the 14th Symposium on Engineering Aspects of MHD, Tullahoma, Tennessee, April, 1974, University of Tennessee, 1974, p. 11.31-11.33.

[71]Wallet, A., and Gruber, A., "Process for the Manufacture of Sulfur and Nitric Acids from Waste Flue Gases," Proceedings of the Second International Clean Air Congress, Washington, D.C., 1970, New York, 1971, p. 217-232.

Chapter 2

Calculation of MHD Channel Flows

2.1. Introduction

The present stage of development of MHD energy conversion raises various new theoretical problems related to flow calculations. Previously (until about the end of 1960), most attention was concentrated on solving a broad class of problems in modeling, so as to investigate the principal physical phenomena in the channel, and on developing the foundations of engineering methods of calculating MHD channel flows. At the present time, the most important problems include improving the accuracy of engineering methods of calculating and numerically modeling real flows in the channel, in order to obtain data required in the design of various types of MHD generators. At the same time, the design of MHD generators frequently leads to a new class of problems that have not been previously analyzed by numerical modeling. Examples of such problems are various aspects of diagonal MHD channels and variational aspects of channel electrodynamics.

A useful approach in investigating electrodynamic problems in an MHD channel is based on that developed in applied magnetohydrodynamics, whereby the general system of equations is separated into electrodynamic and hydrodynamic parts. Solution of axial and transverse electrodynamic problems makes it possible to solve such practical problems as optimal structure configuration and design of loading schemes for segmented channels. Methods based on parametrization of numerical solutions and application of variational calculations are quite effective in such applications.

General, two-dimensional problems of magnetohydrodynamics can be solved numerically, primarily in the inviscid approximation. The most interesting aspect is the investigation of the effect of channel design and loading schemes on the gasdynamic parameters. Important data are obtained about the influence of spatial gasdynamic effects on the overall electrical characteristics of the MHD generator.

Solutions of two-dimensional MHD problems are also extremely useful for the development and improvement of engineering methods of calculating MHD channel flows. The present development of engineering calculation methods is directed, on one hand, toward expansion of the class of real effects taken into account in the quasi-one-dimensional approximation and, on the other hand, toward increasing their reliability, by comparing the results of calculations with the large amount of the available experimental data acquired on the MHD facilities now in operation.

2.2. Axial (Longitudinal) Problems of Electrodynamics

We shall investigate the electrodynamics in MHD channels by the established approach of separating the electrodynamic part from the general system of magnetohydrodynamic equations, and reducing it to a two-dimensional approximation. The foundation of such an approach and the determination of the limits of its applicability have been developed, and are most thoroughly described in a monograph [1].

A great deal of theoretical data on axial electrodynamic problems in MHD channels are available. The results obtained make it possible to solve such problems as the effect of finite segmentation on the characteristics of the MHD generator, including various techniques of connecting electrodes [2-4] and the determination of characteristics of current and potential distributions in the end regions of MHD channels, resulting from a change in the boundary conditions for the channel walls and variation of the fringe magnetic field [5-7]. The solution of similar new problems is directed, primarily, toward a search (in a certain sense) for an optimal structural configuration and loading schemes, as well as an investigation of situations arising during the experiment, especially uneven loading of a segmented MHD channel.

2.2.1. Variational Problem of Electrodynamics. Obtaining optimum solutions providing maximum output parameters of the generator and satisfying various requirements imposed by the operating conditions of the channel in an MHD facility, electrical outlay, and channel construction are important parts of investigations of the electrodynamics of MHD channels. Variational calculation methods are effective for this purpose. The class of variational problems under consideration includes various additional conditions, nonholonomic coupling, isoperimetric conditions, etc., and solution of these problems must provide an arbitrary extremum of functionals. A key element in formulating such problems is the sequence of steps of constructing the first variation of the generalized functional and of obtaining Euler's equations, required boundary conditions, and conditions on surfaces of discontinuities.

2.2.1.1. Formulation of the Variational Problem

We shall consider the optimum potential and current distributions in an arbitrary fixed volume, V, bound by surface, S. It is assumed that the following flow parameters in volume V are given:

flow rate $\mathbf{v} = \mathbf{v}(x,y,z)$, magnetic field $\mathbf{B} = \mathbf{B}(x,y,z)$, electrical conductivity of plasma $\sigma = \sigma(x,y,z)$, Hall parameter $\beta = \beta(x,y,z)$.

The equations for the current density, $\mathbf{j}$, and potential, ϕ have the following form:

$$\mathrm{div}\mathbf{j} = 0 \ , \tag{2.1}$$

$$\mathbf{j} + \beta(\mathbf{j} \times \mathbf{B})/B = \sigma\,(-\mathrm{grad}\ \phi + \mathbf{v} \times \mathbf{B}) \ . \tag{2.2}$$

The boundary conditions on S are not specified and are established when solving the variational problem that seeks the extremum of the power functional

$$N = \int_V \mathbf{j}\ \mathrm{grad}\ \phi\ dV \tag{2.3}$$

subject to the isoperimetric condition

$$N - \eta A = 0 \ , \tag{2.4}$$

where the work performed by the electromagnetic force

$$A = -\int_V (\mathbf{j} \times \mathbf{B})\mathbf{v}\ dV \ ; \tag{2.5}$$

and η is the electrical efficiency of the generator. In addition to the isoperimetric condition (2.4) satisfying the requirement that maximum power in the channel is reached at a given efficiency, it is possible to utilize other isoperimetric conditions and various types of limitations. In the case when the isoperimetric condition is not utilized, solution of the variational problem provides the absolute maximum of the channel power.

We shall seek a solution of the variational problem for an arbitrary extremum using Lagrangian multipliers. Consider the generalized functional

$$J = N + \nu(N - \eta A) + \int_V \lambda \, \mathrm{div}\, \mathbf{j} \, dV ,$$

where ν = const. and $\lambda = \lambda(x,y,z)$ are the Lagrange multipliers.

For convenience in subsequent calculations, let's introduce the following vector:

$$\mathbf{i} = \sigma\left(-\mathrm{grad}\,\lambda + (1+\nu(1-\eta))(\mathbf{v} \times \mathbf{B})\right) + \beta\,(\mathbf{i} \times \mathbf{B})/B . \qquad (2.6)$$

Using (2.2) through (2.6), the expression for J can be transformed into the following form:

$$J = \int_V \left(-\frac{(1+\nu)\sigma}{1+\beta^2} \left((-\mathrm{grad}\,\phi + \mathbf{v} \times \mathbf{B})^2 + \frac{\beta^2}{B^2} (\mathbf{B}\,\mathrm{grad}\,\phi)^2 \right) + \mathrm{div}\,\lambda\mathbf{j} - \mathbf{i}\,\mathrm{grad}\,\phi + \mathbf{i}\,(\mathbf{v} \times \mathbf{B}) \right) dV . \qquad (2.7)$$

Consider the first variation of the functional, J, in order to obtain Euler's equations and the boundary conditions,

$$\delta J = \int_V \left(-\frac{2(1+\nu)\sigma}{1+\beta^2} \left((\mathrm{grad}\,\phi - \mathbf{v} \times \mathbf{B}) - \frac{\beta^2}{B^2} (\mathbf{B}\,\mathrm{grad}\,\phi)\,\mathbf{B} \right) \mathrm{grad}\,\delta\phi + \mathrm{div}\,\lambda\delta\mathbf{j} - \mathbf{i}\,\mathrm{grad}\,\delta\phi \right) dV .$$

Performing integral transformations and utilizing Ohm's law (2.2) in the form

$$\mathbf{j} = \sigma((-\mathrm{grad}\ \phi + \mathbf{v} \times \mathbf{B})$$

$$- \beta\ (-\mathrm{grad}\ \phi + \mathbf{v} \times \mathbf{B}) \times \mathbf{B}/B$$

$$- \beta^2\ (\mathbf{B}\ \mathrm{grad}\ \phi)\ \mathbf{B}/B^2)/(1+\beta^2)\ ,$$

we obtain the following expression

$$\delta J = \int_V \mathrm{div}\ \Big(\mathbf{i}-2\ (1+\nu)\ \mathbf{j}$$

$$- \frac{2(1+\nu)\beta\sigma}{(1+\beta^2)B}\ (-\mathrm{grad}\ \phi + \mathbf{v} \times \mathbf{B})\times\mathbf{B}\Big)\ \delta\phi dV$$

$$+ \oint_S \Big(\lambda\delta j_n - \Big(\mathbf{i}-2(1+\nu)\mathbf{j}$$

$$- \frac{2(1+\nu)\beta\sigma}{(1+\beta^2)B}\ (-\mathrm{grad}\ \phi + \mathbf{v} \times \mathbf{B})\times\mathbf{B}\Big)\mathbf{n}\delta\phi\Big) dS\ , \tag{2.8}$$

where $\mathbf{n}$ is the normal to the surface S. Recalling the principal lemma of calculus of variations and assuming an arbitrary variation of potential $\delta\phi$ in volume V, we obtain Euler's equation from which $\lambda(x,y,z)$ can be determined,

$$\mathrm{div}\ \Big(\mathbf{i}- \frac{2(1+\nu)\beta\sigma}{(1+\beta^2)B}\ (-\mathrm{grad}\ \phi + \mathbf{v} \times \mathbf{B}) \times \mathbf{B}\Big) = 0\ , \tag{2.9}$$

where vector $\mathbf{i}$ is connected with $\lambda(x,y,z)$ through relation (2.6).

The boundary conditions for equations (2.1) and (2.9) are determined from the surface integral in expression (2.8). The surface integral in (2.8) includes variations $\delta\phi$ and δj_n at the boundary of region S. Because, in the variational problem under consideration, the boundary conditions for electrical quantities are not given, the relationship between the potential and the normal component of current density at surface S cannot be determined before solving the problem. Therefore, their variations, $\delta\phi$ and δj_n, are independent. Thus, assuming that any arbitrary potential and current distributions can exit on the surface S, we obtain

$$(\mathbf{i}-2(1+\nu)\mathbf{j}-2(1+\nu)\beta\sigma(1$$
$$+\ \beta^2)^{-1}\ B^{-1}\ (-\text{grad}\ \phi$$
$$+\ \mathbf{v} \times \mathbf{B})\times\mathbf{B})\mathbf{n} = 0,\ \lambda = 0(x,y,z \in S)\ . \qquad (2.10)$$

From (2.9) and (2.10) it follows that when $\beta = 0$, the problem or λ can be separated from the problem for ϕ.

In the case when the problem formulation requires that the surface, S, contain areas with specified current and potential distributions (e.g. insulators and conductors at a fixed potential), one chooses either the first or the second of the two boundary conditions (2.10) for equation (2.9). This procedure is valid because the boundary conditions in the above-noted areas required for the solution of equation (2.1) are given, and either $\delta j_n = 0$ or $\delta\phi = 0$.

Representation of the first variation of the generalized functional, J, in the form (2.8) makes it possible not only to obtain Euler's equation and the boundary conditions, but also to formulate conditions at the discontinuity surfaces. For example, let volume V contain a discontinuity surface, S′. First, the usual electrodynamic conditions, e.g., continuity of the normal component of the current density and the tangential component of the electric field,[1] are fulfilled on S′, i.e.,

$$\{j_n\} = 0 \quad \text{and} \quad \{\text{grad}_\tau \phi\} = 0\ . \qquad (2.11)$$

Additional conditions are obtained by considering variation of δJ (2.8). For this purpose, it is sufficient to divide the volume V into two volumes, V_1 and V_2, separated by surface S′, and to represent as a sum of two

[1] It is assumed that the potential on S′ is continuous. Otherwise, as was shown in Ref. [1], the continuity condition for E_τ may not be fulfilled.

functionals J_1 and J_2, determined for V_1 and V_2, respectively. In this case, in addition to the term in the right side of (2.8), the expression for δJ also contains an additional integral over the surface S′,

$$\int_{S'} \left\{ \lambda \delta j_n - \left(\mathbf{i} - 2(1+\nu)\mathbf{j} - \frac{2(1+\nu)\beta\sigma}{(1+\beta^2)B} (-\mathrm{grad}\ \phi + \mathbf{v} \times \mathbf{B}) \times \mathbf{B} \right) \mathbf{n} \delta\phi \right\} dS \ .$$

Applying (2.11) to the above integral leads to the following expressions:

$$\{\lambda\} = 0 \ ,$$

$$\left\{ i_n - \frac{2(1+\nu)\beta\sigma}{(1+\beta^2)B} ((-\mathrm{grad}\ \phi + \mathbf{v} \times \mathbf{B}) \times \mathbf{B})_n \right\} = 0 \ . \qquad (2.12)$$

The formulation of the variational problem is now completed. A system of equations, (2.1), (2.2), (2.6), (2.9), subject to boundary conditions (2.10) and conditions for matching (2.11) and (2.12), have been obtained for ϕ and λ.

It should be noted that representation of the generalized functional J in the form (2.7) is not unique. For example, it is possible to write J directly as a sum of two (volume and surface) integrals. In this case, variation of the first integral will give Euler's equation, and variation of the second, the boundary conditions. The form of equations obtained in this manner will differ from equations (2.9) and (2.10).

In investigating electrodynamics of MHD channels, it is frequently expedient to obtain solution of a problem in terms of the current stream function, $\boldsymbol{\psi}$, connected with current density $\mathbf{j}$ through the relation $\mathrm{rot}\ \boldsymbol{\psi} = \mathbf{j}$. Let us formulate the variational problems for finding $\psi(x,y,z)$. In this case, the initial equations are

$$\text{rot}\ \mathbf{E} = 0\ , \tag{2.13}$$

$$\text{rot}\ \psi = \sigma(\mathbf{E} + \mathbf{v} \times \mathbf{B}) - \beta\,(\text{rot}\ \psi \times \mathbf{B})/B\ . \tag{2.14}$$

The generalized functional J has the form

$$J = N + \mu(N = \eta A) + \int_V \Xi\ \text{rot}\ \mathbf{E}\ dV\ ,$$

where $\mu =$ const. and $\Xi = \Xi(x,y,z)$ are the Lagrangian multipliers.

Introducing auxiliary vector

$$\text{rot}\ \Xi = \sigma(e - (1+\mu(1-\eta))\ (\mathbf{v} \times \mathbf{B}))$$

$$+ \beta\,(\text{rot}\ \Xi \times \mathbf{B})/B\ , \tag{2.15}$$

J can be represented in the form

$$J = \int_V (-(1+\mu)\sigma^{-1}\ (\text{rot}\ \psi)^2 + e\ \text{rot}\ \psi$$

$$- \text{div}\ (\Xi \times E) - (\mathbf{v} \times \mathbf{B})\ \text{rot}\ \Xi\ dV\ .$$

We obtain the following expression for the first variation of the functional

$$\delta J = \int_V (\delta\psi\ \text{rot}(e-2(1+\mu)\sigma^{-1}\text{rot}\ \psi))\ dV$$

$$+ \oint_S ((\Xi \times \mathbf{n})\delta\mathbf{E}$$

$$+ ((e-2(1+\mu)\sigma^{-1}\text{rot}\ \psi) \times \mathbf{n})\delta\psi)\ dS\ , \tag{2.16}$$

that leads to Euler's equation and the boundary conditions:

$$\mathrm{rot}(e-2(1+\mu)\sigma^{-1}\mathrm{rot}\ \boldsymbol{\psi}) = 0 \ ; \qquad (2.17)$$

$$(e-2(1+\mu)\sigma^{-1}\mathrm{rot}\ \boldsymbol{\psi}) \times \mathbf{n} = 0 \ ,$$

$$\Xi \times \mathbf{n} = 0 \quad (x,y,z \in S) \ . \qquad (2.18)$$

The conditions on the surface of discontinuity S′ have the form:

$$\{(e-2(1+\mu)\sigma^{-1}\ \mathrm{rot}\ \boldsymbol{\psi}) \times \mathbf{n}\} = 0 \ ,$$

$$\{\Xi \times \mathbf{n}\} = 0 \ . \qquad (2.19)$$

Thus, equations (2.13)-(2.15), (2.17) and the boundary conditions (2.18) as well as the conditions for matching (2.19) are available for determination of $\boldsymbol{\psi}$.

In conclusion, it should be noted that the approach to the formulation of the variational problem described here allows generalization to the case, when the boundaries of the region within which the solution is being sought, as well as the magnetic field and flow parameter distribution, can be varied. Naturally, in this case, the formulation of the problem becomes more complicated because of inclusion of new equations and limitations that have to be satisfied by the quantities being varied. Two-dimensional variational electrodynamic problems formulated in a manner similar to that used here have been investigated in [8-10].

2.2.1.2. Variational Problem of the End Effect in an MHD Channel

Consider the two-dimensional variational problem of current and potential distributions in the transverse plane of an MHD channel generating maximum power output at a specified electrical efficiency. Let a stream of anisotropically conducting fluid with conductivity σ flow at a constant velocity, $\mathbf{v} = \{u, 0, 0\}$ in a two-dimensional, infinitely long channel of height h. The magnetic field and the Hall parameter are one-dimensional, i.e., $\mathbf{B} = \{0,0,B(x)\}$ and $\beta = \beta_o B(x)/B_o$, respectively. It will be assumed that the flow field and the magnetic field distribution are specified and are not varied. However, the potential and current distributions in the channel can be varied arbitrarily.

Let's solve the problem for a magnetic field distribution in the form of a step function:

$$B(x) = \{B_o = \text{const. when } x>0; \ 0 \text{ when } x<0\} .$$

It is expedient to obtain the solution in terms of the current stream function ψ [10]. For this purpose, equations (2.13)-(2.19) can be used, setting $\psi = \{0, 0, \psi\}$ and $\Xi = \{0, 0, \Xi\}$ and $\partial/\partial z = 0$. Further transformations will be made for dimensionless quantities, reducing ψ and Ξ to $\sigma u B_o h$ and x and y to h. Then, from equations (2.13)-(2.15) and (2.17), we obtain

$$\Delta\psi = 0, \ \Delta\Xi = 0 \quad \text{when} \quad -\infty<x<0 ,$$

$$0<x<\infty, \ |y|<1/2 . \tag{2.20}$$

The boundary conditions follow from relations (2.18). It is necessary to take into account the fact that in a problem dealing with the end effect, one is given the asymptotic current density and electric field distribution when $x = \pm\infty$. For this formulation of the problem we have

$$\mathbf{j} = 0 \quad \text{when } x = -\infty, \quad |y| \leq 1/2 ,$$

$$j_x = 0, \ j_y = -(1-\eta) \quad \text{when} \quad x = \infty, \ |y| \leq 1/2 .$$

This indicates that when $x = \pm\infty$, variation of ψ and **E** satisfy conditions

$$\delta\psi(-\infty, y) = 0, \quad \delta\psi(\infty, y) \neq 0 ,$$

$$\delta E_y \ (\pm\infty, y) = 0 .$$

It should be noted that, as a result of the fact that $\delta\psi(\infty, y)$ is independent of y when $x = \infty$, condition (2.18) has to be replaced with an integral condition following from the general expression for the variation of the functional (2.16):

$$\int_{-1/2}^{1/2} (2(1+\mu)\psi_x - (1+\mu(1-\eta))$$

$$- \Xi_x + \beta\Xi_y)_{x=\infty}\, dy = 0 \,. \qquad (2.21)$$

The variation of ψ and E_x along the horizontal channel walls is not equal to zero. Therefore, the boundary conditions can be written in the form:

$$\psi_x = 0,\ \psi_y = 0,\ \Xi = 0 \text{ when } x = -\infty \,,$$

$$|y| \leq 1/2 \,,$$

$$\psi_x = 1 - \eta,\ \psi_x = 0,\ \Xi = 0 \text{ when } x = \infty \,,$$

$$|y| \leq 1/2 \,,$$

$$2(1+\mu)\psi_y - \Xi_y = 0, \quad \Xi = 0 \text{ when } -\infty < x < \infty \,,$$

$$y = \pm\, 1/2 \,. \qquad (2.22)$$

Relations (2.22) include the asymptotic condition $\Xi(\pm\,\infty,y) = 0$. If one assumes that Ξ is sought in the same class of functions as ψ, then this condition follows directly from the Laplace's equation $\Delta\Xi = 0$ and the homogeneous condition for Ξ at the channel walls, $\Xi(x, \pm 1/2) = 0$. It can be easily seen that the assumption that Ξ is uniform when $x \to \pm\infty$ provides the asymptotics of solution for $\psi\,(x,y)$.

From (2.21), taking into account asymptotic relations for ψ and Ξ, we obtain the Lagrangian multiplier

$$\mu = (2\eta - 1)/(1-\eta) \,.$$

Let us now establish the conditions for matching solutions in the channel section $x = 0$, where the magnetic field has a discontinuity. From (2.11) and (2.19), we obtain

$$\psi\,(-0,\, y) = \psi(+0,\, y),\ \Xi(-0,\, y) = \Xi(+0,\, y) \,,$$

$$\psi_x\,(-0,\, y) = \psi_x(+0,\, y) + \beta\psi_y(+0,\, y) - 1 \,,$$

$$2(1+\mu)\psi_x(-0,\ y) - \Xi_x(-0,\ y)$$

$$= 2(1+\mu)\psi_x(+0,\ y) - \Xi_x(+0,\ y)$$

$$+ \beta\Xi_y(+0,\ y) - (1+\mu)(1-\eta)\ . \qquad (2.23)$$

Proceeding to solve the problem specified by (2.20), (2.22), (2.23), note the properties of $\psi(x,y)$ and $\Xi(x,y)$, following from the boundary conditions and continuity of these functions:

$$\Xi(-x,y) = \Xi(x,y),\ \psi(-x,y) = \psi(x,y)$$

$$- (1-\eta)x\ . \qquad (2.24)$$

In this case, it is sufficient to find a solution for the right half of the channel. The boundary conditions at $x = +0$ can be obtained by multiplying the third condition (2.23) by $2(1+\mu)$ and subtracting from it the fourth condition (2.23), taking into account properties of (2.24):

$$2(1+\mu)\beta\psi_y - \beta\Xi_y + 2\Xi_x - 2(1+\mu)\eta = 0$$

$$\text{when } x = +0,\ |y|<1/2\ . \qquad (2.25)$$

Taking into account boundary conditions (2.22), the boundary conditions on the horizontal walls can be represented in the form

$$2(1+\mu)\beta\psi_y - \beta\Xi_y + 2\Xi_x = 0 \text{ when } x>0\ ,$$

$$y = \pm\ 1/2\ . \qquad (2.26)$$

It is easy to establish that boundary conditions (2.25) and (2.26) define the following analytic function in the half strip $x>0$, $|y|<1/2$

$$F(z) = 2(1+\mu)\beta\psi_y - \beta\Xi_y + 2\Xi_x$$

$$+ i(2(1+\mu)\beta\psi_x - \beta\Xi_x - 2\Xi_y)$$

in the form

$$F(\omega) = \frac{2(1+\mu)\eta}{\pi i} \int_{-1}^{1} \frac{dt}{t-\omega} + iC$$

$$= \frac{2(1+\mu)\eta}{\pi i} \ln \frac{\omega-1}{\omega+1} + iC , \qquad (2.27)$$

where $\omega = \sin \pi iz$ is the conformal transformation mapping the half strip $x > 0, |y| < 1/2$ into the upper half plane, $iC = F(\infty) = 2i(1+\mu) \times (1-\eta) \beta$, $z = x+iy$.

From (2.27), the value of the imaginary part of F(z) at the boundaries of the half strip $x>0$, $|y|<1/2$ is:

$$2(1+\mu)\beta\psi_x - \beta\Xi_x - 2\Xi_y$$

$$= 2(1+\mu) ((1-\eta)\beta - \eta\pi^{-1}\ln((1-t)/(1+t))) ,$$

$$t = -\sin \pi y \text{ when } x = +0, \ |y|<1/2 ;$$

$$2(1+\mu)\beta\psi_x - \beta\Xi_x - 2\Xi_y$$

$$= 2(1+\mu)((1-\eta)\beta - \eta\pi^{-1}\ln((t-1)/(t+1))) ,$$

$$t = \mp \operatorname{ch}\pi x \text{ when } x>0, \ y = \pm 1/2 . \qquad (2.28)$$

To write out the solution for the Lagrangian multiplier $\Xi(x,y)$, first determine the boundary condition for Ξ when $x = +0$, by substituting the first expression in (2.28) into the third condition (2.23), taking (2.24) into account.

$$\Xi_y(+0, y) = \frac{4(1+\mu)\eta}{\pi(4+\beta^2)} \ln \frac{1-t}{1+t} ,$$

$$t = -\sin\pi y \text{ when } |y|<1/2 .$$

For the boundary condition on the horizontal channel walls, $\Xi(x, \pm 1/2) = 0$, the solution for $\Xi(x,y)$ is represented by an analytical function

$$f(z) = \Xi_x - i\Xi_y = \frac{4(1+\mu)\eta}{i\pi(4+\beta^2)} \ln \frac{\omega-1}{\omega+1} ,$$

$$\omega = \sin\pi iz \text{ when } x>0 . \qquad (2.29)$$

Thus, the variational problem is solved. Combining (2.27) and (2.29) we obtain a solution for $\psi(x,y)$:

$$\psi_y + i\psi_x = (F(z) - (2-i\beta)f(z))/2(1+\mu)\beta$$

$$= \frac{(2-i\beta)\eta}{\pi(4+\beta^2)} \ln \frac{\omega-1}{\omega+1} + i(1-\eta) ,$$

$$\omega = \sin\pi iz \quad \text{when} \quad x>0 . \tag{2.30}$$

Note that, when $\beta = 0$, (2.30) becomes a known solution for the optimum generator, characterized by a discontinuous magnetic field and a constant current density at the electrodes [8].

From (2.30), the distributions ψ_x and ψ_y on the channel walls and along the section $x = 0$ are obtained,

$$\psi_x(x,\pm 1/2)$$

$$= (1-\eta) \mp \frac{\beta\eta}{\pi(4+\beta^2)} \ln \frac{\text{ch}\,\pi x+1}{\text{ch}\,\pi x-1} \text{ when } x>0 ,$$

$$\psi_y(x,\pm 1/2) = \pm \frac{2\eta}{\pi(4+\beta^2)} \ln \frac{\text{ch}\,\pi x+1}{\text{ch}\,\pi x+1} ,$$

$$\psi_x(+0,y) = (1-\eta)$$

$$+ \frac{2\eta}{4+\beta^2} \left(1 - \frac{\beta}{2\pi} \ln \frac{1+\sin\pi y}{1-\sin\pi y}\right) ,$$

$$\psi_y(0,y) = \frac{\eta}{4+\beta^2} \left(\beta + \frac{2}{\pi} \ln \frac{1+\sin\pi y}{1-\sin\pi y}\right) , \tag{2.31}$$

and also the distribution of dimensionless ϕ_x and ϕ_y:

$$\phi_x(x, \pm 1/2) = \beta(1-\eta)\mp$$

$$\mp \frac{(2+\beta^2)\eta}{\pi(4+\beta^2)} \ln \frac{\text{ch}\,\pi x+1}{\text{ch}\,\pi x-1} \text{ when } x>0 ,$$

$$\phi_y(0,y) = - \frac{2\eta}{4+\beta^2} \left(1- \frac{\beta}{\pi} \ln \frac{1+\sin\pi y}{1-\sin\pi y}\right) . \tag{2.32}$$

Figure 2.1a shows the potential distribution on the channel walls when $y = \pm 1/2$ for an optimum MHD generator (continuous curves) and a generator with constant current density on electrodes [7] (dashed curves). The Hall parameter β is set equal to one. The efficiency, η, specifying the homogeneous component of the potential distribution when $x > 0$, is chosen to be 2/3 for both solutions. Figure 2.1b shows the current density distribution on channel walls under the same conditions.

The channel has to be properly loaded in order to have the specified potential and current distributions. As an example, consider a diagonal loading scheme. The dashed lines in Figure 2.2, corresponding to the equipotentials, indicate commutation of ideally segmented electrodes at the same potential. A distributed load is connected to each equipotential electrode pair (or to each equipotential). Figure 2.2 shows the load current density curve when the load is connected to the anode side of the channel. The load current density, j_h, is determined from the following expression:

$$j_h(x) = -j_y(x,1/2) \mp \frac{j_y(x',\pm 1/2)E_x(x,1/2)}{E_x(x',\pm 1/2)} ,$$

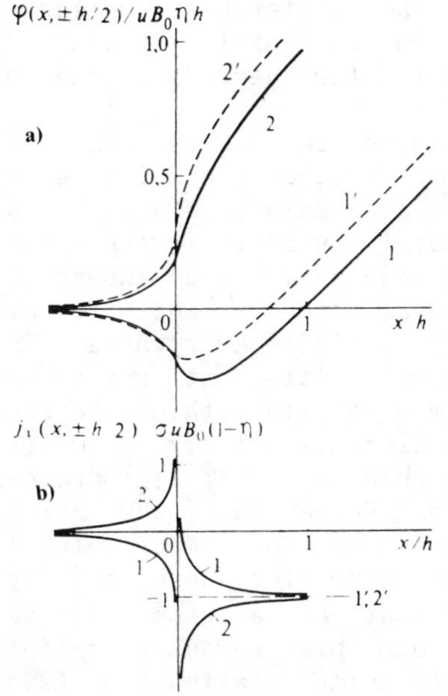

Fig. 2.1 The potential distribution (a) and current density distribution (b) on channel walls for $\beta = 1$ and $\eta = 2/3$: 1,2--variational solution; 1′,2′--solution at constant electrode current density; 1,1′-$y = 1/2\ h$; 2,2′-$y = -1/2h$

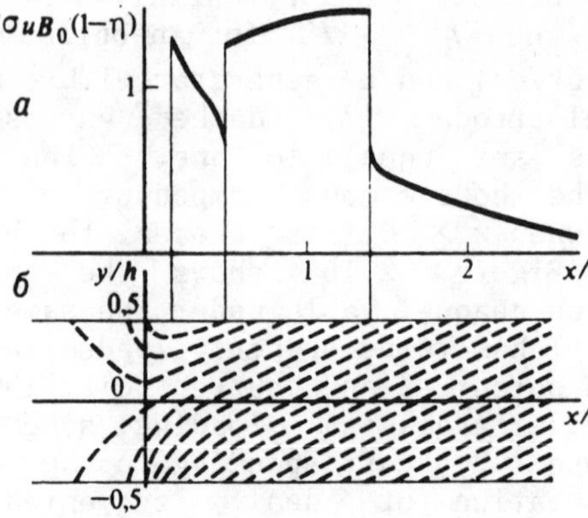

Fig. 2.2 Distribution of load current density fed to the anode wall (a) and plot of equipotentials (b) in an optimal MHD channel

where x varies along the anode wall from the point of the potential minimum ($E_x(x,1/2) = 0$) to ∞, and x' corresponds to the position of the equipotential with respect to $(x,1/2)$ point on the anode or cathode channel wall. The factor $E_x(x,1/2)/E_x(x',\pm 1/2)$ in the formula for j_h takes into account the difference in the number of equipotentials approaching points x and x' or, in other words, the difference in the degree of channel segmentation at these points.

From the curve in Figure 2.2, it follows that a distributed input of the operating current to the optimum channel takes place mainly at the end sector of finite length. The total current to the channel corresponds to the generating mode and, even though the current density on the channel wall becomes infinite at $x = 0$, the load current density is finite everywhere. The distribution of j_h has two discontinuities at points at which the equipotentials that originate within the cross section of the magnetic field discontinuity, $x = 0$ (see Fig. 2.2), are terminated. When $x \to \infty$, j_h decreases rapidly and approaches zero. For example, when $x = 4$, $j_h = 0.0019$.

It is interesting to compare the overall characteristics of the generator described by the variational solution with that of a generator with fixed current density [7]. For this purpose, calculate the power, N, and the work, A, using relationships (2.31) and (2.32):

$$\frac{N}{\sigma(uB_0h)^2} = \eta(1-\eta)\left(\frac{L}{h} -\right.$$

$$- \frac{16}{(4+\beta^2)\pi^3} \sum_1^\infty \frac{1}{(2n-1)^3}\Bigg),$$

$$\frac{A}{\sigma(uB_o h)^2} = (1-\eta)\frac{L}{h}$$

$$+ \frac{16}{(4+\beta^2)\pi^3} \sum_1^\infty \frac{1}{(2n-1)^3},$$

$$L/h >> 1, \sum_1^\infty (2n-1)^{-3} = 1.0518, \tag{2.33}$$

where L is the distance from the origin measured along the channel.

The terms in (2.33) that contain a series determine the dimensionless corrections to the power output and work due to the end effect. In particular, we have the following expression for ΔN:

$$\frac{\Delta N}{\sigma(uB_o h)^2} = -\frac{16\eta(1-\eta)1.0518}{(4+\beta^2)\pi^3}.$$

The continuous line in Figure 2.3 shows a plot of $\Delta N/\Delta N|_{\beta=0}$. The dashed curve in this figure shows the relative power loss in a channel with constant current density on the electrodes [7]. A comparison of the two curves shows the advantages of the variational solution.

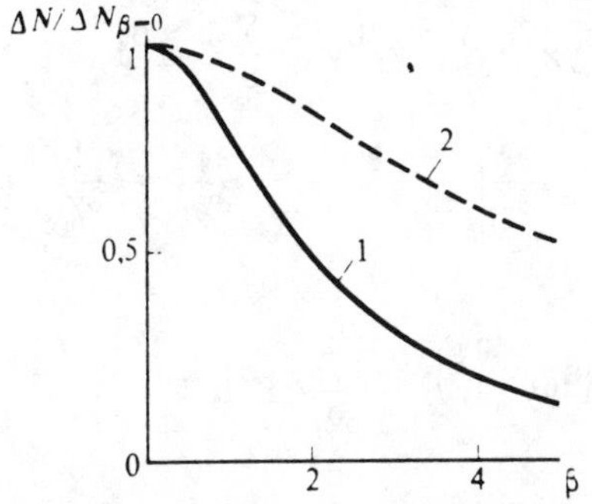

Fig. 2.3 Relative power losses in the channel caused by end effects: 1--variational solution, 2--solution on assumption of constant current density on electrodes

2.2.2. Numerical Solution of Linear Electrodynamic Problems. In addition to the methods of the calculus of variation used to solve the various applied problems of optimizing MHD generators, methods based on parametric representation of solutions under investigation are quite useful and effective. Numerical solutions in parametric forms are especially promising because only numerical solutions can be used to investigate electrical fields and currents in a channel at close to real channel geometry, loading scheme and magnetic field, gasdynamic flow parameter distributions, and plasma properties.

Parametric representation of the numerical solution provides a general solution for a channel, independent of its loading scheme. This solution includes a number of arbitrary parameters determined from additional relationships describing external electrical circuits and loading modes, or representing design of the optimum loading scheme of the channel. Numerical solution in a parametric form makes it possible to obtain any channel characteristics, e.g., voltage-current characteristics of electrodes, groups of electrodes, etc.

2.2.2.1. Problem Formulation and Parametrization of Numerical Solution

The two-dimensional current and potential distributions in the axial plane of the channel in a magnetic field, $\mathbf{B} = \{0,0,B(x,y)\}$, are described by Maxwell's equations (2.1) and (2.13) and the generalized Ohm's law (2.2) that can be represented in the form

$$j_x = a_{xx}E_x + a_{xy}E_y + \sum_0^n b_{xi}p_i \,,$$

$$j_y = a_{yx}E_x + a_{yy}E_y + \sum_0^n b_{yi}p_i \,; \tag{2.34}$$

$$\frac{\partial}{\partial x}\left(a_{xx}\frac{\partial\phi}{\partial x}\right) + \frac{\partial}{\partial x}\left(a_{xy}\frac{\partial\phi}{\partial y}\right) + \frac{\partial}{\partial y}\left(a_{yx}\frac{\partial\phi}{\partial x}\right)$$

$$+ \frac{\partial}{\partial y}\left(a_{yy}\frac{\partial\phi}{\partial y}\right) = \sum_0^n f_i p_i \,,$$

$$\text{where} \quad f_i = \frac{\partial b_{xi}}{\partial x} + \frac{\partial b_{yi}}{\partial y} \,, \tag{2.35}$$

where a_{ji} and b_{ji} are functions of coordinates, p_i are parameters independent of the coordinates, the meaning of which will be explained below.

In considering a class of linear problems, the boundary conditions will be represented in a sufficiently general form, i.e., in the form of a linear relationship between the potential and the normal component of the current density at the boundaries of the region, e.g.,

$$\alpha_1 \phi + \alpha_2 \frac{\partial \phi}{\partial \tau} + \alpha_3 \frac{\partial \phi}{\partial n} = \sum_{o}^{n} \beta_i p_i , \qquad (2.36)$$

where α_i and β_i are functions of coordinates.

In view of the linearity of equations (2.34)-(2.36), the solution of the boundary problem under consideration can be written in the following form:

$$\phi = \sum_{o}^{n} p_i \Phi_i , \qquad (2.37)$$

where Φ_i is a solution of the problem (2.35)-(2.36), subject to the condition that $p_i = 1$ and the remaining parameters $p_o, \ldots, p_{i-1}, p_{i+1}, \ldots, p_n$ are equal to zero. Solutions Φ_i are standard, depending only on the plasma properties and the design of the channel walls, and are independent of the operating mode of the channel. Thus, the potential, ϕ, as presented in (2.37) is a parametric solution which is convenient for the purpose of optimizing and studying the off-design modes of the operation of an MHD channel.

To explain the essence of the parameters p_i, consider the problem of the potential distribution in a channel with one pair of electrodes; this problem is well known and described in various publications [1].

In the case of a working medium with homogeneous isothropic conductance, if the medium flows in a channel of infinite length and height, h, with symmetrically located electrodes of length, ℓ, the parameters are introduced,

$$p_o = u_o B_o , \; p_1 = 1/2V_y ,$$

where u_o and B_o are maximum values of the flow velocity and magnetic field induction; V_y is the difference of the potentials between electrodes. In this case the coefficients in the equations (2.35) and (2.36) are as follows:

$$a_{xx} = a_{yy} = \sigma; \ a_{xy} = a_{yx} = 0 \ ;$$

$$b_{yo} = -\sigma q(x,y); \ b_{xo} = b_{x1} = b_{y1} = 0 \ ;$$

$$f_o = -\sigma \frac{\partial q}{\partial y} \ ; \quad f_1 = \alpha_2 = 0 \ ;$$

$$\beta_o = \begin{cases} 0, \\ -q(x,y); \end{cases} \quad \alpha_1 = \begin{cases} 1, \\ 0; \end{cases} \quad \beta_1 = \begin{cases} \mp 1, \\ 0; \end{cases}$$

$$\alpha_3 = \begin{cases} 0 & \text{at } |x| \leq 1/2\ell, \ y = \pm 1/2h, \\ 1 & \text{at } |x| > 1/2\ell, \ y = \pm 1/2h; \end{cases}$$

$$q(x,y) = u(x,y)\ B\ (x,y)/u_o B_o \ .$$

Thus, for the standard functions Φ_i, which form a parametric solution of (2.37), the following problems can be formulated:

$$\Delta\Phi_o = -\frac{\partial q}{\partial y} \quad \text{at } |x| < \infty, \ |y| < 1/2h \ ;$$

$$\Phi_o = 0 \quad \text{at } |x| \leq 1/2\ell, \ y = \pm 1/2h \ ;$$

$$\frac{\partial \Phi_o}{\partial y} = -q \quad \text{at } |x| > 1/2\ell, \ y = \pm 1/2h \ ;$$

$$\Delta\Phi_1 = 0 \quad \text{at } |x| < \infty, \ |y| < 1/2h \ ;$$

$$\Phi_1 = \mp 1 \quad \text{at } |x| \leq 1/2\ell, \ y = \pm 1/2h \ ;$$

$$\frac{\partial \Phi_1}{\partial y} = 0 \quad \text{at } |x| > 1/2\ell, \ y = \pm 1/2h \ .$$

Solutions of these problems are independent of the operating mode of the channel. Solution Φ_o can be associated with the potential in the channel with short-circuited electrodes and Φ_1 with the potential of a channel with a stationary medium. It should be noted that the parameters p_i are not unique and can be chosen in accordance with the purpose of numerical experiments.

Because the boundary problem of finding $\psi(x,y)$ is written in the same manner as problems (2.35) and (2.36), analogous representation in parametric form is also possible for the stream function, ψ. Therefore,

$$\psi = \sum_{0}^{n} p_i \Psi_i \ , \tag{2.38}$$

where Ψ_i is an analogue of solution Φ_i (see (2.37)). In solving a single boundary problem simultaneously for ϕ and ψ, the values of p_i in solutions (2.37) and (2.38) are the same.

The following example illustrates the technique of determining parameters p_i and the advantages of representing solutions of the boundary problems in the form (2.37) and (2.38). Let a channel with a known magnetic field and flow parameter distribution contain m electrodes. The problem is to determine the influence of electrode commutation on the output characteristics of an MHD generator. Consider the following three most typical cases:

a. All electrode potentials are given (i.e., the boundary conditions (2.36) are given in explicit form),

b. Electrode loading satisfies m relationships between potentials ϕ_i and currents $\Delta\psi_i$ for the i-th electrode, i.e.,

$$G_k(\phi_i,\ \Delta\psi_i) = 0$$

$$(i = 1,\ldots,n;\ k = 1,\ldots,m;\ n = m)\ ; \tag{2.39}$$

In the example of a channel with a single pair of electrodes, a specific form of (2.39) is determined by the type of the loading device connected to the channel. For example, when a channel is loaded with resistance R,

$$\Delta\psi_a = \Delta\psi_k = (\phi_k - \phi_a)/R\ ,$$

where the subscripts a and k refer to the anode and cathode, respectively.

c. The number of relationships, such as (2.39) is smaller than the number of electrodes m (n > m) and the remaining necessary conditions must be determined from the extremum (maximum) power output of the MHD generator.

Parameters p_i are defined in the following manner:

$$P_o = 1,\ P_i = \phi_i\ (i=1,\ldots,\ n)\ .$$

The coefficients of p_i in (2.35) and (2.36) are defined as follows:

$$b_{xo} = \sigma\beta uB/(1+\beta^2),\ b_{yo} = -\sigma uB/(1+\beta^2)\ ,$$

$$b_{xi} = b_{yi} = 0\ \ (i=1,\ldots,\ n)\ ,$$

$$\beta_i = \{1\ \ \text{when}\ \ x,y \in \Gamma_i\ ;$$

$$0\ \ \text{when}\ \ x,y \in \Gamma_1,\ldots,\ \Gamma_{i-1},\ \Gamma_{i+1},\ldots,\ \Gamma_n\}\ ,$$

$$\beta_o = 0\ ,$$

where Γ_i is section of the boundary of the area occupied by the i-th electrode. This definition of parameters p_i and their coefficients in the right side of equations (2.35) and (2.36) defines Φ_i (i=1,..., n) as a solution of the problem in which the i-th electrode potential is 1 and the remaining electrode potentials are zero.

In order to obtain a solution in case (a), it is sufficient to substitute the values of parameters p_i into solutions of (2.37) and (2.38). In case (b), it is first necessary to solve the algebraic system of equations (2.39) for p_i. And, finally, in case (c), it is necessary to solve the system of equations given below by (2.40) jointly with the relationships such as those given by (2.39).

$$\frac{\partial J}{\partial p_i} = 0\ \ (i=1,\ldots,\ n)\ , \tag{2.40}$$

where

$$J = N + \sum_{1}^{m} \lambda_k G_k \quad (m<n) ;$$

$$N = \sum_{1}^{n} \phi_i \Delta\psi_i = \sum_{i,j=1}^{n} \gamma_{ij} p_i p_j ; \qquad (2.41)$$

and λ_k are Lagrangian multipliers.

It should be noted that, in the absence of any limiting conditions (2.39), i.e. when $m = 0$, the optimal solution obtained using procedure (2.40), (2.41) coincides with the solution of the variational problem (see 2.2.1.) for a channel with a specified wall structure.

Hence, the parametric solution of the boundary electrodynamic problem obtained previously can be used for a complete analysis of the operation of an MHD channel, including partial operating modes with different types of limiting relationships and optimization of output characteristics.

2.2.2.2. Optimal Loading of an MHD Channel

Consider the problems of determining optimum loading modes in a channel with continuous electrodes and in the end section of a segmented MHD channel. The channel flow is assumed to be fixed and the flow parameter distribution given in accordance with quasi-one-dimensional model of the flow. Solution of the problems for ϕ and ψ will be obtained in parametric form. To solve each standard boundary problem, use the finite-difference approximation of the differential equation (2.35) obtained by writing the integral equation of current conservation

$$\oint_{\Gamma} j_n d\ell = 0$$

for each Eulerian cell. Solution of the finite-difference analogue of problem (2.35), (2.36) is obtained using the matrix dispersion method [11], modified to represent solution in the form (2.37), (2.38).

<u>Channel with continuous electrodes</u>. Consider first, loading the simplest MHD generator with nonsegmented electrodes. The channel geometry and the magnetic field profile are shown in Fig. 2.4. The gasdynamics of the channel is calculated from the quasi-one-dimensional

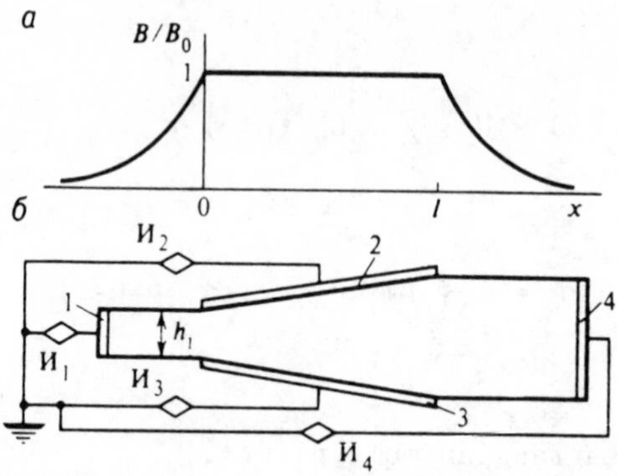

Fig. 2.4 Channel geometry (a) and the magnetic field profile (b); 1-4 - electrodes, I_1-I_4 - load (inverters)

equations, taking into account the thermal and dynamic boundary layers at the walls. The design point was selected to be the operating mode characterized by maximum power output

$$N^o = V_y^o I_y^o ,$$

with the following inlet parameters: Mach number M_1 = 2.3, Reynolds number $Re_1 \sim 10^7$, relative temperature of channels walls $T_\omega/T_1 = 0.65$, relative thickness of the boundary layers $2\delta_1/h_1 = 0.2$, MHD interaction parameter

$$S_1 = \sigma_1 B_o^2 h_1 / \rho_1 u_1 = 0.3; \ \beta_1 \simeq 1 .$$

Velocity and stagnation enthalpy profiles in boundary layers of thickness δ were assumed to be power functions raised to the one-seventh power. The load current was calculated from the integral

$$I_y^o = \int_o^L j_y dx$$

subject to the condition $E_x = 0$ when $0 \leq x \leq L$. The quasi-one-dimensional quantities

$$N^o, \ I_y^o, \text{ and } V_y^o$$

were used as scale factors in two-dimensional electrodynamic calculations.

The channel boundaries were formed by four insulators (horizontal sectors) and four conductors (inclined and vertical sectors). The vertical electrodes model the influence of the nozzle and diffuser adjacent to the channel, to which a load can be connected. Each of the four electrodes is loaded independently. The load potential could assume any arbitrary value.

The parametric solution of the problem (2.35), (2.36) was obtained for $p_0 = 1$, $p_i = \phi_i$ (i=1 through 4), where ϕ_i is the potential of the i-th electrode. The principal integral characteristics of the channel are given in Table 2.1 [12]. The first line in Table 2.1 represents the channel parameters obtained from two-dimensional calculations using boundary conditions applicable to quasi-one-dimensional calculations. In terms of the relationships (2.39), the boundary conditions are

$$G_1 = \phi_1, \ G_2 = \phi_2 + 0.5, \ G_3 = \phi_3 - 0.5, \ G_4 = \phi_4 \ .$$

Compared with quasi-one-dimensional calculations, the two-dimensional computations reveal a substantial decrease in power. The main reason for this effect is the appearance of considerable transverse leakage currents in the end regions. Fig. 2.5a shows the current distribution (continuous curves) and the potential distribution (dashed

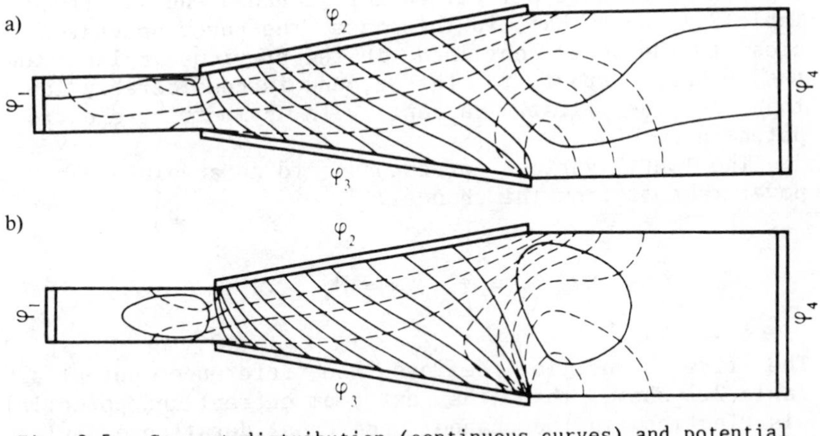

Fig. 2.5 Current distribution (continuous curves) and potential distribution (dashed curves) in the channel; a - $\phi_1 = \phi_4 = 0$, $\phi_2 = -\phi_3 = -0.5$; b - $\phi_1 = 0$, $\phi_3 - \phi_2 = 1$, ϕ_4 is determined from the condition $\Delta\psi_1 = \Delta\psi_4 = 0$

curves) in the channel. The leakage current flows to the nozzle and the diffuser when they are grounded. It increases with increasing Hall parameter.

The results for the second variant in Table 2.1 were obtained for the following conditions.

$$G_1 = \phi_1, \; G_2 = \phi_2 - \phi_3 + 1, \; G_3 = \Delta\psi_1, \; G_4 = \Delta\psi_4 \;,$$

that indicate that, at a specified potential difference between electrodes 2 and 3, the potentials of electrodes 3 and 4 are such that no leakage current flows to electrodes 1 and 4. As can be seen from Table 2.1, elimination of leakage current substantially increases MHD channel power output. Figure 2.5b shows the current and potential distributions for the variant for which calculations were made. The loading scheme of the channel considered is characterized by asymmetric Faraday electrode potentials ($\phi_2 \neq -\phi_3$). Hall voltage approximately 1.5 times greater than the Faraday voltage, $V_y = \phi_3 - \phi_2$, appears at the end electrodes. The third variant was calculated subject to the conditions

$$G_1 = \phi_1, \; G_2 = \phi_4, \quad N = \sum_1^4 \phi_i \Delta\psi_i = N_{max} \;,$$

i.e., Faraday electrode potentials that provide maximum channel power were determined for grounded end electrodes. Table 2.1 shows that, in this case, the power generated is close to the power obtained in the previous variant, and the leakage current to the ground is considerably lower than in the first variant with symmetric electrode potentials.

The fourth variant corresponds to an absolute maximum power takeoff from the channel,

$$G_1 = \phi_1, \quad N = N_{max} \;.$$

The first condition defines the reference potential. Table 2.1 shows that the extremum current and potential distributions in the channel under consideration are close to those when loading a channel with compensated leakage current (second variant). It is interesting to note that complete elimination of the Hall current flow to the

Table 2.1

Integral Parameters of an MHD Channel with Continuous Electrodes

Loading Mode	ϕ_1	ϕ_2	ϕ_3	ϕ_4	$\Delta\psi_1$	$\Delta\psi_2$	$\Delta\psi_3$	$\Delta\psi_4$	N
Electrode potentials are symmetric, diffuser is grounded	0	-0.5	0.5	0	-0.03	-1.04	0.6	0.47	0.82
Electrodes are connected ed to the same load, diffuser is grounded	0	-0.25	0.75	1.51	0	-0.96	0.96	0	0.96
Electrode potentials are optimal, diffuser is insulated from the ground	0	-1.2	-0.01	0	-0.11	-0.78	0.67	0.22	0.93
Electrode and diffuser potentials are optimal (absolute power maximum)	0	-0.47	0.72	1.01	-0.01	-0.84	0.73	0.12	1.04
Anode and nozzle, cathode and diffuser are shorted, channel voltage is specified	0	0	1	1	0.04	-1.06	0.73	0.29	1.02
Anode and nozzle, cathode and diffuser are shorted, channel voltage is optimal	0	0	1.12	1.12	0.05	-0.97	0.67	0.25	1.03

ground does not provide absolute maximum power output. Optimal loading of a channel with continuous electrodes assumed that an axial load is connected while maintaining sufficiently high Hall potential. In view of this, it is interesting to consider series connection of electrodes 1-4 to the load.

The last two variants in Table 2.1 represent the results of the calculations for a single load scheme, when the load is connected to shorted electrode pairs 1-2 and 3-4. For variant 5, calculations were performed subject to the conditions

$$G_1 = \phi_1, \; G_2 = \phi_2, \; G_3 = \phi_3 - 1, \; G_4 = \phi_4 - 1 \; ,$$

and variant 6, for optimal electrode voltage

$$G_1 = \phi_1, \; G_2 = \phi_2, \; G_3 = \phi_3 - \phi_4, \; N = N_{max} \; .$$

These results show that, for series-connected electrodes, the channel power output practically does not differ from the absolute maximum power output variant (4).

Thus, taking into account two-dimensional electrodynamic effects reveals substantial specific characteristics of channel loading for channels with continuous electrodes. A manifestation of these effects is also possible in channels with segmented electrodes, particularly along sectors with sharply defined nonuniformities along the channel length, e.g., in fringe magnetic field areas characterized by nonuniform load connections. It should also be noted that achievement of a maximum power output is not the only practically-useful optimization criterion. Limitations imposed on the nature of local distributions, and absolute values of electrical quantities based on technological considerations and channel construction reliability, can also serve as requirements for the optimal operating mode of the channel.

End sector of a segmented channel. Even though the end effects have been widely investigated and the characteristics of the current potential distributions in the end regions of the MHD channel are adequately understood, a number of applied problems in present-day technology require special analysis. Such problems include investigations of the electrodynamics of the end

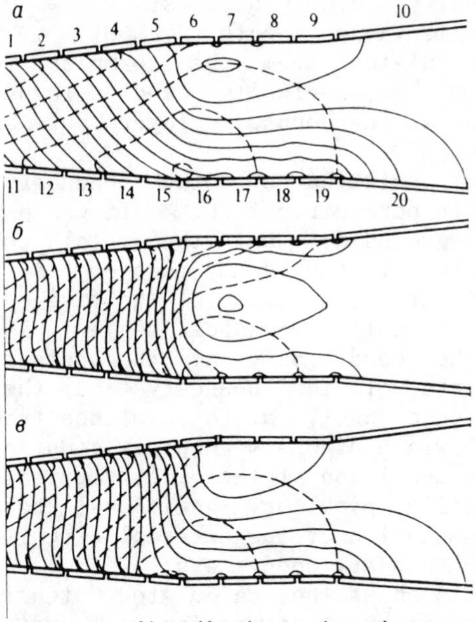

Fig. 2.6 Current distributions (continuous curves) and potential distributions (dashed curves) at the output sector of the channel: a - constant electrode voltage, total axial current is not zero; b - total axial current is zero; c - same as b, with anode 5 grounded; 1 to 20 - electrodes

sectors of channels under actual operating conditions and a search for optimal design solutions and loading schemes for the end sectors of MHD channels being developed. Two end effect problems applicable to MHD channels of the U-25 installation [13, 14] will be considered here.

Consider the end region of an MHD channel (Figure 2.6) formed by a section of the diverging active sector of the channel (electrodes 1-5, 11-15), outlet segmented dummy section of constant cross section (electrodes 6-9, 16-19), and the diverging subsonic diffuser (electrodes 10, 20). The channel is located in a transverse magnetic field that is constant along the active channel sector and decreases exponentially, as $\exp(-x/h_1)$, from the inlet of the segmented dummy sector of constant cross section. The flow in the channel is subsonic (the flow is calculated using quasi-one-dimensional approximation [13]), with the following parameter values at the inlet into the region under consideration: core flow temperature $T_1 = 2600$ K, core flow rate $U_1 = 850$ m/s, pressure $p_1 = 0.12$ MPa, wall temperature $T_w = 1800$ K, conductivity in the core flow $\sigma_1 = 10$ S/m, Hall parameter $\beta_1 = 1.5$, magnetic field along the active sector $B_o = 2$ T, channel height $h_1 =$

1.2 m, boundary layer thickness $\delta_1 = 0.15$ m. The electrode segmentation length $\ell_o = 0.3$ m. The boundary layer was calculated from the formulas for the various criteria. The velocity and stagnation enthalpy distributions in the boundary layer were assumed to obey the "1/7 law."

Numerical solution of the boundary problem was constructed in parametric form. It was assumed that the diffuser is grounded, that the total current to each segment-module of the dummy sector is zero, and the potential difference at each of the five outlet electrode pairs in the active channel segment is given. In addition, the condition that the potential difference between points of the boundary and the inside cross section of the channel, at a distance from the boundary equal to the width of one segment is constant, was used as the boundary condition at the left vertical boundary. In solving specific problems, additional relations, determined by commutation of the external circuit, were used to complete the boundary conditions.

Figure 2.6 shows the calculated potential and current distributions (current stream lines) for different loading schemes of the outlet sector of the channel. When electrode potentials of five electrode segments at constant voltage at opposite electrodes (see Fig. 2.6a) are given explicitly in terms of (2.39), the conditions can be written in the form

$$G_i = \phi_i - p_i \quad (i=1,\ldots,\ 5,11,\ldots,\ 15)\ .$$

In this case, i.e., an almost random loading scheme, the channel operates inefficiently because of considerable transverse current flow. Figure 2.6b shows the distribution of electrical parameters when the total transverse current in the channel at specific voltages and equal currents at opposite electrodes is zero, i.e., when the following relations apply.

$$G_i = \phi_i - \phi_{i-10} + V_y \quad (i=11,\ldots,\ 15)\ ,$$

$$G_j = \Delta\psi_j - \Delta\psi_{j-10} \quad (j=11,\ldots,\ 15)\ .$$

The generator parameters are considerably greater than in the previous case and are close to the optimal characteristics. The current distribution in the end sector of the channel is practically asymmetric. The output equipotentials diverge uniformly from the center of the current vortex, toward the anode and cathode walls.

Figure 2.6c shows the current and potential distributions in a malfunction situation, when the last anode of the active sector (5) is grounded, i.e., when a change is introduced into the functional relationship of the previous case (instead of the condition

$$G_{15} = \Delta\psi_{15'} - \Delta\psi_5 ,$$

it is assumed that $G_5 = \phi_5$). It is interesting to note that, in this case, intense current flows along the short-circuited loop between the grounded electrode and the diffuser, whereas hardly any current flows to the last cathode. The distribution of electrical parameters in the area of the segmented dummy sector section is close to that shown in Fig. 2.6a. Thus, a local disturbance of the outer circuit in the channel can lead to considerable electrical interactions with the components of the MHD installation loop adjacent to the channel.

Now consider the influence of the loading scheme of the end sector of a diagonal MHD generator on its parameters. The channel geometry and the magnetic field profile along the x axis are shown in Fig. 2.7. The channel parameters and the variation of the magnetic field B(x) are close to these quantities for the generator of the U-25 facility. The inlet zone of the channel under consideration consists of a short-circuited initial sector, modeling the equipotential nozzle, and a segmented

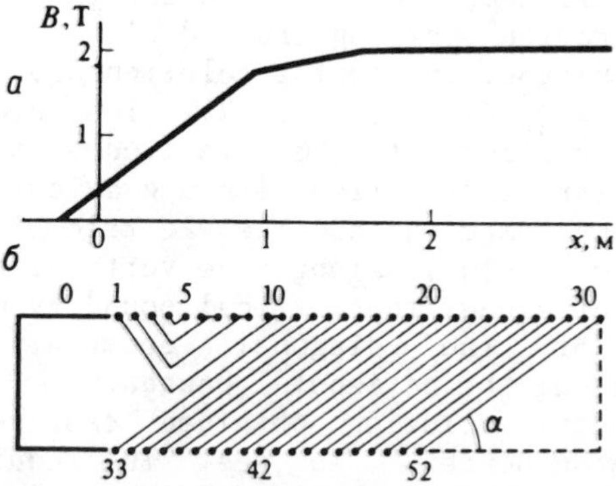

Fig. 2.7 Magnetic field profile (a) and the geometry of the inlet sector (b) of a diagonal MHD channel, 0-52 electrodes

sector containing diagonally connected electrodes. The aim of the investigation is to determine commutation techniques and methods of feeding load current to the end electrodes, meeting the requirements of optimum current and potential distributions in the channel operating at high efficiency and reliability. In the case of frame channels, this is also equivalent to determining the optimal construction. Various techniques of designing and constructing end sectors of diagonally conducting wall channels are known at the present. The most characteristic types are rectilinear diagonal modules [15, 16] and curved modules [17, 18]. Single- and multiple-load connecting techniques correspond to these two design schemes. Optimal operating conditions exist for each of these schemes.

Similarly to the previous problem, the channel flow is specified on the basis of quasi-one-dimensional calculations. The flow parameters at the inlet into the area for which calculations will be performed are: $T_1 = 2600$ K, $V_1 = 900$ m/s, pressure $p_1 = 0.1$ MPa, $T_w = 1800$ K, $\sigma_1 = 7.23$ S/m, mobility of electrons $\mu_e = 0.65\ T/2600\ m^2/V \times s$, $\delta_1 = 0.04$ m. The geometrical dimensions of the channel are as follows: h = 0.8 m; channel width normal to the plane of the diagram, 1 m; length of the short-circuited initial sector $\ell = 0.5$ m; length of an electrode module $\ell_o = 0.1$ m; ratio of the electrode length with the insulator length, 9. Thirty-two electrode modules are located on the upper (anode) channel wall and twenty electrode modules on the lower (cathode) wall. The short-circuited inlet electrodes and the left vertical edge form an equipotential element. An effective boundary condition is specified at the boundaries of the region under consideration.

A parametric numerical solution for the channel sector under consideration is constructed in the form (2.37), (2.38). The parameters of the solution are the electrode potentials $p_i = \phi_i$ ($i = 0,1,\ldots,52$). The potential at the boundary to the right of the electrodes is given in the form of a linear distribution with a gradient equal to the average Faraday component of the electric field along the uniform channel sector, along the vertical boundary, and with a gradient along the vertical boundary multiplied by the tangent of the diagonal commutation angle of electrodes, along the horizontal boundary.

The parametric solution obtained can be analyzed in the following sequence. To meet the requirements of a sufficiently high electrical efficiency in the uniform channel sector ($\eta \gtrsim 0.6$), let the commutation angle of the electrodes along the diagonal be 36° and let the electrodes on the anode wall, that cannot be commutated with cathodes (see Fig. 2.7), be short-circuited in accordance with the two methods of constructing the end zones mentioned in the foregoing discussion.

First, consider a channel with a single load, when electrodes 1-11 on the anode wall are short-circuited with the inlet equipotential component to which the load is connected. In this case, (2.39) assumes the form

$$G_0 = \phi_0 - p_0 ,$$

$$G_i = \phi_0 - \phi_i \quad (i=1,\ldots, 11) ,$$

$$G_k = \phi_k - \phi_{k+21} = \Delta\psi_k - \Delta\psi_{k+21} \quad (k=12,\ldots, 31) .$$

At sufficiently small load currents, the potential distribution on the electrodes of the anode walls of the inlet sector of the MHD channel (Fig. 2.8a) is monotonic. At very high (almost short-circuit) load currents, the initial segmented sector of the wall is characterized by the appearance of a minimum on the potential curve caused by the flow of very high currents in the zone of below-normal induction.

Now consider the multi-load schemes. Analysis of two-dimensional electric fields in the fringe magnetic field zone indicates that under optimal conditions, the potential distribution at the inlet sector of the anode wall must have a potential minimum, whereas the potential distribution on the cathode wall must be monotonic. Special electrode commutation at the inlet sector of the anode wall is necessary to achieve this type of potential distribution. An example of such commutation is shown in Fig. 2.7. The following conditions are fulfilled for such an electrode connection scheme:

$$G_o = \phi_o - \rho_o ,$$

$$G_1 = \phi_1 - \phi_0 = \phi_1 - \phi_{11} = \Delta\psi_1 - \Delta\psi_{10} - \Delta\psi_{11} ,$$

$$G_2 = \phi_2 - \phi_8 = \phi_2 - \phi_9 = \Delta\psi_2 - \Delta\psi_8 - \Delta\psi_9 ,$$

$$G_3 = \phi_3 - \phi_6 = \phi_3 - \phi_7 = \Delta\psi_3 - \Delta\psi_6 - \Delta\psi_7 ,$$

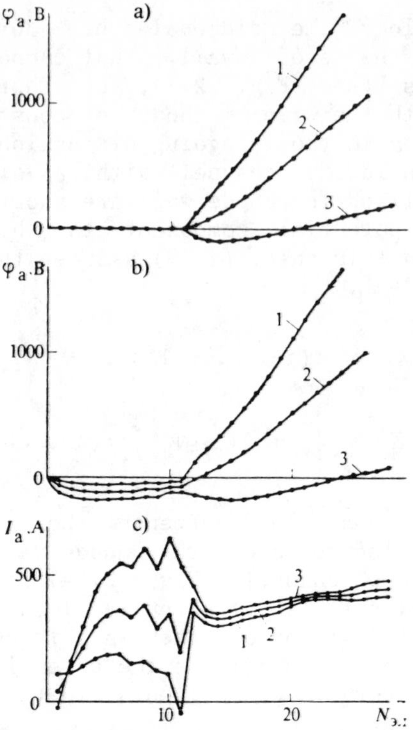

Fig. 2.8 Potential distributions (a,b) and current distribution (c) on anode wall electrodes at the inlet channel sector for different loading schemes: a - single-load scheme with shorted electrodes 0-11, b,c - distributed load scheme with groups of shorted electrodes (1,10,11; 2,8,9; 3,6,7; 4,5). Curves 1-3 are numbered in accordance with increasing load current

$$G_4 = \phi_4 - \phi_5 = \Delta\psi_4 - \Delta\psi_5 ,$$

$$G_i = \phi_i - \phi_{i+21} = \Delta\psi_i - \Delta\psi_{i+21} \quad (i = 12,\ldots,31) .$$

Now consider channel loading according to the scheme shown in Fig. 2.7. It is assumed that the distributive load is connected to the inlet equipotential component and to the electrodes 1-4. The currents to the loads connected to electrodes 1-4 are assigned in the following manner. For each total load current I_H, the current in each load is assumed to be proportional to the sum of currents flowing through electrodes short-circuited as shown in the diagram in Fig. 2.7. These currents were calculated in the previous case, when all electrodes 1-11 were short-circuited. The proportionality constant is assumed to be 0.75. The calculated potential and current distributions on the anode wall electrodes are shown in Fig. 2.8b and Fig. 2.8c. The chosen loading mode of the channel provides potential distribution on the anode wall with a maximum point (see Fig. 2.8b). It should be noted that the nature of the potential distribution on unloaded electrodes (12-31) is the same as in the case of a single load scheme. The current distribution on the loaded electrodes displays peaks as a result of current redistribution over shorted electrodes, caused by the Hall effect. The current distribution on the unloaded electrodes depends weakly on the load current. This can be attributed to the diagonal commutation angle of electrodes used in this example.

In conclusion, compare the integral parameters of the channel end sector for the loading schemes considered. Calculations show that close to the optimum total load current, I_H = 4000 A, the power output from the multi-load channel exceeds the power output from a single-load channel by 0.25 MW. The power generated by a uniform channel sector 1-m long is 2.65 MW.

Analysis shows that the diagonal scheme of connecting electrodes makes it possible to achieve a specified optimal loading regime. However, each new loading regime requires its own electrode commutation scheme. A fixed electrode switching scheme in a variable load mode can result in current or voltage overloading of certain sectors of the end zone. A promising way to avoid such undesirable effects is utilization of schemes in which the position of current takeoff end zones varies with the load [16, 19].

2.2.2.3. The Effects of Nonuniform Loading of a Segmented MHD Channel

Nonuniform loading problems are an important aspect of segmented MHD channels with individual control of loads of individual sectors, or where loading current distribution along sectors is attained by means of various devices, such as ballast resistors. Considerable current and potential nonuniformities that enhance electrical breakdown and destruction of components may occur in the channel when the load is distributed unevenly along the channel sectors. Electrodynamic effects of nonuniform loading are not one-dimensional; their analysis requires application of numerical methods.

Problem formulation. Consider the electrical parameters of individual modules in the active sector of a

Faraday MHD channel with single and multiple connection of electrode pairs to an external load. Two of the earlier published papers [20, 21] are closest to the subject being considered.

The active section of an MHD generator channel consists of electrode insulator modules. Loading devices, i.e., inverters, the control of which maintains either constant current (current control mode) or constant voltage (voltage control mode) in the external circuit are connected to the electrodes. The electrodynamics of an MHD channel for two electrode commutation schemes (Fig. 2.9) [22] will be investigated here. In the first scheme, each module is loaded through its own inverter (Fig. 2.9a). In the second scheme, two modules with their upper and lower electrodes are connected through equal ballast resistors R_b, loaded through a common inverter (Fig. 2.9b). The boundary problem described by (2.34)-(2.36) is solved for rectangular regions, the vertical boundaries of which are subject to the periodicity condition

$$\phi(o,y) = \phi(L,y) - V_x \ ,$$

$$\psi(o,y) = \psi(L,y) - \sum_i I_i \ , \qquad (2.42)$$

where V_x is the Hall voltage and

$$\sum_i I_i$$

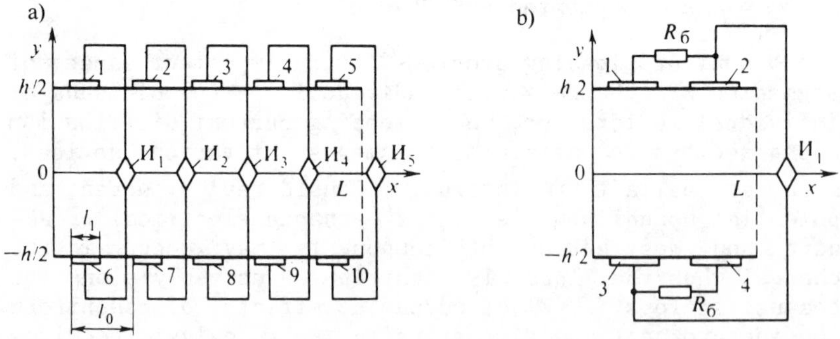

Fig. 2.9 Loading connecting scheme for the active sector of a segmented MHD channel: a - independent loading of each electrode module, b - pair loading of the electrode section with ballast resistors, 1-10 - electrodes, I_1-I_5- - inverters

is the total Faraday current in the channel sector of length L. The constant V_x is determined from the condition that the total Hall current at the vertical boundaries is equal to zero:

$$I_x = \int_{-h/2}^{h/2} j_x(o,y)dy = 0 \; . \qquad (2.43)$$

Constant potential and no current flow conditions apply to the horizontal sectors of the boundary region, consisting of electrodes and insulators. A uniform magnetic field and a gas flow that is nonuniform in the transverse direction are given inside the region. The distributions of the flow rate, u, and its temperature, T, are specified by the following relations:

$$u(y) = U\eta(y), \; T(y) = T_w + (T_\infty - T_w)\eta(y) \; ,$$

$$\eta(y) = \begin{cases} 1 & \text{when } h/2-|y| \geq \delta_T, \\ ((h/2-|y|)/\delta_T)^{1/7} & \text{when } \delta_\pi \leq h/2-|y| < \delta_T, \\ (\delta_\pi/\delta_T)^{1/7}(h/2-|y|)/\delta_\pi & \text{when } h/2-|y| < \delta_\pi, \end{cases}$$

where T_∞ and U are the core flow temperature and flow rate, respectively; T_w is the channel wall temperature; δ_T is the thickness of the turbulent boundary layer, δ_π is the thickness of the laminar sublayer and h is the channel height.

The numerical values of the gasdynamic parameters used in the calculations are: $\beta = 1.5$, $\delta_T = 0.125$ m, $\delta_\pi = 0.0063$ m, $T_w = 1800$ K, B = 2T, u = 850 m/s, p = 0.1 MPa, $T_\infty = 2700$ K, corresponding to conductivity in the core flow $\sigma_\infty \simeq 20$ S/m and a Mach number of about 0.9 for combustion products of hydrocarbon fuels. The height of the region h = 1 m and the length of the region and electrode and insulation gap dimensions varied depending on the degree of channel segmentation.

Nonuniform loading of channel modules. Consider loading modes of a sector of the channel under consideration when, for any reason, the load on the central module varies independently of the load on the other modules that varies similarly on all but the central module. Experimentally, such nonuniform channel loading was investigated on the U-25 facility [23]. It was determined that the no-load voltage of the individually controlled module is lower than the no-load voltage of the jointly controlled channel sector.

Let the current and voltage of all but the central electrode modules be denoted I and V and the analogous quantities for the central module be denoted I* and V*. Let the electrode potential, potential at the grid nodes at the vertical boundary of the region and the Hall voltage, V_x, be the parameters of solution $p_i (i = 1,\ldots,n)$. Then, p_i are determined from relationships (2.42) and (2.43) on the condition that the currents as well as voltages at opposed electrodes are equal.

Figure 2.10 shows the current and potential distributions in a five-module channel at V = 850 V and for values of V* that correspond to three loading modes: short circuit, uniform loading of all modules (V = V*), and no-load for the central module. The sequence of loading modes discussed is equivalent to obtaining a voltage current characteristic of the central module with voltage control of inverters of the other modules. At a specified channel geometry and nonuniform loading, the influence of the central module on outer modules is insignificant, whereas interaction of the central module with adjacent modules alters considerably both the current line and equipotential configurations and load current of these modules.

Now, consider the voltage-current characteristic of the central pair at constant and equal voltages at the remaining pairs. In view of the linearity of the problem and the fact that the individual voltage-current characteristic coincides with total voltage-current characteristic when I* = I and V* = V, we have:

$$I^* = (V - V^*)/r_v^* + (V_{x.x} - V)/r , \qquad (2.44)$$

where $V_{x.x}$ is the total no-load voltage (operation at $I = I^* = 0$, $V = V^* = V_{x.x}$); r_v^* is the internal resistance of the independently controlled electrode module at a constant voltage on the remaining modules, r is the internal resistance of a single electrode module during joint synchronous control of all segments.

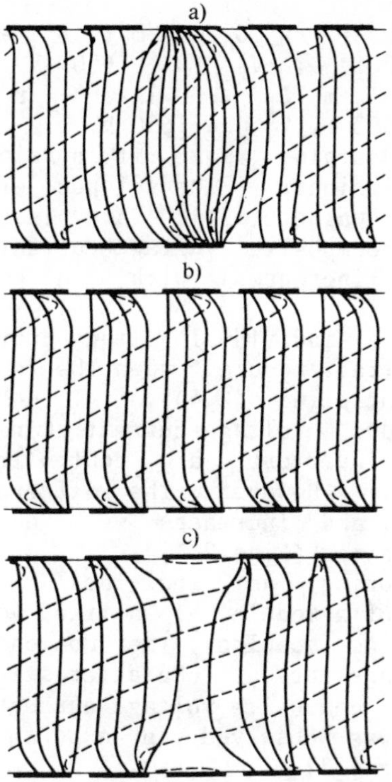

Fig. 2.10 Current distributions (continuous lines) and potential distributions (dashed lines) for nonuniform loading of a segmented MHD channel: a - short-circuited central module, b - uniform loading, c - no-load operation of the central module

Analogously, the voltage-current characteristic of the central pair, when constant current is maintained on the remaining electrode pairs, can be written in the form

$$V^* = r_I^*(I-I^*)-rI+V_{x.x} \,, \qquad (2.45)$$

where r_I^* is the internal resistance of the independently controlled electrode module at constant current on the remaining modules.

The relationships (2.44) and (2.45) also determine the total voltage-current characteristic (when $I^* = I$ and $V^* = V$)

$$V = -rI+V_{x.x} \,. \qquad (2.46)$$

Parametric solution of the boundary problem under consideration makes it easy to calculate the four constants r, r_v^*, r_I^*, and $V_{x.x}$ in formulas (2.44)-(2.46). For a 5-module channel (length of a basic module ℓ_o = 0.4 m, electrode length (ℓ_1 = 0.367 m), r_v^* = 0.146 Ω, r_I^* = 0.150 Ω, r = 0.206 Ω, $V_{x.x}$ = 1650 V. For a 15-module channel (ℓ_o = 0.12 m, ℓ_1 = 0.11 m), r_v^* = 0.260 Ω, r_I^* = 0.288 Ω, r = 0.600 Ω, $V_{x.x}$ = 1650 V.

The results obtained make it possible to reach the following conclusions. The slopes of the individual voltage-current characteristics are less steep than the slope of the composite voltage-current characteristic. The difference increases with increasing degree of segmentation.

The slope of individual voltage-current characteristics of the central module, when the remaining inverters are voltage controlled (2.44), is less steep than the slope of individual voltage-current characteristics, when the remaining inverters are controlled for constant current (2.45). However, the difference between the slopes is small, and increases with increasing degree of channel segmentation (Fig. 2.11).

When measuring an individual voltage current characteristic of a module at a fixed voltage or current on the remaining modules, the no-load voltage of an individual voltage-current characteristic will be lower than that of a composite voltage-current characteristic. This behavior was observed in experiments on segmented Faraday channels [23].

When controlling loads in the central module, the electric parameters of adjoining modules vary. Therefore, the loading mode of these modules when the inverters are voltage-controlled differs from that when the inverters are current-controlled. In the case when inverters are

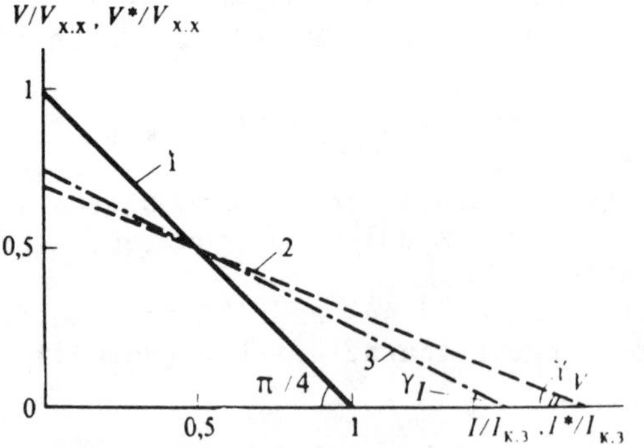

Fig. 2.11 Voltage-current characteristics of electrode modules: 1--composite voltage-current characteristic (V = V*, I = I*); 2,3--individual voltage-current characteristic of an electrode module at a constant voltage (V = 1/2, V* = var), and constant current (I = 1/2, I* = var at the remaining modules); γ_I = arctan (r_I^*/r); γ_v = arctan (r_v^*/r)

voltage-controlled, the current in modules at fixed voltage at sufficiently high values of the load factor changes sign, and the inverter begins to operate in a mode where power is supplied by the grid. When inverters are current-controlled, all modules always generate electric power.

In conclusion, consider the practical problem of selecting the method of controlling the load during the experiment. When uniform load distribution along the channel length is achieved by use of inverters with limited accuracy in controlling specified parameters, current control of inverters is to be preferred. This conclusion follows directly from comparison of the slopes of individual and composite voltage-current characteristics. Because the slope of an individual voltage-current characteristic is less steep than that of a composite curve (see Fig. 2.11), then assuming that current and voltage control have the same relative accuracy, the lowest overload of the channel segment that deviated from nominal loading takes place when current control is used to regulate the inverter.

Loading the channel with ballast resistors. Consider a uniformly loaded channel section with a single loading device connected through ballast resistors to two electrode modules (see Fig. 2.9b). Using ballast resistors makes it possible to decrease the number of loads in the channel, while maintaining a uniform Hall potential distribution. This technique is widely used in experiments and provides positive results [17, 24]. Because the boundary conditions for this problem are similar to those of the previous case, the same quantities will be used for parameters of the solution. In addition to periodicity conditions (2.42) and (2.43), parameters p_i must satisfy additional conditions that depend on the external circuit. These include a specified voltage (or current) on electrodes connected to the inverter, equality of currents on crosswise located electrodes (due to symmetry of the problem), and the following relationship between the voltage and current at the ballast resistor ($V_b = R_b I_b$). Thus, solution of this problem is completely determined by assigning two characteristics of the external circuit: voltage V (or current I) of the inverter and resistance R_b.

To construct the integral channel parameters, the voltage-current characteristic of the inverter and the variation of V_b and V with I and R_b, first consider the dependences of the above voltages on the inverter current,

I, and the ballast resistor current, I_b, that, as a result of the linearity of the problem, can be represented in the form:

$$V = aI + a_1 I_b + V_{x.x} ,$$

$$V_b = bI + b_1 I_b ,$$

$$V_x = cI + c_1 I_b , \qquad (2.47)$$

where $V_{x.x}$ is the no-load voltage.

Substituting $V_b = R_b I_b$ into the second equation in (2.47) leads to the following relationship between I_b and I:

$$I_b = bI/(R_b - b_1) .$$

Using this equation, relationships in (2.47) can be rewritten in the following form:

$$V = (a + a_1 b/(R_b - b_1))I + V_{x.x} ,$$

$$V_b = bR_b I/(R_b - b_1) ,$$

$$V_x = (c + c_1 b/(R_b - b_1))I . \qquad (2.48)$$

From (2.48) it follows that the integral parameters of the channel vary nonlinearly with resistance R_b. What's more, uniform current and voltage distributions on channel electrodes can always be achieved at a certain value, R_b^o that is independent of the loading mode. The following relationships apply in such a (matched) mode:

$$I_b^o = 1/2I^o, \; V^o = (a + 1/2a_1)I^o + V_{x.x} ,$$

$$V_b^o = 1/2V_x^o = 1/2R_b^o I^o . \qquad (2.49)$$

Comparing (2.49) and (2.48) it can be seen that the coefficients b and b_1, c and c_1 are linearly independent because they satisfy the relation

$$R_b^o = 2b + b_1 = c + 1/2c_1 \ .$$

Thus, characteristics (2.48) contain six independent constants. Numerical solution of the boundary problem under consideration (for $\ell_o = 0.12$ m and $\ell_1 = 0.11$ m) provides the following values for these constants:

$$a = -0.427 \ \Omega, \ b = 0.146 \ \Omega, \ c = 0.079 \ \Omega \ ,$$

$$a_1 = 0.697 \ \Omega, \ b_1 = -0.220 \ \Omega, \ V_{x.x} = 1650 \ V \ .$$

The voltage-current characteristics $V(I,R_b)$ (Fig. 2.12) and V_b (I,R_b), $V_x(I,R_b)$ (Fig. 2.13) were constructed for these values of the constants. The characteristics $V(I,R_b)$ demonstrate the limits of variation of the slope of the inverter voltage-current characteristic as a function of resistance R_b.

The range of variation of transverse voltage, V_b, between electrodes connected through a ballast resistor exceeds the range of variation of the Hall voltage along a two-module sector. The highest overvoltages can arise between these electrodes when ballast resistors are mismatches. In this respect, the most dangerous case occurs when the ballast resistor is disconnected ($R_b = \infty$).

Figures 2.12 and 2.13 make it possible to trace the influence of ballast resistance, R_b, on the character-

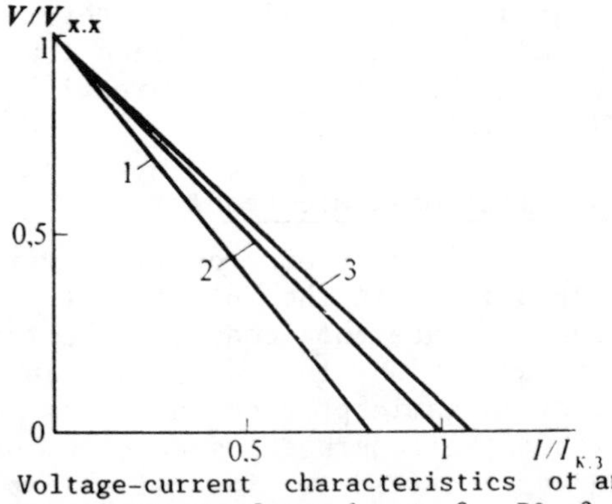

Fig. 2.12 Voltage-current characteristics of an inverter at various ballast resistance values: 1 - ∞, 2 - R_b^o, 3 - 0

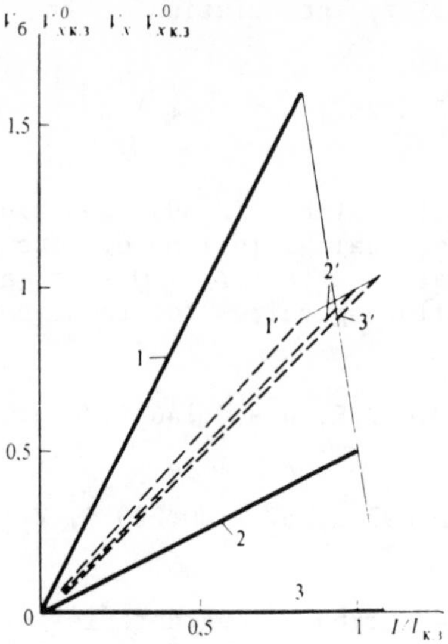

Fig. 2.13 Variation of voltage on ballast resistors (1-3) and Hall voltage (1′-3′) with inverter current at various ballast resistances: 1,1′ - ∝; 2,2′ - R_b^o; 3,3′ - 0

istics of the inverter and the electrode potential distribution in a specified inverter control mode. For example, let the inverter be voltage-controlled and let the resistance R_b vary from zero to infinity. Then, the inverter current will decrease, the Hall potential will also decrease, whereas the voltage between adjacent electrodes will increase. This variation can be most conveniently observed in Figs. 2.12 and 2.13, when the inverter operates in a short-circuit mode, i.e., when V = 0.

Figure 2.14 shows schematically the variation of the electrode potential distribution at three values of R_b equal to $0, R_b^0$ and ∞. It can be seen that the greatest nonuniformity of the electrode potential distribution occurs when R_b = ∞.

2.3. Traverse Electrodynamic Problems

The electrodynamic problems dealing with current and potential distributions in the plane of the cross section of an MHD channel are of considerable theoretical and practical importance. On the one hand, investigation of transverse electrodynamic problems provides necessary data for the development of engineering calculation methods and for understanding the results of physical experiments. On the other hand, it facilitates the search for and the

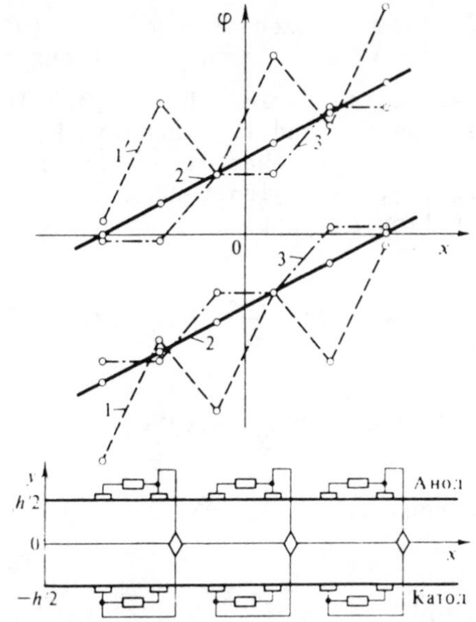

Fig. 2.14 Potential distribution of anode wall electrodes (1-3) and cathode wall electrodes (1′-3′) along the channel length at various ballast resistances: 1,1′ - ∞; 2,2′ - R_b°; 3,3′ - 0

development of the optimal design of an MHD channel.

Present formulations of transverse electrodynamic problems emphasize attaining high MHD channel efficiency. Attempts to achieve this goal are based on methods utilizing variation of the boundary conditions and the configuration of regions in which solutions are constructed. In this approach, not only the electrodynamic, but also gasdynamic and construction advantages of channel models under consideration are taken into account. The most characteristic examples of such problems are either experimental or theoretical investigations of Faraday channels of rectangular cross section with block electrodes [25], hexagonal cross section with V-shaped electrodes [26], circular cross section with arched elec- trodes [27] and frame channels with rectangular, oval, and circular cross sections [28-31].

2.3.1. Problem Formulation. The problem of current and potential distributions in the cross section of an MHD channel will be formulated here by the use of the general methodology developed in [1]. Consider a long cylindrical channel with an arbitrarily shaped cross section and with wall design that allows an arbitrary load distribution along the circumference of the cross section. Assume that the channel flow velocity $\mathbf{v} = \{u(y,x),0,0\}$ and conductivity $\sigma(y,z)$ are known functions of coordinates. The external magnetic field, $\mathbf{B} = \{0,0,B\}$, and the Hall parameter are constant and the magnetic Reynolds number is small. Assuming, also, that the electric field, $\mathbf{E}$, and current density, $\mathbf{j}$, are independent of the x coordinate leads to the following problem:

$$L(\phi) = (\beta E_x - uB)\frac{\partial \ln\sigma}{\partial y} - B\frac{\partial u}{\partial y} \quad (y,z \in D)\ ; \qquad (2.50)$$

$$\mathbf{j} = \sigma(-\mathrm{grad}\ \phi - uB\mathbf{e}_y) - B(\mathbf{j}\times\mathbf{e}_z)\ , \qquad (2.51)$$

$$E_x = \mathrm{const.},$$

where

$$L = \frac{\partial^2}{\partial y^2} + (1+\beta^2)\frac{\partial^2}{\partial z^2} + \frac{\partial \ln\sigma}{\partial y}\frac{\partial}{\partial y} + (1+\beta^2)\frac{\partial \ln\sigma}{\partial z}\frac{\partial}{\partial z}\ .$$

The boundary conditions for equation (2.50) are determined from the relations between the potential and current on the channel walls and can be represented in the following general form:

$$F(\phi, j_n) = 0 \quad (y,z \in \Gamma)\ , \qquad (2.52)$$

where j_n is the normal component of current density on the channel wall, Γ is the boundary of the region D of the channel cross section when x = const.

For channels in which flow parameters vary along their length, formulation of problem (2.50)-(2.52) is approximate. Let us consider conditions for which this approximation is valid. In the approximation used, the term $\partial j_x/\partial x$ in the equation div $\mathbf{j} = 0$ is considered to be small in comparison with the terms $\partial j_y/\partial y$ and $\partial j_z/\partial z$ and is disregarded. Assuming that all components of current density are of the same order of magnitude, we obtain the inequality

$$\left|\frac{\partial j_x}{\partial x}\right| << \left\{\left|\frac{\partial j_y}{\partial y}\right|, \left|\frac{\partial j_z}{\partial z}\right|\right\},$$

provided

$$\alpha/L<<1 , \tag{2.53}$$

where a and L are characteristic linear dimensions of the problem in the transverse and axial directions.

The above is valid for Hall parameter values that are not very large, i.e. when $\beta \lesssim 1$. For larger Hall parameter values, when $\beta \gtrsim 1$, we have the following relations apply for current density components

$$j_x \leq \beta j_y , \quad j_z \sim j_y .$$

In this case, inequality (2.53) becomes more rigorous, i.e.,

$$\beta a/L<<1 .$$

Variation of the transverse components of the electric field along the channel in the electrodynamic equation, rot $\mathbf{E}$ = 0 of (2.13), can be neglected. Analogous analysis of the order of magnitude of derivatives shows that condition (2.53) is fulfilled when $\beta \lesssim 1$, while the weaker condition, $a/\beta L<<1$, applies when $\beta \gtrsim 1$.

Therefore, when the inequality

$$\max\{a/L, \beta a/L\}<<1 \tag{2.54}$$

holds, the electrodynamic problem under consideration is reduced to a two-dimensional problem.

Now, consider the meaning of the characteristic dimensions a and L. The quantities a and L are distances at which **j** and **E** undergo considerable changes. As a rule, in the case of MHD channels, a is its transverse dimension and L is the length at which either the channel geometry or the flow parameters change considerably. L can be associated in the first case with the length of the sector of the channel, and in the second case, with length of the sector at which various effects, such as MHD interaction, heat transfer, friction, etc., become substantial. In the latter case, the characteristic axial dimension of the problem can be smaller than the geometrical length of the active sector of the channel. What's more, assuming L to be the length of the active channel sector, it is necessary to modify condition (2.54) by adding to it characteristic values of dimensionless criteria responsible for the strong manifestation of the above-noted physical effects in the channel. For example, it follows from conservation equations that the following conditions must be fulfilled:

In the presence of strong MHD interaction,

$$\max\{Sa/L, S\beta a/L\} \ll 1 \ ,$$

in the presence of high heat transfer,

$$\max\{4St, 4St\beta\} \ll 1 \ ,$$

in the presence of high friction,

$$\max\{2c_f, 2c_f\beta\} \ll 1 \ ,$$

where St is the Stanton number and c_f is the coefficient of friction.

2.3.2. Variational Problem of the Potential and Current Distributions over the Cross Section of an MHD Channel. The aim of the problem under consideration is determination of the optimal electrodynamic configuration of the active sector of an MHD channel characterized by maximum power output at a specified electrical efficiency. In formulating the problem, it is necessary to obtain, not only Euler's equations and the required boundary conditions, but also relations for the optimum value of the Hall component of the electric field. A comparison of a variational solution with solutions for channels with a known electrodynamic configuration is of interest.

2.3.2.1. Problem Formulation

Consider the problem of optimum potential and current distributions in the cross section of a linear MHD channel [31]. On one hand, the potential and current distributions must satisfy elliptic equation (2.50) and Ohm's law (2.51); on the other hand, they must provide the power functional extremum for power generated per channel unit length. An additional limiting condition is superimposed on the solution, i.e., attainment of a specified electrical efficiency.

The formulation of this problem can be written following the general approach (see Section 2.2.1.1.). This can be done using expressions obtained for variation of the generalized power functional, Euler equations and boundary conditions, taking into account specific distribution of parameters in the transverse electrodynamic problem. In particular, assuming that in (2.9), $\partial i_x / \partial x = 0$, the Euler equation (2.55) is obtained in the form

$$L(\lambda) = (2(1+\nu)\beta E_x - (1+\nu(1-\eta))uB)\frac{\partial \ln \sigma}{\partial y} - (1+\nu(1-\eta))B\frac{\partial u}{\partial y} . \tag{2.55}$$

The boundary conditions follow from (2.10):

$$(2(1+\nu)\phi_y + (1+\nu(1+\eta))uB-\lambda_y)n_y + (1+\beta^2)(2(1+\nu)\phi_z-\lambda_z)n_z = 0 ,$$
$$\lambda = 0 \ (y,z \in \Gamma) . \tag{2.56}$$

In writing expressions (2.55) and (2.56) it was taken into account that $\lambda_x = 0$ as a result of the uniformity of the boundary condition for λ.

The boundary conditions (2.56) correspond to the condition that the surface integral in the expression for δJ (2.8), calculated along the side surface of the channel under consideration, becomes zero. Consider the term being summed in the surface integral (2.8) calculated at the end surfaces of an arbitrary control volume representing a finite channel sector of length Δx. The fact that variation of the potential and the normal components of current density in two cross sections of the channel are related to each other through the relations

$$\delta\phi(x,z) = \delta\phi(x+\Delta x,z) + \delta E_x \Delta x \ ,$$

$$\delta j_x(x,z) = \delta j_x(x+\Delta x,z)$$

will be taken into account.

Performing the transformations, reduce the expression for the sum of the surface integrals, calculated along the channel cross sections x and x + Δx, to the form:

$$\int_D\int \Big(i_x - 2(1+\nu) j_x - \frac{2(1+\nu)\beta\sigma}{(1+\beta^2)B} \big((-\text{grad}\ \phi$$

$$+ \mathbf{v} \times \mathbf{B}) \times \mathbf{B}\big)_x \Big) dy\ dz\ \delta E_x \Delta x \ .$$

This expression leads to the following condition for determining the optimum value of δE_x for an arbitrary variation of the Hall component of the electric field, E_x,

$$\int_D\int \frac{\sigma}{1+\beta^2} \big(2(1+\nu)E_x + (1+\nu(1-\eta))\beta u B$$

$$+ \beta\lambda_y\big) dy\ dz = 0 \ . \qquad (2.57)$$

This condition is equivalent to the total axial current in the channel becoming zero,

$$\int_D\int j_x dy\ dz = 0 \ .$$

In the case of an absolute extremum, when the isoperimetric condition is not used to solve the variational problem (ν = 0), the Hall component of the axial electric field, E_x, determined from (2.57) coincides with the optimum value of the Hall component of the axial field for a rectangular frame channel.

Thus, formulation of the problem is completed. Determination of ϕ and λ requires solution of the boundary problems (2.50), (2.55), and (2.56). Determination of, λ, becomes a separate problem. The Lagrangian multiplier ν is found from condition (2.4).

In solving the problem for ϕ, it is expedient to introduce a new dependent variable,

$$\zeta = 2(1+\nu)\phi - \lambda \ . \tag{2.58}$$

Then the problem (2.50), (2.56) can be written in the form

$$L(\zeta) = -(1+\nu(1+\eta))\left(uB\,\frac{\partial \ln\sigma}{\partial y} + B\,\frac{\partial u}{\partial y}\right) ,$$

$$(\zeta_y + (1+\nu(1+\eta))uB)n_y$$

$$+ (1+\beta^2)\zeta_z n_z = 0 \ (y,z \in \Gamma) \ . \tag{2.59}$$

If ζ is made to correspond to the vector, $g = g_y e_y + g_z e_z$, where

$$g_y = -\frac{\sigma}{1+\beta^2}\,(\zeta_y + (1+\nu(1+\eta))uB) ,$$

$$g_z = -\sigma\zeta_z , \tag{2.60}$$

problem (2.59) assumes the form

$$\frac{\partial g_y}{\partial y} + \frac{\partial g_z}{\partial y} = 0 ,$$

$$g_n = 0 \quad (y,z \in \Gamma) \ . \tag{2.61}$$

It can be shown that condition (2.57) for E_x is identical to the conditions

$$\int_D\!\!\int j_x dy\, dz = 0 \ .$$

Using relationships (2.58)-(2.61), transform condition (2.57) is transformed into the form,

$$\int_D\int (2(1+\nu)j_x = \beta g_y)dy\ dz$$

$$= 2(1+\nu)\int_D\int j_x dy\ dz = 0\ . \qquad (2.62)$$

The relation

$$\int_D\int g_y dy\ dz = 0$$

that applies in view of (2.61) is taken into account in (2.62).

Thus, in the presence of two-dimensional flow nonuniformities when β = const., the channels with optimum characteristics are those with zero total axial current. This conclusion is a natural generalization of well-known results obtained using quasi-one-dimensional flow models.

2.3.2.2. Optimal Potential Distribution in the MHD Channel.

Numerical Solution of the Variational Problem

Solution of the variational problem in the presence of nonuniformities in electrical conductivity and velocity presents considerable difficulties. Therefore, consider, initially, the special case when conductivity in the flow is nonuniform, $\sigma = \sigma(y,z)$, and the flow velocity u = const. In this case, problem (2.61) has a trivial solution $g = 0$ and in view of relations (2.58) and (2.60), potential, ϕ, can be represented in the form

$$\phi = -E_x x - \frac{1+\nu(1+\eta)}{2(1+\nu)}\ uBy + \frac{\lambda}{2(1+\nu)}\ . \qquad (2.63)$$

In view of the fact that, at the boundary of the region, $\lambda = 0$, the equipotentials on channel walls form two-dimensional curves when the flow contains only conductivity nonuniformities. For diagonal-wall MHD channels, this conclusion indicates that an MHD generator with maximum characteristics must be constructed using rectangular frames.

The constants, ν and E_x, in expression (2.63) are determined in accordance with formulation of the variational problem, from conditions (2.4) and (2.57). In order to calculate integrals in relation (2.4), it is convenient to consider a control volume formed by a packet of frames of arbitrary length, Δx, and to represent the expression for the power output, N (2.3), in the form of a surface integral,

$$N = \int_V j \operatorname{grad} \phi \, dV = \oint_S j_n \phi \, dS \, .$$

The side surface of the channel does not contribute any terms to the surface integral for N because the total normal current along the circumference of frames is equal to zero and the potential along the frames is constant. In calculating integrals along the end surfaces stretched over the frames, it is convenient to take into account the fact that the potential difference between points equally distributed over the surfaces $\Delta\phi = -E_x \Delta x$. As a result,

$$N = -\int_D (E_x j_x + E_y|_\Gamma j_y) dy \, dz \, \Delta x \, .$$

Taking into account that

$$\int_D j_x dy \, dz = 0$$

(see condition (2.62)), and determining the Faraday field on the channel walls, $E_{y|\Gamma}$, from the solution (2.63), we obtain

$$N = -\frac{1+\nu(1+\eta)}{2(1+\nu)} uB \int_D j_y dy \, dz \, \Delta x.$$

Because, in the case under consideration, the work

$$A = -uB \int_D j_y dy \, dz \, \Delta x \, ,$$

from condition (2.4), it follows that

$$\nu = (2\eta - 1)/(1-\eta) \, .$$

To determine E_x from condition (2.57), replace $\lambda(y,z)$ with the function,

$$\lambda^{\circ}(y,z) = \frac{\lambda(y,z)(1-\eta)}{2\eta(\beta E_x-(1-\eta)uB)} .$$

Then, from the problem (2.55) and (2.56), taking into account the expression for, ν, we see that $\lambda^{\circ}(y,z)$ is the solution of the problem

$$L(\lambda^{\circ}) = \frac{\partial \ln\sigma}{\partial y} , \quad \lambda^{\circ}|_{\Gamma} = 0 .$$

From this, it follows that $\lambda^{\circ}(y,z)$ is a standard function, independent of the operating mode of the channel, that depends only on the conductivity nonuniformities in the plasma flow. Using the definition of $\lambda^{\circ}(y,z)$, from condition (2.57) it is easy to obtain the following expression:

$$E_x = -\beta uB(1-\eta)/G , \tag{2.64}$$

where

$$G = (1+\beta^2)\langle\sigma\rangle\langle\sigma(1-\partial\lambda^{\circ}/\partial y)\rangle^{-1} - \beta^2 .$$

The angular brackets designate quantities averaged over the channel cross section. In the case of two-dimensional nonuniformities, the G parameter in (2.64) is an analogue of the well-known G factor [32].

In the simplest case of a uniform flow when σ = const. and u = const., the problem (2.55), (2.56) has a trivial solution $\lambda(y,z) = 0$ and the expression for (2.63) assumes the form

$$\phi = \beta uB(1-\eta)x - \eta uBy .$$

In the presence of velocity nonuniformity, the equipotentials on the channel walls are located on a curvilinear surface. It is expected that distortion of this surface in the presence of a thin boundary layer or a turbulent velocity profile will be insignificant.

The nature of the influence of the velocity nonuniformity on the shape of the equipotential frame of the optimum channel can be traced on another simple example, i.e., one-dimensional velocity nonuniformity oriented along the magnetic field vector (the conductivity distribution is arbitrary). Then the problem (2.61) also has a trivial solution and, after determining ν from condition (2.4), potential ϕ can be represented in the form

$$\phi = - E_x x - \eta B \int_0^y u(y)dy + \lambda(1-\eta)/2\eta .$$

Consequently, the equation of the equipotential line on the side surface has the form:

$$x = - \eta B \int_0^y u(y)dy/E_x \quad (x,y \in \Gamma) . \tag{2.65}$$

From (2.65) it follows that within the dynamic boundary layer, the frame angle with the channel axis increases, reaching $\pi/2$ on the wall.

Formulation of problem (2.61) makes it possible to assume that, in the case of a two-dimensional velocity nonuniformity, the relative deviation of equipotentials on the channel walls from a plane parallel to the direction of the magnetic field vector and tangential to the frame at the center of its diagonal parts does not exceed δ^*/a, the ratio of the thickness of displacement of the boundary layer, δ^*, with the characteristic dimension of the channel cross section, a. Thus, in the case of typical velocity nonuniformities in the MHD channel associated with turbulent boundary layers on channel walls, the deformation of the frames of the optimum channel has to be small.

Now consider the variational problem in the presence of two-dimensional velocity and conductivity nonuniformities. The problem (2.50), (2.55), (2.56) shall be solved numerically for the case of an absolute power (N) extremum, when the isoperimetric condition is not utilized, i.e., when $\nu = 0$.

Compare the variational solution with the optimal solution of the electrodynamic problem for a channel with rectangular frames. Solution of the variational problem and a problem for a channel with plane frames were obtained subject to the conditions corresponding to those of the basic model in Section 2.3.4., where the characteristics of an MHD generator with plane frames were determined numerically. The basic model is a channel with rectilinear cross section with the ratio of its sides $a/b = 2$ and $a = 0.5$ m; the structure of velocity and conductivity nonuniformities determined by the velocity and stagnation enthalpy distributions in the turbulent boundary layer in accordance with the "1/7 law"; the ratio of the boundary layer thickness with the channel half width in the direction of the magnetic field $\delta/b = 0.5$; channel wall temperature $T_w = 1800$ K, constant flow parameters outside the boundary layer ($T = 2650$ K, $u = 850$ m/s, $p = 0.1$ MPa, core flow conductivity equal to 18.50 S/m and the conductivity at the wall equal to 0.12 S/m), Hall parameter, $\beta = 1.5$, and magnetic field induction $B = 2T$. The boundary problems (2.50), (2.55), (2.56) were solved by a numerical grid method, using modified matrix dispersion to solve the finite difference equations.

Figure 2.15 shows plots of potential and current density along the circumference of the channel cross section. The difference between ϕ and j_n distributions for the two channels at a given flow nonuniformity is not very large. The variational solution for the frame channel determines the optimum frame configuration (Fig. 2.16), which differs from the rectangular frame mainly by the fact that the angle sectors of the optimum frame circumference are bent upward along the flow at the anode wall and down along the flow at the cathode wall. As a result of the fact that the integral characteristics of the channels being considered (Hall component of the electric field E_x and the average value of the Faraday

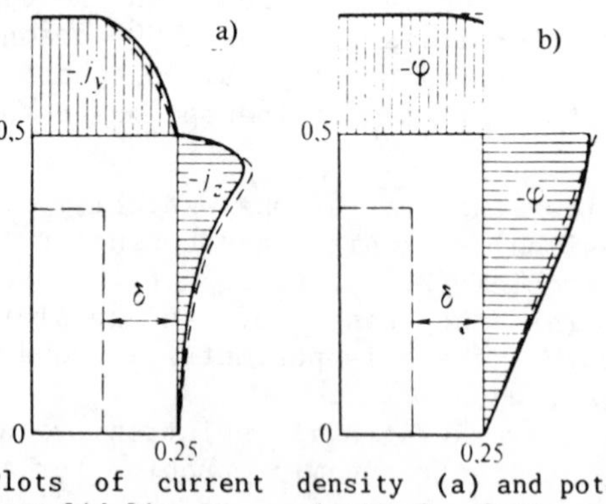

Fig. 2.15 Plots of current density (a) and potential (b) on channel walls: solid line - variational solution; dashed line - solution for a frame channel

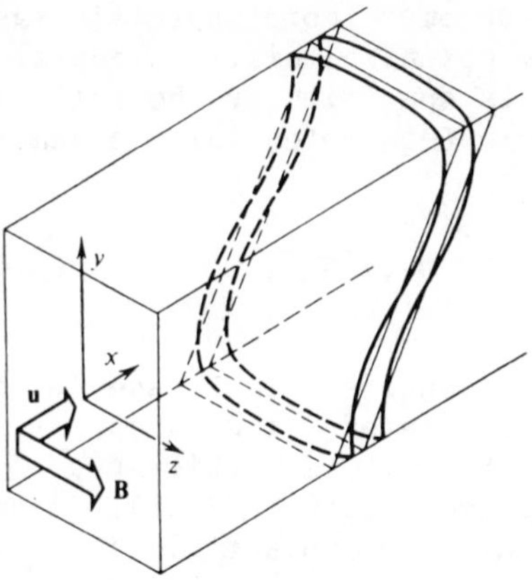

Fig. 2.16 Optimal frame configuration

component of the current density $\langle\langle j_y \rangle\rangle$) coincide, an increase in the integral power achieved by optimizing the shape of the frame is due to redistribution of current along the frame and, for the example being considered, is small (~ 1 %).

The results obtained indicate that, for typical flow rate and conductivity distributions in the MHD generator channel, nearly optimal characteristic conditions for thermal and dynamic turbulent boundary layers are achieved in a rectangular frame channel.

2.3.3. The Influence of Plasma Nonuniformities on the Characteristics of Frame MHD Channels. Let's turn next to an analytical investigation of two-dimensional electrodynamic effects in the cross section of an ideally segmented frame MHD channel with a constant wall angle [29]. Analysis of the characteristics of rectangular frame channels is of considerable practical importance, because frame channels are characterized by simple design and are similar in their output parameters to channels with optimal frame configuration.

For engineering facilities, it is important to determine in general form the dependence of structural and operating modes of frame channels on nonuniformities of the plasma flow in the cross section, and to analyze these characteristics, so as to be able to construct optimal frame channels for comparison with Faraday channels.

2.3.3.1. Problem Formulation

Formulation of the problem of the potential and current distributions in frame MHD channels is identical to the general formulation of the transverse electrodynamic problem (Section 2.3.1.). However, the general boundary condition (2.52) becomes more specific as a result of the fact that the potential distribution along the walls of a frame channel is specified by the following two quantities: the wall angle (equipotentials) α and the transverse electric field, E_x:

$$\phi|_\Gamma = \zeta E_x y, \quad \zeta = \text{ctg}\ \alpha\ (y \in \Gamma) . \tag{2.66}$$

Everywhere below we shall consider the case $\zeta = \text{ctg}\ \alpha = \text{const}$. Unless otherwise stated, the shape of the circumference of the boundary Γ is arbitrary. In analyzing the problem formulated, (2.50), (2.51), (2.66), it is convenient to represent the potential, ϕ, in the form

$$\phi = ((\beta - \zeta)E_x - \langle u \rangle B)\phi_\sigma - \langle u \rangle B \phi_\omega + \zeta E_x y , \tag{2.67}$$

where

$$\langle u \rangle = \iint_D u(y,z)\, dy\, dz \Big/ \iint_D dy\, dz$$

is the velocity averaged over the channel cross section and ϕ_σ and ϕ_ω are solutions of the following boundary problem:

$$L(\phi_\sigma) = \frac{\partial \ln \sigma}{\partial y} , \quad \phi_\sigma|_\Gamma = 0 ; \tag{2.68}$$

$$L(\phi_\omega) = \frac{\partial w}{\partial y} + w \frac{\partial \ln \sigma}{\partial y} , \quad \phi_\omega|_\Gamma = 0 , \tag{2.69}$$

where

$$w = (u - \langle u \rangle)/\langle u \rangle .$$

The functions ϕ_σ and ϕ_ω have the meaning of standard solutions that are independent of the loading mode of the channel and the wall angle. Because standard solutions become zero along the circumference, Γ, they can be equated to the potential distributions over the cross section of a Hall generator. The function, ϕ_σ, determines the potential in the Hall generator in the presence of gas flow at constant velocity, $\langle u \rangle$. The function ϕ_ω describes all effects associated with the nonuniformity of the velocity, and determines the potential in the Hall generators for a profile, $u - \langle u \rangle$, that changes sign, i.e., at zero mean velocity.

2.3.3.2. Integral Nonuniformity Parameters

Consider the average integral power generated in the frame channel in the presence of two-dimensional transverse plasma flow nonuniformities,

$$N_V = -\frac{1}{V}\int_V \mathbf{jE}\, dV = \frac{1}{V}\oint_S j_n \phi\, dS\ .$$

Assuming that volume V is formed by several frames of unit length and noticing that the total normal current along the circumference of the frame (equipotential) is zero, we obtain the following expression for N_V:

$$N_V = -(\langle j_x\rangle - \zeta\langle j_y\rangle)E_x = -\langle j_H\rangle E_x\ , \qquad (2.70)$$

where $\langle j_H\rangle = \langle j_x\rangle - \zeta\langle j_y\rangle$ is the load current averaged over the channel.

The relationship (2.70) shows that the parameter averaged over the channel cross section required for engineering calculations, specific power $N_V = -\langle \mathbf{jE}\rangle$, is determined from the average current density $\langle j_x\rangle$ and $\langle j_y\rangle$. The average volume force, $\mathbf{F} = -\langle \mathbf{j}\mathrm{x}\mathbf{B}\rangle$ can also be expressed in terms of $\langle j_x\rangle$ and $\langle j_y\rangle$. Determine $\langle j_x\rangle$ and $\langle j_y\rangle$ by averaging Ohm's law equations (2.51) and using relations (2.67)-(2.69):

$$\langle j_x\rangle = \frac{\langle\sigma\rangle}{\Xi(1+B^2)}\,(\beta(1+\chi)\langle u\rangle B$$

$$+ (\Xi(1+\beta^2) + \beta(\zeta-\beta))E_x)\ ,$$

$$\langle j_y\rangle = -\frac{\langle\sigma\rangle}{\Xi(1+\beta^2)}\,((1+\chi)\langle u\rangle B + (\zeta-\beta)E_x)\ . \qquad (2.71)$$

Here, are introduced integral nonuniformity parameters Ξ and χ determined from the standard solutions ϕ_σ and ϕ_ω:

$$\Xi = \langle\sigma\rangle \left\langle \sigma\left(1 - \frac{\partial\phi_\sigma}{\partial y}\right)\right\rangle^{-1},$$

$$\chi = \left\langle \sigma\left(w - \frac{\partial\phi_\omega}{\partial y}\right)\right\rangle \left\langle \sigma\left(1 - \frac{\partial\phi_\sigma}{\partial y}\right)\right\rangle^{-1}\ . \qquad (2.72)$$

From (2.71) it follows that Ξ and χ determine the effective conductivity and emf, in that order. By analogy with Faraday channels, in addition to the nonuniformity parameter, Ξ, for frame channels, it is possible to introduce the following parameter related to it.

$$G = \Xi(1+\beta^2) - \beta^2 . \tag{2.73}$$

The G parameter defined in such a manner is a natural generalization of the G-factor to frame channels with two-dimensional, transverse plasma flow nonuniformities.

Using the averaged Ohm's Law (2.71), the average integral specific power, N_V, and the average load current density, $\langle j_H \rangle$, can be represented in the form:

$$N_V = - \langle j_H \rangle E_x$$

$$= \frac{\langle\sigma\rangle\langle u\rangle^2 B^2 (1+\chi)^2 (\beta+\zeta)^2}{(G+\beta^2)(G+\zeta^2)} k(1-k) ,$$

$$\langle j_H \rangle = \frac{\langle\sigma\rangle\langle u\rangle B(1+\chi)(\beta+\zeta)}{G+\beta^2} (1-k) , \tag{2.74}$$

where $k = E_x/E_x^*$ is the load factor and

$$E_x^* = - \langle u\rangle B(1+\chi)(\beta+\zeta)/(G+\zeta^2)$$

is the transverse component of the electric field during the no-load mode.

In order to determine the electrical efficiency of the channel (η), it is necessary to calculate the average specific mechanical power, $A_V = - \langle(\mathbf{j}\mathrm{x}\mathbf{B})\mathbf{v}\rangle$, in addition to the specific electric power, N_V. In calculating A_V, one deals with new nonuniformity parameters λ and μ, determined from the standard solutions ϕ_σ and σ_ω. The expressions for A_V and $\eta = N_V/A_V$ have the form:

$$A_V = \frac{\langle\sigma\rangle\langle u\rangle^2 B^2 \lambda}{G+\beta^2} \left(1 + \mu + \frac{(1+\chi)(\beta^2-\zeta^2)}{G+\zeta^2} k\right) ,$$

$$\eta = \frac{(1+\chi)^2 (\beta+\zeta)^2 k(1-k)}{\lambda((1+\mu)(G+\zeta^2)+(1+\chi)(\beta^2-\zeta^2)k)} , \tag{2.75}$$

where

$$\lambda = \left\langle \sigma(1+\omega) \left(1- \frac{\partial \phi_\sigma}{\partial y}\right)\right\rangle$$

$$\times \left\langle \sigma\left(1- \frac{\partial \phi_\sigma}{\partial y}\right)\right\rangle^{-1} ,$$

$$\mu = \left\langle \sigma(1+\omega) \left(\omega- \frac{\partial \phi_\omega}{\partial y}\right)\right\rangle$$

$$\times \left\langle \sigma(1+\omega) \left(1- \frac{\partial \phi_\sigma}{\partial y}\right)\right\rangle^{-1} . \quad (2.76)$$

Thus, determination of the average integral parameters of a frame MHD channel

$$(\mathbf{F},\ N_V,\ A_V,\ \eta \text{ and } Q_V = \langle j^2/\sigma\rangle = A_V - N_V) ,$$

requires the knowledge of the four nonuniformity parameters Ξ, χ, λ and μ defined by (2.72) and (2.76). In constructing the quasi-one-dimensional flow theory for a frame channel, two-dimensional solutions can be utilized only to determine $\mathbf{F}$ and N_V. In this case, it is sufficient to calculate only Ξ and χ.

To consider certain properties of the nonuniformity parameters, it is useful to write out the expression for joule dissipation:

$$Q_V = \langle\sigma\rangle\langle u\rangle^2 B^2 (\lambda((1+\mu)(G+\zeta^2)$$

$$+ (1+\chi)(\beta^2 - \zeta^2)k)$$

$$- (1+\chi)^2(\beta+\zeta)^2 k(1-k))/(G+\beta^2)(G+\zeta^2) . \quad (2.77)$$

In order that joule dissipation, Q_V, not be negative at arbitrary values of k and ζ, it is necessary and sufficient that the following inequalities be fulfilled:

$$\Xi \geq 1 ,$$

$$4\lambda(1+\mu) \geq (\lambda+\chi+1)^2 + (\lambda-\chi-1)^2/G . \quad (2.78)$$

The first inequality in (2.78) is equivalent to the condition

$$G \geq 1 \ . \tag{2.79}$$

Absolute inequalities (2.78), (2.79) usually apply in the presence of nonuniformities. Equalities are possible only in exceptional cases. For example, $\Xi = 1$ when σ is independent of y.

Let's analyze the expression for the average integral specific power (2.74). At fixed k, maximum N_V is achieved when

$$(\zeta_{opt})_k = G/\beta \ . \tag{2.80}$$

When both k and ζ vary, the absolute maximum power is achieved when k = 1/2 and $\zeta = G/\beta$. In this case

$$N_{Vmax} = 1/4\langle\sigma\rangle\langle u\rangle^2 B^2 (1+\chi)^2/G \ . \tag{2.81}$$

From (2.80), it follows that the optimal wall angle is independent of the velocity profile and is determined only by conductivity nonuniformity. Velocity nonuniformity affects only N_{Vmax}. Even though in the general case the sign of χ is arbitrary, the condition $\chi > 0$ is fulfilled when cold boundary layers are present, i.e., when σ and u decrease toward the walls. Therefore, calculations made assuming a uniform velocity profile that is equal to its average value $\langle u\rangle$, should underestimate the power.

Under conditions at which the absolute maximum power is achieved (k = 1/2, $\zeta = G/\beta$), the average current along the channel becomes zero, i.e.,

$$\langle j_{xopt}\rangle = 0 \ . \tag{2.82}$$

The optimum value of E_x is given by the expression

$$E_{xopt} = -\ 1/2\beta\langle u\rangle B(1+\chi)/G \ . \tag{2.83}$$

To show that E_{xopt} coincides with the Hall component of the electric field along the channel (see 2.3.2.), consider equations (2.55) and (2.57). Representing λ in the form

$$\lambda = (2(1+\nu)\beta E_x - (1+\nu)1-\eta))\langle u\rangle B)\lambda_\sigma$$

$$- (1+\nu(1-\eta))\langle u\rangle B\lambda_\omega ,$$

reduce the problem (2.55), (2.56), to two standard boundary problems

$$L(\lambda_\sigma) = \frac{\partial \ln \sigma}{\partial y} , \quad \lambda_\sigma|_\Gamma = 0 ; \tag{2.84}$$

$$L(\lambda_\omega) = \omega \frac{\partial \ln \sigma}{\partial y} + \frac{\partial_\omega}{\partial y} , \quad \lambda_\omega|_\Gamma = 0 , \tag{2.85}$$

where $w = u/\langle u\rangle - 1$.

Because problems (2.84), (2.85), coincide with problems (2.68) and (2.69), they can be solved by replacing λ_σ with ϕ_σ and λ_ω with ϕ_ω. Then, from (2.57), using definitions of the nonuniformity parameters (2.72), one can obtain the following expression:

$$E_{xvar} = - \frac{1+\nu(1-\eta)}{2(1+\nu)} \beta \langle u\rangle B (1+\chi)/G ,$$

from which it follows that when $\nu = 0$ (this condition guarantees an absolute extremum of the power functional), $E_{xopt} = E_{xvar}$. It should be noted that, in view of the fact that average current along the channel becomes zero, it follows directly from Ohm's law that

$$\langle j_{yopt}\rangle = \langle j_{yvar}\rangle .$$

Thus, in a channel with a constant wall angle, optimized with respect to the absolute maximum power, the average parameters $\langle j_x\rangle$, $\langle j_y\rangle$ and E_x determining the power coincide with the analogous channel parameters described by the variational solution. However, the channel power obtained by solving the variational problem is higher than the power from a rectangular frame channel. This can be attributed to the fact that a change from rectangular frames to their optimal configuration results in an increase in the average load current, $\langle j_H \rangle$.

An important problem in the development and investigation of MHD generators is determination of optimal operating modes of a generator at a specified electrical efficiency, η, (2.75). In the general case, determination of the optimal values of ζ and k at which maximum power is achieved at a fixed efficiency is quite complicated. Therefore, consider only the case of a uniform velocity profile, u = const. When u = const., $\omega \equiv \phi_\omega \equiv 0$ and, consequently, $\chi = \mu = 0$ and $\lambda = 1$. Then, from relations (2.74), (2.75), simple expressions are obtained for the optimum characteristics

$$(\zeta_{opt})_\eta = G\eta/\beta(1-\eta) ,$$

$$(k_{opt})_\eta = \frac{\beta^2(1+\eta)^2+G\eta^2}{\beta^2(1-\eta)+G\eta} ,$$

$$(N_{opt})_\eta = \langle\sigma\rangle u^2 B^2 \eta(1-\eta)/G ,$$

$$(\langle j_{x|opt}\rangle)_\eta = 0 . \tag{2.86}$$

It should be noted that, when u = const., the variational problem considered in Section 2.3.2. has a solution describing a rectangular frame channel. Therefore, expressions (2.86) have a solution of the variational problem in terms of the standard solution ϕ_σ that determines the nonuniformity parameter Ξ (or the G-factor).

From (2.75) and (2.86), it follows that, when u = const., the maximum efficiency, $\eta = 1$, can be achieved only in the limiting case, when $\zeta = \infty$, k = 1. Maximum electrical efficiency that exists at a fixed finite wall angle is given by the following expression:

$$(\eta_{opt})_\zeta = \frac{\beta+\zeta}{\beta-\zeta}(1-2(k_{opt})_\zeta) , \tag{2.87}$$

where

$$(k_{opt})_\zeta = \frac{G+\zeta^2}{\beta^2-\zeta^2}\left(\left(1+\frac{\beta^2-\zeta^2}{G+\zeta^2}\right)^{1/2}-1\right) .$$

From (2.87), it follows that $(\eta_{opt})_\zeta$ is a monotonically decreasing function of the G parameter. Therefore, a decrease of the G parameter leads not only to an increase in power and efficiency of the MHD generator at fixed ζ and k (as can be seen from expressions (2.74), (2.75)), but also to an increase in the maximum electrical efficiency in the case when only ζ is fixed. Figure 2.17 is a plot of the dependence of η_{opt} on the wall angle α = arccot ζ for various values of β and two characteristic values of G (G = 1 when the flow is uniform and G = 1.2 when the conductivity distribution in the flow corresponds to the flow conditions in subsonic open-cycle MHD generators). It can be seen from this figure that high electrical efficiency is achieved in frame channels at sufficiently low wall angles.

Investigations of relations (2.74), (2.75) show that, in the case of a nonuniform velocity profile u = u (y,z), the value η = 1 cannot be reached at any wall angle. (Because of their length, the calculations required will not be given here.) Relations (2.71)-(2.83), (2.86), (2.87), remain valid, even for a rectangular Faraday channel, when equipotential electrodes are located in a plane inclined to the channel axis at an angle, α, that is determined by the generator loading conditions. For a Faraday channel parameter ζ in relations (2.71)-(2.83), (2.86), (2.87) must be replaced with parameter

$$\zeta_F = \text{ctg}\ \alpha = -\langle E_y\rangle/E_x\ .$$

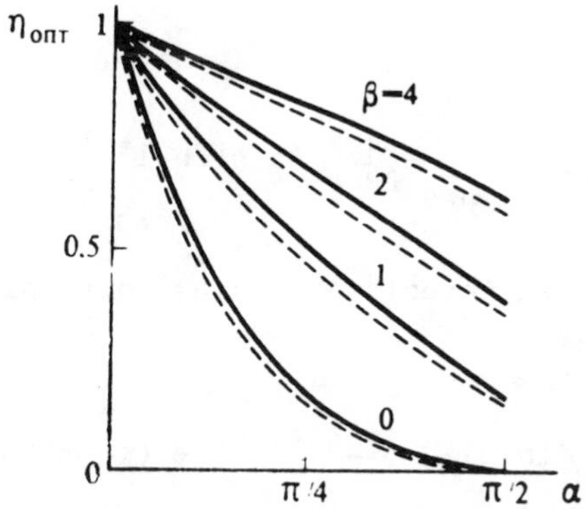

Fig. 2.17 Variation of the maximum efficiency with the wall angle for various values of β and G: _____, -G = 1; _ _ _, G = 1.2

Thus, the only change is in the formulation of the standard boundary problems. In the case of a Faraday channel, instead of boundary problems (2.68), (2.69), it is necessary to solve the following problems:

$$L(\Phi_\sigma) = \frac{\partial \ln\sigma}{\partial y} , \quad \Phi_\sigma\Big|_{\Gamma_1} = 0, \quad \frac{\partial \Phi_\sigma}{\partial z}\Big|_{\Gamma_2} = 0 .$$

$$L(\Phi_\omega) = \frac{\partial \omega}{\partial y} + \omega \frac{\partial \ln\sigma}{\partial y} , \quad \Phi_\omega\Big|_{\Gamma_1} = 0 ,$$

$$\frac{\partial \Phi_\omega}{\partial z}\Big|_{\Gamma_2} = 0 , \tag{2.88}$$

where Γ_1 and Γ_2 are the electrode and insulating sectors of the duct, respectively. The functions Φ_σ and Φ_ω define potential Φ in accordance with formula (2.67):

$$\Phi = ((\beta - \zeta_F)E_x - \langle u\rangle B)\Phi_\sigma - \langle u\rangle B\Phi_\omega + \zeta_F E_x y .$$

2.3.3.3. Rectangular Channel with One-Dimensional Conductivity Nonuniformity

First, consider the simple case of one-dimensional conductivity nonuniformity, $\sigma = \sigma(y)$, in order to obtain a qualitative picture of the dependence of the integral nonuniformity parameter on the flow parameters. In this case, solution of the standard problem (2.68) in the region $|y| \le a$, $|z| \le b$ can be sought in the form of a series

$$\phi_\sigma(y,z) = a \sum_{n=0}^{\infty} A_n\left(\frac{y}{a}\right)\cos(2n+1)\frac{\pi z}{2b} . \tag{2.89}$$

The expansion coefficients A_n (y/a) are solutions of the boundary problems

$$A_n'' + (\ln\sigma)' A_n' - \gamma_n^2 A_n = \delta_n (\ln\sigma)' ,$$

$$A_n(1) = A_n(-1) = 0 , \tag{2.90}$$

where

$$\gamma_n = (1+\beta^2)^{1/2}\pi(2n+1)a/2b \ ;$$

$$\delta_n = 4(-1)^n/\pi(2n+1) \ .$$

For certain special profiles σ (y), the solution of problems (2.90) can be represented by tabulated data. Consider one such case, when σ (y) is given in the form

$$\sigma(y/a) = \sigma_o \cos^2(\omega y/a) \ , \qquad (2.91)$$

where ω = const. $\in [0, \pi/2]$.

Profile (2.91) models a conductivity drop near the cold electrodes. In this case, the coefficients assume the following form:

$$A_n = \frac{2\omega\delta_n}{\gamma_n^2} \operatorname{tg} \frac{\omega y}{a} \left(1 - \frac{\sin\omega \sin\mu_n(\omega y/a)}{\sin\mu_n\omega \sin(\omega y/a)}\right)$$

when $n \leq n_o$,

$$A_n = \frac{2\omega\delta_n}{\gamma_n^2} \operatorname{tg} \frac{\omega y}{a} \left(1 - \frac{\sin\omega \operatorname{sh} \nu_n(\omega y/a)}{\operatorname{sh} \nu_n\omega \sin(\omega y/a)}\right)$$

when $n > n_o$, (2.92)

where

$$\mu_n = (1-\gamma_n^2/\omega^2)^{1/2}; \ \nu_n = (\gamma_n^2/\omega^2-1)^{1/2} \ ;$$

$$n_o = E(\omega b/\pi a(1+\beta^2)^{1/2} - 1/2) \ ;$$

E = E(p) is a whole multiple of p. Assuming (2.91) applies, the conductivity averaged over the cross section $\langle\sigma\rangle$ is given by the formula

$$\langle\sigma\rangle = 1/2\sigma_o(1+\sin 2\omega/2\omega) \ .$$

The nonuniformity parameter calculated from formula (2.72), using (2.89) and (2.92), has the form

$$\Xi = (1-\sigma_o \Psi/\langle\sigma\rangle)^{-1} , \qquad (2.93)$$

where

$$\Psi = \frac{2\omega^2 b^2}{3a^2(1+\beta^2)} \Big(1+ \frac{192 b^2 \omega}{\pi^6 a^2(1+\beta^2)}$$

$$\times \Big(\sum_{o}^{n_0} ((1+\mu_n^2)\sin 2\omega - 4\mu_n \sin^2 \omega \operatorname{ctg} \mu_n \omega)$$

$$\times (2n+1)^{-6} + \sum_{n_o+1}^{\infty} ((1+\nu_n^2)\sin 2\omega$$

$$- 4\nu_n \sin^2 \omega \operatorname{cth} \nu_n \omega)(2n+1)^{-6}\Big)\Big) .$$

Consider the limiting cases:

when $(1 + \beta^2)a^2/b^2 \to 0$

$$\Xi = 1/2\omega^{-1} \operatorname{tg} \omega (1+\sin 2\omega/2\omega) ; \qquad (2.94)$$

when $(1 + \beta^2)a^2/b^2 \to \infty$

$$\Xi \cong 1 + \frac{4\omega^2}{3} \frac{1-\sin 2\omega/2\omega}{1+\sin 2\omega/2\omega} \frac{b^2}{a^2(1+\beta^2)}$$

$$- \frac{512 S\omega^2 \sin^2\omega}{1+\sin 2\omega/2\omega} \left(\frac{b}{a\sqrt{1+\beta^2}}\right)^3 + 0 \left(\frac{b^4}{a^4(1+\beta^2)^2}\right) , \qquad (2.95)$$

where

$$S = \sum_{n=o}^{\infty} \pi^{-5} (2n+1)^{-5} .$$

Next, compare parameters of a frame channel under consideration with characteristics of a Faraday channel. In the case of a Faraday channel, solution of the problem (2.88) for a one-dimensional conductivity profile is independent of β and z and has the simple form:

$$\phi_\sigma^F = y - \int_0^y \sigma^{-1}\langle\sigma^{-1}\rangle^{-1}\, dy \; .$$

Consequently

$$\Xi^F = \langle\sigma\rangle\langle\sigma^{-1}\rangle = 1/2\omega^{-1} \text{ tg } \omega \, (1+\sin 2\omega/2\omega) \; .$$

This expression coincides with (2.94). The result obtained indicates that, when $(1 + \beta^2)a^2/b^2 \to 0$, i.e., when the channel geometry is such that the distance between the electrode walls is much less than the distance between the side walls, parameters of a frame channel approach parameters of a Faraday channel. This conclusion is to be expected because when $a/b \ll 1$, current flow into the side walls in a frame channel becomes insignificant.

A reverse situation occurs when $(1 + \beta^2)a^2/b^2 \to \infty$. It follows from (2.95) that, in this case, $\Xi \to 1$ and the integral parameters of the channel at a fixed Hall parameter are determined only by the mean conductivity $\langle\sigma\rangle$. It should be noted that the limiting case for (2.93) when $(1 + \beta^2)a^2/b^2 \to \infty$ can be interpreted either as an increase in the ratio of the sides, a/b, at a fixed β, or as an increase in the Hall parameter at a fixed ratio, a/b. In the first case, the characteristics of the generator at $a/b \gg 1$ really depend only on the average conductivity $\langle\sigma\rangle$. In the second case, the result is different because the G parameter, which depends on β, appears in expressions for the integral parameters.

Consider the ratio of the mean integral specific power of a Faraday channel with that of a frame channel, N_V^F/N_V. From (2.86) it follows that

$$\frac{(N_{V_{opt}}^F)_\eta}{(N_{V_{opt}})_\eta} = \frac{G}{G^F} = \frac{\Xi(1+\beta^2)-\beta^2}{\Xi^F(1+\beta^2)-\beta^2} \; ,$$

$$\frac{(N^F_{V_{opt}})_\eta}{(N_{V_{opt}})_\eta} \cong$$

$$\cong 1\left(+\frac{4\omega^2 b^2}{3a^2}\ \frac{1-\sin 2\omega/2\omega}{1+\sin 2\omega/2\omega}\right)\ \frac{1}{(\Xi^F-1)\beta^2} \ .$$

Thus, for the conductivity nonuniformity assumed and at large Hall parameter values, the gain in the power output of a frame channel can be quite substantial in comparison with a Faraday channel. At the same time, when the conductivity nonuniformities are substantial, but when parameter $\omega \to \pi/2$ and Ξ and $\langle\sigma\rangle/\sigma_o$ remain finite (see Fig. 2.18), in the case of a Faraday channel $\Xi^F \to \infty$ and

$$(N^F_{V_{opt}})_\eta \to 0 \ .$$

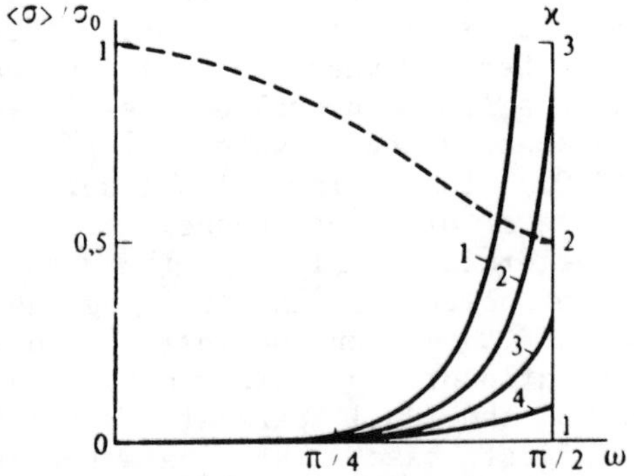

Fig. 2.18 The dependence of dimensionless mean conductivity, $\langle\sigma\rangle/\sigma_o$ (dashed lines), and the nonuniformity parameter, Ξ (continuous line), on ω for a channel with a uniform conductivity profile, at various values of a $\sqrt{1+\beta^2/b}$: 1 - 0, 2 - 1, 3 - 2, 4 - 4

<u>2.3.4. Numerical Analysis of Electrodynamics in the Cross Section of a Frame MHD Channel</u>. An important practical problem in investigations of frame channel electrodynamics is determination of the influence of the flow structure and design parameters of the channel on its characteristics. Determination of the specifics of the spatial current and potential distributions in these channels is also of interest. Numerical solution of the electrodynamic problem and its representation in

parametric form make it possible not only to provide answers to these questions but, also, to determine the integral parameters of the nonuniformity under real operating conditions of frame channels, required when performing engineering calculations.

2.3.4.1. Problem Formulation and the Method of Solution

Consider the problem of the current and potential distributions in the cross section of an MHD channel with plane, ideally thin frames in the presence of two-dimensional conductivity and flow-velocity nonuniformities [28, 30]. The equations and boundary conditions for this problem have been set forth here; see [2.50], [2.51], and [2.66]. Excluding special cases (see, for example, Section 2.3.3.3.), the problem under consideration does not have an analytical solution. Therefore, analysis of the characteristics of an MHD channel for real velocity and plasma conductivity distributions requires numerical solution of the problem.

Analysis of numerical solutions is usually quite complicated. Therefore, it is expedient to represent the numerical solution in a parametric form. In addition to the Hall parameter β, the problem under consideration is characterized by two other parameters: the wall angle α and the electric field along the channel, E_x.

Let the parameter of the numerical solution be

$$p = (\mathrm{ctg}\ \alpha - \beta) E_x a \tag{2.96}$$

and represent the potential in the form

$$\phi = \Phi_* + \beta E_x y\ , \tag{2.97}$$

where Φ_* is the solution of the following boundary,

$$L(\Phi_*) = -\frac{B}{\sigma}\ \frac{\partial(\sigma u)}{\partial y}\ ,\quad \Phi_*|_\Gamma = \frac{py}{a}\ . \tag{2.98}$$

In view of the linearity of the problem (2.98), its solution may be represented in the form

$$\Phi_* = \Phi_o + p\Phi_p\ , \tag{2.99}$$

where Φ_o and Φ_p are solutions of boundary problems

$$L(\Phi_o) = -\frac{B}{\sigma}\frac{\partial(\sigma u)}{\partial y}, \quad \Phi_o|_\Gamma = 0 ,$$

$$L(\Phi_p) = 0, \quad \Phi_p|_\Gamma = y/a . \tag{2.100}$$

Solutions of the boundary problems (2.100) are independent of the loading mode and design characteristics of the channel. Therefore, having obtained solutions Φ_o and Φ_p, one can automatically determine the electrical parameters of the channel for any E_x and ctg α.

Having solutions Φ_o and Φ_p, it is possible to calculate parameters

$$\Xi = \frac{\langle\sigma\rangle}{a}\left\langle\sigma\frac{\partial\Phi_p}{\partial y}\right\rangle ,$$

$$\chi = \left\langle\sigma\left(\frac{\partial\Phi_o}{\partial y} + uB\right)\right\rangle / \langle u\rangle Ba\left\langle\sigma\frac{\partial\Phi_p}{\partial y}\right\rangle - 1 ,$$

specifying the integral characteristics of a channel required for the development of a quasi-one-dimensional flow theory for a frame channel.

The boundary problem (2.98) was solved numerically by means of a grid method, utilizing a modified dispersion technique to solve the finite difference equations (the main features of the method were given above). This method makes it possible to obtain simultaneously two linearly independent solutions, Φ_o and Φ_p, and two analogous solutions, Ψ_o and Ψ_p, for the current stream function, ψ. These solutions and the results of preliminary processing of conductivity and velocity distributions are used to determine $\langle j_x\rangle$ and $\langle j_y\rangle$ and the average integral specific power, $N_v = -(\langle j_x\rangle - \text{ctg}\,\alpha\,\langle j_y\rangle)\,E_x$.

2.3.4.2. Rectangular Channel

Now consider a frame MHD channel with sides 2a (distance between electrode walls in the direction of the y axis) and 2b (distances between side walls in the direction of the z axis). In view of the symmetry of the problem, calculations will be made for one (upper right) quarter of the channel, subject to the following natural boundary conditions at the symmetry axes:

$$\frac{\partial \Phi_*(y,0)}{\partial z} = 0, \quad \Phi_*(0,z) = 0 .$$

The velocity and stagnation enthalpy profiles will be given in accordance with the "1/7 law" for the turbulent boundary layer and will be assumed to be linear for the laminar sublayer:

$$u(y,z) = U\psi_u(y,z) ,$$

$$\psi_u = \eta(y)\xi(z) ,$$

$$i_o(y,z) = i_{o\infty}\psi_i(y,z) ,$$

$$\psi_i = (1-i_\omega/i_{o\infty})\eta\xi + i_\omega/i_{o\infty} , \qquad (2.101)$$

where

$$\eta = \begin{cases} 1 & |y| \leq a-\delta, \\ ((a-|y|)/\delta)^{1/7} & \delta_1 < a-|y| < \delta, \\ v(a-|y|) & a-|y| \leq \delta_1, \end{cases}$$

when

$$\xi = \begin{cases} 1 & |z| \leq b-\delta, \\ ((b-|z|)/\delta)^{1/7} & \delta_1 < b-|z| < \delta, \\ v(b-|z|) & b-|z| \leq \delta_1, \end{cases}$$

δ and δ_1 are the thickness of the turbulent boundary layer and the laminar sublayer

$$(\delta_1 = 194\ Re_x^{-0.7}\delta) ,$$

respectively.

In the formulation given, the principal parameters of the problem are the boundary layer thickness, δ, ratio of the wall temperature to the core flow temperature (the latter did not vary), Hall parameter, β, and the geometrical factor b/a. The set of the dimensional nonvarying parameters was selected in such a way that conductivity in the core flow was approximately 20 S/m and Mach number, $M \simeq 0.9$. The dependence of the conductivity on the temperature was given by the following formula:

$$\sigma(p,T) = 2.05 \cdot 10^5 T^{0.75} \exp(-2.515 \cdot 10^4/T) p^{-0.5} ,$$

that describes adequately the temperature variation of conductivity observed in the experiments using combustion products of hydrocarbon fuels seeded with 1 % potassium [33]. However, the absolute conductivity exceeds the experimental data by approximately 40 %.

The range of variation of the main parameters was chosen to correspond to typical conditions found in an MHD generator operating on combustion products:

$$0 \leq \delta/a \leq 0.25; \quad 1400 \text{ K} \leq T_w \leq 2200 \text{ K} ;$$

$$0.5 \leq \beta \leq 5; \qquad 0.5 \leq b/a \leq 2 .$$

Let us note certain qualitative characteristics of the current and potential distributions obtained (Fig. 2.19). First, current streamline distributions exhibit

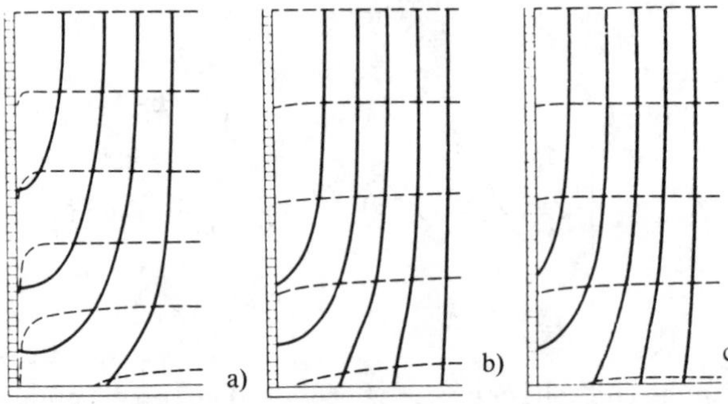

Fig. 2.19 Current distributions (continuous curves) and potential distributions (dashed curves) in the cross section of a frame channel for b/a = 0.5, a = 0.5 m, $\delta/a = 0.25$, ctg $\alpha = 2$, $\beta = 1.5$, u = 850 m/s, $\sigma_\infty = 18.5$ S/m, B = 2T, $E_x = -400$ V/m, V = -400 V, and different wall temperature: a = 2200 K, b = 1800 K, c = 1400 K

considerable two-dimensional behavior and considerable part of the current flows to the side walls (from 20 to 70 % of the transverse current along the channel axis, when the temperature changes from 2200 to 1400 K).

Secondly, except for the near-wall layers, the potential distribution in the cross section is similar to that given by the "parallel equipotential" model [34]. At low wall temperatures, practically all of the potential difference between the axis of the cross section and the side wall takes place in the low-conductivity, near-wall layer that is considerably thinner than the turbulent boundary layer.

For comparison, Figure 2.20 shows the current and potential distributions in the cross section of a Faraday channel with ideally insulating walls. Note that the latter example provides a good basis for using parallel equipotential models (as well as the parallel current streamline models [35]) in engineering methods of calculating Faraday channels.

The most important parameter of an MHD channel is the mean current density, $\langle j_y \rangle$, in terms of which the mean force and specific power output are expressed. Taking (2.96)-(2.99), into account, $\langle j_y \rangle$ can be represented in the form

$$\langle j_y \rangle - \frac{1}{1+\beta^2} \left(\left\langle \sigma \frac{\partial \Phi_o}{\partial y} \right\rangle + \langle \sigma u \rangle B \right.$$

$$\left. + (\mathrm{ctg}\ \alpha - B)\, E_x a \left\langle \sigma \frac{\partial \Phi_p}{\partial y} \right\rangle \right) = \langle j_{oy} \rangle + p \langle j_{py} \rangle .$$

Table 2.2 gives the following nonuniformity parameters for a frame channel:

$$- \frac{1+\beta^2}{\langle \sigma u \rangle B} \langle j_{oy} \rangle = \left\langle \sigma \frac{\partial \Phi_o}{\partial y} \right\rangle / \langle \sigma u \rangle B + 1 =$$

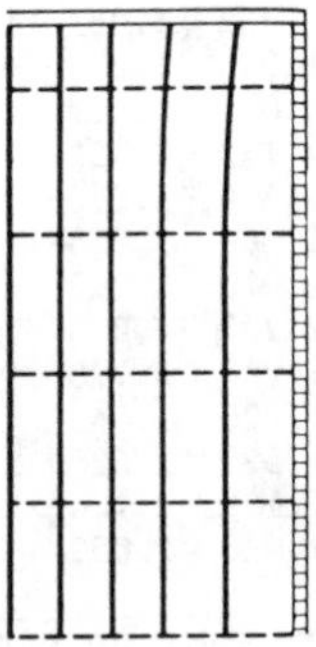

Fig.2.20 Current distribution (continuous curves) and potential distributions (dot-dashed curves) in the cross section of a Faraday channel at $b/a = 0.5$, $\delta/a = 0.25$, $\beta = 1.5$, $T_w = 1800$ K.

$$= \frac{\langle\sigma\rangle\langle u\rangle(1+\chi)}{\langle\sigma u\rangle\Xi} = \frac{1}{\Xi^{\chi}} ,$$

$$- \frac{(1+\beta^2)a}{\langle\sigma\rangle} \langle j_{py}\rangle = \left\langle \sigma \frac{\partial\Phi_p}{\delta y} \right\rangle a/\langle\sigma\rangle \frac{1}{\Xi} ,$$

$$G = \Xi(1+\beta^2) - \beta^2 .$$

For comparison, Table 2.2 also shows the following parameters:

$$\Xi^F = \langle\langle\sigma\rangle\rangle \, \langle\langle\sigma\rangle_z^{-1}\rangle_y \text{ and } G^F = \Xi^F(1+\beta^2) - \beta^2 ,$$

that determine characteristics of a Faraday channel, described by the parallel equipotential model, and also the results of calculations of nonuniformity parameters, based on an approximate model:

$$1/\Xi_m^{\chi},\ 1/\Xi_m,\ G_m$$

(see 2.5.5.4.). The data in Table 2.2 indicate that, within the range considered ($0.5 \leq \beta \leq 5$), the internal resistance of a frame channel is lower than that of the Faraday channel and that this difference increases with increasing β. This conclusion is in agreement with previous results according to which the power output of a frame channel in the presence of a conductivity nonuniformity is higher than that of an analogous Faraday channel.

Table 2.2

Nonuniformity parameters of frame and Faraday MHD Channels with a Nonuniform Flow

δ/α	b/α	T_w,K	β	$1/\Xi^{\chi}$	$1/\Xi^{\chi}_m$	$1/\Xi$	$1/\Xi_m$	$1/\Xi^F$	G	G_m	G^F
0.25	0.5	1800	1.5	0.967	0.992	0.973	0.971	0.947	1.090	1.097	1.182
0.125	0.5	1800	1.5	0.980	0.979	0.981	0.981	0.972	1.063	1.063	1.093
0.0625	0.5	1800	1.5	0.988	0.993	0.990	0.989	0.985	1.033	1.036	1.049
0.125	0.5	1800	0.5	0.976	0.987	0.979	0.977	0.972	1.027	1.029	1.036
0.125	0.5	1800	3.0	0.984	0.994	0.986	0.985	0.972	1.142	1.152	1.288
0.125	0.5	1800	5.0	0.986	0.997	0.991	0.991	0.972	1.236	1.236	1.769
0.125	1.0	1800	1.5	0.975	0.987	0.980	0.978	0.972	1.066	1.073	1.093
0.125	2.0	1800	1.5	0.973	0.985	0.977	0.975	0.972	1.076	1.083	1.093
0.25	0.5	1400	1.5	0.726	0.649	0.725	0.636	0.398	2.232	2.860	5.916
0.25	0.5	1600	1.5	0.929	0.937	0.932	0.918	0.852	1.237	1.290	1.564
0.25	0.5	2200	1.5	0.991	1.012	0.994	0.987	0.990	1.020	1.043	1.033
0.5	1.0	1600	1.5	0.857	0.880	0.861	0.838	0.756	1.525	1.628	2.049
0.25	0.5	1800*	1.5	0.890	0.917	0.961	0.963	0.946	1.129	1.124	1.185
0.25	0.5	2200*	1.5	0.932	0.953	0.991	0.983	0.989	1.027	1.053	1.036

*Electrode wall temperature at the side (diagonal) wall temperature of 1400 K

Most of the data in Table 2.2 were acquired at the same electrode and side (diagonal) wall temperature of the MHD channel. The stagnation enthalpy profile in the channel cross section at different wall temperatures was given in the form

$$\psi_i = \eta\xi + (1-\eta\xi)\,(i_{\omega y} + i_{\omega z})/2i_{o\infty}$$
$$+ (\xi-\eta)\,(i_{\omega} - i_{\omega z})/2i_{o\infty},$$

where $i_{\omega y}$ and $i_{\omega z}$ are the gas enthalpy at central sectors of electrode and side walls, respectively, and η and ξ are profile functions (see formula 2.101).

Cooling of the side walls increases the G-factor for both frame and Faraday channels. At the same time, the influence of cooling of the side walls of frame channels increases with increasing flow nonuniformity, i.e., it increases when the difference between the core flow temperature and the electrode wall temperature increases.

2.3.4.3. Oval Channel

The main distinguishing feature of the electrodynamics of frame channels is the fact that the electric current is fed to the whole surface of the channel walls. In this regard, it is interesting to analyze the influence of the shape of the cross section of a rectangular frame channel on the integral parameters of an MHD generator. It has been noted that, in the case of a rectangular frame channel, a decrease in the ratio of the channel width with channel height (b/a) leads to an increase of integral characteristics of an MHD generator (for example, of power output and mean transverse current). It can also be assumed that, when the channel cross section is finite, its geometrical shape will also influence the generator characteristics. In this case, a search for the optimum shape of the channel section assumes that the corresponding variational problem can be solved.

In view of the difficulties of formulating a variational problem that takes into account variation of the shape of the cross section (region of solution), this discussion is limited to analysis of solutions of direct electrodynamic problems of current and potential distributions in channel cross sections of different shapes. The parameters of a rectangular channel calculated in section 2.3.4.2. will be used as a standard of comparison. The influence of the shape of the channel on its output parameters will be considered, using as an example a family of ellipses with different ratios of semi-major with semi-minor axes. For all variants under consideration, it was assumed that the cross-sectional area of channel A is equal to the area of a rectangular frame channel. The profiles of gasdynamic parameters in elliptic channels are given by analogy with a rectangular frame channel with a uniform core flow and turbulent boundary layers. Inside the channels the variation of gasdynamic parameters for velocity and stagnation enthalpy distributions depends on the distance from the wall along a normal, in accordance with the "1/7 law." The thickness of the boundary layer is constant and equal to 0.125 m. The principal flow parameters are assumed to be the same as in section 2.3.4.2.

The problem (2.98) was solved using parametric representation of the numerical solution in the form given by (2.99). The optimal values of E_x and ctg α were determined in accordance with (2.80) and (2.83).

Figure 2.21 shows current and potential distributions in the upper right-hand quarter of a cross section of an elliptical channel, with symmetry taken into account.

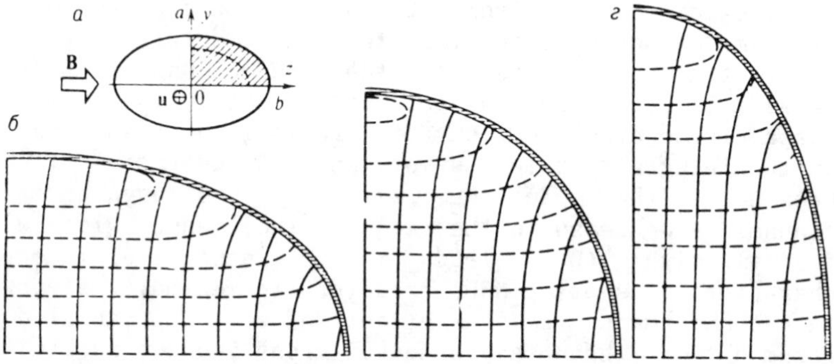

Fig. 2.21 Cross section of a frame MHD channel (a) and current distributions (continuous lines) and potential distributions (dashed lines) along the cross section at ctg $\alpha = 2$, $\beta = 1.5$, $T_w = 1800$ K at the following values of b/a: a - 1.8, b - 1, c - 0.556

The most important parameters of MHD channels of different configuration are summarized in Table 2.3. It can be seen that, for the conditions considered, parameters of an elliptical frame channel exceed considerably parameters of a rectangular frame channel (in particular, power output per unit length, N_L, by up to 15 %). It should be noted that, from the point of view of hydraulic losses to friction and heat transfer, elliptical channels (and to a larger degree circular channels) have obvious advantages over rectangular channels.

Hence, it is possible to assume that, with the electrodynamics of frame channels and the hydraulic

Table 2.3

Output Parameters of Frame MHD Channels of Various Configurations with Nonuniform Flow

Geometry	a, m	$\langle\langle\sigma\rangle\rangle A$, S/m	E_x, V/m	Vy, V	N_L MW/m
Rectangle	0.5	1.999	-1082	-395.4	1.141
Circle	0.395	1.973	-1210	-315.9	1.272
Ellipse	0.526	1.978	-1253	-420.9	1.321
Ellipse	0.297	1.946	-1130	-235.0	1.162

where (b/a = 0.556, b/a = 1.8)

characteristics of channels of different shape taken into account, elliptical channels with cross sections sufficiently close to circular shape are the most preferred type.

2.4. Two Dimensional Inviscid Flows in an MHD Channel

2.4.1. Problem Formulation. Inviscid and thermally nonconducting gas flow in a plane MHD channel at $Re_m \ll 1$ is described by the following system of equations:

$$\frac{\partial \rho}{\partial t} + \text{div}\ \rho \mathbf{v} = 0\ , \tag{2.102}$$

$$\rho \frac{\partial \mathbf{v}}{\partial t} + \rho(\mathbf{v}\nabla)\mathbf{v} = -\ \text{grad}\ p + \mathbf{j} \times \mathbf{B}\ , \tag{2.103}$$

$$\rho \frac{\partial}{\partial t}\left(\varepsilon + \frac{1}{2} v^2\right) + \rho(\mathbf{v}\nabla)\left(\varepsilon + \frac{1}{2} v^2\right) = -\ \text{div}\ p\mathbf{v} + \mathbf{jE}\ , \tag{2.104}$$

$$\varepsilon = \varepsilon(p, \rho)\ , \tag{2.105}$$

$$\text{div}\ \mathbf{j} = 0\ , \tag{2.106}$$

$$\mathbf{j} = \sigma(\mathbf{E} + \mathbf{v} \times \mathbf{B}) - \beta(\mathbf{j}\times\mathbf{B})/B\ , \tag{2.107}$$

$$\sigma = \sigma(p, \rho)\ , \tag{2.108}$$

$$\beta = \mu_e(p, \rho)B\ . \tag{2.109}$$

In equations (2.102)-(2.109), the magnetic field vector, **B**, is assumed to be a known function of coordinates and time. The flow region, D, in the x,y

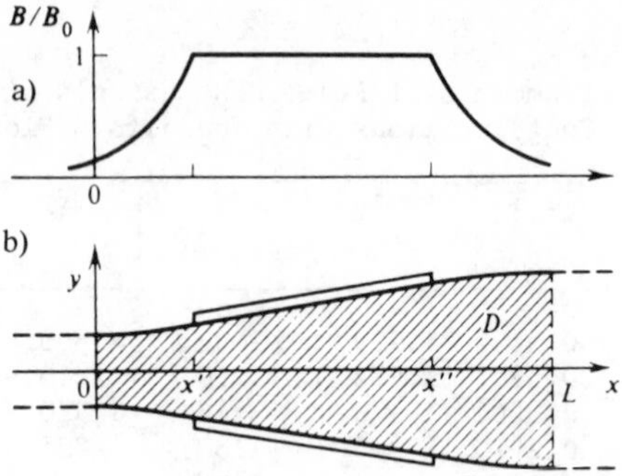

Fig. 2.22 Magnetic field profile (a) and the geometry (b) of an MHD channel

plane (Fig. 2.22) is also assumed to be known (Fig. 2.22). The plane channel is formed by impenetrable upper and lower walls bound by inlet and exit cross sections (x = 0; L). The magnetic field is directed along the z axis, $\mathbf{B} = \{0, 0, B(x)\}$.

The coordinates of the cross sections at the inlet and exit are chosen to be sufficiently far from the active sector of the channel that the MHD interaction at the inlet into the channel and exit from it do not exert substantial influence on the flow. In this case, adequate accuracy is obtained by the use of standard gasdynamic initial and boundary conditions at the boundaries of region D [36]. The electrodynamic boundary conditions are specified in accordance with the design features of the channel walls and the loading scheme [1].

Consider techniques that are most frequently used to formulate gasdynamic boundary conditions. When the flow is achieved by connecting an infinitely large source with constant gas parameters to the channel, the stagnation enthalpy, i_o, and entropy, s, at the inlet into the channel are assumed to be known (instead of entropy, one can use any of its functions, for example, stagnation pressure). If one also assumes that a sufficiently long cylindrical sector is located between the source and the channel, it can further be assumed that a laminar (quasi-developed [1]) flow with a constant static pressure and an absence of a transverse velocity component in the flow is achieved at the channel inlet. In this case, the boundary conditions at the channel inlet can be written in the following form:

when $M_1 < 1$,

$$i_o = i_o(y), \ s = s(y), \ \nu = 0 \quad \text{when} \quad x = 0 \ ;$$

when $M_1 > 1$,

$$i_o = i_o(y),\ s = s(y),\ \nu = 0,\ u = u(y) \qquad (2.110)$$

when $x = 0$.

The condition $\nu = 0$ can be replaced with any other condition satisfying the properties of a quasi-developed flow, e.g., $\partial u/\partial x = 0$. Because when $M_1 > 1$ the flow at the inlet into the channel is completely determined, any three gasdynamic parameters (or their combinations) can be used instead of i_o, s, and u.

When the gas parameters at the source depend on the flow mode in the channel, various functional relationships between flow parameters can be used instead of conditions $i_o = i_o(y)$ and $s = s(y)$ at the channel inlet. For example, if the source models a combustion chamber with gas parameters that depend on the flow rate characteristics of the compressor and fuel nozzles, the boundary conditions at the inlet into the channel can be given in the form

when $M_1 < 1$,

$$i_o = i_o(y),\ p_o = p_o(i_o, m),\ \nu = 0 \quad \text{when } x = 0\ ;$$

when $M_1 > 1$,

$$i_o = i_o(y),\ p_o = p_o(i_o, u, m),\ u = u(y)\ ,$$

$$\nu = 0 \quad \text{when } x = 0\ , \qquad (2.111)$$

where m is the gas flow rate.

The boundary condition at the channel exit is given only when subsonic flow areas exist in the outlet section of the channel. For example, in the case of a subsonic flow into the surrounding medium at a fixed pressure p_∞, it can be assumed approximately that

$$p = p_\infty = \text{const. when } x = L\ . \qquad (2.112)$$

The boundary condition (2.112) is rigidly fulfilled if a quasi-developed flow exists between the active sector of the channel and its exit sector. In this case, any other equivalent condition can be utilized instead of (2.112) [37], e.g., $\partial u/\partial x = 0$, the fulfillment of which along the subsonic sector of the exit section may be required if supersonic zones appear at the channel exit. The no-flow condition,

$$\nu_n = 0 ,$$

is fulfilled at impenetrable walls.

The initial conditions are given in accordance with the chosen stationary flow mode. The boundary conditions for electrical parameters are determined by the channel wall construction, its loading mode and the electrical properties of the MHD facility subsystems adjacent to the channel at the inlet and exit. It is usually assumed that channel walls consist of insulators and electrodes. The no-flow of electrical condition applies at the insulators

$$j_n = 0 , \qquad (2.113)$$

while a relation between the current and potential that depends on the electrode design and the loading scheme of the channel is specified at the electrodes

$$F(j_n, \phi) = 0 . \qquad (2.114)$$

Examples of the simplest realization of relation (2.114) are two frequently encountered cases: ϕ = const. for continuous electrodes and j_n = const. for ideally segmented electrodes. Analogous relationships are given at the inlet to the channel and the exit from it. In particular, when the nozzle or diffuser adjacent to the channel are insulated from each other and are not grounded, the condition $j_n = 0$ is fulfilled in the end sections of the channel.

2.4.2. Supersonic MHD Flows. The system of equations (2.102)-(2.109) can be solved only approximately or numerically. Approximate methods based on linearization of equations (2.102)-(2.107) have been described in detail elsewhere [1]. It should be noted that linearization is a very effective method of investigating two-dimensional MHD flows. In particular, this method was used in the first investigation of the structure of the supersonic flow of an isotropically conducting gas at the inlet into the channel with continuous electrodes [38].

Numerical methods of solution are based on separation of the system of equations (2.102)-(2.109) into "hydrodynamic," (2.102)-(2.105), and "electrodynamic," (2.106), (2.107), subsystems that are solved separately. A complete solution is found within each subsystem by iteration. Gasdynamic methods [36, 39, 40] are used to solve the hydrodynamic problem, in which the field of electrodynamic variables is given for each iteration cycle. The electrodynamic problem, in which the field of hydrodynamic variables is given, is solved by means of various finite-difference techniques of solving elliptic equations [39].

Consider a flow in a plane MHD channel of constant height h. The channel is located in a magnetic field, B(x), specified by the relation

$$\frac{B(x)}{B_o} = \begin{cases} \exp \alpha(x-x')/h & \text{when } 0 \leq x < x', \\ 1 & \text{when } x' \leq x \leq x'', \\ \exp \alpha(x''-x)/h & \text{when } x'' < x \leq L. \end{cases} \qquad (2.115)$$

The electrodes in the channel are located within sectors $x' \leq x \leq x''$. The gas properties (2.105), (2.108), (2.109), are described by the following relations

$$\varepsilon = RT/(\gamma-1) ,$$

$$\sigma = \nu p^{-1/2} T^{3/4} \exp(-J/T) ,$$

$$\beta = \mu p^{-1} T^{1/2} B . \qquad (2.116)$$

At a fixed channel geometry and specified properties of the working fluid, the flow is characterized by the following set of dimensionless parameters: Mach number M_1; MHD interaction parameter

$$S = \frac{\sigma_1 B_o h}{\rho_1 u_1} ;$$

Hall parameter β_1; load factor expressed in terms of the current

$$I = \int_0^L j_y(x, 1/2h)\, dx/\sigma_1 u_1 B_o h$$

or voltage, $k = V/u_1 B_o h$; magnetic field decrement α, parameter J/T_1 and the adiabatic index γ. The subscript "1" denotes the quantities at the channel inlet (at $x = 0$).

The first numerical analysis of a supersonic flow in an MHD channel, subject to conditions (2.115), (2.116), was performed as reported in [41, 42]. The authors conducted a detailed investigation of the gasdynamics of the inlet and exit of a supersonic flow of an isotropically conducting gas in a nonuniform magnetic field in a channel with insulating walls. Flow parameter fields were obtained for channels with electrodes, making it possible to reach important conclusions. As a rule, a flow of isotropically conducting gas ($\beta = 0$) is sufficiently close to being one-dimensional. The largest transverse nonuniformities of flow parameters are observed at the exit sector of the channel, where an intense current swirl occurs in a discontinuous magnetic field ($\alpha = \infty$). A subsonic flow region with continuous supersonic-to-subsonic transition in this zone can be formed when the MHD interaction parameter is sufficiently large ($S \sim 1$).

Taking the Hall parameter into account exerts considerable influence on the flow structure. When $\beta \simeq 1$, the transverse electromagnetic force toward the cathode, as a result of the Hall current, leads to the appearance of a substantial pressure gradient at the channel walls. The axial asymmetry of the flow increases because of concentration of the current at the edges of the electrodes, caused by the Hall effect. Because, at the entrance to the channel, the current is concentrated at the edge of the anode, the gas in this zone stagnates more intensely than near the cathodes. As a result, the flow first deviates toward the cathode wall and then is sequentially reflected from the electrode walls. The pressure gradient at the walls is oscillatory along the channel axis.

An analysis of the integral characteristics of supersonic MHD generators, based on numerical solution of the problem (2.102)-(2.110), has been made [43]. Channels with both continuous and ideally segmented electrodes were

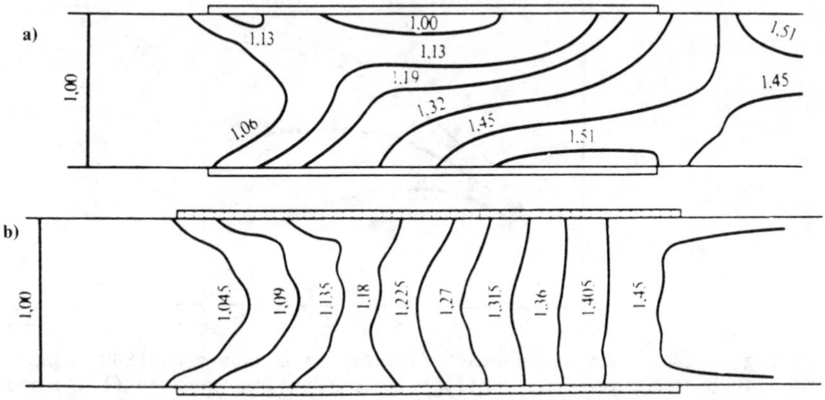

Fig. 2.23 Pressure distribution in a supersonic MHD channel with continuous electrodes (a) and with perfectly segmented electrodes (b): ($M_1 = 2.5$, $S = 0.1$, $\beta_1 = 1$)

considered. In the case of segmented electrode channels, specified current density, j_y = const., was used as the boundary condition. Noticeable differences in local parameter distributions were observed in these channels. The pressure distribution was calculated in a channel with continuous electrodes (Fig. 2.23a) and in an ideally segmented channel (Fig. 2.23b) [43] for the following conditions: $M_1 = 2.5$; $S = 0.1$; $\beta_1 = 1$; $I = 1.8$; $J/T_1 = 10$; $\alpha = 0$; $(x'' - x')/h = 3$; $\gamma = 1.14$.

Flow nonuniformities caused by current flow along the channel and electrical end effects are observed in a channel with continuous electrodes. The pressure distribution in a channel with ideally segmented electrodes is close to being one-dimensional. Insignificant transverse nonuniformities are caused by the end effect.

Analysis of distributions of other gasdynamic variables indicates that the most significant deviation from nonuniformities occurs in the case of pressure and density, and the velocity and enthalpy distributions are characterized by relatively small nonuniformities in the cross section of the channel. The latter fact forms the basis for using uniform velocity and enthalpy distributions in quasi-one-dimensional calculations of the inviscid core flow in the presence of transverse current in the channel [44].

The integral parameters of generators with continuous segmented electrodes are shown in Figure 2.24. The maximum power takeoff in generators of both types is achieved at exit Mach numbers close to one. Subsonic zones appear at the exit when the total current is $I = 3.2$

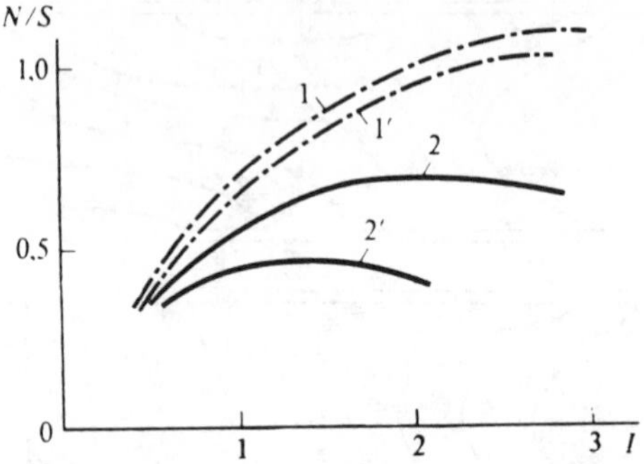

Fig. 2.24 Loading characteristics of a supersonic MHD channel with a constant cross section with perfectly segmented (1,1′) and continuous electrodes (2,2′), 1,2 - two-dimensional calculation, 1′,2′ - hydraulic approximation neglecting the end effects

for an ideally segmented channel and I = 2.8 for a channel with continuous electrodes.

The two-dimensional and quasi-one-dimensional calculations have been compared in [43]. In the quasi-one-dimensional model, it was assumed that the hydrodynamic variables of the canonical flow are uniform over the channel cross section. The electrodynamic terms in quasi-one-dimensional equations were calculated by two methods: (1) using one-dimensional Ohm′s law equations and neglecting edge effects, and (2) on the basis of solution of the two-dimensional equations (2.106), (2.107), with one-dimensional field of hydrodynamic variables. Figure 2.24 shows the loading characteristics of channels calculated in a rigid one-dimensional approximation (curves 1′ and 2′). The difference between the quasi-one-dimensional and two-dimensional calculations, resulting from the fact that the edge effects were neglected in the one-dimensional electrical model, is most significant for a channel with continuous electrodes. The use of two-dimensional distributions of electrodynamic variables increases considerably the accuracy of quasi-one-dimensional calculations (there is a difference of several percent between these and accurate two-dimensional calculation for characteristics averaged over the cross section and integral parameters).

Numerical analysis for different initial conditions has shown that satisfactory agreement between two-dimensional and quasi-one-dimensional calculations that take the edge effects into account is achieved when the following condition is fulfilled:

$$M^2 S_x \leq 1 \, , \tag{2.117}$$

where S_x is the MHD interaction parameter calculated from the current along the flow (e.g., for a channel with continuous electrodes, $S_x = \beta S/(1 + \beta^2)$). When condition (2.117) is fulfilled, the change in pressure in the channel cross section is relatively little. This is also verified by comparing terms of the transverse projection of the equations of motion (2.103)

$$\Delta_y p/p \lesssim M^2 S_x \ .$$

Under conditions when relation (2.117) is no longer valid, the flow structure is so strongly distorted by the transverse electromagnetic force that the quasi-one-dimensional flow model must be modified by introducing a transverse pressure gradient [44].

Transverse flow nonuniformities in the channel are caused not only by the effect of the transverse component of the electromagnetic force, but also by the initial disturbances of the flow at the channel inlet. One such possible disturbance is the flow slant at the channel inlet, most frequently resulting from the technology. An analysis of the influence of the initial slant of a supersonic flow on the structure of the flow and the output parameters of the MHD generator has been performed [45]. The investigation was conducted by numerically solving the problem formulated similarly to another study [43], at the following values of the governing parameters: $M_{1x} = 2.5$; $S_1 = 0.2$; $\beta_1 = 2$; $J/T_1 = 10$; $\gamma = 1.14$; $L/h = 8$; $(x'' - x')/h = 6$; $\alpha = 2$.

Boundary conditions at the channel inlet included uniform profiles of the axial component of velocity and of thermodynamic parameters, as well as the transverse velocity component that varies along the channel height,

$$v(0,y)/u_1 = v_o \cos(\pi y/h) \ .$$

Solution of this problem for a channel with continuous electrodes has shown that an increase of transverse nonuniformities and pressure pulsations at the channel walls is observed in the presence of an initial flow slant toward the anode ($v_o > 0$). This occurs because the increase in pressure near the anode wall at the inlet sector of the channel, caused by concentration of the current at the edge of the electrode because of the Hall effect, amplified by additional flow stagnation caused by an increase of the transverse flow velocity. An inverse effect is observed when the flow slant is toward the cathode. The force exerted on the flow, and the dynamic interaction with it are in antiphase, causing the flow in the channel to become considerably more uniform.

The change in the flow structure in the presence of a slant influences the output channel parameters. Figure 2.25 shows the dependence of power reduced to

$$\sigma_1 u_1^2 B_o^2 h^2$$

on the slant parameter ν_o (the tangent of the angle between the velocity vector and the channel axis, at the center of the inlet section at a constant load current, I = 0.9). The maximum power output is observed at negative values of ν_o, i.e., in the presence of an initial flow slant toward the cathode. Even though, in the inviscid approximation, the flow slant exerts only a weak influence on the integral characteristics of the MHD

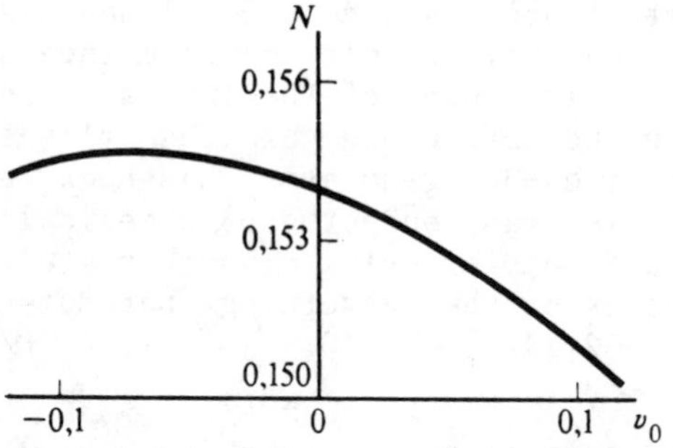

Fig. 2.25 Variation of the dimensionless power output of a supersonic MHD generator with the initial flow slant

channel, the change in flow structure it causes can exert substantial influence on the development of the viscous boundary layer at the channel walls and, as a result, on the output parameters of a generator. In particular, it can be expected that the initial slant, providing dynamic interaction with the flow opposed to that caused by the electromagnetic force interaction, will hinder boundary layer separation at the cathode wall.

2.4.3. Subsonic MHD Flows. In view of the fact that many MHD facilities, e.g., U-02 and U-25 (Institute of High Temperatures), operate in nominal subsonic modes, analysis of two-dimensional subsonic flows in an MHD channel is also of considerable practical interest. Most important is determination of the influence of two-dimensional effects on the flow structure and parameters of the MHD generator [46].

Consider a flow in a channel of constant height, h, located in a magnetic field (as shown in Figure 2.22). The channel electrodes located along a sector between cross sections x′ and x′′ are either continuous or ideally segmented. Therefore, in the first case, the operating mode of the generator is determined by the voltage between the electrodes (V_y) or the load current (I) and, in the second case, by parameter k that controls distribution of the normal component of current density in the form

$$j_y = -\frac{1-k}{h}\int_0^h \sigma u B \, dy \ . \tag{2.118}$$

The remaining boundary sectors of the region of calculation outside the electrodes are insulators and are governed by condition (2.113).

Consider flow modes in the channel when the flow is everywhere subsonic and the parameters given are the inlet stagnation enthalpy, i_{01}, gas flow rate, m, and the exit static pressure p_∞, i.e., when boundary conditions (2.111) and (2.112) are fulfilled at the channel inlet and outlet, in that order. The choice of such operating modes of the channel corresponds to control of the compressor systems of the MHD facility for a constant flow rate. In this case, the inlet channel pressure will depend on the degree of loading.

The (2.102)-(2.109) system of equations with the given boundary conditions were solved by the relaxation method, using intersystem iteration. Electrical parameter distributions are determined during each relaxation stage of the solution of the gasdynamic subsystem, defined to be $n\tau$ (τ is the time step and n is the number of steps). The gasdynamic equations are integrated in the same manner as reported by Isakov [47], using a scheme proposed by Godunov [48]. The electrodynamic equations are integrated applying the method of axial-transverse dispersion [49], using a modification [43]. The entropy, s_1, or equivalently, the channel inlet stagnation pressure, are chosen so that stable stationary flow satisfies the given flow rate, m, at fixed i_{01} and p_∞.

Dimensionless parameters were used in numerical integration of the system of equations (2.102)-(2.109). The scaling factors were as follows: linear dimension, $h_* = h$; velocity $u_* = ((\gamma - 1)\, i_{01}/\gamma)^{1/2}$; density, $\rho_* =$

m/u_*h_*; time, $t_* = h_*/u_*$; pressure, $p_* = \rho_*u_*^2$; gas enthalpy, $i_* = u_*^2$; temperature, $T_* = u^2/R$ (R is the gas constant); electric density, $j_* = \sigma_*u_*B_*$; electric field strength, $E_* = u_*B_*$. The characteristic value of B_* was determined by the maximum value of magnetic induction and σ_*, by relation (2.108), $\sigma_* = \sigma(p_*, \rho_*)$. The quantities given were chosen so that the flow in the channel under consideration would correspond to flow in large MHD generators: the height of the channel, h = 1 m; total specific enthalpy of the incident flow, $i_{01} = 6.6 \times 10^6$ j/kg; gas flow rate per one meter channel width, m = 100 kg/s; outlet pressure, $p_\infty = 0.1$ MPa; adiabatic index, $\gamma = 1.15$; relative molecular mass, $\mu_o = 27$; magnetic field strength in the electrode zone, $B_o = 2$ T; magnetic field logarithmic decrement, $\alpha = 2$; the electrode length, $x'' - x' = 3h$; total channel length, L = 7h. The constants, v, J and μ, in equations (2.116) were chosen using the experimental data obtained elsewhere [33]. If, in formula (2.116), $[\sigma]$ = S/m, [p] = MPa, [T] = K, then $v = 1.68 \times 10^2$, $J = 2.6 \times 10^4$, $\mu = 0.123 \times 10^{-2}$.

Solution of the problem under consideration was obtained for a channel with continuous electrodes at different load currents I. The MHD interaction parameter, S, calculated from the flow parameters in an average channel cross section was determined to be 0.4 and the Hall paramater, $\beta = 1.2$.

The distributions of electrical parameters reveal the characteristics noted above (see section 2.4.2.). The field of the gasdynamic variables is characterized by Mach number, velocity, and temperature distributions fairly uniform over the channel cross sections and substantially nonuniform pressure and density distributions at high load current. The transverse electromagnetic force, $F_y - j_xB$, leads to the appearance of pressure differences at the electrode walls, δp. The variation of δp along the channel length is not characterized by pulsations observed in a supersonic flow. The pressure gradient at the channel walls, δp, increases smoothly from zero at the inlet, reaches a maximum at the electrode sector, and decreases to zero at the channel exit.

Increasing the load and the Hall parameter leads to an increase in δp. If, at I = 0.5, δp at the center of the electrode zone is 7 % of the outlet pressure (p_∞), then, at I = 0.75, δp = 10 % of p_∞. A comparison with calculations for $\beta = 0.6$ has shown that, in the short-circuit operating mode, δp at $\beta = 1.2$ is two times as large as at $\beta = 0.6$. This effect can be important in improving engineering calculation methods [44] based on quasi-one-dimensional flow models, because the conductivity nonuniformity resulting from the transverse pressure gradient has an effect on the internal resistance of the generator.

The joint influences of forces and joule dissipation on the subsonic flow leads to flow acceleration and a corresponding decrease in pressure along the channel. The pressure gradient in the channel increases with increasing load current, I. In view of the foregoing, the channel inlet pressure increases with increasing electrical load, provided the outlet flow rate and pressure remain constant. The other parameters vary in a similar manner (Fig. 2.26).

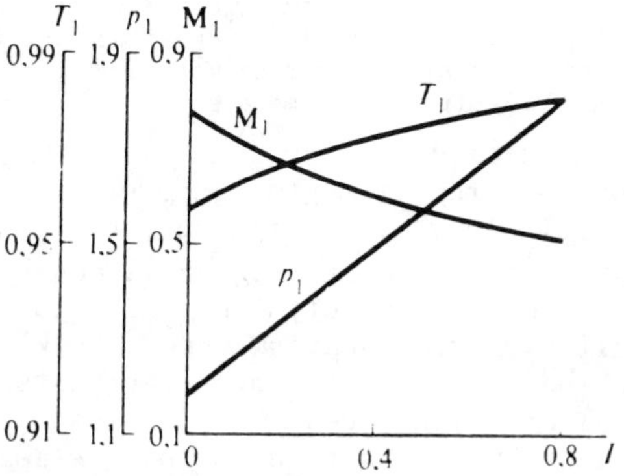

Fig. 2.26 Variation of the inlet parameters of the flow with the load current in a subsonic MHD channel with continuous electrodes for $\beta_* = 1.2$

The voltage current characteristics of the channel (Fig. 2.27) are close to being linear dependences. However, the internal resistance changes significantly with increasing I, because of characteristics of the flow. The flow stagnation is accompanied by an increase in the channel inlet pressure and temperature (see Fig. 2.26), as well as an increase in the average pressure and temperature along the flow region. On one hand, this leads to a decrease in the local values of Hall parameter and, on the other, to an increase in conductivity such that the effective conductivity increases. It should be noted that transverse nonuniformities in the flow resulting from the transverse electromagnetic force, F_y, appear at large values of I. Transverse nonuniformities should cause a decrease in the effective electrical conductivity of plasma. However, their influence is relatively small and variation of the internal resistance of the channel with I depends primarily on restructuring

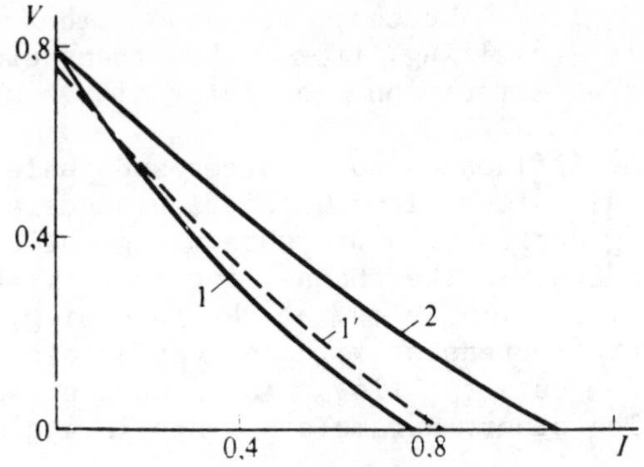

Fig. 2.27 Voltage-current characteristics of a subsonic MHD channel with continuous electrodes at different Hall parameter values: 1,1′ - 1.2; 2 - 0.6; 1,2 - two-dimensional computation; 1′ - quasi-one-dimensional approximation for two-dimensional distributions of electrical parameters

of the flow in axial direction.

Variation of the integral power output, N, and electrical efficiency, η, with current extracted from electrodes shows that increasing β leads not only to a decrease in the power output, but also to a decrease of electrical efficiency. What's more, peak efficiency is shifted into the region of lower currents (Fig. 2.28). The electrical efficiency near the no-load operating mode is characterized by a sharp drop that is associated with joule losses caused by eddy currents.

Analysis of the solution of the problem for a channel with segmented electrodes at which condition (2.118) is fulfilled, indicates that there is hardly any Hall current in the channel, and that the distributions of all gasdynamic parameters over the cross section of the channel are close to being uniform. In this case, the one-dimensional flow model that takes into account electrical edge effects describes the flow with an accuracy of up to 1 %. From Fig. 2.28, it follows that the power output and efficiency of a channel with segmented electrodes exceeds considerably the efficiency and power output of a channel with continuous electrodes. An increase in efficiency of a channel with segmented electrodes is primarily the result of an increase in effective electrical conductivity of plasma. The same effect is also observed in a channel with continuous electrodes, when the Hall parameter is decreased (see Fig. 2.28).

The influence of the two-dimensional nature of the flow gasdynamics on the output parameters of the generator can be traced by comparing solutions obtained from the

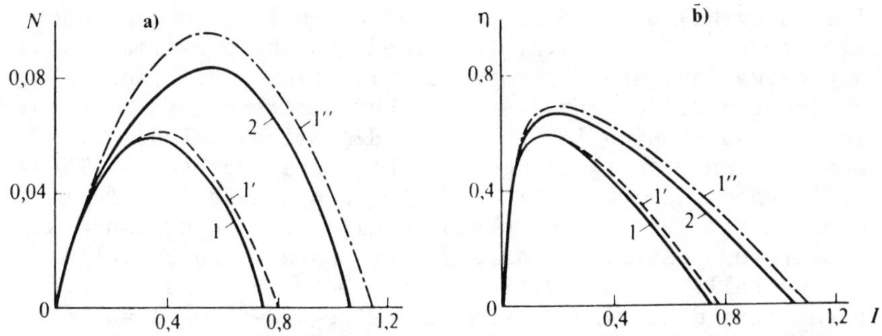

Fig. 2.28 Load characteristics (a) and variation of the electrical efficiency (b) with current for MHD channels with continuous electrodes (1,1′,2) and with ideally segmented electrodes (1′′): 1,1′,1′′ - $\beta_* = 1.2$; 2 - $\beta_* = 0.6$; 1,1′′,2 - two-dimensional computations; 1′ - quasi-one-dimensional approximation for two-dimensional distributions of electrical parameters

two-dimensional flow model with those from one-dimensional model. Consider, for example, one-dimensional flow in a channel on the assumption that

$$u = u(x),\ \nu = 0,\ p = p(x),\ \rho = \rho(x)\ ,$$

$$(\mathbf{j} \times \mathbf{B})_x = \langle j_y \rangle_y B(x),\ (\mathbf{j} \times \mathbf{B})_y = 0\ ,$$

$$(\mathbf{jE}\rangle = \langle \mathbf{jE} \rangle_y\ . \tag{2.119}$$

The corresponding system of one-dimensional gasdynamic equations describing the flow was obtained from (2.102)-(2.105), taking (2.119) into account. The values of $\langle j_y \rangle_y$ and $\langle \mathbf{jE} \rangle_y$ are calculated by averaging $\mathbf{j}(x,y)$ and $\mathbf{E}(x,y)$ distributions over the channel cross section obtained by integrating the two-dimensional electrodynamic equations (2.106)-(2.109). This one-dimensional model takes into account nonuniformities of electrical parameters associated with the end effects and conductivity anisotropy.

A complete solution of the one-dimensional problem was obtained by iteration, with the two-dimensional electrodynamical problem solved, during each step, for a fixed one-dimensional distribution of gasdynamic parameters. The quantities $\langle j_y \rangle_y$ and $\langle \mathbf{jE} \rangle_y$ were determined by averaging and the one-dimensional gasdynamic equations were then solved for constant values of $\langle j_y \rangle_y B$ and $\langle \mathbf{jE} \rangle_y$.

The integral parameters for a channel with continuous electrodes at $\beta = 1.2$, obtained in the one-dimensional approximation, are represented in Figs. 2.27 and 2.28. Noticeable differences between parameters determined from the two-dimensional and one-dimensional models are observed only in the range of high load currents. These differences can be attributed to a decrease in effective conductivity because of nonuniformity of the pressure and temperature distributions over the channel cross section.

Overall, the differences between the integral parameters determined from the two-dimensional and one-dimensional gasdynamic models are small. From this it can be concluded that, for subsonic flows at Hall parameters on the order of unity, the influence of transverse flow nonuniformities on the parameters of the MHD channel is insignificant. This conclusion is in agreement with the foregoing estimate (2.117).

2.5. Quasi-One-Dimensional (Hydraulic) Approximation and Engineering Methods of Calculating Flow Parameters in an MHD Generator Channel

The quasi-one-dimensional (hydraulic) approximation forms the foundation of most of the present-day engineering methods of calculating flow parameters in various types of MHD generator channels [1, 13, 34, 35, 44, 50-53]. This can be attributed to those requirements that have to be satisfied by engineering methods and to possibilities presented by quasi-one-dimensional calculations.

The principal aim of the engineering calculations is to obtain quantitative data for the design of facilities and, also, for analysis of parameters of the existing experimental, pilot, and commercial-scale MHD installations. In this connection, engineering methods must describe with necessary completeness and accuracy the integral parameters of an MHD generator channel (the important integral parameters are the total power output of the MHD generator, efficiency, total pressure loss in the channel, total heat losses, transverse potential difference, etc.). Other requirements imposed on engineering methods are acceptably accurate calculations of flow parameter distributions along the channel length, such as the static pressure, heat losses into the wall, working current density, electric field strength, and simplicity of calculations, which is necessary in engineering practice.

To satisfy these requirements, any method should to a considerable extent take into account the characteristic properties of the class of flows under consideration. For flow conditions in a linear MHD generator channel with working fluid consisting of combustion products of organic fuels seeded with an alkali metal, the principal dimensionless parameters fall within the following ranges: Reynolds number, Re = 10^6-10^7; magnetic Reynolds number Re_m << 1; Mach number, M $\lesssim$ 0.3-2.5; MHD interaction parameter, S = 0.1-10; Hall parameter, $\beta \lesssim$ 0.3-5. In addition, one should note the channel geometry, specifically the large ratio of the channel length, L, to the characteristic transverse dimensions, 2a (L/2a $\simeq$ 10 >> 1) and a relatively weak variation of the cross-sectional area.

Under such conditions, the flow structure in the MHD generator channel is characterized by sharp transverse velocity, temperature and electrical field nonuniformities (primarily because of the viscous and thermal boundary layers) and a relatively weak variation of these nonuniformities along the channel length. This makes it possible to describe adquately the flow parameter distributions over the cross section, with the aid of a limited number of certain integral quantities that depend only on the transverse coordinate. In this sense, the flow along the axial coordinate is "quasi-one-dimensional."

Considerable nonuniformity of gasdynamic parameters outside the boundary layers may arise in the end regions of the channel, if the local MHD interaction parameter becomes large, or in the regular part of the channel at high values of the transverse component of current. As will be shown, all of these characteristics can be taken into account, to a varying degree, within the framework of quasi-one-dimensional approximation.

2.5.1. The Basic Equations of the Quasi-One-Dimensional Approximation. Construction of a system of equations in a quasi-one-dimensional approximation requires introduction of a certain fictitious canonical flow into the analysis that, to a necessary degree, maintains some properties of the real flow that are important from the point of view of a specific formulation of a problem. Consider one possible approach to construction of such a system of equations [1, 34, 44].

The basis of any quasi-one-dimensional system of equations is formed by conservation laws for the mass, momentum, and energy flows. For the class of flows under consideration, the most natural way to express these laws is in terms of a system of ordinary differential equations with the coordinate along the channel axis as the independent variable, and the mass, momentum, and energy flows as functions being sought. Visualize a steady-state flow of electrically conducting gas in a rectangular channel of a linear MHD generator with impenetrable walls, the length, L, of which is much longer than the characteristic transverse dimension 2a. In this case, the system of principal quasi-one-dimensional equations can be written in the form [1][2]

$$\frac{d}{dx}(A\rho U\alpha_1) = 0\ ,$$

$$\frac{1}{A}\frac{d}{dx}(A(\rho U^2\alpha_2 + p\alpha_4)) =$$

$$= F_{wx} + F_x - \frac{1}{A}\oint_{\ell} p(\ell)\mathrm{ctg}(\mathbf{n},\ \mathbf{e}_x)d\ell\ ,$$

$$\frac{1}{A}\frac{d}{dx}(A\rho U i_o\alpha_3) = -\ Q_w - N_V\ ,$$

$$\frac{d}{dx} I_x = -\oint_{\ell} j_n(\ell)\frac{d\ell}{\sin(\mathbf{n},\mathbf{e}_x)}\ ,$$

$$\frac{d}{dx}\Phi = -\ E_x\ . \qquad (2.120)$$

where

$$F_{wx} = -\frac{1}{A}\oint_{\ell}\tau_w d\ell,\ F_x = \frac{1}{A}\int_A (\mathbf{j}\mathrm{x}\mathbf{B})_x dA\ ,$$

$$Q_w = \frac{1}{A}\oint_{\ell} q_w d\ell,\ N_V = -\frac{1}{A}\int_A (\mathbf{jE})dA\ ,$$

$$I_x = \int_A j_x dA\ ,$$

$$\alpha_n = \frac{1}{A}\int_A \frac{\rho(x,y,z)u^n(x,y,z)}{\rho U^n}\, dA\ (n = 1.2)\ ,$$

2 For a wide class of flows (2.120) the system of equations is exact. Conditions that specify when this is not the case are described in detail in [1].

$$\alpha_3 = \frac{1}{A}\int_A \frac{\rho(x,y,z)u(x,y,z)i_\circ(x,y,z)}{\rho U_{\ell\circ}} dA ,$$

$$\alpha_4 = \frac{1}{A}\int_A \frac{p(x,y,z)}{p} dA , \qquad (2.121)$$

where $i_\circ = i + |u|^2/2$, F_{wx} and F_x are components along the channel of specific (per unit volume) friction and electromagnetic force, Q_w is heat losses to the walls, N_v is the electric power output, ℓ is the coordinate of the circumference of the cross section; U, p, $i_\circ$, and ρ are the velocity, pressure stagnation enthalpy and density of the working fluid along the axis of the channel; I_x is the total current along the channel, Φ is the potential of the channel axis relative to the ground, E_x is the axial (Hall) component of the electric field.

The thermodynamic properties of the working fluid are assumed to be known:

$$\rho_\ell = \rho(i_\ell, p_\ell), \; T_\ell = T(i_\ell, p_\ell) , \qquad (2.122)$$

where $\rho\ell$, $i\ell$, $p\ell$, $T\ell$ are the local parameters of the working fluid.

Under the assumptions made, the system of equations (2.120) is exact. However, in the general case it cannot be integrated directly because of its incompleteness. In addition to equations (2.121) and (2.122), closure of this system of equations requires knowledge of the relationship between the force and energy interaction of the electromagnetic field and channel walls with the mass, momentum, and energy flows in terms of which the system of equations (2.120) is written. An accurate representation of such relationships would imply that an exact solution of the complete spatial problem can be obtained and, in view of the insurmountable mathematical difficulties, is at the present unrealistic.

Introduction of approximate relationships to achieve closure of the (2.120) system of equations determines the structure of the canonical flow that models the real flow. From a practical point of view, to investigate some properties of the flow in the MHD generator within the framework of the hydraulic approximation, it is necessary to choose a combination of closure relations that would, on one hand, provide an adequate description of the characteristics of a real flow within the framework of the canonical flow introduced by these relationships and, on the other hand, considerably simplify mathematical calculations. At the present time, a system of ordinary differential equations represents the acceptable level of complexity of mathematical apparatus for lengthy calculations typical of engineering problems. The principal aim of engineering calculations is to determine the integral characteristics of an MHD generator channel (total power output, efficiency, pressure drop in the channel, loading scheme parameters, et al.), and also distributions along the channel of gasdynamic and electrical parameters of the flow (static pressure, heat flows to the wall, electric field, current density, et al.).

Hence, success in applying the quasi-one-dimensional approximation is determined in each specific case by the degree to which the integral parameters and real flow parameter distributions of interest to specific analysis are maintained upon transformation to canonical flow. In this connection, the structure of the canonical flow must reflect the principal phenomena determining the flow parameters.

2.5.2. Gasdynamic Structure of the Model. At high Reynolds numbers, friction and heat transfer processes in the channel must be taken into account only in thermal and dynamic boundary layers on channel walls. In a general case, the thicknesses of the dynamic and thermal boundary layers differ both along the same wall and along different walls. Starting with these concepts, it is possible to introduce a canonical flow that is nonuniform with respect to the hydrodynamic parameters. The structure of this canonical flow corresponds to an inviscid and thermally nonconducting core flow (with uniform stagnation enthalpy and axial velocity distributions) and boundary layers on channel walls. In this case, the axial velocity and stagnation enthalpy distributions in the channel cross section can be described by the following profile functions:

$$\psi_u = u(x,y,z)/U = \eta_u \xi_u \ ,$$

$$\psi_i = i_o(x,y,z)/i_o = \eta_i \xi_i + (1+\eta_i \xi_i)(i_{wy}-i_{wz})/2i_o +$$

$$+ (\xi_i - \eta_i)(i_{wy}-i_{wz})/2i_o \ , \qquad (2.123)$$

where i_{wy}, i_{wz} are the enthalpy at the temperature of the electrode and insulating walls (it is assumed that the temperature of opposite walls is the same). In relations (2.123), the profile functions of axial velocity and stagnation enthalpy are approximately represented as bilinear forms of dimensionless profiles η_u (x,y), η_i (x,y), ξ_u (x,z), ξ_i (x,z) along the OY and OZ axes and the cross section of the channel.

These profiles will be represented here by power functions that provide an adequate approximation to velocity and stagnation enthalpy distributions across the boundary layer, satisfactory for engineering applications. Thus, in the general case, assuming that the boundary layer parameters and the exponents in the power function approximation for distributions at different channel walls are different, let's introduce an axial velocity profile along the OY axis in one of many possible forms:

$$\eta_u(x,y) = \begin{cases} ((a-y)/\delta^+_{uy})^{1/n^+_{uy}} & \text{when } y \geq a-\delta^+_{uy}, \\ 1 \quad \text{when} \quad a-\delta^+_{uy} > y > -a+\delta^-_{uy}, & \\ ((a+y)/\delta^-_{uy})^{1/n^-_{uy}} & \text{when } y \leq -a+\delta^-_{uy} \end{cases} \tag{2.124}$$

and, in a completely analogous manner, profiles ξ_u (x,z), η_i (x,y), ξ_i (x,z).

Within the framework of the model of the boundary layer of finite thickness, the quantities δ^+_{uy}, δ^-_{uy}, δ^+_{iy} (see Fig. 2.29) in relations (2.124) represent dimensions of regions with sharp manifestation of viscosity and thermal conductivity associated with the thickness of the boundary layers. One way to determine the variation of thickness of the boundary layers along the channel is by means of integral equations for the boundary layer that represent the same conservation laws of mass, momentum, and energy flows written for the boundary layer [54]. In particular, equations for the dynamic and thermal boundary layer's thicknesses at one of the channel walls can be written in the form

$$\frac{d}{dx}(\rho U^2 \delta^{**}_u) + \rho U \delta^*_u \frac{dU}{dx} = \tau_w + \int_0^{\delta_u} (F_x(\delta_u) - F_x(y))dy , \tag{2.125}$$

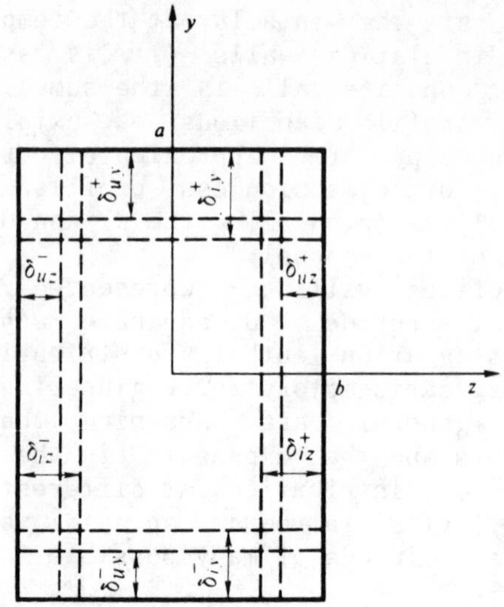

Fig. 2.29 Structure of the flow in the channel cross section

$$\frac{d}{dx}\left(\rho U \Delta i\, \delta_i^{**}\right) + \rho U \frac{di_o}{dx} \int_o^{\delta_i} \left(\frac{\mathbf{jE}}{(\mathbf{jE})_{y=\delta_i}} - \psi_u \frac{\rho(y)}{\rho(\delta_i)} \right) dy = q_w , \qquad (2.126)$$

where

$$\delta_u^* = \int_o^{\delta_u} \left(1 - \frac{\rho(y)}{\rho(\delta_u)}\, \psi_u\right) dy ;$$

$$\delta_u^{**} = \int_o^{\delta_u} \psi_u \frac{\rho(y)}{\rho(\delta_u)} (1-\psi_u)\, dy ;$$

$$\delta_i^{**} = \int_o^{\delta_i} \psi_u \frac{\rho(y)}{\rho(\delta_u)} \frac{i_o(\delta_i) - i_o(y)}{\Delta i}\, dy ;$$

$$\Delta i = i_o(\delta_i) - i_w; \quad F_x = (\mathbf{j} \mathrm{x} \mathbf{B})_x . \qquad (2.127)$$

For simplicity, the wall angle of the channel in relations (2.125)-(2.127) is zero. At known dependences of the frictional stress, τ_w, on the wall and heat flux density into the wall, q_w, as well as the electrodynamic terms in equations (2.125) and (2.126), on the predominant flow parameters, the system of equations (2.123)-(2.127) completely determines the gasdynamic structure of the canonical flow, i.e., it is then possible to calculate profile coefficients (2.121) in each channel cross section, required to express the mass, momentum, and energy laws in terms of the axial velocity, static pressure, and total enthalpy.

When the axial gradients of gasdynamic parameters in the core flow are relatively small, the size of the near-wall nonuniformity (boundary layer thickness) can be calculated from equations (2.125), (2.126), that correspond to a nongradient flow of incompressible fluid past a rectangular plate [44]. In this case

$$\delta_u = \text{const.} \times Re_x^{-0.2}, \quad \delta_i = \text{const.}\ \delta_u Pr^{-2/3}. \tag{2.128}$$

It was taken into account that $n_u = n_i = 7$.

In conclusion, it should be noted that, in calculating internal resistance of the channel, which to a considerable degree depends on the conductivity (temperature) distribution over the cross section, it is useful to modify profiles (2.124) by introducing intervals corresponding to linear distributions of velocity and total enthalpy in the laminar layer (see, for example, (2.101)).

<u>2.5.3. Calculation of Friction and Heat Transfer.</u> Calculation of quantities describing the influence of friction and heat transfer on the flow both in the system of equations (2.120) and in integral equations for the boundary layer will now be discussed. It should be noted that a close relationship exists between the distributions of parameters across the boundary layer and the heat flux and frictional stresses at the channel wall. This relationship depends on the physical meaning of friction and heat transfer processes during the flow of gas stream past a surface. Consequently, any approximations of distributions of parameters across the boundary layer and stresses resulting from friction and heat flux at the wall cannot be introduced independently. In particular, power approximation for the velocity and enthalpy profiles in the boundary layer is borrowed from the so-called two-

layer model of a turbulent boundary layer [54]. In this model, the region of tangential stresses near the surface affected by the gas flow is separated into two subregions. One of these, in direct contact with the wall, is the laminar sublayer zone with a linear distribution of axial velocity, and the other a turbulent core of a boundary layer with a velocity distribution in the form of a power function. The matching condition for the sublayers actually determines the axial velocity gradient at the surface and, at the same time, frictional stresses proportional to this gradient.

The structure of the thermal boundary layer is introduced on the basis of the model accepted for the dynamic layer, within the Reynolds analogy for frictional and heat transfer processes in turbulent layers. The power approximation and the expressions for the coefficients of friction and heat transfer following from it on the basis of a two-layer model of the boundary layer adequately describe the experimental data throughout a wide range of values of the governing parameters for a nongradient flow of an incompressible fluid past a flat plate, in the absence of MHD interaction. In particular, for Reynolds numbers most characteristic of flow conditions in an MHD channel ($Re \simeq 10^7$), the coefficient of friction and heat flux into the wall are given by the expressions:

$$c_f^o = 2\tau_w^o/\rho U^2 = 0.0576 Re_x^{-0.2},$$

$$St^o = q_w^o/\rho U(i_o - i_w) = 1/2 c_f^o Pr^{-0.75}. \qquad (2.129)$$

However, surface roughness, compressibility, and nonisothermal behavior of the boundary layer and transverse velocity gradients in the external flow can substantially change the values of the coefficients of friction and the dimensionless heat flux into the wall. Many methods have been developed to take into account the influence of the above-noted characteristics on friction and heat transfer processes in the turbulent boundary layer. The most important aspect of one of these methods is the introduction of relative corrections to the simplest relationships established for a nongradient, incompressible flow past a rectangular plate [54]. Local value of the momentum thickness loss, δ_u^{**} (or energy thickness loss δ_i^{**}), is used as a linear dimension. This partially takes into account the influence of previous history of the flow on the local values of friction and heat flux. Broad experimental verification of the relationships developed for pure gasdynamic flows indicates that the expressions for the coefficient of friction and the dimensionless heat flux obtained in this manner are sufficiently accurate. On this basis, it will be assumed that, disregarding the influence of the electromagnetic field, the expressions for the coefficient of friction and heat transfer in an MHD generator channel can be written in the form

$$c_f = \Psi c_f^o = \Psi \cdot 0.0252(\mathrm{Re}^{**})^{-0.25},$$

$$\mathrm{St} = \Psi_S \mathrm{St}^o = \Psi_S \cdot 0.0126)\mathrm{Re}_i^{**})^{-0.25} \mathrm{Pr}^{-0.75}, \tag{2.130}$$

where Ψ and Ψ_s are corrections compensating for compressibility, nonisothermal behavior, and velocity gradient in the external flow [54]. The local values of the Reynolds number in (2.130), calculated from the momentum thickness losses and total energy thickness losses, may be calculated using integral equations for momentum (2.125) and energy (2.126) for the dynamic and thermal boundary layers. It will be assumed that the velocity and enthalpy distributions are described by power functions such as (2.124).

Consider, qualitatively, the possible influence of MHD interaction on the friction and heat transfer coefficients in an MHD generator channel. Note that electrodynamic effects in the boundary layer depend considerably on the direction of the wall relative to the external magnetic field vector. Assuming ideal electrode segmentation, the current density component normal to the electrode wall that is parallel to the magnetic field may be, with sufficient accuracy, assumed to be constant across the boundary layer. Then, the second term in the right side of (2.125) approaches zero, eliminating direct influence of the electromagnetic field on the dynamic boundary layer parameters. At the same time, joule dissipation in highly-cooled, near-wall regions of the boundary layer of low electrical conductivity (under typical conditions in MHD generators, the difference between electrical conductivity near the wall and that at the external boundary of the boundary layer may be two or more orders of magnitude) leads to a considerable heat release inside the boundary layer.

Complete heat release through joule dissipation may be comparable with convective heat flux to the wall. Consequently, one can expect considerable influence of the current flow through the boundary layer at the electrode wall on the temperature distribution in the boundary layer and density of heat flux into the wall (in the frame channel, analogous conditions may arise even on the side wall extending across the external magnetic field). As can be seen from equation (2.126), the thickness of the boundary layer depends considerably on the heat release in this layer. Therefore, utilization of the integral equation for the boundary layer energy (2.126), assuming validity of power function representation for the enthalpy distribution makes it possible to take into account the effect described above qualitatively. Nevertheless, the problem of taking this effect into account qualitatively within the framework of the boundary layer model described above requires meticulous experimental and theoretical verification.

In a certain sense, the reverse situation exists on the side wall that is normal to the external magnetic field. The principal component of the current density is directed parallel to the wall. Because the corresponding electric field component hardly varies across the boundary layer, current density depends mostly on the electric conductivity and velocity and, consequently, changes rapidly away from the boundary of the boundary layer toward the wall. This leads to a sharp nonuniformity of the current density (in particular, the component of current density under consideration may even change sign) and, therefore, to nonuniformity of the ponderomotive force. In this connection, the second term in the right side of equation (2.125) may reach values comparable with those of terms on the left side of equation (2.125), and thus play an important part in determining the thickness of the dynamic boundary layer and frictional stress. The influence of one of the terms in the energy equation (2.126) on the integral characteristics of the boundary layer on the side wall will apparently be considerably smaller than on the electrode wall. Similar to the previous case of the electrode wall, it can be assumed that, within the framework of integral relations of the boundary layer, it is possible to take these effects into account, at least quantitatively. Improvement of integral characteristics of the boundary layer when using equations (2.125), (2.126), requires introduction of profile functions of velocity and entropy more exact than power functions, as well as an improvement of dependences of St and c_f on the governing parameters of the flow. This information may be obtained from, among other things, analysis of exact solutions of two-dimensional equations of the boundary layer.

With these comments in mind, introduce additional compensation factors k_ξ^0 and k_α^0 into the expressions for the coefficient of friction and the dimensionless heat flux. These factors should take into account the influence of the electrodynamic effects on friction and heat transfer. Hence, assume that

$$c_f = k_\xi^0 \Psi c_f^0 = k_\xi c_f^0 , \tag{2.131}$$

$$St = k_\alpha^0 \Psi_S St^0 = k_\alpha St^0 . \tag{2.132}$$

A method of determining correction factors is by processing the experimental data or by obtaining the exact solution of the boundary layer equations.

2.5.4. Electrodynamic Flow Model for an MHD Generator Channel. A quasi-one-dimensional model of electrodynamics of an MHD generator channel must provide acceptable accuracy in calculating the integral electrodynamic characteristics of the channel; loading circuit parameters, the amount of heat scattered by the nonideal channel insulation, the force and energy interaction of the electromagnetic field with the working fluid flow, and the distribution of current density and the electric field along the channel length.

The distributions of electrodynamic parameters in the active volume of the channel are determined by the boundary conditions at the walls and in the inlet and exit channel sections, external conditions and distributions of electrophysical properties, and gasdynamic parameters of the working fluid in the channel. Because the channel electrodynamics is closely associated with the gasdynamic parameters, the electrodynamic model must rely on the gasdynamic flow model introduced above.

It is assumed that the current and potential distributions are defined by equations (2.1) and (2.2), that the external magnetic field is known (only one component of the magnetic field vector, $\mathbf{B} = \{0, 0, B(x)\}$, will be considered).

Constructing the model and obtaining the required integral parameters will follow the approach developed in [55, 34, 35], which is based on a compromise between limitations of the assumptions made and the possibility of obtaining fairly simple expressions for the quantities sought by successively averaging equations (2.1) and (2.2) across the channel cross sections.

First, consider a linear, ideally segmented MHD generator channel with a rectangular cross section. Assume that the required gasdynamic flow parameter distributions are known and have central symmetry. The nonideal insulation of the channel walls in transverse and axial directions will be modeled by an infinitely thin surface layer ($\delta_\pi \rightarrow 0$) of surface conductivity σ_π with the relationship between the electric field and the current density in this layer given by Ohm's law

$$\mathbf{j} = \hat{\sigma}_\pi \mathbf{E}_w \,, \qquad (2.133)$$

where

$$\hat{\sigma}_\pi = \begin{Vmatrix} \sigma_{\pi x} & 0 & 0 \\ 0 & \sigma_{\pi y} & 0 \\ 0 & 0 & \sigma_{\pi z} \end{Vmatrix}$$

is the antisotropic conductivity in the layer and where the Hall effect is assumed to be absent.

The whole conducting region of the channel cross section, i.e., the flow region and the layer of surface conductivity adjacent to it is considered when averaging equations (2.1) and (2.2). The dimensions of this expanded region shall be denoted by an apostrophe (2a′, 2b′, A′ = 2a′b′). It is assumed that, outside the surface conductivity layers, the boundary condition specifying that the normal component of the current density is equal to zero applies at the side insulating walls, the boundary condition specifying that the tangential component of the electric field in the plane of the channel is equal to zero applies at the electrode wall, i.e., that $j_n = 0$ on the insulator and $E_\tau = 0$ on the electrode.

The external load, r_H, is connected to each pair of opposed electrodes and is characterized by specific (per unit channel length in axial direction) load current, i_H. In addition, leakage resistance, r, from the electrode to the ground is included in the load circuit. The channel geometry and the external electrical circuits for the case under consideration are shown in Fig. 2.30. Certain general relations will now be obtained.

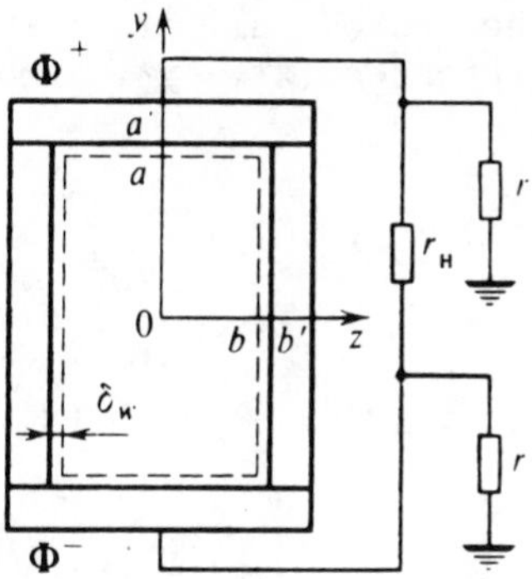

Fig. 2.30 Cross section of the channel and the external electrical circuit

Selecting the surface bounding the flat layer of the channel volume located between two arbitrarily close channel cross sections to be the closed integrating surface in equations (2.1), (2.2), gives

$$\frac{dI'_x}{dx} = \frac{d}{dx}\int_{F'} j_x dA = -\,i^+ + i^- ,$$

where

$$i^{\pm} = i_y^{\pm} \mp i_x^{\pm}\,\frac{da}{dx}\ ;\ i_y(y) = \int_{-b'}^{b'} j_y dz\ ;$$

$$i_x(y) = \int_{-b'}^{b'} j_x dz;\quad i_y^{\pm} = i_y(\pm a');\ i_x^{\pm} = i_x(\pm a')\ .$$

The loading conditions under consideration (see Fig. 2.30) can be written

$$i^{\pm} = i_H \pm \Phi^{\pm}/r$$

and in the final form,

$$\frac{dI'_x}{dx} = -\,\frac{\Phi^+ + \Phi^-}{r}\ .$$

Electrical power taken off per unit length of a channel may be determined in a similar manner:

$$N'_L = \frac{d}{dx}\int_{A'} j_x \phi dA + V_y i_H + \frac{(\Phi^+)^2 + (\Phi^-)^2}{r}\ ,$$

where $V_y = \Phi^+ - \Phi^-$ is the electrode voltage and Φ^+ and Φ^- are electrode potentials relative to the ground.

Note that the potential at a certain point of a channel with coordinates (x^*, y^*, z^*) can be represented in the form

$$\phi(x^*,y^*,z^*) = \phi(0,0,0) - \int_0^{x^*} E_x(x,0,0)dx$$

$$- \int_0^{y^*} E_y(x^*,y,0)dy - \int_0^{z^*} E_z(x^*,y^*,z)dz ,$$

and, in particular,

$$\Phi^{\pm} = \Phi(x) - \int_0^{\pm a'} E_y(x,y,0)dy$$

$$= \Phi(x) - \int_0^{\pm a} E_y(x,y,0)dy \pm \Delta V^{\pm} ,$$

where

$$\Phi(x) = \phi(0,0,0) - \int_0^{x} E_x(x,0,0)dx .$$

Sometimes it is more convenient to use the last equation written in the following form

$$\frac{d\Phi(x)}{dx} = - E_x(x,0,0) \equiv - E_x .$$

The transverse component of the ponderomotive force is given by the expression

$$AF_x = \int_A (\mathbf{j}\mathrm{x}\mathbf{B})_x dA = \int_A j_y B \, dA ,$$

and when $B = B(x)$

$$AF_x = B \int_{-a}^{a} (i_y(y) - 2\sigma_{wy} E_y(y,b'))dy ,$$

where, here and below, $\sigma_{wi} = \sigma_{\pi i}\delta_\pi$, and for simplicity, it is assumed that $\sigma_{wy}(b') = \sigma_{wy}(-b^-)$. Thus, the current and potential distributions (or the electric field) over the channel cross section have to be known in order to calculate the integral parameters of the flow.

By analogy with the previous procedure of constructing a gasdynamic flow model, a set of profile functions is now introduced:

$$\varepsilon_x = E_x(x,y,z)/E_x(x,0,0) \ ,$$

$$\varepsilon_y = E_y(x,y,z)/E_y(x,y,0) \ ,$$
$$\nu_y = i_y(x,y)/i_y(x,0) \ .$$

Then, after averaging components of equations (2.1), (2.2) taking equation (2.133) into account (the sequence of averaging is described in detail in [34]),

$$\frac{1}{A} I'_x = E_x\langle\sigma_x\rangle + UB\langle\delta\psi_u\rangle - i_y\langle\nu\beta\rangle/2b,$$

$$- V_y/2a = i_y r_i/2a + UB\langle\sigma\psi_u\rangle - E_x\langle\beta_x\rangle \ ,$$

$$E_y(x,y) = \left(i_y \nu_y + UB \int_{-b'}^{b'} \frac{\sigma\psi_u dz}{1+\beta^2}\right.$$

$$\left. - E_x \int_{-b'}^{b'} \frac{\beta\sigma\varepsilon_x dz}{1+\beta^2}\right) \Big/ \int_{-b'}^{b'} \frac{\sigma\varepsilon_y dz}{1+\beta^2} \qquad (2.134)$$

where

$$\langle\sigma_x\rangle = \frac{1}{A} \int_{-a'}^{a'} \left(\int_{-b'}^{b'} \frac{\sigma\varepsilon_x dz}{1+\beta^2} \right.$$

$$\left. + \int_{-b'}^{b'} \frac{\sigma\beta\varepsilon_x dz}{1+\beta^2} \int_{-b'}^{b'} \frac{\sigma\beta\varepsilon_y dz}{1+\beta_2} \Big/ \int_{-b'}^{b'} \frac{\sigma\varepsilon_y dz}{1+\beta^2}\right) dy \ ;$$

$$\langle\sigma\psi_u\rangle = \frac{1}{A} \int_{-a'}^{a'} \left(\int_{-b'}^{b'} \frac{\sigma\beta\psi_u dz}{1+\beta^2} \right.$$

$$\left. - \int_{-b'}^{b'} \frac{\sigma\psi_u dz}{1+\beta^2} \int_{-b'}^{b'} \frac{\sigma\beta\varepsilon_y dz}{1+\beta^2} \Big/ \int_{-b'}^{b'} \frac{\sigma\varepsilon_y dz}{1+\beta^2}\right) dy \ ;$$

$$\langle\nu\beta\rangle = \frac{1}{2a} \int_{-a'}^{a'} \nu_y \left(\int_{-b'}^{b'} \frac{\sigma\beta\varepsilon_y dz}{1+\beta^2} \Big/ \int_{-b'}^{b'} \frac{\sigma\varepsilon_y dz}{1+\beta^2}\right) dy \ ;$$

$$r_i = \int_{-a'}^{a'} \nu_y \left(\int_{-b'}^{b'} \frac{\sigma\varepsilon_y dz}{1+\beta^2}\right)^{-1} dy \ ;$$

$$\langle\sigma\psi_u\rangle = \frac{1}{2a}\int_{-a'}^{a'}\left(\int_{-b'}^{b'}\frac{\sigma\psi_u dz}{1+\beta^2}\Big/\int_{-b'}^{b'}\frac{\sigma\varepsilon_y dz}{1+\beta^2}\right) dy\ ;$$

$$\langle\beta_x\rangle = \frac{1}{2a}\int_{-a'}^{a'}\left(\int_{-b'}^{b'}\frac{\sigma\beta\varepsilon_x}{1+\beta^2}\,dz\Big/\int_{-b'}^{b'}\frac{\sigma\varepsilon_y dz}{1+\beta^2}\right) dy\ . \tag{2.135}$$

Assuming, for example, that the axial current, I_x', and the electrode voltage, Vy, in a specified cross section are known, the closure relationships can be represented in the form

$$E_x = (I_x' r_i/A - V_y\langle\beta\nu\rangle/2b$$

$$- UB(\langle\sigma\psi_u\rangle\langle\beta\nu\rangle a/b + r_i\langle\delta\psi_u\rangle))/G\ ,$$

$$i_y = (I_x'\langle\beta_x\rangle/(\langle\sigma_x\rangle A) - V_y/2a$$

$$- UB(\beta_x\rangle\langle\delta\psi_u\rangle/\langle\sigma_x\rangle + \langle\sigma\psi_u\rangle))2a\langle\sigma_x/G\ , \tag{2.136}$$

where

$$G = \langle\sigma_x\rangle r_i - \langle\beta_x\rangle\langle\beta\nu\rangle a/b\ .$$

Strictly speaking, the profiles, ε_x, ε_y, and ν_y, are not independent. In particular, the following condition must be satisfied,

$$i_H \mp \frac{\Phi^{\pm}}{r} = \nu_y(x,\pm a')i_y \mp i_x^{\pm}\frac{da}{dx}\ . \tag{2.137}$$

In general, the concept of profile functions (such as those introduced above) is that there exists a fairly broad class of conditions for which profiles introduced describe the distribution of quantities in a real flow, with an accuracy acceptable for specific applications.

Considering one of the characteristic classes of flows in an MHD generator channel, the principal characteristics of distributions of electrodynamic parameters over the channel cross section may be formulated (at least qualitatively).

Assume that the characteristic length over which flow parameters (velocity, pressure, and temperature) and the external and boundary conditions change is the channel length, L. In this case, the characteristic length of variation of electrodynamic parameters in the direction of the flow will also be channel length L. It is assumed that, under conditions existing in an MHD generator channel,

$$i_x^* \sim \sigma^* \beta^* U^* B^*, \quad i_y^* \sim \sigma^* U^* B^* ,$$

$$E_x^* \sim \beta^* U^* B^*, \qquad E_y^* \sim U^* B^* ,$$

where, here and below, asterisks denote characteristic values of parameters.

Represent the transverse specific current, $i_y(x,y)$ in the form,

$$i_y(x,y) = i_y(x) - \int_0^y \frac{\partial i_x}{\partial x} \, dy .$$

Taking into account the foregoing comments, the variation of the transverse current in the cross section can be estimated from the relationship

$$\Delta_y i_y^* \sim a^* \beta^* i_y^* / L . \tag{2.138}$$

Thus, when $\alpha^* \beta^* \ll L$, from the estimate (2.138), it can be assumed that

$$\nu_y = 1, \text{ i.e., } i_y(x,y) = i_y(x) .$$

On the other hand, assuming that $i_y(y)$ is a monotonic function of the transverse coordinate, (2.137) can be used to evaluate the same quantity

$$\frac{\Delta_y i_y^*}{i_y^*} \sim \frac{\Phi^*}{r^* i_y^*} \lesssim \frac{E_x^* L}{r^* i_y^*} \lesssim \frac{\beta^* L}{\sigma^* r^* b^*} . \tag{2.139}$$

A comparison of (2.138) with (2.139) provides the limit on leakage resistance between the electrodes and the ground:

$$\sigma^* r^* a^* b^* / L^2 \gg 1 \ .$$

Using the same assumptions, evaluate the changes of the transverse component of the electric field E_x along the cross section. Represent E_x in the form

$$E_x(x,y,z) = E_x(x) + \int_0^y \frac{\partial E_y}{\partial x} dy + \int_0^z \frac{\partial E_z}{\partial x} dz \ .$$

From this,

$$\Delta_x E_x^* \lesssim \max\{E_y^* a/L, E_z^* b/L\} \ .$$

Under conditions considered $E_z^* \lesssim F_y^*$, therefore when $a \simeq b$,

$$\Delta_y E_x^* / E_x^* \lesssim a^* / L\beta^* \ .$$

Hence, when $\beta^* \gtrsim 1$ and $a\beta^*/L \ll 1$, it can be assumed that

$$\varepsilon_x = 1, \text{ i.e., } E_x(x,y,z) = E_x(x) \ .$$

Proceeding to evaluate the change in the transverse component of the electric field along the external magnetic field, it can be assumed, initially, that the transverse conductivity in the surface leakage layer along the walls is zero, $\sigma_{wy} = 0$. Numerous solutions of the two-dimensional electrodynamic problem of the potential and current distributions over the channel cross section have been obtained for such a case, under very general assumptions of the nature of the gasdynamic and electrodynamic flow parameter distributions. It follows from the potential and current distributions over the channel cross section (see Fig. 2.20) that one can assume with sufficient accuracy, that

$$\varepsilon_y = 1, \text{ i.e., } E_y = E_y(x,y) \ . \qquad (2.140)$$

In the case of nonideal insulation of side walls in the traverse direction, a variation of the y component of the electric field, $\Delta_z E_y^*$, along the direction of the external magnetic field vector may be caused by the shunting effect of the nonideal insulation wall ("local shunting," see [55]). The quantity $\Delta_y E_y^*$ may be determined in the following manner:

$$\Delta_z E_y^* \sim \left(\frac{\partial E_y}{\partial z}\right)^* b^* \sim b^* \left(\frac{\partial E_z}{\partial y}\right)^* \sim \frac{b^* j_z^*}{a^* \sigma^*} ,$$

where

$$j^*_z \sim \left(\frac{\partial}{\partial y}(\sigma_{wy} E_y)\right)^* \lesssim \frac{\sigma_{wy}^* E_y^*}{a^*} .$$

From this,

$$\Delta_z E_y^* / E_y^* \lesssim \sigma_{wy}^* b^* / \sigma^* a^{*2} ,$$

i.e., when $\sigma_{wy}^* b^* / \sigma^* a^{*2} \ll 1$, assumption (2.140) may be extended to the case of nonideal insulation of side walls.

It should be noted that the estimate given here does not apply in regions of sharp variation of conductivity in the transverse direction, i.e., primarily near the angles of the channel. Modeling assumption (2.140) in this case leads to a local above-normal current leakage along side walls.

Thus, modeling assumptions $\varepsilon_x = \varepsilon_y = \nu_y = 1$ is fulfilled, provided:

$$a^* \beta^* / L \ll 1, \; \beta^* \gtrsim 1, \; \sigma^* r^* a^* b^* / L^2 \gtrsim 1 ,$$

$$\sigma_{wy}^* b^* / \sigma^* a^{*2} \ll 1 .$$

The principal integral characteristics for the case of $\varepsilon_x = \varepsilon_y = \nu_y = 1$ can be represented in a somewhat simplified form.

The electrical power takeoff from a channel unit length is given by the expression,

$$N_L' = - E_x I_x' + i_H V_y + ((\Phi^+)^2 + (\Phi^-)^2)/r , \qquad (2.141)$$

and, subject to an additional condition, $\Phi^{\pm} = \Phi(x) \pm V_y/2$,

$$N_L' = - E_x I_x' + i_H V_y + (4\Phi^2 + V^2)/2r .$$

Electrical power taken off from the flow (see system (2.120)) is partially dissipated in the surface layer, δ_π, i.e.,

$$N_L = N_L' + \delta N_L \;, \tag{2.142}$$

where

$$\delta N_L = - i^+ \Delta V^+ - i^- \Delta V^- -$$
$$- \Pi \sigma_{wx} E_x^2 - 2 \int_{-a}^{a} \sigma_{wy} E_y^2 \, dy \;,$$

(Π is the circumference of the cross section, and where in the previous expression, it was assumed, for simplicity, that σ_{wx} is constant along the whole circumference).

The total transverse ponderomotive force and the specific power output are represented in the form

$$F_x = 2B \left(a i_y - \int_{-a}^{a} \sigma_{wy} E_y dy \right) /A \;,$$

$$N_V = N_L/A \;. \tag{2.143}$$

The total transverse current in the flow, I_x, is given by the following expression:

$$I_x = I_x' - \sigma_{wx} \Pi E_x \;. \tag{2.144}$$

It should be noted that the following boundary conditions must be satisfied at the input (x = 0) and the output (x = L) of the channel.

$$I_x'(0) = I_H(0) - \Phi(0)/R_1 \;,$$

$$I_x'(L) = I_H(L) + \Phi(L)/R_2 \;,$$

where $I_H(0)$, $I_H(L)$, are currents in the load switched to the channel ends and R_1, R_2, are the leakage resistances to the ground from the channel ends.

Thus, we have derived relationships determining the governing integral and local channel parameters: power output at the load and at the leakage resistors, integral force along the channel length, total axial current in the channel, and transverse current and transverse electric field distribution over the cross section. These quantities are sufficient for closure of the general quasi-one-dimensional system of equations.

2.5.5. Incorporating Certain Real Effects. The results obtained make it possible to formulate a completely closed (in a mathematical sense) quasi-one-dimensional flow model of an MHD generator channel. A mathematical description of the model includes the mass, momentum, and energy flow conservation laws (the energy conservation law is written in terms of the total enthalpy) and electrodynamic equations (2.120).

This system of five first-order ordinary differential equations requires boundary conditions at the inlet (x = 0) and outlet (x = L) of the channel. The following set of relationships may be used as such boundary conditions at the outlet on the assumption that the flow is subsonic:

$$m = \int_A \rho u dA|_{x=o}, \quad H_o = \int_A \rho u i_o dA|_{x=o} ,$$

$$p(L) = p_2 ,$$

$$I_x'(0) = I_H(0) - \Phi(0)/R_1 ,$$

$$I_x'(L) = I_H(L) + \Phi(L)/R_2 , \tag{2.145}$$

where m is the flow rate of the working fluid, H_o is the total enthalpy, p_2 is the static pressure at the channel outlet.

When the flow at the channel outlet is supersonic, the condition $p(L) = p_2$ must be replaced with one of the gasdynamic parameters at the channel inlet (e.g., Mach number). Additional equations are required for closure of the system, in addition to the five basic equations given above. In particular, integral equations for the boundary layer (2.125), (2.126), or relations replacing them, e.g., (2.128), can be used to describe the gasdynamic nonuniformities. The structure of nonuniformities in the channel cross section is determined by profile functions such as (2.128).

Expressions for the friction and heat transfer coefficients, relations (2.129) and (2.130), or (2.131) and (2.132), are used in the governing and boundary layer equations to describe friction and heat transfer processes and their influence on the flow parameters. Electrodynamic relationships (2.132), (2.133), and expressions (2.141)-(2.144) are used for closure of the governing and boundary layer equations. It is assumed that the channel geometry, external magnetic field distribution, and loading scheme parameters, as well as the thermodynamic equation of state and the dependence of the electrophysical properties of the working fluid on thermodynamic parameters, are given.

Mathematical description of such a basic model is based on the assumption that electrode segmentation is ideally fine; current losses to the channel side walls are negligibly small; transverse parameter gradients of the flow and of internal and boundary conditions are sufficiently small (in particular, that one can neglect the end effects associated with large external magnetic field gradients, as well as variability of loading conditions at the channel inlet and exit and possible channel constriction at the electrodes). Under real conditions, some of the listed assumptions are not fulfilled. Accordingly, it is necessary to consider the possible influence on the flow characteristics of phenomena that were neglected, and to analyze the feasibility of a more or less accurate way of taking them into account within the framework of the basic model formulated.

2.5.5.1. The Effect of Finite Segmentation

Fairly simple model electric field and current distributions near the segmented wall have been used [25, 56, 57] to estimate the influence of finite electrode and insulating wall segmentation. The influence of finite segmentation was described in terms of parameters of the quasi-one-dimensional theory, using certain effective variables associated with transverse leakage and near-electrode potential drop. It will be shown here that such a representation is actually possible.

For simplicity, consider a two-dimensional electric field and current density distributions due to electrode wall segmentation. Assume that the variation of flow parameters within a single segment is negligibly small and that electrode loading (boundary) conditions for a group of adjacent electrode pairs are identical. In this case, it can be assumed that the electric field and current density distributions in the region corresponding to the central electrode pair are periodic along a sector of length, ℓ, equal to the geometric segmentation period. No limitations are imposed on the type of variation of quantities across the channel. It will be assumed that the tangential component of the electric field at electrodes and the normal component of current density at interelectrode insulators are equal to zero.

Under these assumptions, the components of the averaged generalized components of Ohm's law are:

$$j_x = \sigma E_x - \beta j_y ,$$

$$j_y = \frac{\sigma}{1+\beta^2} (E_y - uB + \beta E_x) .$$

Integrating with respect to x between the limits of one segment gives

$$\langle j_x \rangle_\ell \ell = \sigma \langle E_x \rangle_\ell \ell - \beta \langle j_y \rangle_\ell \ell ,$$

$$\langle j_y \rangle_\ell \ell = \frac{\sigma}{1+\beta^2} (\langle E_y \rangle_\ell \ell - uB\ell + \beta \langle E_x \rangle_\ell \ell) .$$

In view of the periodicity, $I_y = \langle j_y \rangle_\ell \ell$ and $V_x = - \langle E_x \rangle_\ell \ell$ are independent of the transverse coordinate and represent the total current to the electrode and the transverse (Hall) voltage at the interelectrode gap, respectively. Taking this into account and performing second integration with respect to y, gives

$$\int_{-a}^{a} \int_{0}^{\ell} j_x dx\, dy = - 2a \langle \sigma \rangle_\ell V_x - 2a \langle \beta \rangle I_y ,$$

$$2aI_y \left\langle \frac{1+\beta^2}{\sigma} \right\rangle_y = \int_{-a}^{a} \int_{0}^{\ell} E_y dx\, dy - 2a \langle uB \rangle_y \ell - 2a \langle \beta \rangle_y V_x .$$

Changing the order of integration,

$$\int_{-a}^{a} \int_{0}^{\ell} j_x dx\, dy = \int_{0}^{\ell} \int_{-a}^{a} j_x dy\, dx = \langle I_x \rangle_\ell \ell = (I_x + \Delta I) \ell ,$$

$$\int_{-a}^{a}\int_{o}^{\ell} E_y \, dx \, dy = \int_{o}^{\ell}\int_{-a}^{a} E_x \, dy \, dx$$

$$= - \langle V_y \rangle_{\ell} \ell = - (V_y + \Delta V)\ell ,$$

where I_x is the specified transverse current in the channel, V_y is the electrode voltage, and ΔI and ΔV are corrections to the "quasi-dimensional" parameters due to finite segmentation. It should be noted that I_x and V_y are defined by the relations

$$I_x = \int_{-a}^{a} j_x \, dy|_{x=o}, \qquad V_y = - \int_{-a}^{a} E_y \, dy|_{x=1/2} .$$

In the general case, corrections ΔI and ΔV are functions of geometric segmentation, distributions of electro-physical properties of the working fluid and loading conditions. The dependence of ΔI and ΔV on loading conditions is responsible for the arbitrariness of the disruption of the effect of finite segmentation in terms of the near-electrode drop and the transverse leakage current. In view of the linearity of the problem in terms of the loading parameters V_y and I_x, the solution of the two-dimensional problem of finite segmentation has the form

$$\phi = \Phi_0 + \Phi_1 V_y + \Phi_2 I_x ,$$

$$\psi = \Psi_0 + \Psi_1 V_y + \Psi_2 I_x .$$

Consequently, corrections ΔI and ΔV can be represented in an analogous form,

$$\Delta V = a_0 + a_1 V_y + a_2 I_x , \quad \Delta I = b_0 + b_1 V_y + b_2 I_x .$$

Since when $V_y = V_{x.x} = - 2a \langle uB \rangle$ and $I_x = 0$, $\Delta V = 0$, $\Delta I = 0$, the expression for ΔI and ΔV can be represented in the form,

$$\Delta V = a_1 (V_y - V_{x.x}) + a_2 I_x ,$$

$$\Delta I = b_1 (V_y - V_{x.x}) + b_2 I_x . \qquad (2.146)$$

The coefficients a_1, a_2, b_1, b_2, may be determined from the numerical solution of the finite segmentation problem and approximated, using the results of these solutions, by sufficiently simple relations as functions of geometrical segmentation parameters and integral characteristics of the flow structure. It can be readily seen that the dependence of the effective near-electrode potential drop, ΔV, and the effective leakage current, ΔI, on the load parameters (V_y and I_x) is due to the Hall effect. The latter is associated with the diagonal terms in expressions (2.146).

Therefore, because the loading parameters are either given or determined from the basic system of quasi-one-dimensional equations for each section, the influence of finite segmentation can be taken into account with the framework of the "basic" model by using a suitable approximation of expressions (2.146). It should be noted that, in this case, all mode parameters become variables averaged over the length of a single segment in axial direction.

2.5.5.2. Discharge Constriction at the Electrode

Under conditions existing in combustion-driven MHD generators at moderate current densities ($j \simeq 1$ A/cm^2) that is characteristic, as a rule, of most industrial applications, passage of current through the near electrode region of the channel is accompanied by constriction. Arc formation processes in the near-electrode regions exert considerable influence on the duration of electrode operation, structural integrity, and, obviously, on the integral parameters of the MHD generator. A possible method of describing characteristics of an MHD generator channel in the arc-discharge operating mode will now be considered. Separate analysis of the arc mode discharge, in a chapter that deals with quasi-one-dimensional approximation, is necessitated by the breakdown of principal assumptions of the basic model in the discharge constriction zone.

This analysis shall deal with an MHD channel of constant rectangular cross section, 2a and 2b, placed in a uniform external magnetic field, $\mathbf{B} = (0,0,B)$. The two opposed channel walls oriented normal to the magnetic field vector, $\mathbf{B}$, are ideal insulators, and the other two are uniformly segmented electrode walls with the segmentation period ℓ (the length of the electrode-insulator module along the OX axis).

Assume that the flow parameters (flow rate $\mathbf{u} = \{u, 0, 0\}$, conductivity σ, and the Hall parameter β) do not vary in the transverse direction, and that the electrode loading conditions are the same for all electrode pairs located in the sector under consideration, of length L that is at least greater than the characteristic transverse length, 2a. In this case, assuming an arbitrary variation of flow parameters (u, σ, β), the distribution of electrodynamic parameters, electric field **E** and current density **j**, are functions that are periodic along the transverse coordinate, x, with period, ℓ, equal to the geometrical period of the electrode wall segmentation, i.e.,

$$\mathbf{E}(x,y,z) = \mathbf{E}(x+\ell,y,z) \; ,$$

$$\mathbf{j}(x,y,z) = \mathbf{j}(x+\ell,y,z) \; .$$

Assume that current constriction near electrodes is caused by formation of high conductivity regions, D_i, that are small in comparison with the volume D under consideration. For simplicity, assume that these high conductivity regions (arc channels) are cylinders of known radii ($r_a \ll c$) and height ($h_{ai} \ll a$), the axes of which form a specified angle, α_i, with the plane of the electrode. Assume that conductivity in the arc channel is much higher than conductivity at any point outside it. What's more, the total current of each arc, I_{ai}, is constant along the arc channel.

It will be assumed that the usual form of the generalized Ohm's law is valid throughout the whole region under consideration.

$$\mathbf{j} + \beta/B\cdot\mathbf{j} \times \mathbf{B} = \sigma(\mathbf{E}+\mathbf{u}\mathrm{x}\mathbf{B}) \; .$$

Consider an elementary volume, D, bound by two channel cross sections, the distance between which is equal to the segmentation period of the electrode wall. In accordance with the concept of a constricted discharge conductivity distribution $\sigma\,(x,y,z)$ and Hall parameter distribution β (x,y,z) in the region D under consideration will be introduced, in the form

$$\sigma = \sigma_o(y,z) + \delta(r)\sigma_a \; ,$$

$$\beta + \beta_o(y,z) + \delta(r)\beta_a \; , \qquad (2.147)$$

where

$$\delta(r) = \begin{cases} 1 \text{ when } r \in D_i, \\ \qquad (i = 1,2,\ldots,n_a^-,\ldots,(n_a^+ + n_a^-)); \\ 0 \text{ when } r \in \left(D - \sum_i D_i\right) \end{cases}$$

$$r = \{x,y,z\}; \qquad \sigma_a >> \sigma_o(y,z) ;$$

$\sigma_o(y,z)$ and $\beta_o(y,z)$ are the conductivity and Hall parameter in a diffuse mode of operation of channel electrodes, n_a^- and n_a^+ are the number of arcs at the lower and upper electrodes, respectively. Using assumptions (2.147), the generalized Ohm's law may be written in the form

$$j_x = (\sigma_o E_x - \beta_o j_y)(1-\delta(r)) + \delta(r) j_{xa},$$

$$j_y = \frac{\sigma_o}{1+\beta_o^2}(E_y - uB + \beta_o E_x)(1-\delta(r))$$

$$+ \delta(r) j_{ya} , \tag{2.148}$$

where the current density components in the arc channel, D_i, are given by the following expressions:

$$j_{xa} = \sigma E_x - \beta j_y ,$$

$$j_{ya} = \frac{\sigma}{1+\beta^2}(E_y - uB + \beta E_x) .$$

Now, introduce

$$I_y = \int_{-b}^{b}\int_{x_1}^{x_1+\ell} j_y dx\, dy. \quad V_x = -\int_{x_1}^{x_1+\ell} E_x\, dx ,$$

$$V_y = -\int_{-a}^{a} E_y\, dy, \quad I_x = \int_{-b}^{b}\int_{-a}^{a} j_x dy\, dz ,$$

where x_1 is the coordinate of the upstream edge of the electrode that is part of the elementary volume, D, under consideration, $I_y = I_y^0 = \text{const.}$ is the total transverse (Faraday) current of one channel module, $V_x = V_x^0 = \text{const.}$ is the axial (Hall) potential difference across a segmentation period, $V_y = V_y(x)$. At the surface points of an electrode

$$(x_1 \le x \le x_1 + c),$$

where c is the length of the electrode along the transverse direction), $V_y = V_y^\circ = \text{const.}$ and is the potential difference between electrodes of the segment under consideration, $I_x = I_x(x)$, where $I_x = I_x^\circ = \text{const.}$ is the total axial current in the given segment of the channel on the insulated sectors

$$(x_1 + c < x < x_1 + \ell) .$$

The quantities I_y°, V_y°, V_x°, I_x°, connected by relations for external circuits of an MHD generator, are used in quasi-one-dimensional calculations of the flow in the channel (e.g., see expressions (2.136)).

Now, taking into account these properties of the integral characteristics I_y, V_y, V_x, and I_x, integrate relations (2.148) with respect to x and z, within the boundaries of the elementary volume, D. The following equations are then obtained:

$$I_y^0 = \frac{1}{2b} \int\int E_y dz\, dx \cdot \int \frac{\sigma_0 dz}{1+\beta_0^2}$$

$$+ \int\int \Delta_\sigma E_y dx\, dz - \int\int \frac{\sigma_0 uB}{1+\beta_0^2} dx\, dz$$

$$- V_x^0 \int \frac{\sigma_0 \beta_0}{1+\beta_0^2} dz + \int\int \delta(r)(j_{ya} - j_{y\Phi}) dx\, dz ,$$

$$\int\int j_x dx\, dz = - \int \sigma_0\, dz \cdot V_x^0 - \frac{1}{2b} \int \beta_0\, dz \cdot I_y^0$$

$$- \int\int \Delta_\beta j_y dx\, dz + \int\int \delta(r)(j_{xa} - j_{x\Phi}) dx\, dz ,$$

(2.149)

where $j_{y\Phi}$ and $j_{x\Phi}$ are components of a fictitious current density determined in arc channels, D_i, from the local electric field and current density, using unperturbed ("diffuse") values of conductivity and Hall parameter, and the quantities Δ_σ and Δ_β are given by the following equalities:

$$\Delta_\sigma = \frac{\sigma_o}{1+\beta_o^2} - \frac{1}{2b}\int \frac{\sigma_o dz}{1+\beta_o^2} \quad ,$$

$$\Delta_\beta = \beta_o - \frac{1}{2b}\int \beta_o dz \ .$$

Dividing the first of equalities (2.149) by

$$\frac{1}{2b} \int\int \frac{\sigma_o dz}{1+\beta_o^2} \ ,$$

then integrating (2.149) with respect to y within the limits between the lower and upper electrode walls will finally obtain

$$I_y^o R_y = 2b\ell\,(-V_y^o+\Delta V) + 2b\ell \quad - \mathcal{E}$$

$$- 4ab\bar{\beta}V_x^o + I_a^- R_s^- + I_a^+ R_s^+ \ ,$$

$$\ell(I_x^o+\Delta I_x) = - \langle\sigma\rangle 4abV_x^o - 2b\langle\beta\rangle I_y^o \ , \qquad (2.150)$$

where the quantities introduced have the following meaning:

effective near electrode potential drop

$$\Delta V = (\langle V_y\rangle_{\ell-c} - V_y^o)(1-c/\ell) +$$

$$+ \frac{1}{2b\ell}\int\Big(\int\int \Delta_\sigma E_y dx\, dz \Big/ \frac{1}{2b}\int \frac{\sigma_o dz}{1+\beta_o^2}\Big)\, dy$$

$$- \frac{1}{2b\ell}\int\Big(\int\int \delta(r) j_{y\Phi} dx\, dz \Big/ \frac{1}{2b}\int \frac{\sigma_o dz}{1+\beta_o^2}\Big)\ dy \ ,$$

where

$$\langle V_y \rangle_{\ell-c} = \frac{1}{2b(\ell-c)} \int_{-b}^{b} \int_{x_1+c}^{x_1+\ell} V_y dx\, dz \quad ;$$

effective axial leakage current

$$\Delta I_x = \frac{c}{\ell} (\langle I_x \rangle_c - I_x^o) + \frac{1}{\ell} \int \int \int \Delta_\beta j_y dx\, dy\, dz +$$

$$+ \frac{1}{\ell} \int \int \int \delta(r)(j_{xa} - j_{x\Phi}) dx\, dy\, dz \ ,$$

where

$$\langle I_x \rangle_c = \frac{1}{c} \int_{x_1}^{x_1+c} I_x dx \ ;$$

effective emf

$$= - \frac{1}{2b\ell} \int \frac{\left(\int\int \frac{\sigma_o uB}{1+\beta_o^2} dz\, dx \right.}{\frac{1}{2b} \int \frac{\sigma_o dz}{1+\beta_o^2}} dy \ ;$$

total current fed to the upper (I_a^+) and lower (I_a^-) electrodes along the arc channels:

$$I_a^{\pm} = \sum_i^{n_a^{\pm}} I_{ai}^{\pm} = \pm \frac{1}{h_a^{\pm}} \int_o^{\ell} \int_o^{\pm a} \int_{-b}^{b} j_y \delta(r) dz\, dy\, dx \ ;$$

resistance of high-resistance layers in the presence of a breakdown on the upper (R_s^+) and lower (R_s^-) electrodes. (For simplicity, here and below, it is assumed that arcs on each of the electrodes are the same):

$$R_s^{\pm} = \pm\, 2b \int_{\pm a \mp h_a^{\pm}}^{\pm a} \left(\int_{-b}^{b} \frac{\sigma_o dz}{1+\beta_o^2} \right)^{-1} dy \ .$$

In addition to this, the following relationships are used in equations (2.150):

$$\langle\beta\rangle = \frac{1}{4ab} \int_{-a}^{a}\int_{-b}^{b} \beta_o dz\, dy \ ,$$

$$\langle\sigma\rangle = \frac{1}{4ab} \int_{-a}^{a}\int_{-b}^{b} \sigma_o dz\, dy \ ,$$

$$R_y = 2b \int_{-a}^{a} \left(\int_{-b}^{b} \frac{\delta_o dz}{1+\beta_o^2} \right)^{-1} dy \ ,$$

$$\overline{\beta} = \frac{1}{2a} \int_{-a}^{a} \left(\int_{-b}^{b} \frac{\sigma_o \beta_o dz}{1+\beta_o^2} \Big/ \int_{-b}^{b} \frac{\sigma_o dz}{1+\beta_o^2} \right) dy \ .$$

Relations (2.150) connect four integral parameters of one electrode module of the channel: V_y^0 - electrode voltage, I_y^0 - total electrode current, V_x^0 - axial (Hall) potential difference across one segment, I_x^0 is the total axial (Hall) current in the channel. These quantities are required for closure of flow equations in the channel. For example, if it is assumed that the electrode voltage, V_y^0, and the total transverse current in the channel, I_x^0, are given, relations (2.150) may be represented in explicit form in terms of the remaining pair of unknown parameters: transverse current I_y^0 and axial potential difference V_x^0,

$$I_y^o = \frac{2b\ell\langle\sigma\rangle}{G} \left(- \frac{V_y^o - \Delta V + \mathcal{E}}{2a} + \overline{\beta} \frac{I_x^o + \Delta I_x}{4ab\langle\sigma\rangle} \right) ,$$

$$V_x^o = \frac{\ell\langle\beta\rangle}{G} \left(\frac{V_y^o - \Delta V + \mathcal{E}}{2a} - \frac{1}{\langle\beta\rangle} \frac{I_x^o + \Delta I_x}{4ab\langle\beta\rangle} (G + \langle\beta\rangle\overline{\beta}) \right) , \qquad (2.151)$$

where

$$G = \langle\sigma\rangle\, R_y 2a - \langle\beta\rangle\overline{\beta} - \langle\sigma\rangle(\xi^+ R_s^+ + \xi^- R_s^-)2a \,, \qquad (2.152)$$

$\xi^{\pm} = I_a^{\pm}\ I_y^{o}$ is the total relative current flowing to the electrodes along the arc channels.

Expressions (2.152) are valid in both diffuse and constricted operating modes of electrodes, subject to the condition that the above assumptions on the constricted nature of the discharge are fulfilled. In a diffuse operating mode, ξ^+ and ξ^- become zero and expressions (2.151) become known relations used in quasi-one-dimensional approximation (for example, see expressions (2.136)). In this case, G becomes the G-factor that was introduced for one-dimensional nonuniformities [59], generalized to the case of two-dimensional nonuniformities.

Typically, in the presence of constriction, the structure of relations (2.151) remains the same as in the case of a diffuse operating mode. However, taking constriction into account leads to a change in values of G, ΔV, and ΔI_x in these relations.

For qualitative analysis of the influence of current constriction on the dimensionless effective internal resistance G, it is convenient to rewrite expression (2.152) in the form

$$G = G^{o} + \langle\sigma\rangle((1-\xi^+)R_s^+ + (1-\xi^-)R_s^-)/2a \,, \qquad (2.153)$$

where $G^{o} = \langle\sigma\rangle\ R_y^{o}/2a - \langle\beta\rangle\ \overline{\beta}$ is the G-factor calculated from the resistance of the flow outside the high-resistance layers affected by breakdown:

$$R_y^{o} = R_y - R_s^{+} - R_s^{-}$$

$$= \int_{-a+h_a^-}^{a-h_a^+} \left(\frac{2}{2b}\int_{-b}^{b}\frac{\sigma_o dz}{1+\beta_o^2}\right)^{1} dy \,.$$

Two other terms in the rightB side of relation (2.153) describe direct influence of current constriction on G. If only a small part of the total electrode current, I_y^{0}, flows along the arc channels, i.e., when $\xi^+ \ll 1$ and $\xi^- \ll$

1, G is close to the value for a diffuse discharge mode. As a result, the influence of arc formation on the effective internal resistance is small. When $\xi^{\pm} \simeq 1$, the qualitative influence of constriction on the effective internal resistance is specified by the last two terms in relation (2.153).

It should be noted that $\xi^{\pm}$ depends on both geometrical parameters of the MHD channel (a,b,c,ℓ) and arc channels (r_a, h_a) and on the relation between resistances R_y^0, $R_s^{\pm}$, and $R_a = h_a/\pi r_a^2 \sigma_a$. Even at $\xi^{\pm}$ close to unity and when the numbers of arcs is limited, the terms in the relation (2.153), $(1 - \xi^-)\ R_s^-$ and $(1 - \xi^+)\ R_s^+$ will have a finite value that depends on the resistance of the region of "spreading out" of the current from the arc channel to a uniform distribution away from the electrode. It should be noted that when, $R_s^{\pm} \neq \infty$, as a result of the current flowing in a diffuse mode outside of arc channels, $\xi^{\pm} \neq 1$. This current may be very small $((1 - \xi^{\pm}) \ll 1)$, but it must be taken into account when determining effective internal resistance.

Thus, closure of quasi-one-dimensional relations (2.151) requires only determination of the dependence of ξ on the geometrical parameters of arcs and the discharge gap and the relationship between the total resistance of the gap in the diffuse mode and the resistance of the layer affected by breakdown, i.e., relation of the form

$$\xi = \xi(h_a/a, r_a/a, r_a/r_o, n, R_s R_y) ,$$

where n is the number of arc channels on the electrode that depends on the total current I_y^0 to the electrode, $r_o \simeq (bc/n)^{1/2}$ is the characteristic dimension of the electrode region associated with a single arc channel, c is the length of the electrode along the channel. A more detailed analysis of this model of constricted discharge and a discussion of feasibilities of utilizing this model to describe the current-voltage characteristics are given elsewhere [60, 61].

2.5.5.3. End Effects

Nonuniform current and potential distributions at the MHD generator inlet and exit can arise as a result of nonuniformity of boundary and external conditions (in the transition zone between the loaded and unloaded channel sectors that also frequently coincides with sectors of tapered external magnetic field). Within the framework of the quasi-one-dimensional theory, the influence of the nonuniform current and potential distributions can only be estimated. The nature of the end effects and the possibility of incorporating these effects within the framework of the quasi-one-dimensional approach depend on the specifics of the end zones of the channel and the external magnetic field profile. Therefore, it is rather difficult at the present time to obtain general relationships describing the end effects.

Most of the present-day commercial-scale MHD generator facilities are designed for efficient utilization of the magnetic field generated by an external magnetic system. This is achieved by having the working (loading) sector of the channel located in the fringe zone of the magnetic field, and determining loading from optimum conditions, from one of the principal integral characteristics. Construction and loading of the end sectors of such a channel make it possible to smooth out nonuniformities of the current potential distribution at the channel walls and, therefore, make it easier to operate channel components in the area where the active sector joins the unloaded input (output) sector. In engineering practice, calculation of flows at the end sectors of the channel is usually performed in the same way as for the active sector, using the local values of loading parameters and the magnetic field in quasi-one-dimensional expressions for the electrodynamic terms. Naturally, the two-dimensional electrical effects, the influence of which on the flow may be considerable under certain circumstances, are not taken into account.

Consider a model problem of axial end effect in an MHD channel with constant flow parameters, in the absence of the Hall effect. Solution of the problem for $\beta = 0$ provides an upper limit estimate of the end effects (because they decrease with increasing β) [7, 10].

An infinite channel of constant height, 2a, has insulation walls at $x < 0$ and electrode walls at $x > 0$. The distribution of the magnetic field has the form

$$\frac{B(x)}{B_\circ} = \begin{cases} \exp \mu(x-\ell)/2a & \text{when } x<\ell \ , \\ 1 & \text{when } x \geq \ell \ . \end{cases}$$

A plot of the magnetic field includes two parameters: μ, scattered field attenuation decrement, and ℓ, the distance from the beginning of the electrode zone to the sector with a uniform magnetic field.

Two solutions are known for a segmented channel when $\ell = 0$ and the electrodes are characterized by constant current density [62]:

$$j_y(x,\pm a) = \begin{cases} 0 & \text{when } x<0 \,, \\ -(1-k)\sigma u B_o & \text{when } x\geq 0 \,, \end{cases}$$

and a channel with continuous electrodes at a constant voltage [63]:

$$j_y(x,\pm a) = 0 \qquad \text{when } x<0 \,,$$

$$\phi(x,\pm a) = \mp 1/2V_y \qquad \text{when } x\geq 0 \,.$$

Performing simple calculations, these solutions can be generalized to the case $\ell \neq 0$.

Having solution of the problem, it is possible to calculate corrections to the power output, N, and the work, A, due to two-dimensional effects:

$$\Delta N = N - N_{id}, \quad \Delta A = A - A_{id} \,,$$

where N_{id} and A_{id} are calculated with the use of one-dimensional representations:

for a segmented channel, when $L/2a \gg 1$

$$N_{id} = 2\sigma u^2 B_o^2 a(1-k) \int_o^L \left(k + \frac{B(x)}{B_o} - 1 \right) dx \,,$$

$$A_{id} = 2\sigma u^2 B_o^2 a(1-k) \int_o^L \frac{B(x)}{B_o}\, dx \,;$$

for a channel with continuous electrodes

$$N_{id} = \sigma u B_o V_y \int_o^L \left(\frac{B(x)}{B_o} - \frac{V_y}{2auB_o} \right) dx \,,$$

$$A_{id} = 2a\sigma u^2 B_o^2 \int_o^L \left(\frac{B(x)}{B_o} - \frac{V_y}{2auB_o} \right) \frac{B(x)}{B_o}\, dx \,.$$

Then, two-dimensional corrections for a segmented channel can be represented in the form

$$\frac{\Delta N}{4a^2 \sigma u^2 B_o^2} = -(1-k)\left(\frac{1}{\mu}\left(1-\frac{2}{\mu}\operatorname{tg}\frac{\mu}{2}\right)(1-e^{-\mu \ell/2a})\right.$$

$$\left.-4\mu\sum_1^\infty \frac{1-e^{-\beta_n \ell/2a}}{\beta_n^3(\mu-\beta_n)} - 4\sum_1^\infty\left(\frac{1}{\beta_n^2(\mu+\beta_n)}-\frac{k}{\beta_n^3}\right)\right),$$

$$\frac{\Delta A}{4a^2 \sigma u^2 B_o^2} = -(1-k)\left(\frac{1}{\mu}\left(1-\frac{2}{\mu}\operatorname{tg}\frac{\mu}{2}\right)(1-e^{-\mu \ell/2a})\right.$$

$$\left.-4\mu\sum_1^\infty \frac{1-e^{-\beta_n \ell/2a}}{\beta_n^3(\mu-\beta_n)} + 4k\mu\sum_1^\infty \frac{1}{\beta_n^3(\mu+\beta_n)}\right), \tag{2.154}$$

where $\beta_n = \pi(2n-1)$.

The formulas obtained include a free parameter, ℓ, that when designing MHD generators, may be chosen from various conditions, e.g., assuming maximum power output.

In addition, at a given k, i.e., a specific loading mode, ℓ can be used to make either ΔN or ΔA equal to zero. Apparently, applying more complex loading control techniques in the end zone (e.g., assuming $k = k(x)$ et al.), it's possible to make ΔN and ΔA become zero simultaneously. Hence, in the same manner it is possible to provide optimal output characteristics of an MHD generator subject to various limiting conditions (specified efficiency, maximum allowable currents and voltages on components of the structure, etc.). Experience gained from investigating two-dimensional electric fields in MHD channels shows that, in the optimum design and optimum loading scheme case, corrections for two-dimensional effects are sufficiently small, i.e., when calculating characteristics of the generator, specific aspects of electrodynamics of end effects are taken into account mostly within the framework of the quasi-one-dimensional approximation. To illustrate this, consider a characteristic case of a discontinuous magnetic field, when the electrodes are located only within the tapered magnetic field zone ($\mu = \infty$, $\ell = 0$). Such a configuration was considered in the variational solution (see 2.2.1.) for a generator of maximum power output at a specified electrical efficiency. In this case, from the first equation in (2.154) the following numerical value was obtained for the ratio of correction to the power output due to the two-dimensional effect with the ideal power output of a channel sector of unit length, calculated using one-dimensional relations:

$$\frac{\Delta N}{N_{id}\big|_{L=2a}} = -4\sum_{1}^{\infty}\frac{1}{\pi^3(2n-1)^3} = -0.136 \ .$$

From the variational solution (2.33), it follows that, in the presence of the Hall effect, this value decreases proportionally to the factor $(1+(\beta/2)^2)^{-1}$.

For a channel with continuous electrodes, two-dimensional corrections can be written in the form

$$\frac{\Delta N}{4a^2\sigma u^2 B_o^2} = \frac{V_y}{2auB_o}\left(\frac{1}{\mu}\left(1 - \frac{1}{\sqrt{\pi}}\,\Gamma\left(\frac{\mu}{2\pi}+\frac{1}{2}\right)\Big/\Gamma\left(\frac{\mu}{2\pi}+1\right)\right)e^{-\frac{u}{l}} - \frac{V_y}{\pi u B_o}\ln 2\right),$$

$$\frac{\Delta A}{4a^2\sigma u^2 B_o^2} = \frac{\Delta N}{4a^2\sigma u^2 B_o^2} + \frac{16}{\pi^3}\,e^{-\frac{2\mu\ell}{a}}\sum_{1}^{\infty} nA_n^2 - \frac{1}{2\mu}e^{-\frac{2\mu\ell}{a}} + \left(\frac{V_y}{2anB_o}\right)^2\frac{\ln 2}{\pi},$$

where

$$A_n = \frac{(-1)^n\pi^2\mu^{-1}\cdot 2^{-(1+\mu/\pi)}\,\Gamma(\mu/\pi+1)}{\Gamma(\mu/2\pi+1+n)\,\Gamma(\mu/2\pi+1-n)} \ ;$$

and Γ is the gamma function.

For a channel with a tapered magnetic field ($\mu = \infty$, $\ell = 0$), the relative correction to the power output is given by the expression

$$\frac{\Delta N}{N_{id}\big|_{L=2a}} = -\frac{\ln 2}{\pi}\,\frac{V_y}{2auB_o}\left(1-\frac{V_y}{2auB_o}\right)^{-1}.$$

At high voltages, this quantity is on the order of unity, i.e., under these conditions, a channel with continuous electrodes is far from being optimal and a one-dimensional description of the channel is considerably in error.

Correction quantities, ΔN and ΔA, are usually introduced into the quasi-one-dimensional flow calculations with the aid of proper discontinuity surfaces at the channel inlet and outlet, where the enthalpy-flow and stagnation-pressure discontinuities are determined in terms of ΔN and ΔA [62]. An alternative, more convenient approach to taking end effects into account is to introduce additional terms ΔN and ΔA, reduced to the volumes of the input and output channel sectors, into the conservation equations for momentum and stagnation enthalpy flows (2.120).

In conclusion, note that the distribution of electrodynamic parameters in the end regions of frame channels with diagonally conducting walls are considerably three-dimensional, and the methods of taking them into account within the framework of any quasi-one-dimensional flow models (as well as in more complex cases than those considered here) with plane end effects have not yet been developed.

2.5.5.4. Current Leakage to Channel Side Walls

Current flow directly from the working fluid to the inner side surfaces of the channel can occur in diagonally conducting wall channels or in Faraday channels with nonideal insulation of channel side walls. The leakage current flow to the side walls is caused by nonuniformity of flow parameters in the transverse or axial directions and nonuniformity of boundary conditions associated with finite segmentation, loading nonuniformities, and other, similar effects. Some of these effects will next be estimated and the feasibility of their description considered within the framework of the quasi-one-dimensional model. The current and potential distributions across the channel cross section (2a x 2b) can be described by the following equations:

$$\frac{\partial j_y}{\partial y} + \frac{\partial j_z}{\partial z} = 0 \, ,$$

$$j_y = \frac{\sigma}{1+\beta^2}\left(-\frac{\partial \phi}{\partial y} - uB + \beta E_x\right) ,$$

$$j_z = -\sigma\frac{\partial \phi}{\partial z} \, , \quad E_x = \text{const.} \qquad (2.155)$$

Integrating equation (2.155) along the z axis,

$$\frac{d\langle j_y\rangle_z}{dy} + \frac{j_{zw}}{b} = 0 \ , \qquad (2.156)$$

$$\langle j_y\rangle_z = - \left\langle \frac{\sigma}{1+\beta^2} \frac{\partial \phi}{\partial y} \right\rangle_z - \left\langle \frac{\sigma u}{1+\beta^2} \right\rangle_z B$$

$$+ \left\langle \frac{\sigma\beta}{1+\beta^2} \right\rangle_z E_x \ , \qquad (2.157)$$

$$\phi_o - \phi = \int_o^z \frac{j_z}{\sigma} dz \ . \qquad (2.158)$$

where $\langle ...\rangle_z$ is the mean value along the z axis, $j_{zw} = j_z\,(y,b)$, $\phi_o = \phi(y,0)$, $\phi = \phi(y,z)$, and the fact that $j_z\,(y,0) = 0$ has been taken into account.

Combining equations (2.156)-(2.158) leads to the equation:

$$\frac{d}{dy}\left(\left\langle \frac{\sigma}{1+\beta^2} \right\rangle_z \frac{d\phi_o}{dy}\right)$$

$$- \frac{d}{dy} \left\langle \frac{\sigma}{1+\beta^2} \frac{\partial}{\partial y} \int_o^z \frac{j_z}{\sigma} dz \right\rangle_z - \frac{j_{zw}}{b}$$

$$+ \frac{d}{dy}\left(\left\langle \frac{\sigma u}{1+\beta^2} \right\rangle_z B - \left\langle \frac{\sigma\beta}{1+\beta^2} \right\rangle_z E_x\right) = 0 \ .$$

Estimating the second and the third terms in this equation, one finds that the order of magnitude of their ratio is equal to $b^2/a^2\ (1+\beta^2)$. Assuming that $b^2 \ll a^2\,(1+\beta^2)$, and neglecting the second term in comparison with the third, gives an approximate equation for $\phi_o(y)$:

$$\frac{d}{dy}\left(\left\langle \frac{\sigma}{1+\beta^2} \right\rangle_z \frac{d\phi_o}{dy}\right) - \frac{j_{zw}}{b}$$

$$+ \frac{d}{dy}\left(\left\langle \frac{\sigma u}{1+\beta^2} \right\rangle_z B - \left\langle \frac{\sigma\beta}{1+\beta^2} \right\rangle_z E_x\right) = 0 \ .$$

$$(2.159)$$

This equation contains an unknown variable, $j_{zw}(y)$. Determination of this unknown requires the use of boundary conditions on the side walls of the channel with z = b and certain additional modeling assumptions. Consider a channel with a diagonally conducting, ideally segmented wall. Boundary condition (2.66) is fulfilled at such a wall:

$$\phi_w = E_x y \operatorname{ctg} \alpha.$$

The wall potential, ϕ_w, is connected with the plasma potential along the y axis by the following relation (see relation (2.158)):

$$\phi_o - \phi_w = \int_o^b \frac{j_z}{\sigma} dz , \qquad (2.160)$$

from which it follows that, to determine the relation between ϕ_o and j_{zw}, it is necessary to know the dependence of j_z on z. Assume that the variation of j_z along z is described by the function

$$j_z \equiv \begin{cases} 0 & \text{when} \quad |z|<b-\delta_1 , \\ j_{zw} & \text{when} \quad b-\delta_1 \leq |z| \leq b, \end{cases}$$

where δ_1 is a running parameter, the value of which is on the order of the thickness of the boundary layer. In this case, a relation between ϕ_o and j_{zw} has the form

$$\phi_o - \phi_w = j_{zw} \int_{b-\delta_1}^{b} \frac{dz}{\sigma} .$$

Using relation (2.160) and the boundary condition for ϕ_w, equation (2.159) can be transformed into the following form [28, 64]:

$$\frac{d}{dy}\left(< \frac{\sigma}{1+\beta^2} >_z \frac{d\Phi}{dy}\right) - \Phi/b \int_{b-\delta_1}^{b} \frac{dz}{\sigma}$$

$$+ \frac{d}{dy}\left(< \frac{\sigma u}{1+\beta^2} >_z B - \left(< \frac{\sigma\beta}{1+\beta^2} >_z\right.\right.$$

$$\left.\left. - < \frac{\sigma}{1+\beta^2} >_z \operatorname{ctg} \alpha\right) E_x \right) = 0 , \qquad (2.161)$$

where $\Phi = \phi_o - \phi_w$.

In view of the symmetry of the distribution of the potential relative to the z axis and the fact that the potential on the electrode wall is constant, Φ (y) satisfies homogeneous boundary conditions

$$\Phi(0) = \Phi(a) = 0 .$$

Thus, formulation of the model is completed. Note that an analogous model differing only by the type of the boundary condition on the insulating wall has been previously considered [65]. Let us construct an approximate analytical solution of the problem for Φ (y). It will be assumed that $\sigma = \sigma(y,z)$ and $u = u(y,z)$ distributions over the channel cross section, corresponding to a turbulent flow with boundary layer of thickness δ and a uniform flow in the flow core, are specified. For simplicity, assume that β = const. For the flow core, equation (2.161) becomes uniform and has the following solution:

$$\Phi(y) = C \sinh \Delta y \quad \text{when} \quad |y| \leq a-\delta ,$$

where

$$\Delta = \sqrt{1+\beta^2} \left(\int_o^b \sigma(0,z\ dz \int_{b-\delta_1}^{b} \frac{dz}{\sigma(0,z)} \right)^{-1/2} .$$

Let's extend this solution over the whole interval $|y| \leq a$ and determine the constant C by equating C sinh Δa to the potential drop in the near electrode layer of thickness δ_2, when a transverse current of density $\langle j_y(a,z) \rangle_z$ flows through it. Then the following approximate formula is obtained for the current density averaged over the cross section [28, 64]:

$$\langle\langle j_y \rangle\rangle = - \frac{\langle \sigma(0,z) \rangle_z}{1+\beta^2} \left(\frac{\langle \sigma(0,z) u(0,z) \rangle_z B}{\langle \sigma(0,z) \rangle_z} \right.$$

$$- (\beta - \text{ctg}\alpha)\, E_x \left(1 - \langle \sigma(0,z) \rangle_z \int_{a-\delta_2}^{a} \frac{dy}{\langle \sigma(y,z) \rangle_z a} \right.$$

$$\left. \times \left(1 - \langle \sigma(0,z) \rangle_z \int_{a-\delta_2}^{a} \frac{\varepsilon dy}{\langle \sigma(y,z) \rangle_z a} \right)^{-1} \right) , \tag{2.162}$$

where $\varepsilon =, \Delta a \coth \Delta a$.

The results of the calculations using formula (2.162) for the case $\delta_1 = b$ and $\delta_2 = \delta/8$ are in satisfactory agreement with the solution of the two-dimensional problem (see Table 2.3). Calculation of the characteristics of a frame MHD channel using formulas for a Faraday channel, leads to significant differences and serves as a lower estimate of the characteristics of a frame channel.

Analyzing a Faraday channel with an insulating wall of finite conductivity, it will be assumed that the conducting region of the wall consists of a thin layer with surface conductivity, σ_w. The boundary condition at the wall is fulfilled then by the following relation [1]:

$$\frac{\partial}{\partial y}\left(\sigma_w \frac{\partial \phi_w}{\partial y}\right) = - j_{zw} . \quad (2.163)$$

The closure assumption concerning the distribution of j_z along z will be represented in the form

$$j_z = j_{zw} z/b . \quad (2.164)$$

In order to obtain finite results, introduce a conductivity distribution over the channel cross section in terms of a step function

$$\sigma(y,z) = \begin{cases} \sigma_0(z) & \text{when} \quad |y|<a-\delta , \\ \sigma_1(z) & \text{when} \quad a-\delta\leq|y|\leq a . \end{cases} \quad (2.165)$$

In this case, taking into account (2.160), (2.163)-(2.165), equation (2.159) is transformed into the form (u = const., β = const.),

$$\frac{d^2\Phi}{dy^2} - \gamma^2\Phi = 0 , \quad (2.166)$$

where

$$\gamma^2 = \frac{1+\beta^2}{\langle\sigma\rangle_z \langle z\sigma^{-1}\rangle_z b}\left(1 + \frac{1}{1+\beta^2}\,\frac{\langle\sigma\rangle_z b}{\sigma_w}\right) .$$

The function, $\Phi(y)$, satisfies uniform boundary conditions $\Phi(0) = \Phi(a) = 0$ and, when $y = a - \delta$, to the following matching conditions: $\Phi(a-\delta+0) = \Phi(a-\delta-0)$, $j_y(a-\delta+0, 0) = j_y(a-\delta-0, 0)$. Taking these conditions into account, solution of equation (2.166) has the form

$$\Phi(y) = j_{yo} \frac{a(1+\beta^2)(\Xi-1)/\sigma_o}{1+Q},$$

$$j_y(y) = j_{yo}\left(1 - \frac{QJ(y)}{1+Q(\mathrm{ch}\ \gamma_o(a-\delta)-1)}\right), \tag{2.167}$$

where

$$Q = \frac{(a/b)^2(1+\beta^2)(\Xi-1)}{\gamma(a-\delta)\mathrm{sh}\ \gamma_o(a-\delta)};$$

$$J(y) = \begin{cases} \mathrm{ch}\gamma_o y & \text{when } y \le a-\delta, \\ \mathrm{ch}\ \gamma_o(a-\delta) & \text{when } y > a-\delta; \end{cases}$$

$$\phi_o(y) = \phi_o(0) - (1+\beta^2)\int_o^y \frac{j_y}{\sigma}dy - (uB-\beta E_x)y;$$

$$\phi_w(y) = \phi_o(y) - \Phi(y); \quad \Xi = \langle\sigma\rangle\langle\sigma^{-1}\rangle.$$

In writing relations (2.167) it was assumed that

$$\sigma_o(z) = \sigma_o = \text{const.}, \ \sigma_1(z) = \sigma_1 = \text{const.},$$

$$\sigma_1/\sigma_o \ll 1, \ (1+\beta^2)\sigma_w/\sigma_o b \ll 1,$$

$$\gamma_o(a-\delta) \ll 1, \ \gamma_1\delta \ll 1.$$

Expressions (2.167) include two parameters: j_{yo} and E_x, the values of which have to be determined from the loading conditions. For example, for a Faraday channel it can be assumed that the electrode voltage, V_y, and the mean axial current density, $\langle j_x \rangle$, are specified. In this case, $j_{y\circ}$ and E_z are determined from the relationship

$$\phi_o(a) - \phi_o(0) = 1/2V_y, \; \langle\sigma\rangle E_x - \beta\langle j_y\rangle = \langle j_x\rangle \; .$$

It is easy to verify that, even in the presence of considerable surface conductivity, $((1+\beta^2)\sigma_w/\sigma_0 b \simeq 1)$, a simpler model of parallel equipotentials introduced [34, 55] to take into account the effect of local shunting of the working current in the channel by the surface conducting walls, provides satisfactory results for both integral characteristics, $\langle\langle j_y\rangle\rangle$, $\langle\langle \mathbf{jE}\rangle\rangle$, and for distributions of principal parameters everywhere except a small region near the discontinuity of the model conductivity. A comparison of distributions of the y component of the electric field in the wall (E_{yw}), obtained using two different models shows that, in the case of constant surface conductivity, (σ_{wo} = const.), the results obtained using a simplified model of parallel equipotentials may be made to approach exact results with the aid of a simple additional assumption that has been used [66] to simplify finite relationships:

$$\sigma_w = \sigma_{wo}\,\sigma(y,0)/\sigma(0,0) \; .$$

2.5.5.5. Transverse Pressure Gradient

Considerable flow asymmetry relative to the channel axis can be expected in an MHD generator channel flow in the presence of transverse current when the Hall parameter exceeds unity [42, 43]. The characteristic two-dimensional effects arising in the presence of flow asymmetry were considered in 2.4. for an inviscid flow. The feasibility of taking this effect into account within the framework of quasi-one-dimensional calculations was discussed elsewhere [44]. Analysis of the two-dimensional distributions of gasdynamic and electrodynamic parameters of the flow at moderate Mach numbers ($M \lesssim 2$) and MHD interaction parameters ($S < 0.25$) shows that one can with reasonable accuracy utilize the following approximate expression for the static pressure distribution across the channel:

$$\psi_p = 1 - \frac{1}{p(x)} \int_o^y j_x B \, dy, \quad \frac{\partial\psi_p}{\partial z} = 0 \; ,$$

$$\alpha_p = \frac{1}{A} \int_A \psi_p dA \; .$$

Transverse velocity and total enthalpy profile functions can still be represented by expressions (2.123). Then, except for the expression for α_p, the system of quasi-one-dimensional equations maintains its standard form.

In calculating the thickness of the boundary layers along the electrode walls, one should use the values of static pressure on the wall as the external flow parameters, whereas one can apparently use the static pressure averaged over the cross section when calculating the boundary layers on insulating walls.

The influence of the transverse pressure gradient on the integral parameters of the MHD generator and the asymmetry of development of boundary layers on the anode and cathode walls can be illustrated by calculating the supersonic flow in an ideally segmented Hall MHD generator [44]. A channel was considered with linearly increasing interelectrode distance and a constant distance between insulating walls. The area of the cross section at the channel inlet (A_1) was 0.6 m by 0.6 m and the cross section area at the exit (A_2) was 0.6 m by 0.96 m. The total length of the channel, L, was 6 m and the active section of the channel extended between $1 \leq x \leq 5$. The magnetic field was assumed to be constant in the active zone of the channel and exponentially decreasing in the inlet and exit sections of the channel. The initial conditions at the channel inlet were: Mach number, M_1 = 2.5; static temperature, T_1 = 2500 K; flow rate, $\dot{m}$ = 60 kg/s; momentum thickness loss at the inlet, δ_1^{**} = 0.13×10^{-2} m, and wall temperature, T_w = 1200 K. It was assumed that the working fluid obeyed the equation of state of a perfect gas with the adiabatic index $\gamma = 1.14$. The boundary layers at the electrode walls were taken into account. Their thickness was calculated from equations (2.125) on the assumption that the thickness of the thermal layer at each wall is equal to the thickness of the dynamic layer.

Voltage-current characteristics of the MHD generator were calculated with the transverse pressure gradient taken into account and, also, disregarding the transverse pressure gradient (Fig. 2.31). The differences between these curves can be attributed to a substantial increase of the effective internal resistance of the generator because of the nonuniformity of the static pressure profile across the channel. At a specified transverse current, an increase in the internal resistance leads to a decrease in the transverse current and, consequently, to a decrease of transverse potential difference and total power output.

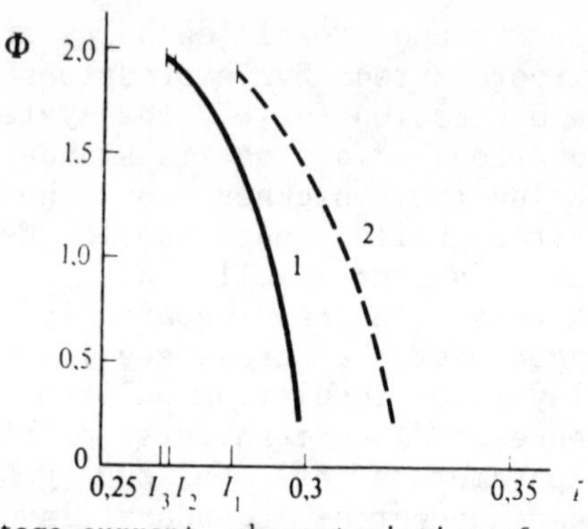

Fig. 2.31 Voltage-current characteristics of a supersonic Hall MHD generator, calculated taking the transverse pressure gradient into account (1) and neglecting the transverse pressure gradient (2)

The potential distribution along the channel axis,

$$\Phi(x) = - \int_0^x E_x dx ,$$

is shown in Figure 2.32. The dashed curve in this figure shows the potential distribution at $\alpha_p = 1$, calculated at the same total transverse current. The drop in the total power output due to nonuniformity of static pressure, which is proportional in this case to the Hall potential difference, is approximately 20 %.

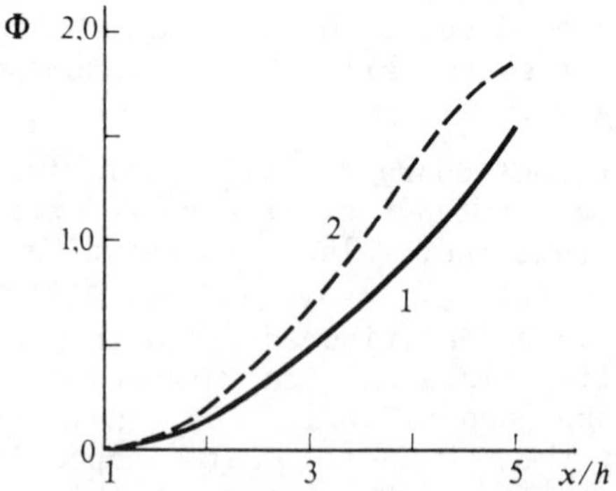

Fig. 2.32 Axial potential distribution in a supersonic Hall MHD generator, calculated with the transverse pressure gradient taken into account (1) and with the transverse pressure gradient neglected (2)

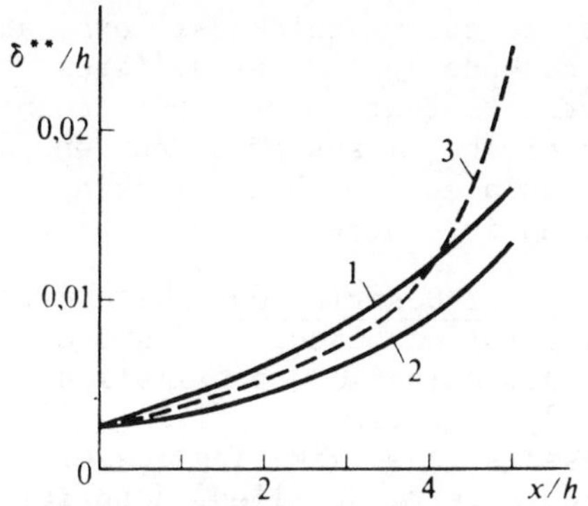

Fig. 2.33 Development of the boundary layer at the electrode walls in a supersonic Hall MHD generator with the transverse pressure gradient taken into account (1,2), with the transverse pressure gradient neglected (3): 1 - at the anode wall, 2 - at the cathode wall

Figure 2.33 shows the variation of the momentum thickness loss along the channel axis at the anode and the cathode. For comparison, this figure also shows variation of the momentum thickness loss in the flow calculated for $\alpha_p = 1$. It can be seen that the transverse pressure gradient causes considerable difference between the integral characteristics of boundary layers at the anode- and cathode-walls and qualitatively changes the nature of the boundary layer growth in comparison with the constant transverse pressure case. It should be noted that, in addition to direct influence of the variation of internal resistance of the generator on its output parameters, the transverse pressure gradient can lead to considerable variation of flow parameters, with a possibility of boundary layer separation.

Figure 2.32 shows regions of separated flows determined from criteria used in standard gasdynamics [67]. Inclusion or exclusion of the transverse pressure gradient leads to different qualitative conclusions. At a specified current, $I \leq I_1$, calculations for $\psi_p = 1$ lead to a separated flow, although, in reality (calculations for $\psi_p \neq 1$), the channel in the region $I_1 > I > I_2$ can still operate in a nonseparated mode (see Fig. 2.33). At a specified total current, $I < I_3$, the separation first occurs at the anode wall. At a specified potential difference, the current that can be extracted from the channel in the nonseparated mode is lower than the current calculated subject to the conditions $\psi_p = 1$.

In any case, in the example considered, the boundary layer separation criteria based, in one way or another, on the use of the momentum thickness loss, are satisfied at the anode and cathode walls at different values of total transverse current that does not coincide with the critical value of the transverse current at which occurs separation calculated without taking the transverse pressure gradient into account.

2.5.6. Level of Description. Within the framework of the modeling assumptions made, the system of equations of the quasi-one-dimensional approximation (2.120)-(2.127), (2.131), (2.132), (2.136), (2.141)-(2.143), (2.145), actually represents the limiting case. However, in a practical sense, a very detailed description of the flow is frequently unnecessary. Replacement of any equations of the model by simplified equations requires, as a result, more or less substantial corrections to the other equations of the model, in order to satisfy a natural requirement of agreement pertaining to accuracy and detail of description of various physical processes that determine flow parameters of interest in an MHD generator channel.

As the simplest, "lowest" level of the quasi-one-dimensional approximation, consider next a system of equations the left-hand sides of which formally correspond to a homogeneous canonical flow. In this case, assuming that the influence of nonuniformity of the gasdynamic parameters in the cross section of the channel, or the momentum and total enthalpy flows, can be taken into account in a very simple manner (i.e., with an accuracy on the order of δ^2/a^2, where δ is the nonuniformity factor, the average of the product is equal to the product of the averages), it will be assumed that

$$\alpha_1 = \frac{1}{A\rho U}\int_A \rho u \, dA, \quad \alpha_2 = \alpha_1 \langle \psi_u \rangle ,$$

$$\alpha_3 = \alpha_1 \langle \psi_i \rangle, \quad \alpha_p = 1 .$$

Then, the gasdynamic part of the general system of equations assumes the form

$$\frac{d}{dx}(\rho U \alpha_1 A) = 0 \quad \text{or} \quad \langle \rho u \rangle A = m ,$$

$$\frac{1}{A}\frac{d}{dx}(m\langle u \rangle) + \frac{dp}{dx} = F_x - k_\xi c_f^0 \frac{\langle \rho u \rangle \langle u \rangle}{2}\frac{\Pi}{A} ,$$

$$\frac{1}{A}\frac{d}{dx}(m\langle i_o \rangle) = -N_V - k_\alpha St^o \langle \rho u \rangle (i_o - i_w)\frac{\Pi}{A} ,$$

where m is the flow rate and $\langle u \rangle$, $\langle i_o \rangle$ are the velocity and total enthalpy averaged over the cross section, respectively.

In this case, the structure and dimensions of the near-wall velocity and temperature nonuniformities, as well as the friction and heat exchange coefficients, are given by the following simple expressions:

$$\delta(x) = 0.37\ \bar{x}\ Re_{\bar{x}}^{-0.2},\quad \delta_i(x) = 0.37\ \bar{x}\ Re_{\bar{x}}^{-0.2} Pr^{-2/3},$$

$$c_f^0 = 0.0576\ Re_{\bar{x}}^{-0.2},\quad St^0 = 1/2\ c_f^0\ Pr^{-0.75},$$

where $\bar{x} = x + 4A(0)/\Pi(0)$.

It will be assumed that, for the accuracy required, the thermodynamic and transfer properties of the working fluid at all levels of the quasi-one-dimensional description must correspond to the working fluid under consideration.

In determining the simplest level of electrodynamic equations, we shall begin with the assumptions that the Hall parameter depends weakly on the temperature, and that the modifications of the "base" model are minor. We shall also utilize the assumption already made about the electrodynamics of the channel. In this case, the basic electrodynamic relations are simplified considerably and can be represented in the form

$$\varepsilon_x = \varepsilon_y = \nu_y = 1,\ (1+\beta^2)\sigma_{wy}/\langle\sigma\rangle b \ll 1\ ,$$

$$\langle\sigma_x\rangle = \langle\sigma\rangle + \sigma_{wx}\Pi/A,\ \langle\nu\beta\rangle = \langle\beta\rangle_x = \beta\ ,$$

$$\langle\sigma\psi_u\rangle = \langle\psi_u\rangle,\quad \langle\delta\psi_u\rangle = 0\ ,$$

$$\langle j_y\rangle = -\langle\sigma\rangle\left(\langle u\rangle B + \frac{V_y - \Delta V}{2a} - \beta\frac{\langle j_x\rangle}{\langle\sigma\rangle}\right)/G\ ,$$

$$E_x = -\beta\left(\langle u\rangle B + \frac{V_y - \Delta V}{2a} - \frac{\langle j_x\rangle}{\langle\sigma\rangle\beta}(G+\beta^2)\right)/G\ ,$$

$$G = (\Xi-1)\beta^2 + \Xi, \quad \Xi = \langle\langle\sigma\rangle_z^{-1}\rangle_y \langle\langle\sigma\rangle_z\rangle_y ,$$

$$\Delta V = \Delta V^+ + \Delta V^-, \quad F_x = \langle j_y\rangle B ,$$

$$N_V = - \langle j_x\rangle E_x - \langle j_y\rangle (V_y-\Delta V)/2a .$$

The coefficients in Ohm's law must be averaged over the channel cross section using profile functions (2.123) and specified variations of conductivity, σ (p, T), and Hall parameter, $\beta/B = \beta_o (p, T)/B_o$, with temperature and pressure, i.e.,

$$T = T(i,p), \quad i = i_o - u^2/2, \quad u = \psi_u U ,$$

$$U = \langle u\rangle/\langle\psi_u\rangle, \quad i_o = \psi_i \langle i_o\rangle/\langle\psi_i\rangle ,$$

$$i(y,z) = i_o(y,z) - 1/2u^2(y,z) =$$

$$= \psi_i \langle i_o\rangle/\langle\psi_i\rangle - 1/2\psi_u^2 \langle u\rangle^2/\langle\psi_u\rangle^2 .$$

Greater accuracy can be attained by utilizing more rigid electrodynamic relations, (2.134)-(2.136), and successive improvement of calculations of the structure of the gasdynamic flow. Improvement in accuracy is required in cases such as when calculating strongly nonuniform flows (with thick boundary layers) and, also, when it is necessary to take into account the influence of the nonideal channel insulation.

Hence, the next level of the electrodynamic part of the model are relations (2.134)-(2.136). The model of parallel equipotentials ($\varepsilon_y = 1$) is assumed to be valid in every conducting region of the channel cross section (both the flow and surface conductivity layer, σ_w), subject to the condition $\sigma_{wy} \ll \langle\sigma\rangle_F b$ and the use of an additional model assumption on the nature of the variation of surface conductivity, σ_{wy} along the y axis:

$$\sigma_{wz} = \sigma_{wo}\,\sigma(y,0)\,\sigma(0,0) .$$

Calculation of flows with parameters that depend substantially, according to preliminary estimates, on certain effects disregarded at the second level of approximation requires addition of corrections that can be done as described in 2.5.5. Because the effects of finite segmentation and current constriction in the near-electrode layer were evaluated only in terms of certain effective parameters of the basic quasi-one-dimensional model, they must be calculated or evaluated starting with the non-one-dimensional calculations, or from the experimental data.

When taking into account the transverse pressure gradient in the channel, it is necessary to use finite or differential equations to calculate the development of the boundary layer along the length of the channel, separately for each wall (anode, cathode, and insulating). It is also necessary to reject the assumption of symmetry in the distribution of gasdynamic parameters along the channel cross section, which is effectively used in averaging the coefficients in Ohm's law in all other cases.

It should be noted that the use of a nonuniform profile function for the static pressure, $\psi_p \neq 1$, requires iteration in calculating gasdynamic distributions and in averaging electrodynamic variables over each channel cross section.

A switch to calculations of channels other than the segmented Faraday channel can be made within the framework of the basic model described, by using other closure equations for channel electrodynamics. For example, the following relationship between the axial and transverse components of the electric field is used in the case of a series (diagonal wall) channel:

$$\langle E_y \rangle = - E_x \operatorname{ctg} \alpha , \tag{2.168}$$

where α is the angle between the "outer" equipotential and the channel axis (wall angle in the frame channel). In addition, it is necessary, in the case of a single-load channel loading scheme, to use the condition of conservation of the total current in the load,

$$I_H = A \left(\langle j_x \rangle_A - \langle j_y \rangle_A \operatorname{ctg} \alpha \right) - E_x \sigma_{wx} \Pi , \tag{2.169}$$

which for an arbitrary load distribution becomes the differential equation,

$$\frac{dI_H}{dz} = - i_H^+ - i_H^- .$$

In calculating frame channels with "cold" side walls, when the resistance of the boundary layers at these walls exceeds considerably the resistance of the boundary layers at the electrode wall, it is possible to limit one's self to the calculation scheme used for diagonal wall channel ((2.168), (2.169)). Expressions (2.162) can be used when the resistance of the boundary layers at the side, diagonally-conducting walls is comparable to, or smaller than the resistance of the boundary layers at the electrode walls.

Improvement in the accuracy of calculating the gas-dynamic structure of the flow is achieved by sequential use of a more detailed description of friction and the heat exchange processes at the channel walls. In particular, one can suggest the following sequence. Instead of a finite equation for the thickness of the dynamic boundary layer, δ, one can use a differential equation for the momentum thickness (2.125). The thickness of the thermal boundary layer, δ_i is determined from the finite equation (2.128). In this case, the friction coefficients and the Stanton number are calculated from (2.129), (2.130), taking into account correction factors to compensate for the difference between the flow conditions from the standard conditions (flow past an incompressible fluid plate) [54]. In the general case, additional correction factors should be introduced to take into account the influence of the electrodynamic processes on friction and heat transfer. Thus,

$$c_f^o = k_\xi^o \Psi c_f^o = k_\xi c_f^o, \quad St = k_\alpha^o \Psi_s St^o = k_\alpha St^o ,$$

$$c_f^o = c_f^o(Re^{**}), \quad St^o = St^o(Re_i^{**}) .$$

The energy equations for the thermal boundary layer at the electrode and side walls (2.126) are used for a more detailed description of the influence of heat transfer on the flow in the MHD generator channel. In this case, as well as in those immediately foregoing, the Stanton number is also calculated from the enthalpy thickness, δ_i^{**}, determined by integrating (2.126) jointly with the general system of quasi-one-dimensional equations.

Finally, a complete system of integral equations for the boundary layer, i.e., two for each electrode channel wall and two for the boundary layers at the insulating (side) wall, must be used when the transverse pressure gradient exerts considerable influence, and when it is necessary to take more fully into account the influence of this effect on friction and heat transfer in the channel. As already noted, in this case the external flow parameters for each of the boundary layers are determined from the static pressure at the wall. However, it should be noted that, for an almost developed flow (boundary layer thickness is almost equal to the channel half-width), such an approach requires special justification.

A fairly effective method of closure of the principal quasi-one-dimensional system of equations, (2.120), by means of profile functions determined by solving equations of a two-dimensional boundary layer [68, 69] has been recently developed. In addition to greater reliability of profile functions, the connection between local coefficients of frictions, c_f, and the Stanton number, St, and the governing parameters of the flow is automatically determined during the solution.

Selection of one or the other level of description must be determined in accordance with the purpose of the specific calculation. Even from general considerations, it is clear that use of very accurate equations for a certain physical process does not guarantee an increase in accuracy of the overall quasi-one-dimensional calculation. On the other hand, it is expedient to use modeling of the process under consideration sufficiently detailed to reveal the parametric influence of a specific physical process on the flow parameters as a whole. Furthermore, even the simplest level of description of other physical processes may provide the necessary accuracy of calculating the "background" upon which are superposed the effects of the physical process of interest.

Characteristic examples include the two-dimensional electrodynamic problems discussed in the first two chapters. For example, investigation of two-dimensional effects in the plane of the channel cross section with diagonally-conducting side walls, their dependences on gasdynamic parameters that determine the flow structure that, in this case, determines the nature of two-dimensional current and potential distributions, does not require calculation of the flow throughout the whole channel, because a priori evaluations reliably determine the region of application of the results. These results, expressed in terms of parameters of the quasi-one-dimensional theory, can be used in the corresponding quasi-one-dimensional calculation, instead of a direct solution of the two-dimensional boundary problem for each cross section.

2.5.7. Experimental Data Processing Using the Quasi-One-Dimensional Approximation. The degree of correspondence of the chosen canonical flow with a specific class of flows must be checked by analyzing the experimental data and comparing these data with the results of calculations obtained by use of the quasi-one-dimensional model. In addition, in making such a comparison, one also selects or improves the empirical constants or functional dependences in the equations of the quasi-one-dimensional model.

Thus, comparison of the results of calculations using the quasi-one-dimensional model with the experimental data justifies the adequacy of the model accepted for the class of flows under consideration, and indicates the limits of applications of the model. Without such a comparison, one cannot ascertain the accuracy of calculations of flow parameters for the conditions specified, even when calculations are based on the use of a quasi-one-dimensional model that can be made as complex as desirable (complex in the sense of taking into account various physical factors). On the other hand, comparison can provide justification for the use of sufficiently simple models for calculating the flow under consideration with acceptable accuracy.

When the correspondence of the quasi-one-dimensional model with the class of flows under consideration is established, processing of the experimental data by means of this model provides a reliable tool for the determination of parameters not measured directly during the experiment.

As an example, consider processing of the experimental data obtained on different channels (1D, R, RM).

Consider the simplest, "lowest" level of the quasi-one-dimensional model to be the canonical flow. The thermodynamic properties of the working fluid (combustion products seeded with alkali metals) were given in the form of tabulated data, calculated by use of the technique described elsewhere [73]. The dependence of conductivity and Hall parameter on the thermodynamic parameters of the working fluid were represented in the form,

$$\sigma = k_\sigma f(c)\ \sigma_o\ ,$$

$$\sigma_o(T,\ p) = 4.45\cdot 10^5 T_1^{3/4}\ \exp\{-\ 27\ 200/T_1\}p^{-0.5},$$

$$\beta = (0.14+0.2\cdot 10^{-3}T)\cdot 10^5 B/p\ , \qquad (2.170)$$

where $T_1 = T - 80$, σ_o (T, p) corresponds to combustion products seeded with K_2CO_3 at 1 % potassium concentration by mass. According to report [33], f(c) takes into account the variation of seed concentration [33], and the factor, k_σ, is introduced (by analogy with k_ξ and k_α) to account for the possible difference between the conductivity in the channel and that calculated using relations (2.170). The set of parameters to be measured for use in calculations as the initial data and for comparison with the results of calculations is chosen on the basis of the structure of the theoretical model and the set of quantities measured during the experiment. In particular, the load current, $I_H(x)$, serves as the initial data for Faraday channels, and the axial components of the electric field, $E_x(x)$, or the Faraday voltage, $V_y(x)$, are used as the initial data for diagonal channels. Various parameters are used to establish relationships between calculated and experimental data. For example, the axial component of the electric field, $E_x(x)$; Faraday voltage, $V_y(x)$, and the pressure, p(x), are chosen for Faraday channels. The load current, $I_H(x)$, and the pressure, p(x), are chosen for diagonal channels. The magnetic field distribution, B(x); geometrical parameters of the channel, a(x) and b(x); wall temperature, T_w, as well as the boundary conditions (flow rate, m, and the stagnation enthalpy, i_{01}, at the input, and the pressure, p_2, at the output), are used as the initial data in calculations. The potentials of all frames or electrodes relative to the ground, load currents, magnetic field, static pressure distribution along the channel, are measured during the experiment on the U-25 facility. The injection rate of fuel, oxidizer, and seed used, as well as a number of other parameters, make it possible to determine adequately the total enthalpy at the inlet into the channel.

Hence, the experimental data specify all of the parameters required for calculations. In addition, in view of the excess of data, it is possible to compare independently the experimentally determined distributions with calculated distributions. Processing of the experimental data acquired on Faraday and diagonal channels of the U-25 facility shows that the empirical quantities can be divided into three groups, based on the physical meaning and model interpretation.

The first group includes quantities (k_ξ and k_α) that depend weakly on the design and operating mode of the channel. In principle, these quantities depend on the flow mode. However, for the class of flows under consideration, the effect of this dependence on the results of the calculations falls within the limits of the experimental accuracy and the degree of agreement between the experimental and theoretical distributions.

The second group is formed by quantities that depend primarily on the design characteristics of the channel (segmentation size, electrode materials, condition of insulation, - ΔV, σ_w. The agreement of calculations with the experimental data is reached at specific values of these quantities that are constant for a specific channel.

The third group includes quantities that actually are a function of the flow mode and of design characteristics. For example, the conductivity correction coefficient, k_σ, depends not only on the structure of the flow, but also on seed injection method.

Obviously, the results of processing the experimental data using models containing as few as possible of the empirical quantities that include, predominantly, quantities from the first and second groups, are of the greatest practical value. From the point of view of the foregoing classification, separation of the quantities in the first and second groups and determination of their values is the aim of processing the experimental data, whereas determination of the values of the quantities of the third group is only a means of comparison. Within the framework of the model, numerical values of the quantities from the first and the second groups characterize the physical processes or peculiarities of the design of the channel.

For example, consider the choice of value of k_ξ. The coefficient k_ξ is chosen from operating modes in the absence of MHD interaction, when the calculated static pressure distribution along the length of the channel agrees with the experimental distribution. Figure 2.34 shows typical static pressure distributions in the R channel. The data scatter observed is characteristic of all channels, and may be attributed to measurement error as well as characteristics of the flow. For the flow mode under consideration, within a certain range of measurement of k_ξ, the degree of agreement between the calculated curves and the experimental curves is consistent.

It should be noted that the value, $k_\xi = 2$, from this range provides good agreement of the results of calculations with the experimental data for both pure subsonic operating modes and for a weakly supersonic ($M_{max} \simeq 1.15$) flow with a density jump (Fig. 2.35). Variation of k_ξ in the range $1.85 \leq k_\xi \leq 2.15$ does not lead to noticeable disagreement between calculations and experimental data.

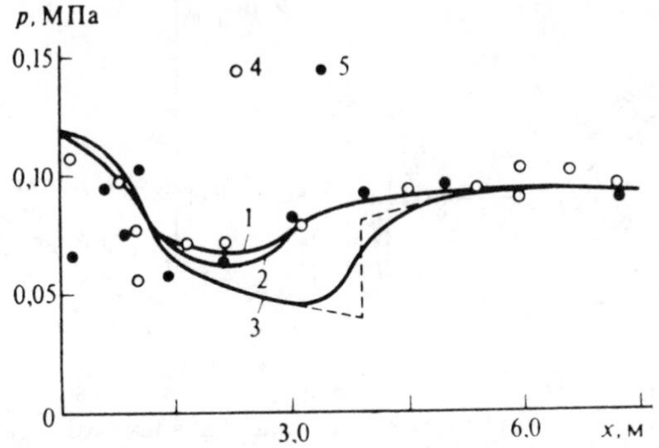

Fig. 2.34 Calculated (1-3) and the experimental (4,5) static pressure distributions in the R channel at a flow rate of 26 kg/s and at the following values of k_ξ: 1-2.25; 2-2; 3-1.75; 4,5, respectively, at the cathode and anode walls

When processing experimental data acquired on the 1D and the RM channels, it was possible to choose a range of values of k_ξ that provides agreement between the results of the calculations and the experimental data for various modes. Furthermore, $k_\xi = 2$ turned out to be useful for a description of flows without MHD interaction in all channels under consideration. An exception is the case of

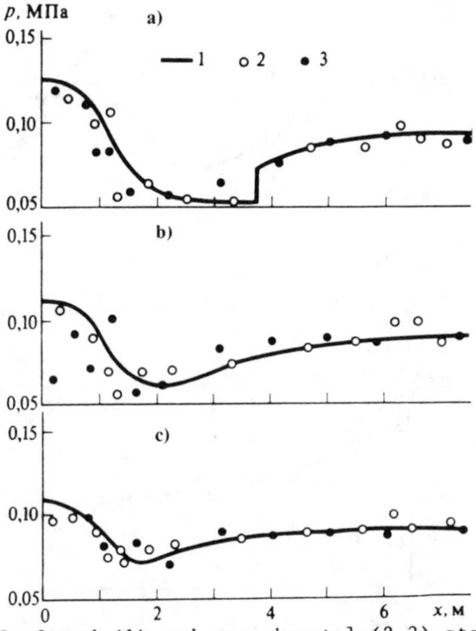

Fig. 2.35 Calculated (1) and experimental (2,3) static pressure distributions in the R channel at $k_\xi = 2$ at the following flow rates (in kg/s): a-26, b-22.5, c-21.5; 2,3, on the cathode and anode walls, respectively

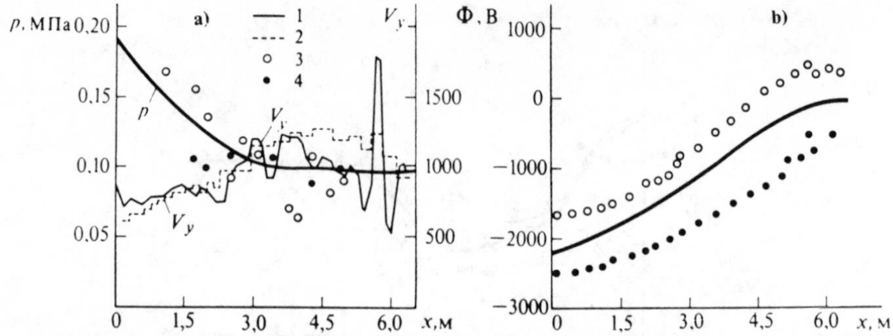

Fig. 2.36 Static-pressure and Faraday-voltage distributions (a), Hall and electrode potentials (b) in the 1-D channel (run 58, measurement 218), 1 = results of the calculations, 2-4 = experimental data

stagnation of supersonic flows with supercritical values of the Mach number ($M > M_{cr} \approx 1.3$). As the Mach number increases, the supersonic flow induces boundary layer separation before the density jump [67, 73]. In this case, the flow behind the jump (or system of jumps) is not described by this model.

Variation of the value of the coefficient k_α exerts a weak influence on the static pressure distribution. In

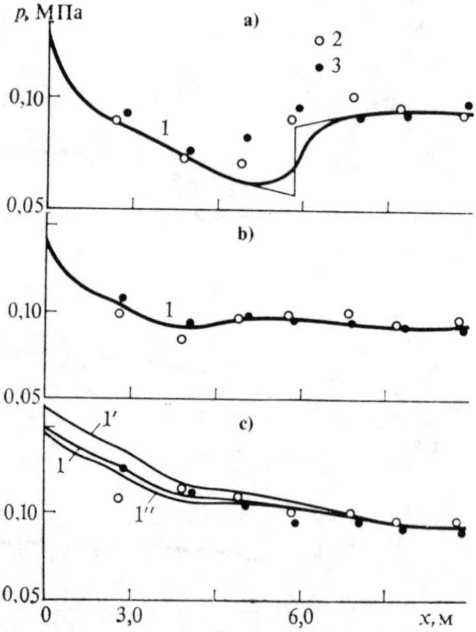

Fig. 2.37 Static-pressure distribution in the RM channel during run 62, measurements 682(a), 768(b), and 1905(c): 1,1′,1″ - calculations for k_σ = 0.977, 1.087, and 0.915, respectively; 2,3, experimental data

this connection, the choice of k_α is made difficult by the fact that the temperature regime of the walls and conditions of heat exchange at them are known only approximately. As a result of this and the well-known fact that the heat transfer coefficient depends weakly on the flow conditions, k_α was assumed to be equal to one during analysis applicable to all modes.

The value of k_ξ determined in operating modes in the absence of MHD interaction was used in calculations of flows in loaded channels. A completely satisfactory agreement of calculations with the experimental data (Figs. 2.36 and 2.37) occurs at different levels of MHD interaction. On the basis of this, it is concluded that, in the model used, flows considered k_ξ depend only weakly on the level of the MHD interaction.[3]

The empirical parameters, ΔV, σ_w, and k_σ, are determined during processing of data acquired during flow modes accompanied by MHD interaction. The near-electrode potential drop in the 1-D Faraday channel, associated with the near-wall surface effects, was estimated to be approximately 50 V [69]. Apparently, this parameter is somewhat lower for the R and RM frame channels. Hence, ΔV was assumed to be 50 V for the 1-D channel and 25 V for the R and RM channels. It can be seen from Table 2.4 and Fig. 2.36 that satisfactory agreement between results of the calculations and the experimental data was obtained for the 1-D Faraday MHD channel.

Processing of the experimental data shows that, at a specified value of ΔV, practically a single pair of values, σ_w and k_σ, corresponds to each pair of values of the power output and the Hall voltage. From Table 2.4 it can be seen that, for the 1-D channel, σ_w is the same for various modes. Therefore, it can be assumed that, within the framework of the models used, σ_w determined characterizes the nonideal behavior of the insulating properties of the 1-D channel. The power losses during characteristic operating modes corresponding to $\sigma_w = 0.7$ S may reach 10-20 %.

When processing experimental data acquired on frame channels, i.e., at specified voltage distributions $V_y(x)$, variation of k_σ leads to a change of $j_y(x)$ and, consequently, of $p(x)$. Therefore, in each case, pressure correlation may be attained for a certain and, as shown experimentally (see Fig. 2.37), fairly narrow range of

---

[3] A more complete analysis of the dependence of k_ξ on the structure of the flow, MHD interaction, etc., is given in Reference 72.

Table 2.4

Experimental and Calculated Characteristics
of the 1D Faraday Channel

Channel Parameters	Run 47, measurement 92		Run 58, measurement 218	
	Experimental	Theoretical	Experimental	Theoretical
Magnetic field B, T	2.0	2.0	1.85	1.85
Total load current I_H, kA	12.9	-	19.2	-
Hall voltage V_x, kV	2.16	2.37	2.25	2.26
Power output N, MWt	11.8	10.7	18.4	17.4
k_σ coefficient	-	1.0	-	1.55
Wall conductivity σ_w, S	-	0.7	-	0.7

Table 2.5

Experimental and Calculated Characteristics
of R and RM Diagonal Channels

Channel Parameters	R Channel, run 41			RM Channel, run 62		
	Meas. 1664	Meas. 3624	Meas. 3658	Meas. 682	Meas. 768	Meas. 1905
Magnetic field B, T	1.32	2.0	2.0	0.89	1.65	2.0
Hall volage V_*, kV	0.83	1.16	1.18	1.12	1.45	2.81
Power output N, MWt	0.85	1.51	3.11	1.81	4.95	8.12
Calculated power N, MW	0.78	1.48	3.15	2.12	4.83	8.66
k_σ coefficient	0.455	0.457	0.665	0.805	0.833	0.977
Wall conductivity σ_w,S	0.1	0.1	0.1	0.0	0.0	0.0

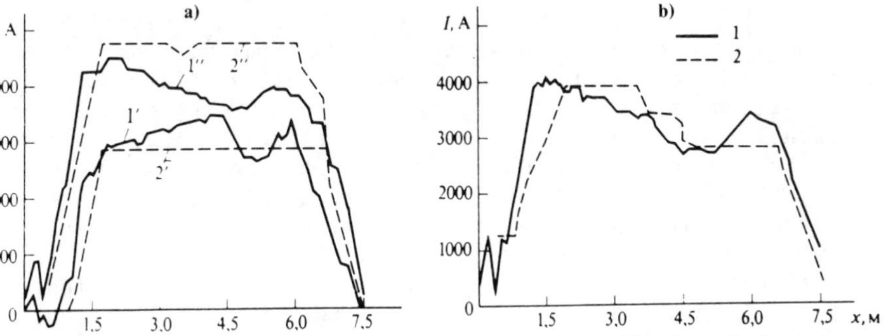

Fig. 2.38 Load current distribution in the RM channel during run 62, measurements 682 (1′,2′) and 768 (1′′,2′′)(a) and 1905(b): 1, results of the calculations: 2, experimental data

values of k_σ. As can be seen from Table 2.5 and Figs. 2.38 and 2.39, an adequate agreement was obtained between the results of calculations and the experimental data.

Similar to the case of a Faraday channel, the values of k_σ that lead to an agreement between the calculated and experimental data for various flow modes differ from each other, whereas the values of σ_w in each channel are constant.

The σ_w for frame channels (σ_w = 0.1 S for the R channel and σ_w = 0 for the RM channel) indicate that, within the framework of the model used, the influence of nonideal insulation of the structure is considerably less than in the 1-D Faraday channel (σ_w = 0.7 S), e.g., for the RM channel one can neglect nonideal behavior of the insulation.

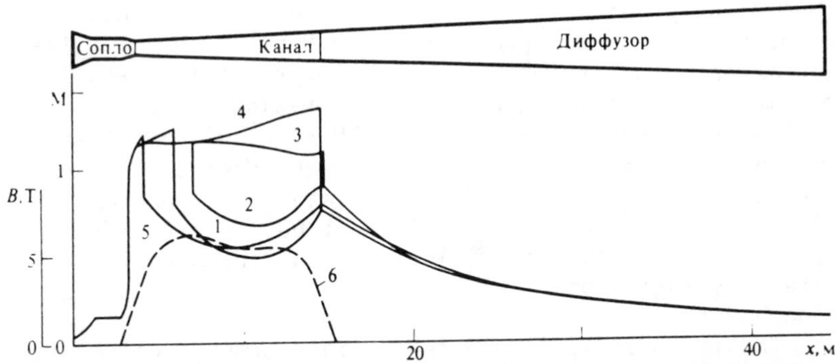

Fig. 2.39 Geometry of the flow-through sector, Mach number distribution (1-5), and magnetic field profile (6) in a diagonal channel of a commercial-scale MHD generator for nominal combustion products flow rate of 190 kg/s when the load is changed; sequence of modes 1-5 corresponds to increasing load current, 3 - nominal mode

2.5.8. Calculation of a Flow in a Channel of a Commercial-Scale MHD Generator. When calculating a flow in an MHD generator, it is necessary to take into account conditions at which the operating mode of the channel is established when the channel is part of the thermal loop of an MHD power plant. These are specified by other assemblies of the electric power plant, compressor systems, air preheaters, combustion chamber, inverter subsystem, steam generator, et al. On the other hand, the influence of the channel on the operation of those assemblies quite frequently must be taken into account. Inverse coupling of the channel with certain components of equipment is difficult to formalize in the form of finite relations. In many cases, it becomes necessary to calculate jointly the operation of the channel and other components of the equipment. The latter is most valid for the nozzle and diffuser. When performing variant calculations of a flow in a channel, it is practically impossible to specify the universal characteristics of the nozzle and the diffuser before calculations. Therefore, determination of the flow mode and characteristics of the MHD channel must be performed with the use of the results of a joint calculation with a nozzle and diffuser.

Consider an example of such a calculation under conditions existing in a commercial-scale 500-MW MHD power plant [76]. Calculations must be performed for the gasdynamic loop, from the inlet into the nozzle to the diffuser outlet. The flow rate of the working fluid and the stagnation enthalpy at the inlet into the nozzle, as well as the static pressure at the diffuser outlet, are used as boundary conditions. These parameters are specified by characteristics of the rest of the equipment, especially the air preheater, combustor, and the steam generator. In the 500-MW MHD power plant scheme [76], parameters determining equipment operation are as follows: combustion products flow rate, 190 kg/s; maximum pressure in the combustion chamber, 1 MPa; steam generator pressure, 0.1 MPa. The fuel is natural gas preheated to 773 K, the oxidizer is the air enriched by oxygen to 30 % by weight, and its excess is 0.95, and the seed injection rate (dry potassium) is 2 % by mass of the combustion products flow rate.

The two principal problems encountered in calculating flows in an MHD generator channel are construction of nominal operating modes and determination of partial operating modes. Construction of nominal flow modes in an MHD generator is reduced to selection of the optimum geometry of the flow-through sector and the load distribution along the channel axis. This is attained by solving the optimization problem in which the governing parameters are the velocity distribution functions u(x) and the electrical efficiency function $\eta(x)$. The magnetic field distribution V(x) and the loading scheme of the channel are specified.

The nominal flow mode in an MHD generator channel with diagonal commutation of electrodes and axially distributed load ($\alpha = 30^0$) is supersonic with smooth stagnation along the channel length [77]. The electrical efficiency distribution in the channel corresponds to a boundary maximum, i.e., maximum value of $\eta(x)$ that is possible in a diagonal loading scheme, is reached in each cross section of the channel. Electrical efficiency increases along the length of the channel, with the mean value of $\eta(x)$ being 0.8. A direct jump occurs at the channel outlet. The subsonic flow stagnates in the diffuser.

When the loading mode changes from nominal toward the no-load mode (see curves 2, 1 in Fig. 2.39), the jump is displaced upward along the flow. When loading changes toward the short-circuit mode, the flow in the channel is first accelerated (curve 4) and the amplitude of the jump in the diffuser increases. After this, the jump is displaced sharply toward the channel inlet (curve 5). This evolution of the flow in the diagonal channel is associated with characteristics of its electrodynamics.

The loading characteristics of the channel at various combustion products flow rates (Fig. 2.40) are in the shape of a loop, i.e., two loading modes with various loading points and, generally speaking, with different power outputs correspond to a single value of efficiency.

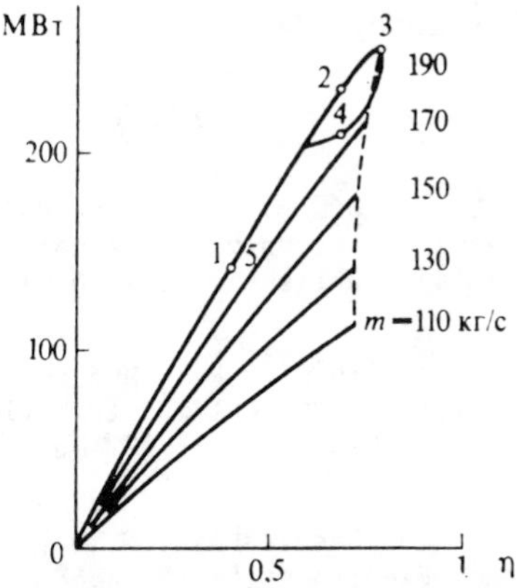

Fig. 2.40 Variation of the power output of a commercial-scale MHD generator with the electrical efficiency averaged over the channel length at various combustion products flow rates; sequence of points 1-5 corresponds to an increase in the loading current

However, noticeable splitting of the loading curve is observed only at the nominal flow rate (m = 190 kg/s) at high values of η. In the case of lower flow rates, the difference in power outputs at various current but the same electrical efficiency is insignificant.

Computation of loading characteristics at various flow rates is required for the determination of partial operating modes of the generator and the power plant. Analysis shows that the most economical method of controlling the load of an MHD power plant is to decrease combustion products flow rate in the channel. All other control methods without changing the flow rates result in a sharp decrease in efficiency of the MHD power plant. In addition, a decrease in the power output of an MHD generator when the nominal flow rate is maintained leads to excessive thermal loading of the steam generator. In the case of a non-nominal combustion products flow rate, the load should be regulated until the generator operates in a maximum power output mode (the only way to attain maximum efficiency of the electric power plant).

The loading characteristics of a diagonal MHD generator at each value of the combustion products flow rate are characterized by a single maximum point at which maximum power output is reached. In Fig. 2.40, these points are connected by a dashed curve that represents the control curve of a diagonal MHD generator under consideration that is part of an MHD power plant. The nature of this curve makes it possible to reach the important conclusion that diagonal MHD generators allow a high degree of power output control, while maintaining high electrical efficiency.

REFERENCES

[1]Vatazhin, A. B., Lyubimov, G. A., and Regirer, S. A., "Magnitogidrodinamicheskiye techeniya v kanalakh" (Magnetohydrodynamic Flows in Channels), Publishing House Nauka, Moscow, 1970.

[2]Hurwitz, H., Jr., Kilb, R. W., and Sutton, G. W., "Influence of Tensor Conductivity on Current Distribution in an MHD Generator," Journal of Applied Physics, Vol. 32, No. 2, 1961, p. 205-216.

[3]Tolmach, I. M., and Yasnitskaya, N. N., "Hall Effect in a Channel with Segmented Electrodes," AN SSSR, Izvestiya, Energetika i transport, No. 5, 1965, p. 91-104.

[4]Oliver, D. A. and Nitchner, M., "Nonuniform Electrical Conductivity in MHD Channel," AIAA Journal, Vol. 5, No. 8, 1967, p. 1424-1432.

[5]Dzung, L. S., "Hall Effect and End Loop Losses of Magnetohydrodynamic Generators" (Russian translation of a paper), Magnetogidrodinamicheskiye generatory electricheskoy energii (MHD Generators of Electric Power), Publishing House VINITI, Moscow, 1963, p. 260-270.

[6]Vatazhin, A. B., "Certain Two-Dimensional Problems of Current Distribution in an Electrically Conducting Medium Moving Along a Channel in a Magnetic Field," Zhurnal prikladnoy mekhaniki i tekhicheskoy fiziki, No. 2, 1963, p. 39-54.

[7]Sutton, G. W., "End Losses in Magnetohydrodynamic Channels with Tensor Electrical Conductivity and Segmented Electrodes," Journal of Applied Physics, Vol. 34, No. 2, 1963, p. 396-403.

[8]Medin, S. A., "End Effect in an MHD Generator with an Optimum Load Distribution," Teplofizika vysokikh temperatur, Vol. 9, No. 6, 1971, p. 1271-1276.

[9]Medin, S. A., and Farber, N. L., "Potential and Magnetic Field Distributions in an Optimal Two-Dimensional MHD Generator," Teplofizika vysokikh temperatur, Vol. 11, No. 2, 1973, p. 390-395.

[10]Medin, S. A., "End Effect in an Optimal Two-Dimensional MHD Generator in the Presence of Conduction Anisotrophy," Magnitnaya gidrodinamika, No. 3, 1974, p. 99-104.

[11]Berezin, I. S., and Zhidkov, N. P., Computational Methods (Metody vychisleniy), Publishing House Fizmatgiz, Moscow, Vol. 2, 1960.

[12]Bityurin, V. A., Burakhanov, B. M., Medin, S. A., and Ponomarev, V. M., "Two-Dimensional Electric Fields in MHD Generator Channels, Optimum Loading Circuits," 14th Symposium on Engineering Aspects of MHD, Tullahoma, U.S.A., 1974, p. III.2.1-III.2.6.

[13]Bityurin, V. A., Burkhanov, B. M., Dronov, Yu. A., et al., "Computational Analysis of Characteristics of the U-25 MHD Generator Installation," Teplofizika vysokikh temperatur, Vol. 12, No. 2, 1974, p. 390-398.

[14]Isayenkov, Yu. I., Iserov, A. D., Kirillov, V. V., et al., "Investigation of the U-25 MHD Generator Channel," Teplofizika vysokikh temperatur, Vol. 12, No. 2, 1974, p. 399-411.

[15]Rosa, R., and Petty, S., "Status Report on the Mark-VI Long-Duration Generator Facility," 13th Symposium on Engineering Aspects of MHD, Stanford, U.S.A., 1973, p. II.7.1-II.7.3.

[16]Brogan, T. R., Hill, J. A., and Aframe, A. M., "A Preliminary Design of a Single Output Variable Wall Angle Window Frame MHD Channel for the U-25 Power Plant," Sixth International Conference on MHD Electrical Power Generation, Washington, D.C., Vol. 1, 1975, p. 267-286.

[17]Pinkhasik, M. S., and Pishchikov, S. I., "Certain Results of Investigations on the U-25 Facility," Sixth International Conference on Magnetohydrodynamic Electrical Power Generation, Washington, D.C., Vol. 1, 1975, p. 167-182.

[18]Barshak, A. E., Bityurin, V. A., Granovskiy, E. A., et al., "Construction of the RM Frame MHD Channel for the U-25 Installation," Third U.S.-U.S.S.R. Colloquium on MHD Energy Conversion, Moscow, 1976, p. 133-143.

[19]Bazarov, G. P., Kufa, E. N., and Medin, S. A., "Commuta-tion of Electrodes at the End Sector of a Series-Connected MHD Generator," Teplofizika vysokikh temperatur, Vol. 15, No. 6, 1977, p. 1276-1283.

[20]Kholshchevnikova, E. K., "The Influence of the Hall Effect on the Characteristics of an MHD Generator with Two Pairs of Electrodes," Zhurnal prikladnoy mekhaniki i tekhnicheskoy fiziki, No. 4, 1966, p. 74-84.

[21]Nemkova, N. G., and Alavidze, G. R., "Flow of Current in a Cross Connected MHD Generator with Four Electrodes," Philosophical Transactions of the Royal Society of London, Series A, Vol. 261, 1967, p. 455-460.

[22]Bityurin, V. A., Burakhanov, B. M., and Medin, S. A., "Numerical Investigation of Nonuniform Loading of a Segmented MHD Channel," Magnitnaya gidrodinamika, No. 1, 1981, p. 93-100.

[23]Bityurin, V. A., Iserov, A. D., Kovbasyuk, V. I., et al., "Certain Results of Investigations of Gasdynamics and Electrodynamics of a Faraday MHD Channel on the U-25 Facility," Sixth International Conference on Magnetohydro-dynamic Electric Power Generation, Washington, D.C., Vol. 1, 1975, p. 183-198.

[24]Sonju, O. K., and Teno, J., "An Experimental and Theoretical Investigation of a High Interaction Combustion-Driven MHD Generator," Fifth International Conference on Electrical Power Generation, Munich, Vol. 1, 1971, p. 199-212.

[25]Bityurin, V. A., "Quasi-One-Dimensional Description of Flow in an MHD Generator Channel," Abstract of Dissertation for Candidate of Physical-Mathematical Sciences Degree, IHT, 1970.

[26]Oliver, D. A., and Lowenstein, A. I., "The Three-Dimensional Electrical Conduction Field in a Slanted Electrode Wall MHD Generator," Fifth International Conference on MHD Electrical Power Generation, Munich, Vol. 2, 1971, p. 225-243.

[27]Fishman, F., "Steady Magnetohydrodynamic Flow Through a Channel of Circular Section," Advanced Energy Conversion, Vol. 4, No. 1, 1964, p. 1-14.

[28]Bityurin, V. A., Burakhanov, B. M., Zhelnin, V. A., et al., "Theoretical Investigation of Two-Dimensional Electric Effects and Development of Engineering Calculation Methods for Frame-Type MHD

Channels," Sixth International Conference on Magnetohydrodynamic Electrical Power Generation, Washington, D.C., Vol. 1, 1975, p. 501-516.

[29]Medin, S. A., and Rutkevich, I. M., "Influence of Two-Dimensional Nonuniformities on Characteristics of Frame MHD Channels," Teplofizika vysokikh temperatur, Vol. 14, No. 6, 1976, p. 1305-1312.

[30]Bityurin, V. A., Burakhanov, B. M., Zhelnin, V. A., et al., "Investigation of Two-Dimensional Electric Effects in a Frame MHD Channel," Teplofizika vysokikh temperatur, Vol. 15, No. 2, 1977, p. 390-398.

[31]Bityurin, V. A., and Medin, S. A., "Selection of the Optimal Configuration of the Frame and Shape of the Cross-Section of a Frame MHD Channel," Teplofizika vysokikh temperatur, Vol. 15, No. 5, 1977, p. 1086-1094.

[32]Rosa, R. J., Magnetohydrodynamic Energy Conversion, McGraw-Hill Book Company, 1968.

[33]Nedospasov, A. V., Poberezhskiy, L. P., and Chernov, Yu. T., Composition and Properties of Working Fluids of Open-Cycle MHD Generators (Sostav i svoystva rabochikh tel MGD generatorov otkrytogo tsikla), Publishing House Nauka, Moscow, 1977.

[34]Bityurin, V. A., and Lyubimov, G. A., "Quasi-One-Dimensional Analysis of Flow in the MHD Generator Channel," Teplofizika vysokikh temperatur, Vol. 7, No. 5, 1969, p. 974-986.

[35]Burakhanov, B. M., Medin, S. A., and Shumyatskiy, B. Ya., "Engineering Method of Calculating Flows in the Channel of a Linear Conduction MHD Generator," Magnetohydrodynamic Method of Generating Electric Power (Magnitogidrodinamicheskiy metod polucheniya elektroenergii), No. 3, 1972, p. 19-33.

[36]Godunov, S. K., Zabrodin, A. V., Ivanov, M. Ya., et al., Numerical Solution of Many-Dimensional Gasdynamic Problems, (Chislennoye resheniye mnogomernykh zadah gasovoy dinamiki), Publishing House Nauka, Moscow, 1976.

[37]Vatazhin, A. B., and Isakova, N. P., "Flow of Conducting Gas Along a Circular Pipe in a Strong Axi-Symmetric Magnetic Field," AN SSSR, Izvestiya, Mekhanika zhidkosti i gaza, No. 5, 1972, p. 145-155.

[38]Barmin, A. A., Kulikovskiy, A. G., and Lobanova, L. F., "Linearized Problem of Supersonic Flow at the Inlet into the Electrode Zone of an MHD Channel," Prikladnaya matematika i mekhanika, Vol. 29, No. 4, 1965, p. 609-615.

[39]Samarskiy, A. A., Introduction to Difference Scheme Theory, (Vvedeniye v teoriyu raznostnykh skhem), Publishing House Nauka, Moscow, 1971.

[40]Belotserkovsky, O. M., and Davydov, Yu. M., "Nonstationary Method of Large Particles for Gasdynamic Calculations," Zhurnal

vychislitel'noy matematiki i matematicheskoy fiziki, Vol. 11, No. 1, 1971, p. 182-207.

[41]Gubarev, A. V., Degtyarev, L. M., Samarskiy, A. A., and Favorskiy, A. P., "Flow of Supersonic Electrically Conducting Gas in a Nonuniform Magnetic Field," AN SSSR, Doklady, Vol. 192, No. 3, 1970, p. 520-523.

[42]Gubarev, A. V., Degtyarev, L. M., Samarskiy, A. A., and Favroskiy, A. P., "Certain Two-Dimensional Effects of Supersonic Flow of Electrically Conducting Gas in MHD Channels," Fifth International Conference on MHD Electrical Power Generation, Munich, Vol. 2, 1971, p. 159-174.

[43]Ponomarev, V. M., "Computation of Supersonic Flow in an MHD Generator Channel," Teplofizika vysokikh temperatur, Vol. 12, No. 3, 1974, p. 619-625.

[44]Bityurin, V. A., Lyubimov, G. A., Medin, S. A., and Ponomarev, V. M., "Engineering Methods of Calculating Flows in MHD Generator Channel," Teplofizika vysokikh temperatur, Vol. 12, No. 4, 1974, p. 817-826.

[45]Likhachev, A. P., "Supersonic Flow in an MHD Generator Channel Slanted at the Inlet," Magnitnaya gidrodinamika, No. 3, 1978, p. 49-56.

[46]Isakova, N. P., and Medin, S. A., "Subsonic Flow in an MHD Generator Channel," Teplofizika vysokikh temperatur, Vol. 16, No. 2, 1978, p. 377-383.

[47]Isakova, N. P., "Solution of the Direct Problem of a Mixed Flow in a Laval Nozzle in the Presence of a Meridional Magnetic Field," AN SSSR, Izvestiya, Mekhanika zhidkosti i gaza, No. 5, 1971, p. 14-20.

[48]Godunov, S. K., Zabrodin, A. V., and Prokopov, G. P., "A Difference Scheme for Two-Dimensional Nonstationary Problems of Gasdynamics and Computation of Flow with a Separated Shock Wave," Shurnal vychislitel'noy matematiki i matematicheskoy fiziki, Vol. 1, No. 6, 1961, p. 1020-1050.

[49]Yanenko, N. N., Method of Fractional Increments of Solving Multi-Dimensional Problems of Mathematical Physics (Metod drobnykh shagov resheniya mnogomernykh zadach matematicheskoy fiziki), Publishing House Nauka, Novosibirsk, 1967.

[50]Sonju, O. K., Teno, J., and Brogan, T. R., "Comparison of Experimental and Analytical Results for 20 MW Combustion-Driven Hall Configuration MHD Generator," 11th Symposium on Engineering Aspects of MHD, Pasadena, 1970, p. 5-10.

[51]Panchenko, V. P., "Quasi-One-Dimensional Analysis and Program for Calculating Flow in an MHD Generator Channel Taking into Account Development of Turbulent Boundary Layers," preprint IAE-2118, 1971.

[52]Breyev, V. V., and Panchenko, V. P., "Quasi-One-Dimensional Method of Calculations for an MHD Generator in a Boundary Layer Approximation," AN SSSR, Izvestiya, Mekhanika zhidkosti i gaza, No. 4, 1974, p. 139-145.

[53]Lyubimov, G. A., "Averaging MHD Flows and Application of the Quasi-One-Dimensional Approximation for Computation of MHD Flows in Channels," AN SSSR, Izvestiya, Mekhanika zhidkosti i gaza, No. 3, 1966, p. 3-11.

[54]Kutateladze, S. S., and Leont'yev, A. I., Heat and Mass Transfer and Friction in a Turbulent Boundary Layer (Teplomassoobmen i treniye v turbulentnom pogranichnom sloye), Publishing House Energiya, Moscow, 1972.

[55]Bityurin, V. A., Kovbasyuk, V. I., and Medin, S. A., "Certain Electrodynamic Effects in the MHD Generator Channel," AN SSSR, Izvestiya, Mekhanika zhidkosti i gaza, No. 3, 1968, p. 3-8.

[56]Bityurin, V. A., and Medin, S. A., "The Influence of Finite Segmentation of the Insulated Wall on the Characteristics of an MHD Generator," Magnetohydrodynamic Facilities (Magnitogidrodinamicheskiye ustanovki), Publishing House Nauka, Moscow, 1975, p. 82-87.

[57]"The U-25 MHD Pilot Plant, Part II, MHD Generator Studies," Scientific-Technical Report No. 242 (Issledovaniya na Ustanovke U-25 [II]. Issledovaniya MGD Generatora, Nauchno-tekhnicheskiy otchet No. 242), IHT, Moscow, 1974.

[58]Zalkind, V. I., Zuyeva, N. M., Kirillov, V. V., et al., "Investigation of the Microarc Operating Mode of Electrodes in Open-Cycle MHD Generator," The First U.S.-U.S.S.R. Colloquium on Energy Conversion, Moscow, 1974, p. 289-304.

[59]Rosa, R., "Hall and Ion Slip Effects in a Nonuniform Gas," Physics of Fluids, Vol. 5, No. 9, 1962, p. 1081-1090.

[60]Bityurin, V. A., and Lyubimov, G. A., "Taking into Account the Effect of Current Constriction at Electrodes on Characteristics of an MHD Generator," Teplofizika vysokikh temperatur, Vol. 17, No. 5, 1979, p. 1069-1080.

[61]Bityurin, V. A., and Lyubimov, G. A., "Models of Arc Discharge on Electrodes of an MHD Generator Channel," Teplofizika vysokikh temperatur, Vol. 17, No. 6, 1979, p. 1299-1308.

[62]Vatazhin, A. B., Medin, S. A., Sokol'skiy, A. G., and Shumyatskiy, B. Ya., "Influence of End Effects on the Thermal Efficiency of MHD Plants," Teplofizika vysokikh temperatur, Vol. 8, No. 2, 1970, p. 403-412.

[63]Vatazhin, A. B., and Regirer, S. A., "Electric Fields in Channels of Magnetohydrodynamic Facilities," Addendum in J. A. Shercliff's book, Electromagnetic Flow Measurement Theory, Russian translation by Publishing House Mir, 1965, p 205-266.

[64]Bityurin, V. A., Lyubimov, G. A., and Medin, S. A., "Calculation of Flows in MHD Channels," Scientific-Technical Report No. 78/1, IHT, Moscow, 1978.

[65]Sheyndlin, A. E., Kirillov, V. V., and Shumyatskiy, B. Ya., "Local Characteristics of a Nonideal MHD Generator and Approximate Methods of Their Computation" (Lokal'niye kharakteristiki neideal'nogo MGD generatora i priblezhenniye metody ikh rascheta), Scientific-Technical Report No. A 77/7 (Nauchno-tekhnicheskiy otchet No. A 77/7), IHT, Moscow, 1977.

[66]Bityurin, V. A., "Problem of the Influence of Leakage Losses on Characteristics of MHD Generators," Teplofizia vysokikh temperatur, Vol. 7, No. 4, 1969, p. 885-889.

[67]Abramovich, G. N., Applied Gasdynamics (Prikladnaya gasovaya dynamika), Publishing House Nauka, Moscow, 1976.

[68]Bityurin, V. A., Zhelnin, V. A., Lyubimov, G. A., and Medin, S. A., "Quasi-One-Dimensional Models of Flow in an MHD Generator Channel Based on Two-Dimensional Equations of a Boundary Layer," AN SSSR, Izvestiya, Mekhanika zhidkosti i gaza, No. 2, 1982, p. 69-77.

[69]Bityurin, V. A., Zhelnin, V. A., and Satanovskiy, V. R., "Numerical Calculation of Quasi-One-Dimensional Flow Models at Various Accuracy Levels of Describing the Boundary Layer," AN SSSR, Izvestiya, Mekhanika zhidkosti i gaza, No. 3, 1982, p. 78-87.

[70]"The U-25 MHD Pilot Plant, Part III., Major Components and the Long Duration R Channel," Scientific-Technical Report No. 72 (Issledovaniya na ustanovke U-25 [III]. Osnovnyye systemy teplovogo kontura i resursnyy kanal R, Nauchno-tekhnicheskiy otchet No. 72), IHT, Moscow, 1975.

[71]"Studies Conducted at the U-25 Facility, Part IV, Experimental Studies of the MHD Generator of the U-25 Facility with the 1D Channel," Scientific-Technical Report No. 54 (Issledovaniya na ustanovke U-25 [IV]. Eksperimental'noye issledovaniye MGD generators U-25 s kanalom 1D, Nauchno-tekhnicheskiy otchet No. 54), IHT, Moscow, 1976.

[72]"Studies Conducted at the U-25 Facility, Part V," Scientific-Technical Report No. A 77/4 (Issledovaniya na ustanovke U-25 [V]. Nauchno-tekhnicheskiy otchet No. A 77/4), IHT, Moscow, 1977.

[73]Yungman, V. S., Gurvich, L. V., and Rtishcheva, N. P., "Composition and Thermodynamic Properties of Methane Combustion Products with Ionizable Seed," Teplofizika vysokikh temperatur, Vol. 4, No. 4, 1966, p. 507-512.

[74]Bityurin, V. A., Zhelnin, V. A., Lyubimov, G. A., and Medin, S. A., "Comparison of the Results of Computations of a Flow in an MHD Generator Channel with the Experimental Data from the U-25 Facility," Teplofizika vysokikh temperatur, Vol. 16, No. 4, 1978, p. 854-867.

[75]Chen, P., Discontinuous Flows, (translation from English), Publishing House Mir, Moscow, Vol. 3, 1972.

[76]Kirillin, V. A., Sheyndlin, A. E., Pishchikov, S. I., et al., "A Few Thoughts Concerning Construction of the First Industrial Facility with an MHD Generator," The First U.S.-U.S.S.R. Colloquium on Energy Conversion, Moscow, 1978, p. 5-19.

[77]Bityurin, V. A., Ivanov, P. P., Koryagina, G. M., et al., "Numerical Simulation of Operation of an MHD Generator in Variable Modes as Part of an MHD Power Plant," Teplofizika vysokikh temperatur, Vol. 20, No. 2, 1982, p. 347-358.

Chapter 3

Characteristics of a Nonideal MHD Generator

3.1. Introduction

An ideal MHD generator may be represented in the following manner. An MHD channel of rectangular cross section is placed into a uniform magnetic field. The gasdynamic flow parameters and the physical properties of plasma are uniformly distributed over the cross section of the channel with the lines of force of the magnetic field directed normal to one pair of channel walls and parallel to the other. The first pair of walls is characterized by ideal insulating properties. Electrodes are located on the other two walls. The Hall effect influences the tranverse current flowing in the MHD channel in a magnetic field, generating transverse emf. To avoid short-circuiting electrode walls in the axial direction, the electrodes must be ideally segmented, i.e., divided along the length of the channel into individual, infinitely-thin (infinitely-short) sectors, ideally insulated from each other. Each pair of such infinitely-small electrodes must be equipped with its own load, electrically insulated from other loads. In order to avoid current flow along the plasma between adjacent electrodes, the electrical load distribution along the channel length must be uniform.

To avoid any potential drop when the current flows across the boundary between the electrodes and the plasma, the electrodes must have ideal emission properties. Finally, there should be no end effects associated with plasma flow at its entrance into and exit from the magnetic field.

In such an MHD generator, the distribution of electrical parameters across the channel cross section will also be uniform. What's more, the local external characteristics of the MHD generator can be described by simple relations of the generalized Ohm's law in terms of projections along the x and y coordinate axis:

$$j_y = -\frac{\sigma}{(1+\beta^2)} uB + \frac{\sigma\beta}{1+\beta^2} E_x + \frac{\sigma}{1+\beta^2} E_y ,$$

$$j_x = \frac{\sigma\beta}{1+\beta^2} uB + \frac{\sigma}{1+\beta^2} E_x - \frac{\sigma\beta}{1+\beta^2} E_y ,$$

where j_x, j_y, E_x, E_y are components of the current density and electric field vectors, respectively; $B = B_z$ is the magnetic field; $u = u_x$ is the plasma flow rate in the MHD channel; σ is the electric plasma conductivity and β is the Hall parameter.

In the absence of the longitudinal loading (longitudinal load current $I_x^H = 0$),[1] the expressions for the external characteristics of the MHD generator become even more simple

$$E_y = K_y uB, \quad j_y = \sigma uB(1-K_y) \ ,$$

$$E_x = \beta uB(1-K_y) \ ,$$

where $K_y \equiv E_y/E_y^{x \cdot x}$ is the transverse electrical load factor (by definition), $E_y^{x \cdot x}$ is the electric field in the open circuit operating mode. For an ideal MHD generator, $E_y^{x \cdot x} = uB$, while $K_y = E_y/uB$.

Unfortunately, an ideal MHD generator cannot be fabricated. As early as the 1960s, experimental investigations of the first MHD generators had shown that their electrical characteristics are considerably below the expected parameters. For small MHD generators, the difference may be by multiples of dozens, or even hundreds. Some experimentalists attributed this to large near-electrode potential drops, others to poor channel insulation, and still others to both effects together. However, no qualitative interpretation of the observed results was provided. Some investigators were of the opinion that small MHD generators are, in principle, unsuited for investigation of MHD energy conversion processes.

United States specialists believe that this problem can be solved by increasing the size of the experimental MHD generators. This belief led to the development of the 1-MW Mark 2 MHD generator [1], then the 30-MW Mark-5 MHD generator [2]. In such MHD generators, the influence of the nonideal behavior of the MHD channel[2] on its characteristics was manifested to a considerably lesser decree. However, this nonideal behavior could not be

[1] Such a generator is frequently referred to as a Faraday MHD generator, a term that will be used here from this point on.

[2] In referring to nonideal (nonideal behavior) of the MHD channel, we refer to a totality of factors depending on the design and structural features of an MHD channel that exert an influence on its characteristics.

correctly estimated from the experimental results obtained on these facilities. This was the result of very short-duration operation (dozens of seconds) of MHD channels in these facilities and strong degradation during the experiments. Investigation of the properties of the channel and their influence on the characteristics of an MHD generator under such conditions was difficult. Furthermore, the required experimental techniques had not yet been developed.

Development of commercial-scale MHD generators requires a capability to adequately predict their characteristics, which, in turn, makes it necessary to investigate operational characteristics of actual MHD channels. Without solving this problem, it was difficult to expect success in developing effective commercial-scale MHD generators.

A systematic investigation of the whole set of problems associated with the development of MHD power facilities began after the U-02 model was brought on line [3]. Investigation of the MHD generator was one of the most important aspects of this work, with particular emphasis placed on investigation of the electrodynamics of real MHD generators.

Experiments performed on the U-02 have quickly demonstrated that the influence of the nonideal behavior of an MHD channel depends considerably on the Hall effect. The "Start" experimental facility was designed to expand the range of work in this direction [4]. This chapter generalizes the results of many years of experimental investigations of the electrodynamics of a real MHD generator, conducted at the Institute of High Temperatures of the Soviet Academy of Sciences on the U-02 and "Start" facilities.

3.2. The Influence of Various Factors on the Characteristics of a Real MHD Generator

Electrodynamics of an ideal MHD generator is very simple. Specific electrical characteristics of an ideal Faraday MHD generator are specified by only four parameters: magnetic field, B; flow rate, u; electrical conductivity of plasma, σ; and Hall parameter, β. Because the design of an MHD channel is ideal, it exerts no influence on operation of the MHD generator. In real MHD channels, the MHD energy conversion process becomes considerably more complex. Additional factors appear, which exert considerable influence on the efficiency of energy conversion. In one way or another, all of them are associated with the design parameters of an MHD channel or with its characteristics. For example, in a real MHD channel the choice of construction materials limits the operating temperature of its walls and, hence, the temperature profiles in the boundary plasma layers. The temperature profiles determine the plasma conductivity profiles, and the latter exert a direct influence on the electrical characteristics of the MHD generator. However, the primary cause of degradation of the characteristics of an MHD generator is the channel wall temperature.

One could give many such examples. The principal factors exerting considerable influence on a real MHD generator, will be considered here.

3.2.1. Nonideal Behavior of an MHD Channel. A hypothetical model of an ideal MHD generator is based on the assumption that the flow parameter distribution over the MHD channel cross section is uniform. However, when viscous fluid (plasma in this case) flows in channels, it adheres to the walls, leading to formation of hydrodynamic boundary layers with a nonuniform velocity distribution that varies between zero and the core flow rate along the thickness of the boundary layer. The nonuniformity of velocity in the boundary layers may exert an influence on operation of an MHD generator. The degree of influence will depend on the relative thickness of the boundary layers and the velocity profile within them. Both are determined by a number of factors, including geometry of the channel and conditions at the walls.

In real MHD channels, the distribution of plasma electrical conductivity over the cross section will also be nonuniform. Construction materials capable of operating at the temperature close to the temperature of the plasma flow in an MHD generator channel do not exist at the present-day level of the technology. Therefore, the walls of real MHD channels need to be cooled. This leads to formation of thermal boundary layers at the walls, in which plasma temperature is considerably below the temperature in the core flow. However, because even a small drop in plasma temperature results in a considerable drop of its electrical conductivity, the presence of "cold" boundary layers on channel walls generates considerable nonuniformity in the distribution of plasma electrical conductivity over the channel cross section. Such nonuniformity of plasma conductivity, as well as velocity nonuniformity, depends on the geometry of the channel and, also, on wall cooling conditions. As will be shown below, under certain conditions plasma nonuniformity can exert strong negative influence on the external electrical parameters of the MHD generator.

Insulation walls of real MHD channels and interelectrode insulators within them are characterized by ideal insulating properties. Therefore, channel walls have finite conductivity in both axial and transverse directions. Electrode segmentation is also nonideal, being limited by the design.

Near-electrode potential drop occurs when current flows across the plasma-electrode boundary. This potential drop depends on the emission properties of the electrode material, its surface temperature, and current density.

Other design features (e.g., special electrode geometry) that may exert an influence on the characteristics of an MHD generator are also possible. All of this must be taken into account in developing generators.

<u>3.2.2. Certain Characteristics of the U-02 and "Start" Experimental Facilities</u>. The U-02 and "Start" facilities have been previously described in the literature [3-7]. Therefore, only those characteristics of those facilities that have a bearing on the solution of

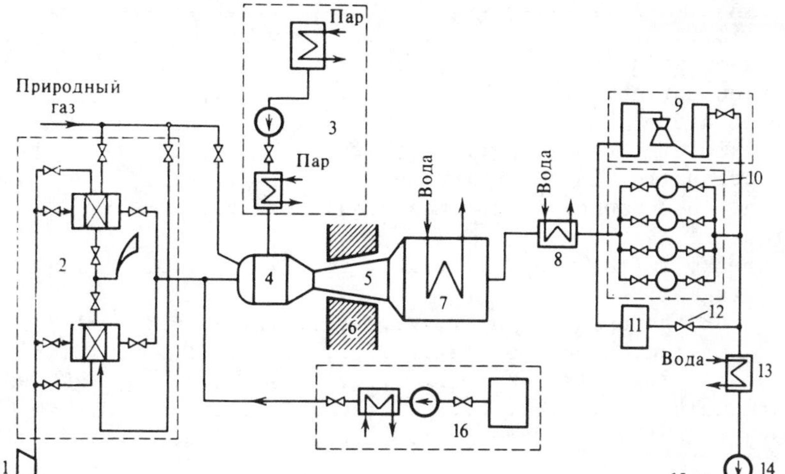

Fig. 3.1 U-02 facility, 1--blower, 2--preheater, 3--seed injection system, 4--combustion chamber, 5--MHD generator channel, 6--magnet, 7--model steam generator, 8--additional heat exchanger for wet seed recovery system, 9--wet seed recovery system, 10--bag filter/dry seed recovery system, 11--electrostatic precipitator, 12--switch, 13--additional heat exchanger for dry seed recovery system, 14--exhauster, 15--exhaust to atmosphere, 16--oxygen enrichment system

the problem under consideration will be discussed here. The U-02 facility, brought on-line in 1964, is the first complex, experimental MHD facility in the world designed for model investigations of all of the major components of the thermal loop of an open-cycle MHD facility (Fig. 3.1 is a diagram of the facility). The following major components were investigated in detail during operation of the U-02 facility: a unique high-temperature air preheater (2) to heat air to 2000 K; combustion chamber (4); seed injection system (3) and seed removal systems (9-11) of three types, and model steam generator (7). However, the most important component under investigation was the MHD generator. The fairly large U-02 experimental facility operates on plasma comprised of combustion products of natural gas and air enriched with oxygen. The principal parameters of the experimental U-02 facility are as follows:

Thermal Loop	
Thermal power output, MW	6
Flow rate, kg/s	
combustion products	1.50
oxygen	0.40
fuel	0.12
gas	0.10
oil	0.02
seed solution	0.05
Oxygen enrichment, %	up to 50
Temperature, ^0C	
oxidizer preheating	1700
adiabatic, in the combustor	2650
at the channel inlet (measured)	2500
Pressure, Pa	
peak, in the loop	$1.7 \cdot 10^5$
stagnation	
at the channel inlet	$1.3 \cdot 10^5$
at the channel outlet	$0.4 \cdot 10^5$
Channel flow rate, m/s	700-900
Mach number	0.60-0.85
Oxygen supply in vessels, t	12
Magnetic System	
Pole piece area, m^2	$0.4 \cdot 3.0$
Warm bore height, m	0.2
Magnetic field, T	1.77
MHD Generator	
Working volume of the MHD generator	
height, m	0.065-0.08
width, m	0.22-0.35
length, m	3.0

No-load voltage, V..............................	300
Short-circuit current of an electrode pair, A...	up to 40
Power output, kW	
pair of electrodes............................	up to 3
total, MHD generator..........................	75
Specific power takeoff, MW/m^3...................	3.3

Thus, the U-02 facility makes it possible to conduct investigations of an MHD channel using adequately large models (up to 3 m long and with up to 80 times 300 mm cross section).

The U-02 facility operates on natural gas using air heated to 2000 K in a high-temperature air preheater (2) as oxidizer. System 16, enriching air with oxygen, is intended to increase plasma temperature and, therefore, its electrical conductivity. Oxygen enrichment in the U-02 facility does not exceed the limits of economic feasibility of future power-generating MHD facilities. Potassium carbonate, which is very likely to be used in a commercial-scale MHD facility, is used as seed injected into the combustion products flow. Thus, the composition, temperature, and other parameters of the plasma products flowing in the U-02 facility are close to those of the working fluid of open-cycle, power-generating MHD facilities. This makes it possible to test construction models of an MHD channel (5) and other thermal loop assemblies under conditions close to the expected natural conditions.

The second advantage of the U-02 facility is its practically unlimited duration of operation. Experiments of 100-200 and even 300 hours duration were conducted on the U-02 facility. Other than the U-25 installation, for a long time none of the facilities known to us possessed such capabilities.

A Mark-6 facility capable of long-duration tests of an MHD generator was constructed only in the middle 1970s by Avco Everett Research Laboratory in the United States.

The U-02 facility is also well suited for investigation of the electrodynamic processes in an MHD generator. An electric field of 1400 V/m, current density that may reach 1.2×10^4 A/m^2 in the short-circuit operation mode, and transverse electric field of up to 1000 V/m are induced in the MHD generator channel at a maximum magnetic field, B = 1.77 T. The potential drop in the space charge layer near electrodes (cathode drop) that strongly masks MHD effects in small MHD channels is of much less importance in the U-02 facilities (it was only 20-30 V or 10 % of the induced emf at hot electrodes). The Hall parameter can exert considerable influence on external characteristics of a nonideal MHD generator, reaching $\beta \approx 2$ at the maximum magnetic field in the channel.

The U-02, well-equipped with diagnostic equipment, makes possible detailed investigations of both the properties of channel structure and their influence on the electrodynamic processes in the channel. Figure 3.2 is a schematic block diagram of an experimental facility

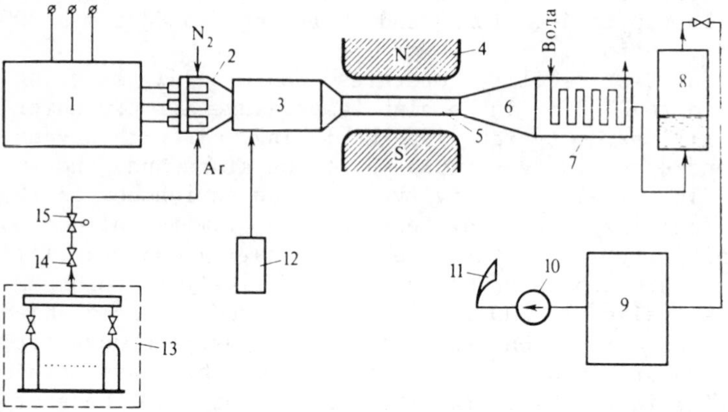

Fig. 3.2 "Start" facility, 1--current stabilizer, 2--three-phase plasmatron, 3--mixing chamber, 4--magnet, 5--MHD generator channel, 6--diffuser, 7--heat exchanger-potassium condenser, 8--bubble-type emulsifier, 9--damping capacitance, 10--vacuum pump, 11--exhaust to atmosphere, 12--preparation and seed injection system, 13--argon-container ramp, 14--shut off valve, 15--regulator

operating on argon-potassium plasma, known as "Start." This facility was developed in order to expand the range of investigations of electrodynamics of a nonideal MHD generator, primarily into the range of larger values of the Hall parameter. The "Start" facility is considerably smaller than the U-02 facility.

Heat Loop

Thermal power output, MW........................	0.5
Flow rate, g/s	
argon..................................	40-100
nitrogen additive.........................	2-5
potassium seed............................	0.1-0.8
Stagnation temperature at the channel inlet, K..................................	1600-3000
Stagnation pressure at the channel inlet, Pa......................................	$(0.3\text{-}0.8)\cdot 10^5$
Mach number at the channel inlet...............	0.5-0.8
Duration of continuous operation, h............	up to 1.2

MST 1.5/90 Magnet	
Magnetic core	C-shaped
Pole piece area, mm^2	160x600
Warm bore diameter, mm	0-180
Magnetic field strength, T (at a diameter of 100 mm)	1.4
Operation	Long-duration
MHD Channel	
Inner cross section, mm^2	19x90
Length, mm	up to 600
Wall temperature, K	
electrode	800-1600
insulating	1000-1900
Electric field strength, V/m	up to 800
Current density, A/m^2	up to $0.3 \cdot 10^4$

The choice of argon-potassium plasma as the working fluid in the "Start" facility can be attributed to its simple composition and its well-known properties. This is extremely important in performing precision laboratory investigations.

The low collision cross sections of electrodes with argon atoms make it possible to obtain high Hall parameter values at a moderate magnetic field (this solves the principal problem of expanding investigations of the MHD generator over a wider range of Hall parameters and, also, provides high plasma conductivity at only a moderate plasma temperature). However, low-collision cross sections have their shortcomings. They generate conditions under which the electron temperature may differ from the temperature of neutrals. This leads to nonequilibrium effects that complicate investigations of an equilibrium MHD generator. Because molecules of nitrogen are characterized by high inelastic loss coefficient, nitrogen is introduced into the argon-potassium plasma to avoid this effect. Addition of 5 to 15 % nitrogen is sufficient to suppress the nonequilibrium effects.

The principal aim in designing the "Start" facility was simplicity of design and operation. Therefore, the device was fabricated in the form of an open loop, shown in Fig. 3.2. The working fluid (argon with an addition of partially purified nitrogen) is fed into the facility from the ramp of the compressed gas containers (13), through the shut-off valve (14), and regulator (15) controlling pressure at the inlet into the gasdynamic loop of the facility. The inert gas is heated in a three-phase ac plasmatron (2) connected to a current stabilizer (1).

Seed in the form of metal potassium is fed into a mixing chamber located between the plasmatron and the MHD channel (5). To attain better mixing, seed is introduced into the inert gas flow in the form of vapor. Vaporization and determination of injection rate takes place in the preparation and seed injection system (12). From the combustion chamber, argon-potassium plasma flow passes into the MHD generator channel located within the gap of a C-shaped magnet (4). The flow exiting the MHD channel is cooled and purified of seed first (preliminary purification) in the heat exchanger (7) and then (fine purification) in the bubble-type emulsifier (8). The gas purged of seed is pumped from the facility by a vacuum pump (10) and is exhausted into the atmosphere.

Experience gained on the "Start" facility has shown that it can be used to perform multistage investigations of the MHD generators operating with both equilibrium and nonequilibrium plasma.

Although parameters of the "Start" facility are relatively low, they make it possible to conduct a wide range of investigations of local electrical characteristics of an MHD generator. The range of some of the important operating parameters of the facility exceeds considerably the possible range of operation of future MHD generators. For example, the Hall parameter in the "Start" facility can vary between 2 and 25 and the relative surface channel wall conductivity, between 10^{-1} and 10^{-3}. By varying wall temperature, it is possible to vary the degree of nonuniformity of plasma conductivity through a wide range.

The measuring equipment of the facility provides a full set of the experimental data required for analysis of the principal phenomena specifying the local characteristics of a real MHD generator. The development of the measuring system and measuring techniques for the "Start" facility was based on the principle that experimental methods must be characterized by "negative degrees of freedom." This refers to the fact that the techniques must not only allow the use of certain freely varying parameters, but must also provide a set of quantities measured in the experiments that allows determination of the parameter being sought by different, independent methods. Conforming to this principle for such a complex object as an MHD generator encounters considerable technical and methodological difficulties. Nevertheless, this was achieved in many cases on the "Start" facility.

The fairly large number of quantities being measured and the broad range of thorough investigations made it possible to perform detailed investigations of characteristics of a real MHD generator and to identify and investigate a series of secondary electrodynamic effects characteristic of real MHD channels.

The U-02 and "Start" facilities complemented each other rather well, making it possible to investigate electrodynamics of an MHD generator under considerably different conditions, including those close to natural.

3.2.3. Transverse Plasma Conductivity Nonuniformity. Experimental investigation of the characteristics of a nonideal MHD generator at the Institute of High Temperatures of the Soviet Academy of Sciences, on the U-02 and "Start" facilities under widely different experimental conditions, has shown that, in the presence of the Hall effect, the strongest negative influence on the specific power output of an MHD generator is exerted by plasma conductivity nonuniformity along the channel, caused by the "cold" boundary layers at the channel electrode walls.

A decrease of the axial current density, j_y, and the specific power, $W = j_y E_y$, of a Faraday MHD generator caused by nonuniformity of transverse plasma conductivity can be explained as follows. When the transverse current intersects plasma layers with considerably different conductivity, $\sigma = \sigma(y)$, the Hall effect-induced transverse emf[3] will be different; i.e., $E_x^* = \beta j_y / \sigma$. As a result, the current circulation appears in the flow even in those cases when the axial current averaged over the cross section is zero. In the boundary layers, where $\sigma < \langle\sigma\rangle$ (here and below the symbol $\langle\ \rangle$ will designate quantities averaged across the channel cross section), the local axial emf exceeds the axial electric field averaged over the cross section ($E_x^* > \langle E_x \rangle$) and the current flows along the direction of the flow ($j_x > 0$). In the core flow, where $\sigma > \langle\sigma\rangle$, the current is directed opposite to the direction of flow ($j_x < 0$). The axial current induces a transverse emf, $E_y^* = \beta j_x / \sigma$. In layers of high conductivity, where $j_x < 0$, E_y^* is directed in accordance with the emf induced by the plasma flow, **U** x **B**, whereas, in layers of low conductivity, it is directed opposite to it. However, because $E_y^* \sim 1/\sigma$, the additional emf in layers of high conductivity is always smaller than the counter-emf induced in layers of low conductivity. The residual counter-emf opposes the flow of the transverse

[3] As is customary in literature on MHD generators, the term emf will refer to specific emf per unit length.

operating current, j_y.

This effect was first predicted theoretically in Reference 8, where the author obtained the following relations between the transverse current density, j_y, and the axial field intensity, E_x, in a Faraday channel ($\langle j_x \rangle = 0$) in the presence of transverse plasma conductivity nonuniformity in the flow:

$$\bar{j}_y = \bar{j}_y^{s.c.}(1-K_y), \quad V_y \equiv \bar{V}_y^{x.x}K_y, \quad \bar{E}_x = \beta\bar{j}_y \, , \qquad (3.1)$$

$$\bar{j}_y^{s.c.} = 1/G_o, \quad \bar{V}_y^{x.x} = 1 \, , \qquad (3.2)$$

where

$$\bar{j}_y \equiv j_y/\langle\sigma\rangle\langle uB\rangle; \qquad \bar{E}_x \equiv E_x/\langle uB\rangle \, ;$$

$$\bar{V}_y \equiv V_y/2a\langle uB\rangle \, ;$$

$\bar{j}_y^{s.c.}$, $\bar{V}_y^{x.x}$ are the dimensionless current density in an operating mode and the no-load voltage, respectively; $G_0 = 1 + \alpha(1 + \beta^2)$; $\alpha = \langle\sigma\rangle\langle 1/\sigma\rangle - 1$ is the degree of plasma conductivity nonuniformity;[4] a is the half-width of the channel; K_y is the electrical load factor; u is the plasma flow rate, and B is the magnetic field induction.

In experiments conducted on the U-02 facility at maximum electrode wall temperature,

$$T_w^e = 2300 \text{ K } (\bar{T}_w^e \equiv T_w^e/T_0 \simeq 0.8 \, ,$$

where T_o is the plasma temperature in the core flow), the degree of plasma conductivity nonuniformity, α, was relatively small, varying between 0.06 and 0.15, as a result of the development of boundary layers along the channel length (ℓ = 3 m). At a maximum magnetic field induction, B = 1.8 T, when the Hall parameter reached β =

4 The nonuniformity of the Hall parameter distribution along the channel cross section, β (y), was also taken into [8]. In this case, $G_o = \langle\sigma\rangle\langle(1 + \beta^2)/\sigma\rangle - \langle\beta^2\rangle$. However, the Hall factor is considerably less sensitive to plasma temperature variation than conductivity. Therefore, it is frequently possible to neglect the variation of the Hall parameter along the thickness of the boundary layer and to utilize the simpler relation (3.2).

2, the plasma conductivity nonuniformity led to a drop of j_y and E_x by a factor of one-and-a-half on the average. The specific power of the MHD generator decreased further by approximately a factor of two, because of the influence of other factors.

When the relative wall temperature decreased to $\bar{T}^e_w$ = 0.7, the degree of plasma conductivity nonuniformity increased approximately by a factor of 2.5-3.[5] The relative current distribution along the channel length,

$$\bar{I}^{s.c.} \equiv I^{s.c.}/I^{s.c.}_{id.}, \quad \text{where } I^{s.c.} \text{ and } I^{s.c.}_{id.}$$

are the transverse current in the short-circuit mode of the MHD generator being investigated and that in an ideal MHD generator, respectively, is shown in Fig. 3.3. A general current drop due to nonideal behavior of the MHD channel was by a factor of 5 at the inlet into the channel, increasing to approximately a factor of 15 at the channel outlet [9]. The plasma temperature drop in the

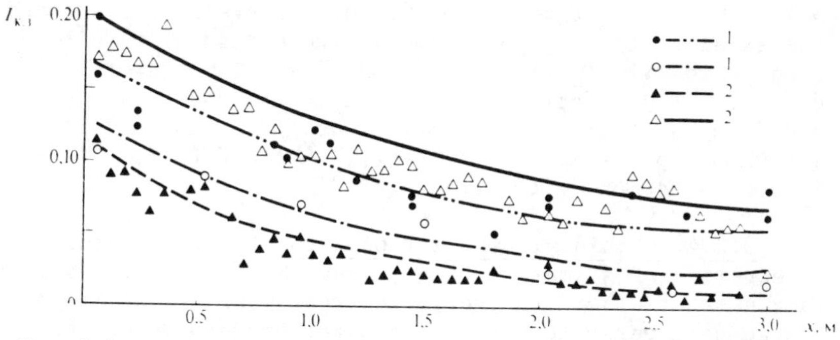

Fig. 3.3 Transverse current distribution along the channel length in the short-circuit operating mode of the U-02 facility, lines--results of the calculations; points--experimental data; 1,1'--run No. 11, T_o = 2450-2500 K and 2200-2250 K, respectively; 2,2'--run No. 15, T_o = 2200-2250 K and 2450-2500 K, respectively

5

Here and below the variation of plasma conductivity nonuniformity, α, will frequently be associated with variation of the relative electrode wall temperature, $\bar{T}^e_w$. This implies that other conditions in the flow remain the same. In a general case, α depends on plasma composition and, for the same composition,

$$\alpha = \alpha\,(\bar{T}^e_w, \bar{\delta}y, U_i/kT_o), \text{ where } \bar{\delta}_y = \delta_y/a\ ,$$

is dimensionless thickness of the boundary layer at the electrode walls, U_i is the ionization potential and k is the Boltzmann constant.

core flow at approximately the same $\overline{T}^e_w$ led to an additional increase of α[6] and a decrease of $\overline{j}_y$ (see curves 1′ and 2).

Analogous behavior was observed in experiments on the "Start" facility. The range of measurements on this facility was considerably broader. In particular, this applied to the range of measurements of the Hall parameter (between 2 and 25). The relative channel wall conductivity, $\overline{\sigma}^w$, also varied across adequately broad limits. The range of variation of the degree of nonuniformity of plasma conductivity on the "Start" facility is practically the same as on the U-02 facility ($0.1 < \alpha < 1$). Parameter α varied primarily as a result of variation of the electrode wall temperature at a practically constant thickness of the boundary layer and constant other gasdynamic conditions in the flow [10].

In experiments performed at adequately high electrode wall temperature ($\overline{T}^e_w = 0.7$) and a relatively small Hall parameter ($\beta = 2$–4), external characteristics of the MHD generator were in sufficiently good agreement with theoretical calculations for an ideal generator (Fig. 3.4). However, at $\beta = 20$–25, the transverse current and the axial field drop by almost two orders of magnitude, even at relatively high electrode wall temperature. The drop of $\overline{T}^e_w$ by a factor of two leads to a decrease of $\overline{j}_y$ and $\overline{E}_x$ by another order of magnitude. Such low values and the nature of variation of $\overline{j}_y$ (β, $\overline{T}^e_w$) and E_x (β, $\overline{T}^e_w$) are in good agreement with theory [8].[7]

3.2.4. Nonideal Behavior of Insulation. According to theoretical estimates (e.g., see Ref. 11), a strong decrease of the transverse current and axial field at large Hall parameter values in a Faraday MHD generator may be caused by nonideal insulation of the MHD channel. As already noted, when $\beta > 0$, the transverse current induces an axial field, $E^*_x = \beta j_y/\sigma$, in the flow. If the channel walls are characterized by finite surface conductivity,

6 At constant $\overline{T}^e_w$, and $\overline{\delta}_y$, α increases with decreasing T_o, because the dependence of plasma conductivity, δ, on temperature increases.

7 In electrodynamics calculations, the velocity and conductivity distributions over the channel cross section are assumed to be specified. In analyzing the experimental data, the velocity conductivity profiles under experimental conditions were determined from gasdynamic computation of the boundary layers, using techniques described in the appendix.

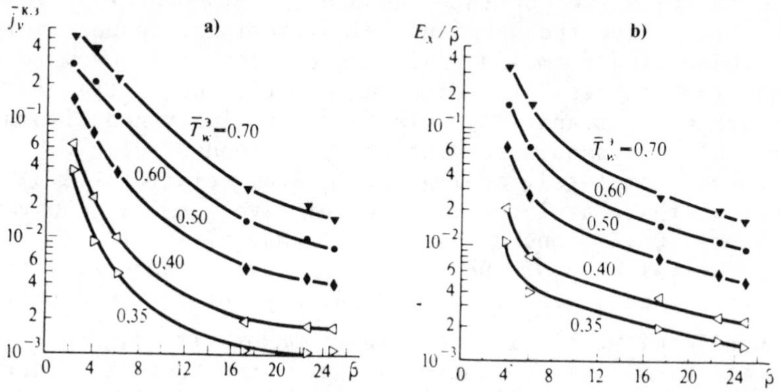

Fig. 3.4 Experimental dependences of transverse current density (a) and transverse field intensity (b) in an MHD channel on the Hall parameter at various channel wall temperatures

axial current with current density $j_x = E_x \sigma_x^w (a + b)/ab = E_x^* \sigma\tilde{\sigma}_x^w)/(1+\tilde{\sigma}_x^w)$, where σ_x^w is the surface channel wall conductivity in the axial direction (a and b are half-width and half-height of the channel cross section), may be short-circuited along them. In this case, $j_x > 0$ and the transverse current under the influence of the Hall effect induces transverse counter-emf $E_y^* = \beta j_x/\sigma = \beta^2 j_y \tilde{\sigma}_x^w/\sigma\,(1 + \tilde{\sigma}_x^w)$ in the flow. Similar to the previous case, this emf lowers the density of the transverse operating current, j_y, and the intensity of the transverse field, E_x, that it induces. When the flow is uniform and the channel wall conductivity is finite, MHD generator characteristics will be described by relations (3.1). However, the expressions for

$$\bar{j}_y^{s.c.} \quad \text{and} \quad \bar{V}_y^{x.x}$$

change somewhat [12]:

$$\bar{j}_y^{s.c.} = (1 + \tilde{\sigma}_x^W)/G_W \,,$$

$$\bar{V}_y^{x.x} = (1 + \tilde{\sigma}_y^W G_W (1 + \tilde{\sigma}_x^W))^{-1} \,, \tag{3.3}$$

where

$$G_W \equiv 1 + \tilde{\sigma}_x^W (1 + \beta^2); \quad \tilde{\sigma}_x^W \equiv \sigma_x^W (a + b)/\sigma ab$$

$$\text{и } \sigma_y^W \equiv \sigma_y^W/\sigma b$$

are the relative surface channel wall conductivities in the transverse and longitudinal directions, respectively.

Comparing formulas (3.2) and (3.3), it is easy to see that the nature of variation $\bar{j}_y(\beta)$, or $\bar{E}_x(\beta)$ in the presence of transverse plasma conductivity nonuniformity ($\alpha > 0$), or transverse channel wall conductivity ($\sigma_x^w > 0$) turn out completely analogous. From relations (3.3), it follows that, at large β, even relatively small transverse channel wall conductivity, σ_x^w, may lead to the same substantial drop of the transverse current and axial field intensity as in the case of transverse plasma conductivity nonuniformity. All of this makes it difficult to distinguish the effects under consideration during experimental investigations of an MHD generator. However, in order to improve the design of the MHD channel, it is necessary to establish the factor responsible for the decrease of its characteristics. Therefore, it is necessary to establish clearly the importance of each factor to the energy conversion process in an MHD channel.

Serious methodological difficulties arise in solving such a problem. These difficulties are associated with measurement of the surface channel wall conductivity under operating conditions. Our experience, as well as that of non-Soviet investigations, have demonstrated that the importance of this parameter depends considerably on conditions in plasma flow and on its walls. Direct measurement of the surface conductivity of channel walls under operating conditions is hindered by the short-circuiting effect of the plasma flow. Indirect methods of estimating the surface conductivity of channel walls are based on the use of various models of processes in an MHD channel and result in deviations of up to two orders of magnitude [13] in σ^w.

A method of measuring channel wall surface conductivity was developed during investigations performed on the "Start" facility. This made it possible to conduct detailed investigation of this parameter as a function of operating conditions of the channel [14].

Analogous investigations were also performed on the U-02 facility, where several different methods were used to measure surface conductivity of channel walls. These techniques included a compensation method that makes it possible to measure the transverse surface conductivity of channel walls directly in the operating channel [15]. Investigations performed on the "Start" facility have demonstrated that the surface channel wall conductivity is strongly affected by seed. The electrical conductivity of dense ceramic in the presence of potassium vapors increased tens and even hundreds of times, even though its temperature in the experiments ($T_w \geq 1200$ K) exceeded considerably the saturation temperature of cesium vapors in the flow ($T_s \simeq 600$ K). Apparently, thin (molecular) conducting films were formed on the surface. This hypothesis was verified by the temperature dependence $\sigma^w(T_w)$, and certain other factors, the analysis of which can be found elsewhere [14]. Figure 3.5 shows a plot of $\sigma^w(T_w)$. It can be seen from this figure that, as the wall temperature increases from 1200 K to 1900 K, the surface wall conductivity drops by almost two orders of magnitude.

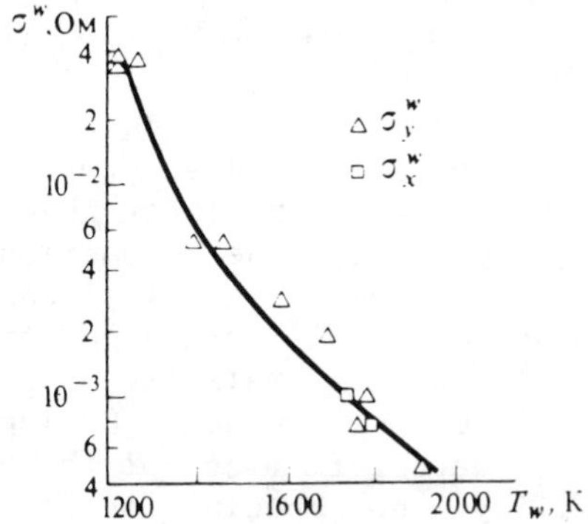

Fig. 3.5 Temperature dependence of the surface conductivity of a ceramic wall in the presence of potassium vapor (p_K = 0.4–1.6 kPa)

Measurements of the surface wall conductivity in transverse and axial directions verified its isotrophy: $\sigma_x^w = \sigma_y^w$.

Uniformity of the $\sigma^w(y)$ distribution in the experiment was controlled by measuring the potential distribution on the wall, $\overline{\Phi}_w(y) = \phi_w/V$. The slope of the wall potential distribution curve, $\phi_w(y)$, is proportional to the surface current density in the wall, j_y^w, and inversely proportional to the surface wall conductivity: $\partial\phi_w/\partial y = j_y^w/\sigma_y^w$. From this it follows that, when $j_y^w(y)$ = const., σ_y^w is uniquely specified by the slope of the curve $\phi_w(y)$.

A typical potential distribution on the wall, $\overline{\Phi}_w(y)$, is shown in Fig. 3.6. An analogous potential distribution was observed in practically all experiments. From the curve, it can be seen that $\overline{\Phi}_w(\overline{y})$ distribution is linear almost throughout the width of the insulating walls $\partial\overline{\Phi}_w/\partial\overline{y}$ = const. This indicates that σ_w is constant practically along the whole surface of the insulating wall exposed to the flow. An exception to this are narrow strips (approximately 1 mm) along the junction of the insulating walls with the electrode walls. A sharp

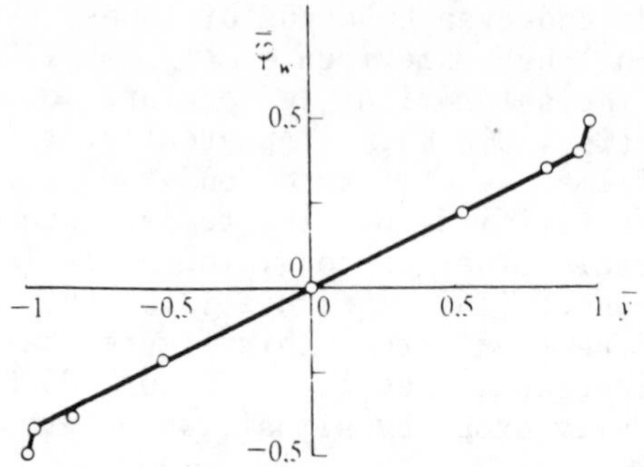

Fig. 3.6 Potential distribution on the insulating wall in the transverse diration in the experiments on measurements of surface conductivity of channel walls

increase in the potential gradient, $\partial\overline{\Phi}_w/\partial\overline{y}$, occurs here, indicating extremely low σ_w in these zones. Formation of nonconducting strips at the junction between the insulating and the electrode walls can apparently be attributed to penetration of pure argon from the mounting space, which is always at small excess pressure, through the junction and into the channel. During investigations of insulation and characteristics of the MHD generator on the "Start" facility, no special importance was assigned to the nonconducting strips on channel walls. According to theoretical estimates available at that time, the principal influence on the characteristics of the MHD generator in the presence of a strong Hall effect should be exerted by the axial wall conductivity, σ_x^w, which was hardly affected by the strips. However, as was revealed later, such strips may exert considerable influence on the generator characteristics. This problem will be considered in more detail in following text.

Overall, investigations of the conductivity of channel walls provided exhaustive data on the nature of its dependence on external conditions, and made it possible to determine the feasibility of controlling this parameter. Other possible leakages and current shorts along the insulation were investigated simultaneously with investigations of channel wall conductivity [14]. Careful measurements were made of the impedance of insulation between metal components of the channel, and their impedance relative to the adjacent loop assemblies and the ground under normal temperature and operating conditions. It was determined that the impedance of insulation is adequately high that the local current leakage could not exert much influence on the characteristics of the MHD generator.

3.2.5. Joint Influence of Plasma Conductivity Nonuniformity and Nonideal Behavior of Insulation. Even at high-channel wall temperature, the maximum σ^w measured in the "Start" facility was so low that the influence of this factor on operation of the MHD generator could be neglected. This made it possible to separate and investigate solely the influence of the plasma conductivity nonuniformity on the MHD generator characteristics. When the wall temperature dropped, the surface wall conductivity increased rapidly enough (see Fig. 3.5). Therefore, it was possible to vary $\tilde{\sigma}_x^w$ throughout a wide range of values ($0.005 < \tilde{\sigma}^w < 0.4$).

Investigation of the MHD generator in this range of $\tilde{\sigma}^w$ variation has shown that the harmful influence of the surface wall conductivity on j_y and E_x during the nonuniform flow is considerably weaker than predicted by theory on the assumption of flow uniformity. Physically, this is quite clear. Degradation of generator characteristics resulting from nonideal insulation is caused by short-circuiting of the axial current proportional to the induced Hall emf along the channel walls. The transverse plasma conductivity nonuniformity decreases considerably the importance of this parameter. Transverse current leakages along the walls decrease correspondingly and their influence on the external characteristics of an MHD generator also decreases.

The foregoing discussion is verified by qualitative estimates of the mutual influence of plasma conductivity nonuniformity and nonideal behavior of insulation. Considering channel wall conductivity as an additional external generator load, i.e., neglecting the feasibility of current flow from the plasma to the insulating walls with finite electrical conductivity ("$j_z = 0$" model). Then, analysis of the influence of the transverse plasma conductivity nonuniformity (see Ref. 8) can be generalized to the case of finite channel wall conductivity. Under the assumptions made, averaging Ohm's law equations provides the following load characteristics for a Faraday MHD generator. Relations (3.1) once again maintain their validity. However, expressions for

$$\bar{j}_y^{s.c.} \text{ and } \bar{V}_y^{x.x}$$

assume a somewhat different form,

$$\bar{j}_y^{s.c.} = (1 + \tilde{\sigma}_x^W)/G \ ,$$

$$\bar{V}_y^{x.x} = (1 + \tilde{\sigma}_y^W G/(1 + \tilde{\sigma}_x^W))^{-1} \ , \qquad (3.4)$$

where

$$G = (1 + \alpha)(1 + \tilde{\sigma}_x^w)(1 + \beta^2) - \beta^2$$

can be considered a parameter of the nonideal behavior of an MHD generator. For uniform flow ($\alpha = 0$), relations (3.4) become expressions (3.3), while for ideal insulation ($\sigma_x^w = \sigma_y^w = 0$) in (3.2).

From relations (3.4) it follows that, in the presence of transverse plasma conductivity nonuniformity, the influence of the nonideal insulation is weakened considerably. For example, when $\alpha = 0.3$ (this corresponds approximately to $\overline{T}_w = 0.6$ and $\overline{\delta}_y = 0.2$), an increase of $\tilde{\sigma}_x^w$ from 0 to 0.1, when $\beta = 10$ lowers $\overline{j}_y$ and $\overline{E}_x$ only by a factor of 1.4, whereas, in the case of a homogeneous flow ($\alpha = 0$), similar change of $\tilde{\sigma}_x^w$ at the same value of β has to lower $\overline{j}_y$ and $\overline{E}_x$ by more than an order of magnitude.

In experiments performed on the "Start" facility, the influence of $\tilde{\sigma}^w$ was even less noticeable than indicated by relations (3.4). Comparison of the experimental data at $\tilde{\sigma}^w = 0.005$ and 0.09 is shown in Fig. 3.7. Deviation of the experimental data reduced to other equivalent conditions is on the order of 15 %. Furthermore,

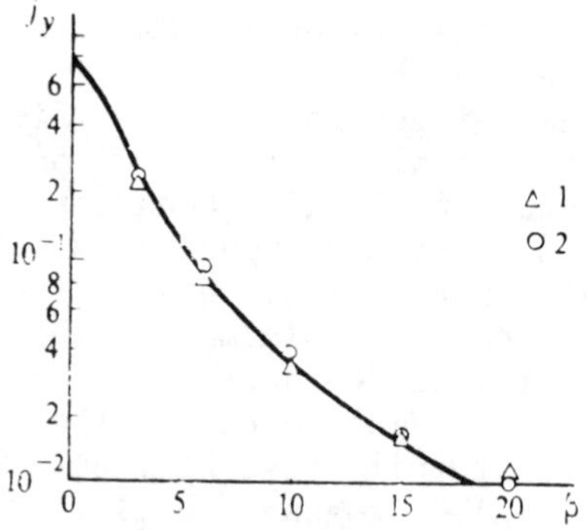

Fig. 3.7 Variation of the transverse current density with the Hall parameter at various values of MHD channel-wall conductivity ($\overline{T}_w = 0.62$), lines--results of the calculations ($\tilde{\sigma}^w = 0$), points - experimental data 1 - $\tilde{\sigma}^w = 0.005$, 2 - 0.9

experimental points at high electrical wall conductivity ($\tilde{\sigma}^w = 0.09$) on the average lie even somewhat higher (approximately by 10 %) than for lower conductivity ($\tilde{\sigma}^w = 0.005$). Obviously, the difference lies beyond the limits of experimental accuracy. However, the tendency of suppressing the harmful effect of surface wall conductivity by the transverse plasma conductivity nonuniformity is revealed clearly in the experiments.

Below, it will be shown that under certain conditions the transverse channel wall conductivity may suppress the negative influence of transverse plasma conductivity nonuniformity, so that the MHD generator characteristics with nonideal insulation walls may turn out to be even better than in a case of ideal insulation.

3.2.6. No-Load Voltage in an MHD Generator. The influence of nonideal insulation on the no-load voltage, $V_y^{x.x}$, is more noticeable. This influence increases with the Hall effect, and, as a result of nonideal insulation, $V_y^{x.x}$ at large β may decrease several times even at a low surface wall conductivity due to nonideal insulation. Variation of $\bar{V}_y^{x.x}$ with β at different $\tilde{\sigma}^w$ is shown in Fig. 3.8. The mechanism responsible for the no-load voltage drop due to nonideal insulation in the presence of the Hall effect is as follows. The initial cause of the drop of $V_y^{x.x}$ in the Faraday MHD generator is the nonideal behavior of the transverse insulation. Neither the axial surface channel wall conductivity nor the transverse plasma conductivity nonuniformity can influence $V_y^{x.x}$.

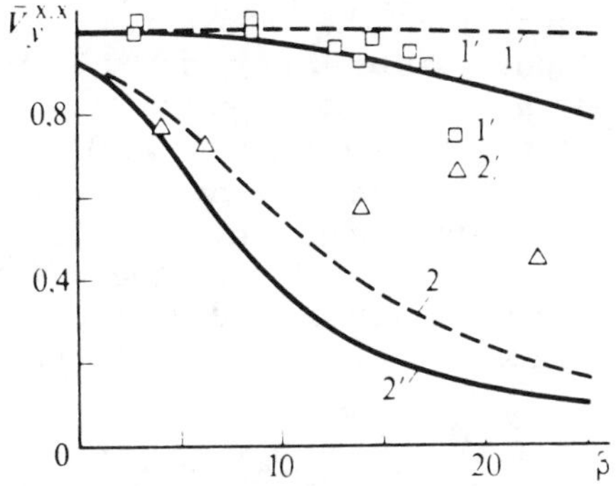

Fig. 3.8 Variation of the no-load voltage on the Hall parameter in the case of nonideal insulation of an MHD channel, lines--results of the calculations; points--experimental data; 1,1′ - $\tilde{\sigma}^w = 0.005$; $\alpha = 0(1)$ and $\alpha = 0.08(1')$; 2,2′ - $\tilde{\sigma}^w = 0.09$, $\alpha = 0(2)$ and $\alpha = 0.08(2')$

Their influence on the characteristics of the generator is associated with the transverse emf in the flow induced by the transverse current. When the transverse insulation during a no-load operating mode is ideal, current density $j_y = 0$. When the external circuit is open, the transverse current in the plasma flow appears only when $\bar{\sigma}_y^w \neq 0$. Furthermore, the effects considered above, associated with the transverse plasma conductivity nonuniformity and nonideal insulation in the direction along the channel, appear in the plasma. At large β, the transverse counter emf induced as a result of these effects exceeds considerably the voltage drop across the dc plasma resistance caused by the passage of the transverse current. This counter emf is responsible for the substantial drop of the no-load voltage, whereas the dc plasma resistance at large β is only of tertiary importance.

Thus, plasma conductivity nonuniformity and nonideal behavior of insulation along the channel in the presence of the Hall effect enhance considerably the negative effect of the nonideal behavior of the transverse insulation on the no-load voltage $V_y^{x\cdot x}$. This apparently is responsible for the more noticeable influence of σ^w on $V_y^{x\cdot x}$, even though the effect itself is considerably weaker than the influence of the nonideal behavior of the MHD channel on j_y and E_x.

Within the framework of the "$j_z = 0$" model, the effects considered are taken into account in (3.4). When insulation is nonideal, the experimental data qualitatively verify the existence of dependence of $\overline{V}_y^{x\cdot x}$ on β. However, the experimentally observed dependence is considerably weaker than the theoretical dependence. This is especially noticeable at moderate σ^w. Hence, at moderate $\tilde{\sigma}^w$ and high β, theoretical values of $\overline{V}_y^{x\cdot x}$ may differ several times from the experimental values.

Here, again is an indication of the fact that the theoretical "$j_z = 0$" model overenhances the negative influence of nonideal insulation on the characteristics of the generator. Explanation of a deviation requires three-dimensional analysis of the electric field in the MHD channel.

<u>3.2.7. Direct Experimental Verification of Current Circulation in a Nonuniform Flow Caused by the Hall Effect.</u> The concept of current circulation in a nonuniform flow under the influence of the Hall effect, resulting in a substantial decrease of current density, j_y, and the electric field intensity, E_x, [8] has not gained immediate acceptance. Practically identical dependences of j_y (or E_x) on the Hall parameter in the presence of a transverse plasma conductivity nonuniformity in the flow ($\alpha > 0$) and nonideal channel wall insulation in the axial direction ($\sigma_x^w > 0$) allowed a non-unique interpretation of the results of experiments in MHD channels with a nonuniform flow and nonideal insulation. The non-unique interpretation was also enhanced by the fact that mutual influence of both factors differed little from interaction of either one of them.

Special proof was required that current circulation generated by the Hall effect actually occurs in a nonuniform flow. The following was a strong argument against this concept. Assuming that the mechanism of the phenomenon considered in 3.2.3. is valid, one may write the following expression for the electric field intensity in the core flow:

$$\bar{E}^o_y = E^o_y/u_oB = 1 + \frac{\beta^2}{G_o}\frac{\langle u\rangle}{u_o}\left(1 - \frac{\langle\sigma\rangle}{\sigma_o}(1 + \beta^{-2})\right)(1 - K_y) \,. \tag{3.5}$$

This expression follows when j_y and E_x from relations (3.1) and (3.2) are substituted into the generalized Ohm's law for the core flow.

From formula (3.5),, it follows that, if

$$1 - \langle\delta\rangle/\sigma_o\rangle\,(1 + \beta^2)^{-1}, \tag{3.6}$$

then $E^o_y > u_oB$, i.e., at sufficiently low plasma conductivity in the boundary layers and at high β, the electric field intensity in the core flow has to be higher than the emf u_oB induced in it. What's more, no matter how paradoxically it may sound, the field intensity in the core flow has to increase with decreasing load factor K_y, even though the mean value $\langle E_y\rangle$ has to drop.

The dependence of E^o_y on β in the short-circuit mode ($K_y = 0$) at various degrees of plasma conductivity nonuniformity is shown in Fig. 3.9. Calculations were performed, using formula (3.5), assuming conditions close to the actual conditions existing in the "Start" facility.

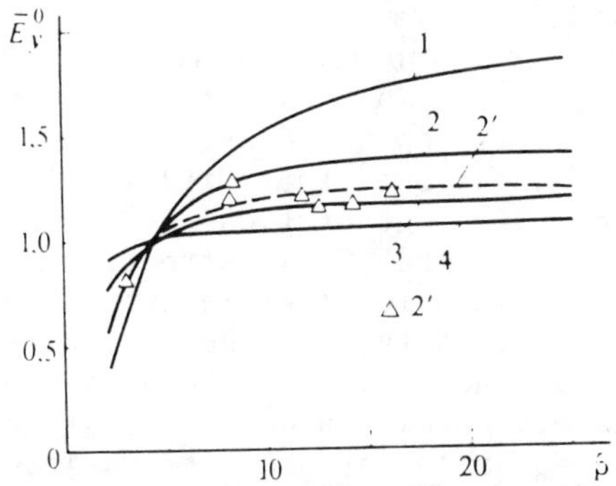

Fig. 3.9 Influence of the Hall effect on the electric field intensity in the core of a nonuniform flow in an MHD channel, lines--results of the calculations; points--experimental data; 1-4 - $\sigma^w_x = 0$, $T^w_y = 0.7$, 0.6, 0.5, and 0.4, respectively; 2′ - $\sigma^w_x = 0.005$, $T^y_w = 0.6$

As can be seen from Fig. 3.9, under these conditions $E_y^\circ > u_oB$ may occur only when $\beta > 5$. It is interesting to note that $\bar{E}_y^\circ$ decreases with decreasing wall temperature, even though the ratio $\langle\sigma\rangle/\sigma_o$ also decreases when the wall temperature drops, and it appears that $\bar{E}_y^\circ$ should actually increase. However, at high β, the drop of $\bar{T}_w$ leads to a sharp increase of the nonuniformity parameter, G_o. This determines the variation of $\bar{E}_w^\circ$.

The fact that E_y° exceeds u_oB as well as the dependence of $\bar{E}_y^\circ$ on β and $\bar{T}_w$ can be easily explained physically. As was shown in 3.2.3., the electric current circulating in the flow due to the Hall effect in the core is directed toward the flow ($j_x < 0$) and induces a transverse emf E_y^* in the direction of the emf u_oB. When the Hall effect increases, the additional emf increases and, beginning at a certain value of β, becomes greater than the potential drop due to the dc resistance of the plasma in the core flow, ΔE_y, during passage of transverse current across the core flow. This leads to the condition $E_y^\circ > u_oB$ in the core flow. Because the system is linear ($E_y^* \sim j_x \sim j_y$ and $\Delta E_y \sim j_y$), the inequality sign does not depend on the transverse current and will remain constant at any $K_y < 1$. Only the difference $E_y^\circ - u_oB$ will change with changing current. The sharpest manifestation of this effect will occur at maximum current, i.e., when $K_y = 0$.

A decrease in E_y° with decreasing $\bar{T}_w$ can be attributed to the fact that the plasma nonuniformity increases with decreasing wall temperature and, for various reasons considered in 3.2.3., the transverse current decreases sharply at high β. At the same time, the Hall field induced by it and Hall current that induces additional emf E_y^* in the core flow also decrease. Apparently, optimal $\bar{T}_w$ exists at which E_y° reaches a peak. However, we have not investigated this problem.

When it was first noticed that, according to theory, $E_y^\circ > u_oB$ may occur in a nonuniform flow in the short-circuit mode [8], it appeared so unlikely that certain researchers have questioned the validity of the hypothesis of current circulation generation in the nonuniform flow in the presence of the Hall effect, as well as all conclusions pertaining to the influence of the plasma conductivity nonuniformity on the characteristics of an MHD generator. At that time, $E_y^\circ > u_oB$ was not observed in any of the experimental investigations and, therefore, there were no reasons to assume it possible. During the first stage of our experiments, we have not observed any cases when the electric field strength in the core flow exceeded the induced emf u_oB, even though condition (3.6) was fulfilled in many of our experiments.

A more detailed analysis has shown that the pattern can be strongly distorted as a result of current leakage along insulation. The Hall current in plasma being short-circuited along the insulating walls flows with the flow and, by inducing a field against the emf, reduces completely the effect under consideration.

The influence of nonideal behavior of the insulation on E_y^o can be easily taken account by substituting j_y and E_x from relations (3.4) into the Ohm's law equation. One then obtains

$$E_y^o = 1 + \frac{\beta^2}{G} \frac{\langle u \rangle}{u_o} \left(1 - \frac{\langle \sigma \rangle}{\sigma_o} (1 + \tilde{\sigma}_x^w)(1 + \beta^{-2}) \right) , \qquad (3.7)$$

$$1 - \frac{\langle \sigma \rangle}{\sigma_o} (1 + \tilde{\sigma}_x^w) > \frac{1}{1+\beta^2} , \qquad (3.8)$$

then $E_y^o > u_o B$.

When $\tilde{\sigma}_x^w = 0$, formula (3.7) becomes (3.5) and condition (3.8) becomes (3.6). From relation (3.7), it can be readily observed that, even for insulation adequate for experimental facilities ($\tilde{\sigma}_x^w = 0.1$), $E_y^o > u_o B$ may occur only at such a high plasma conductivity nonuniformity (low $\overline{T}_w$) that the effect is already so small as to be difficult to observe ($E_y^o/u_o B - 1 < 0.01$). Obviously, this is the reason that this effect was not observed in our first series of experiments ($\overline{\sigma}_x^w \simeq 0.1$).

However, after the wall surface conductivity was successfully decrease to $\tilde{\sigma}_x^w = 0.005$, the field intensity in the core flow in the short-circuit mode began to exceed considerably the quantity $u_o B$. The results of these experiments are shown in Fig. 3.9. The experimental data occur somewhat below the theoretical curve calculated for $\tilde{\sigma}_x^w = 0$, but are in good agreement with the theoretical curve for $\tilde{\sigma}_x^w = 0.005$.

Although E_y^o exceeds $u_o B$ by only a small amount, this fact itself is, in our view, of major importance. Of all the effects in the MHD generator channel, only Hall current circulation may lead to E_y^o exceeding $u_o B$. The relation, $E_y^o > u_o B$, observed in the experiments is direct experimental verification of appearance of current circulation in the nonuniform flow in the presence of the Hall effect.

3.2.8. Other Factors. The plasma conductivity nonuniformity and the nonideal behavior of channel wall insulation considered in the foregoing are two principal factors exerting the strongest influence on the electrical characteristics of a nonideal MHD generator, especially pronounced at high Hall parameter values. The influence of these factors on the electrical characteristics of experimental MHD generators is quite considerable. Apparently, they will also exert noticeable influence on the characteristics of commercial-scale MHD generators. In any case, such a possibility should not be neglected in designing commercial-scale MHD generators.

The nonideal behavior of an MHD channel is not limited to these two factors. Many other factors are directly associated with the design of a real MHD generator (e.g., nonideal electrode segmentation or a near-electrode potential drop) that may also influence the operation of an MHD generator. Apparently, the influence of these parameters will be less substantial in commercial-scale MHD facilities. However, it is quite considerable in experimental MHD generators. For example, in the case of a commercial-scale MHD generator with an induced transverse voltage of 1000-2000 V, the near-electrode potential drop of 50 V will not be so significant. However, in an experimental facility where the induced emf is only 200 V (as in the case in the U-02 facility), 50 V losses will be quite considerable. What's more, they will exert substantial influence on the overall energy conversion process, including those effects that will be of considerable importance in large-scale generators.

Therefore, in conducting experimental investigations of the energy conversion process on test and prototype MHD facilities, one cannot be limited to investigations of only those effects that will be of quite considerable importance in commercial-scale facilities. In order to reach proper conclusions in analyzing the results of experiments on a test facility, it is necessary to take into account all factors exerting influence on the energy conversion process under these conditions. Taking this into account, consider other factors that may exert an influence on the energy conversion process in an MHD channel.

3.2.8.1. Nonideal Behavior of Electrode Segmentation

The physical influence of this factor on the characteristics of an MHD generator in the presence of a Hall effect is close to that of the nonideal behavior of transverse insulation. Because the electrodes are characterized by infinite conductivity, they short-circuit the transverse field induced in each segment, leading to a nonuniform current distribution over the surface of electrodes. At the anodes, the current is concentrated near the front edges of electrodes exposed to the flow; at the cathodes, it is concentrated near the rear edges. A transverse component of current that induces transverse counter emf that lowers the operating current and the transverse voltage of the MHD generator appear in the plasma flow.

This phenomenon was investigated in many theoretical papers (e.g., see Ref. 16), most of which dealt with a uniform flow. The most detailed data on the dependence of the characteristics of an MHD generator with a uniform flow on the degree of electrode segmentation and their length at various β can be found in Reference 17.

The plasma conductivity nonuniformity in the flow suppresses the influence of the nonideal electrode segmentation. This can be attributed to two factors: 1) the transverse plasma conductivity nonuniformity decreases considerably the transverse field in an MHD channel; 2) the presence of a boundary layer with below normal conductivity enhances equalization of the current distribution over the surface of the electrode. Both decrease the transverse current and, therefore, harmful counter emf induced by it.

The foregoing discussion is substantiated by theoretical calculations [18, 19]. However, these calculations are only qualitative, because they are based on a model (linear) conductivity distribution profile in the boundary layer that deviates considerably from the real profile. Unfortunately, these authors know of no experimental work investigating this problem.

Experiments on the "Start" facility were conducted at two different relative degrees of electrode segmentation, $\bar{s} = s/2a$ (0.11 and 0.19). However, no difference was observed in characteristics of the MHD generator. Therefore, in processing the experimental data, the influence of the degree of electrode segmentation was neglected. Apparently, finite electrode segmentation will not exert much of an influence on the operation of large, commercial-scale MHD generators, the degree of electrode segmentation of which will depend on the absence of axial interelectrode breakdown. This phenomenon has not been thoroughly investigated. According to various authors, permissible interelectrode voltage varies between 30 and 150 V. The influence of operating conditions of an MHD channel on this quantity is not yet clear. Even if one assumes a maximum value of 150 V, then at the electric field intensity expected in commercial-scale channels (up to 5 kV/m), the degree of electrode segmentation cannot exceed 30 mm, or $\bar{s}$ = 0.02-0.03. Furthermore, according to maximum estimates for a uniform flow, the decrease of specific characteristics of an MHD generator will not exceed 10 %. It is possible that this problem will generate considerable interest at later stages of the investigation. However, problems associated with interelectrode breakdown, intensively investigated at present, are more urgent.

3.2.8.2. Nonuniformity in Flow Rate and Transverse Plasma Conductivity

The influence of these factors on the electrical characteristics of a generator is not associated with the Hall effect. If a comparison is made at the same mean plasma flow rate across the channel cross section, the flow rate nonuniformity will not affect the current and the voltage in any way, provided the electrical load factor is K_y = idem. However, it is more apropos to conduct a comparison at the same local values of the internal relative efficiency of the MHD generator, η_{oi} = idem. In an ideal Faraday MHD generator, $\eta_{oi} = K_y$. However, in a nonideal MHD generator, $\eta_{oi} < K_y$ is always true. When η_{oi} = idem, the influence of the transverse flow rate nonuniformity on the specific characteristics of the generator will always be negative. This can be attributed to the appearance of reverse transverse current causing an increase in dissipative losses and a decrease in efficiency in that part of boundary layers at the insulating walls, where the local flow rate is less than mean flow rate along the height of the channel. Maintaining it at a constant level requires an increase in K_y and that causes a drop in the specific power (when $K_y \geq 0.5$). However, in a turbulent boundary layer, the flow rate profile is adequately filled, so that, qualitatively, the influence of axial flow rate nonuniformity on η_{oi} will not be so substantial. This is especially true in the presence of the Hall effect that equalizes current distribution nonuniformity along the channel height, caused by the flow rate nonuniformity. A more substantial influence on η_{oi} is exerted by current leakages along insulation walls in the transverse direction. The axial plasma conductivity nonuniformity exerts a considerably weaker influence on operation of an MHD generator than the transverse conductivity. A decrease in conductivity in cold boundary layers at insulating walls quite obviously decreases the overall conductivity of the plasma flow. Transverse working current decreases correspondingly. According to our estimates, when the boundary layer is turbulent, the decrease of the transverse current resulting from nonuniformity of axial plasma conductivity does not exceed 15-20 %. This can be attributed to the temperature profile (which actually determines the conductivity profiles) being more than adequately filled when the boundary layer is turbulent, even in unfavorable cases when the generator channel is long and flat (a/b>>1) and the insulating walls are cold. For commercial-scale MHD generators, this current drop will be substantially smaller.

The influence of the axial conductivity nonuniformity $\sigma(z)$ on the operating current of the generator may be taken into account theoretically, with adequate accuracy, by introducing plasma conductivity averaged over the height of the channel $\langle\sigma\rangle_z$.

Axial conductivity nonuniformity also exerts some effect on the no-load voltage $V_y^{x \cdot x}$, and this effect differs from the previous one by being positive. The plasma conductivity drop near the insulating wall suppresses reverse current flows along the boundary layer. As a result, the no-load voltage increases somewhat. It can be shown that $V_y^{x \cdot x} \sim g^* \equiv \langle u\sigma\rangle_z/\langle\sigma\rangle_z$ and $g^* > 1$. In the case of well-filled profiles, the difference, $g^* - 1$, is only several percent.

By suppressing reverse current flows, nonuniformity in axial plasma conductivity will exert certain positive effects on the efficiency of an MHD generator.

3.2.8.3. Near-Electrode Potential Drop

The term near-electrode potential drop will refer to a potential drop in the space charge at the plasma-electrode boundary. The near-electrode potential drop at the anode is usually assumed to be zero. The cathode potential drop depends on the emissivity of the electrode surface, its temperature, and electrode current density. In small MHD channels, the cathode drop may be of the same magnitude as the induced emf.

Because the electric layer at the electrode surface is quite thin, gasdynamics and electrodynamics of the flow, including the Hall effect, apparently will not exert any substantial influence on the near-electrode potential drop, $\Delta\phi$. Furthermore, the influence of the cathode drop on processes in the flow will be manifested only in terms of the boundary conditions. Relations (3.1) and (3.4) can be considered to be load factors of the flow, from which one can obtain the external load factor of the MHD channel, taking into account that

$$j_y^H = j_y^\Pi, \quad V_y^H = V_y^\Pi - \Delta\phi\,(j_y^H)\ , \tag{3.9}$$

where, in the no-load operating mode, $\Delta\phi(0) = 0$ and $(V_y^H)^{x.x} = (V_y^\Pi)^{x.x}$. Here, $j_y^H = I^H/2bs$ and j^Π are the load current density (I^H is the load current) and current density in the plasma flow, respectively; V_y^H, V_y^Π, are the voltages at the load and at the plasma flow boundaries.

Taking equations (3.9) into account, relation (3.1) in the presence of a cathode potential drop assumes the form

$$\bar{j}_y^H = (\bar{j}_y^\Pi)^{s.c.}\,(1 - K_y - \Delta\bar{\phi}(\bar{j}_y^H))\ ,$$

$$\bar{V}_y^H \equiv K_y(\bar{V}_y^H)^{x.x}\ , \tag{3.10}$$

where $(j_y^\Pi)^{s.c.}$ is the current density in a regime when the mean value of the transverse field intensity in the flow is $\langle E^\Pi \rangle = 0$ (or $V_y^\Pi = 0$) and $\Delta\bar{\phi} \equiv \Delta\phi/V_y^{x.x}$. The values of $(j_y^\Pi)^{s.c.}$ and $V_y^{x.x}$ depend on the electrodynamics of the flow and are independent of $\Delta\phi$. During theoretical analysis performed within the frame of the "$j_z = 0$" model, they can be determined from relations such as (3.4). When solving magnetogasdynamic problems, the cathode potential drop is usually considered to be a specified function of the current density j_y.

When $(\bar{j}_y^\Pi)^{s.c.}$, $\bar{V}_y^{x.x}$, and $\Delta\bar{\phi}(\bar{j}_y)$ are known, expression (3.10) is a transcendental equation, a solution of which may not always be represented analytically, even though it can be found approximately with any desired accuracy in each specific case. Hence, the influence of the cathode potential drop on the external characteristics of an MHD generator can be easily taken into account within the framework of the "$j_z = 0$" model.

As has already been noted in the foregoing discussion, the cathode potential drop may exert considerable influence on the operation of a small-scale MHD generator, and must be taken into account during investigation of the electrodynamic processes in an MHD channel of laboratory facilities. In commercial-scale MHD channels with "hot" ($T_w \simeq 2000$ K) or "semi-hot" ($T_w \simeq 1600$-1700 K) electrodes, the near-electrode potential drop will be on the order of several dozen volts [20] and $V_y^{x \cdot x}$ may reach several kilovolts. Under such conditions, the near-electrode potential drop will not exert substantial influence on the external characteristics of MHD generators. If necessary, it can be taken into account using relation (3.10).

A large potential drop (hundreds of volts) in the near-electrode region[8] may be expected only during operation with "cold" electrodes. However, in this case, they will operate in an arc mode. This case lies outside the scope of the present analysis, which is based on the assumption of diffuse current spreading across the whole channel cross section.

Thus, of all the factors considered, the strongest influence on the specific characteristics of large-scale MHD generators can be expected from the nonuniformity of transverse plasma conductivity, caused by the development of relatively cold boundary layers on electrode channel walls.

A question arises as to whether it is possible to exclude, or at least considerably weaken the harmful effect of this factor. Probably the most effective means is a decrease of conductivity nonuniformity at the expense of a rise of the electrode wall temperature. However, if this parameter is limited by the material properties or other conditions, one can attempt to find other means to decrease or eliminate the influence of plasma conductivity nonuniformity in an MHD channel.

3.2.9. Protrusion of Electrodes into the Flow. Tranverse Current Nonuniformity. Initially, the problem of decreasing the negative influence of plasma conductivity nonuniformity in the boundary layers on electrode walls did not appear to be complex. It was apparent that this problem could be easily solved by power takeoff outside the boundaries of the cold boundary layer.

8 It is also frequently referred to as "near-electrode potential drop," even though, apparently, it would be correct to refer to it as "effective near-electrode potential drop." This can be attributed to the fact that the processes determining it are not limited to the space charge layer, but affect a much broader flow region. They are determined by discharge constriction and current spread out from the arc. These phenomena have not been adequately investigated. Some data on the influence of various factors on the effective near-electrode potential drop during a constricted discharge on electrodes may be found in Ref. 21.

Technically, this could be achieved by protruding electrodes into the flow.

Another way to suppress the harmful effect of plasma conductivity nonuniformity is by means of the so-called S-shaped electrodes, part of the working surface of which extends into the plane of the horizontal ("insulating") walls of the channel. Part of the current to the S-shaped electrodes flows along the lines of force of the magnetic field without inducing undesirable transverse emf. This should enhance equalization of the distribution of the latter across the cross section and, as has been shown here, it is the nonuniformity of the distribution of E_x^* over the cross section that generates effects responsible for a substantial drop of MHD generator parameters.

Five types of electrodes designed to take into account the concepts discussed elsewhere [22] were tested on the "Start" facility. The shape and dimension (millimeters) of the electrodes investigated and the results of experiments on these electrodes are shown in Fig. 3.10. As can be seen, protrusion of electrodes into the flow, or their extension to the horizontal walls, produces a positive effect. However, it proved to be considerably less than expected.

Protrusion of electrodes into the flow does exclude the influence of the plasma conductivity nonuniformity. However, current nonuniformity appears just as harmful in

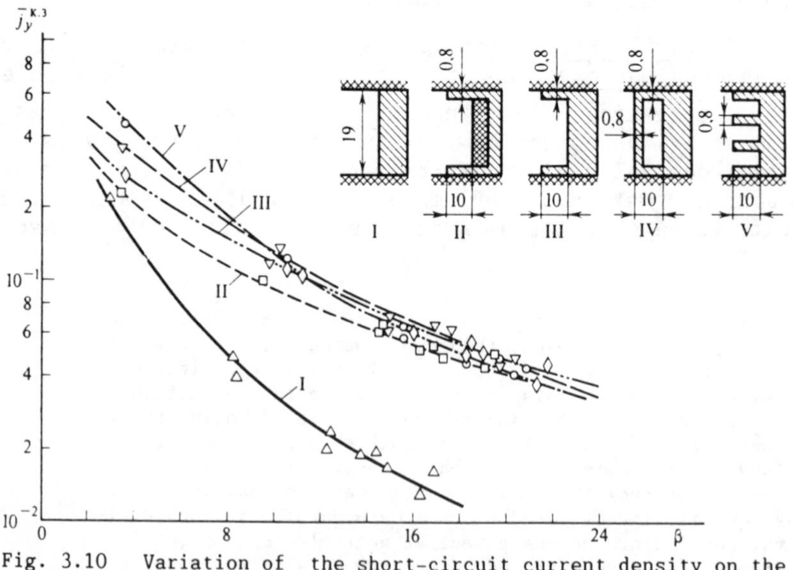

Fig. 3.10 Variation of the short-circuit current density on the Hall parameter for electrodes of various configurations

the presence of the Hall effect as the plasma conductivity nonuniformity. When the electrodes are protruded, transverse current does not penetrate the boundary layers at the electrode walls, and zones are formed where the Hall emf is not induced. As a result, the axial current is short-circuited along these zones (Fig. 3.11), the formation of which is equivalent to degradation of channel insulation in the axial direction.

Taking this into account, an approximate method of calculating MHD channel characteristics with electrodes protruding into the flow (a variety of the "$j_z = 0$" model) was developed [23]. This model is based on the following assumptions.

The temperature of electrodes protruding into the flow a distance y* is equal to the plasma temperature at the same distance from the wall: $T_y = T(y^*)$. The electrodes do not disturb velocity and conductivity distributions in the boundary layer.

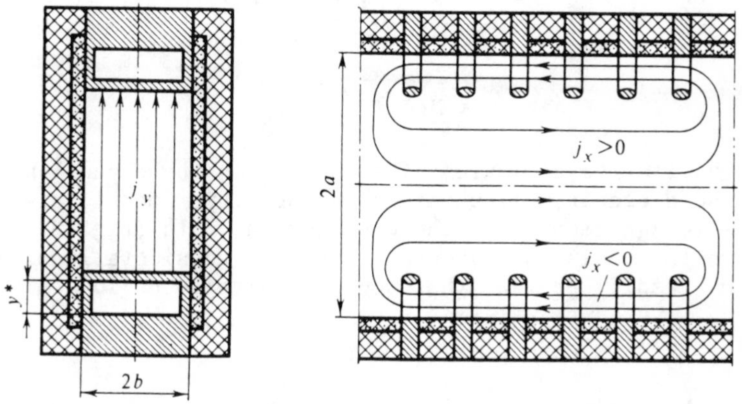

Fig. 3.11 Current circulation in an MHD channel with protruding electrodes

When $|y| < a - y^*$, the transverse current distribution is uniform ($j_y(y)$ = const.), whereas, in the zone between the electrodes and the wall ($|y| > a - y^*$), the current density is $j_y = 0$.

In view of the last condition, the axial field in the zone $|y| > a - y^*$ is not generated, and the transverse current is short-circuited along it. Therefore, this zone performs the function of an external axial load with respect to the rest of the flow, with conductivity

$$\tilde{\sigma}_x^o = b \int_{a-y^*}^{a} \sigma \, dy \; .$$

Based on these assumptions, the external characteristics of an MHD generator with electrodes protruding into the flow can be represented by relations (3.4), by replacing the following quantities in these relations:

$$\tilde{\sigma}_x^w \text{ with } \tilde{\sigma}_x^* \equiv \tilde{\sigma}_x^w + \tilde{\sigma}_x^o \text{ and } \alpha \text{ with } \alpha^* \equiv \langle\sigma\rangle^* \langle 1/\sigma\rangle^* ,$$

where

$$\tilde{\sigma}_x^* = (\sigma_x^w(a + b) + \sigma_x^o)(b\langle\sigma\rangle^*)^{-1} ;$$

$$\langle\sigma\rangle^* \equiv (1-\bar{y}^*)^{-1} \int_0^{1-\bar{y}^*} \sigma \, d\bar{y} ;$$

$$\left\langle \frac{1}{\sigma} \right\rangle^* \equiv (1 - \bar{y}^*)^{-1} \int_0^{1-\bar{y}^*} \frac{1}{\sigma} \, d\bar{y} ;$$

$$\bar{y}^* = y^*/a; \quad \bar{y} = y/a .$$

Qualitative estimates of j_y and E_x in the channel with protruded electrodes, obtained using this model, are in complete agreement with the experimental data (Fig. 3.12).

On the basis of general physical considerations, it can be readily proven that current divergence in a plane

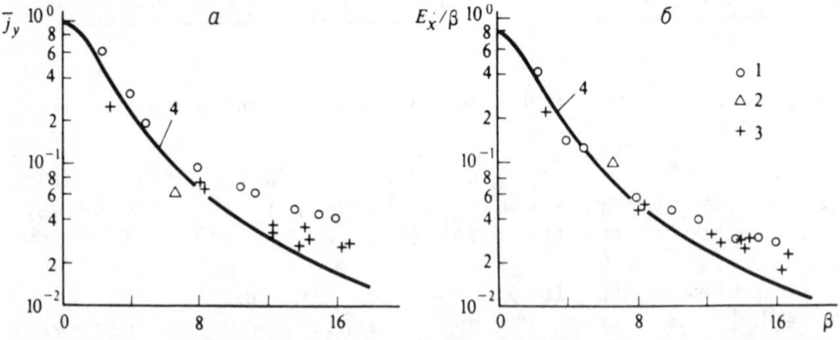

Fig. 3.12 Influence of the Hall parameter on the transverse current density (a) and electric field intensity (b) in the case of electrodes protruding into the flow ($\bar{y}^* = \bar{\delta}$); 1-3 - results of the experiments; 1--T_o = 2260 K, $\bar{T}_w$ = 0.63, $\bar{\delta}$ = 0.23; 2--T_o = 2290 K, $\bar{T}_w$ = 0.66, $\bar{\delta}$ = 0.23; 3--T_o = 2190 K, $\bar{T}_w$ = 0.60, $\bar{\delta}$ = 0.31; 4--results of the calculations; T_o = 2200 K, $\bar{T}_w$ = 0.6, $\bar{\delta}$ = 0.2

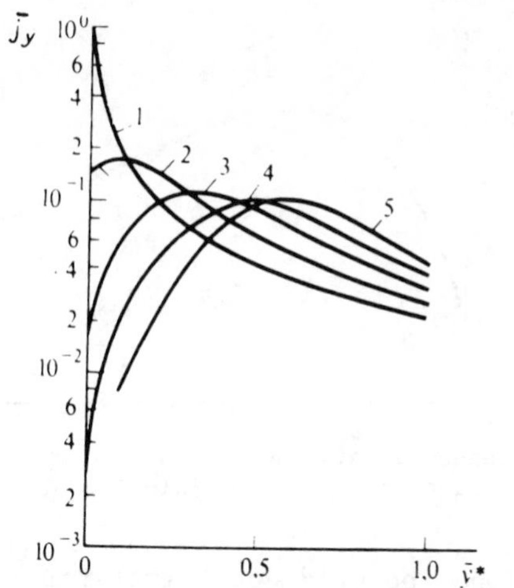

Fig. 3.13 Dependence of the transverse current density on the degree of protrusion of electrodes into the flow at various relative wall temperatures (β = 15, $\bar{\delta}$ = 0.2); 1--1; 2--0.8; 3--0.6; 4--0.4; 5--0.2

flow in the presence of a Hall effect generates current circulation independent of whether the current flows to the horizontal channel walls directly, or through electrodes protruding into the flow. Such current circulation and circulation resulting from below-normal plasma conductivity in the boundary layers are of opposite signs, and may partially compensate each other. Therefore, protrusion of electrodes into the flow improves characteristics of an MHD generator.

However, current nonuniformity may exert a harmful effect. As $\bar{y}^*$ increases, the influence of the plasma conductivity nonuniformity decreases and the current nonuniformity increases. Calculations of $\bar{j}_y(\bar{y}^*)$ at various $\bar{\delta}$, β, and $\bar{T}_w$ show that there is an optimum, $\bar{y}^*_{opt}$, at which $j_y = j_y^{max}$. It turned out that $\bar{y}^*_{opt}$ is strongly influenced by $\bar{T}_w$ and is practically independent of either δ, or β. When $\bar{T}_w = 1$, protrusion of electrodes into the flow only degrades characteristics of an MHD generator. This degradation is quite considerable at large β. However, when $\bar{T}_w < 1$, $\bar{j}_y(\bar{y}^*)$ is nonmonotonic (Fig. 3.13). As the wall temperature decreases, $\bar{y}^*_{opt}$ also increases. The positive effect caused by protrusion of electrodes into the flow also increases with decreasing wall temperature and at large β, and small $\bar{T}_w$ may reach several dozen times that of normal. The calculated dependence $\bar{j}_y(\bar{y}^*)$ is in adequate agreement with the experimental data (Fig. 3.14).

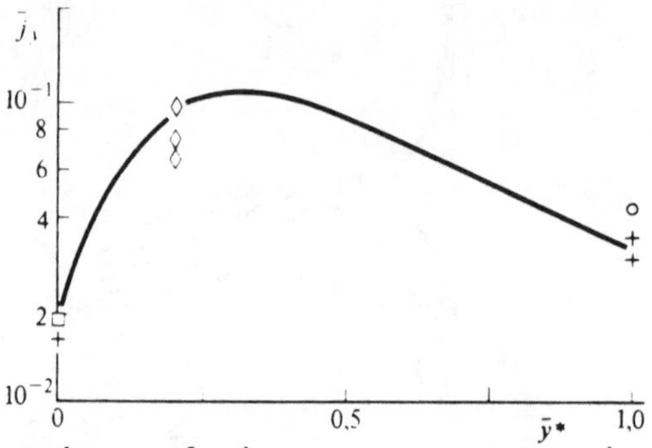

Fig. 3.14 Dependence of the transverse current density on the degree of protrusion of electrodes into the flow at $\bar{T}_w = 0.6$, $\beta = 15$, $\bar{\delta} = 0.2$

It should be pointed out that protruding electrodes strongly disturb the flow and, therefore, will be unlikely to find application. The S-shaped electrodes, which lead to a considerable reduction of the harmful effect of the plasma conductivity nonuniformity in boundary layers (see Fig. 3.10) but cause no disturbance of the flow, are more promising. The electrodynamics of a channel with S-shaped electrodes is more complex, but also depends on the same effects; i.e., with S-shaped electrodes, most of the current is fed to the horizontal sectors of electrodes and, as with protruding electrodes, the current does not reach vertical walls. As a result, the voltage of the induced Hall field is weakened. This suppresses current circulation and harmful effects from the transverse plasma conductivity nonuniformity.

Although protruding electrodes are not especially promising, detailed investigation of an MHD channel with such electrodes has clearly revealed another effect, the influence of the transverse current nonuniformity. It also made it possible to investigate its interaction with the influence of plasma conductivity nonuniformity under simpler conditions. This problem is of primary importance.

3.2.10. Simplest Model that Takes into Account the Current Flow to the Insulation Walls. The transverse current nonuniformity may also appear in the flow as a result of current flow to the horizontal (insulating) walls. Two conditions are necessary for the appearance of current leakage caused by current flow to the insulating walls: 1) the insulating walls must be characterized by surface conductivity not equal to zero, 2) transverse distribution of plasma conductivity in the flow must be nonuniform. In a uniform flow, the potential distribution across the flow will be linear. It will also be linear on the horizontal walls of the channel with a uniform surface conductivity. Consequently, no potential difference between the flow and the horizontal walls that could lead to current leakages from the flow will occur in the z direction. Boundary layers with below-normal plasma conductivity on the electrode walls of the channel are a typical flow nonuniformity generating current leakages to the insulating walls.

The current fed to the walls is partially returned to the flow along the walls. However, the largest part of it flows along the walls to the electrodes and across the electrodes to the load, bypassing boundary layers of low plasma conductivity. Decreasing current density, j_y, in layers with below normal conductivity lowers the transverse emf induced in it and, as a result, weakens the current circulation in the x,y plane and the associated harmful effects.

As a result, the appearance of current on the horizontal walls suppresses the harmful influence of the transverse plasma conductivity nonuniformity in the boundary layers at the electrode channel walls. The "j_z = 0" model does not take this effect into account. One of the first attempts to take into account the influence of current leakage to the horizontal channel walls with finite surface conductivity on the characteristics of the MHD generator [24] dealt with the limiting case of current leakage based on the assumption that the plasma impedance in the z direction is infinitely small ("E_z = 0" model). Under this assumption, the problem of the current density (j_y) distribution and field intensity (E_y) distribution over the cross section of an MHD channel can be solved by quadratures. This solution can be used to obtain the load factor of a Faraday MHD generator. The latter is specified by relations (3.1) with different $\bar{j}_y^{s.c.}$ and $\bar{V}_y^{x.x}$.

For an arbitrary distribution of parameters u(y,z), σ(y,z), and β(y,z) over the flow cross section, the expressions for $\bar{j}_y^{s.c.}$ and $\bar{V}_y^{x.x}$ are fairly complicated [24]. However, they can be simplified considerably, provided the distribution of the parameters of plasma over the channel cross sections can be represented by approximate relations, such as

$$\bar{u}(\bar{y},\bar{z}) \equiv u(\bar{y},\bar{z})/u_{oo} = f(\bar{y})g(\bar{z}) ,$$

$$\bar{\sigma}(\bar{y},\bar{z}) \equiv \sigma/\sigma_{oo} = \mu(\bar{y})\ \nu(\bar{z}) ,$$

$$\beta(\bar{y},\bar{z}) = \text{const.} , \tag{3.11}$$

where u_{oo}, σ_{oo} are the flow rate and plasma conductivity along the channel axis, $\bar{z} = z/b$ are dimensionless coordinates, $f(\bar{y}) = u(\bar{y},0)/u_{oo}$, $g(\bar{z}) = u(0,\bar{z})/u_{oo}$, and $\mu(\bar{y}) = \sigma(\bar{y},0)/\sigma_{oo}$, $\nu(\bar{z}) = \sigma(0,\bar{z})/\sigma_{oo}$ are the flow rate and plasma conductivity profiles in the channel cross section.

3.3. Simplified Model and Methods of Calculating Local Characteristics of an MHD Generator taking into Account Three-Dimensional Current Distribution in Channels

Two models have been considered that allow computations of local electrical characteristics of a nonideal MHD generator. Both models are sufficiently simple and convenient for engineering practice, and make it possible to obtain solutions of the problem in closed form without numerical integration of complex differential equations.

The "$j_z = 0$" model makes it possible to take into account the influence of basic factors controlling operation of a nonideal MHD generator on its local electrical characteristics, i.e., plasma conductivity nonuniformity and the nonideal insulation of the MHD channel walls. Actually, channel wall conductivity is taken into account, as an external load connected between electrodes. The probability of current leakage from the plasma flow directly into insulating walls is not taken into account and the decrease of the influence of the transverse plasma conductivity nonuniformity on the current leakage to the insulating walls is also neglected. Therefore, the "$j_z = 0$" model provides minimum estimates of the characteristics of the MHD generator. This can explain the deviation of the calculated dependences obtained using the "$j_z = 0$" model from the experimental data, referred to in 3.2.5. and 3.2.6.

The "$E_z = 0$" model takes into account current leakage to the insulating walls. However, assuming that, for current leakage, plasma impedance in the vertical direction is zero, the model clearly provides high estimates of current leakage to the insulating walls. As a result, it also provides high estimates of the external characteristics of an MHD generator.

Therefore, "j_z = 0" and "E_z = 0" models can be used to estimate both the upper and lower limits of the MHD generator characteristics. Quite frequently, the difference between the two is small, and either model can be used for engineering calculations. However, the "j_z = 0" model is preferred because of its simplicity. In many cases, the difference between the maximum and minimum is too large and a more careful analysis is required.

An investigation of this problem is of considerable practical interest, especially in connection with the development of MHD facilities intended for operation on ash-laden fuel (coal). Liquid slag film with fairly high surface conductivity will always be present on MHD channel walls. Even more uniform transverse current distribution is to be expected in frame channels of MHD facilities that are promising for uses in commercial-scale MHD generators. At the present time, methods of solving two-dimensional magnetogasdynamic problems, particularly problems dealing with electrical fields across the channel [25, 26] have been developed and widely used, both in the Soviet Union and abroad. However, two-dimensional calculations are quite difficult and, for this reason, are not well suited for large-scale engineering computations at the stage that involves development of prototypes or preliminary optimization of the MHD channel parameters. In engineering practice, one must search for less complex, but sufficiently accurate methods of calculating the characteristics of an MHD generator.

It is possible to construct a relatively simple model of the flow in the y, z cross section of the flow and, using this model, to develop approximate methodology for calculating local characteristics of an MHD generator, taking into account practically all of the principal effects specifying operation of a nonideal MHD generator.

In such a model, in addition to the plasma conductivity nonuniformity, nonideal insulation, and near-electrode potential drop, one must account for current flow to the horizontal walls. This case differs from the "E_z = 0" model in that it is necessary to take into account the dc resistance of plasma in the direction of the current density, j_z, especially in cold boundary layers at insulating walls of the channel. Such a model will be referred to as "$j_z \neq 0$, $E_z \neq 0$."

3.3.1. Construction of the Computational Model. With the appearance of current leakage to the insulating walls, the distribution of electrodynamic parameters in an MHD channel becomes three-dimensional. In this case, the electric field in the flow is described by a well-known system of equations:

$$\mathrm{rot}\,\mathbf{E} = 0 \ (\text{or } \mathbf{E} = -\,\mathrm{grad}\,\phi),\ \mathrm{div}\,\mathbf{j} = 0\ . \tag{3.12}$$

The connection between **E** and **j** is specified by Ohm's law which, for quasi-neutral plasma, can be written in the following form:

$$\mathbf{j} = \sigma\,(\mathbf{E} + \mathbf{u} \times \mathbf{B}) - \Xi\mathbf{j} \times \mathbf{B}\ , \tag{3.13}$$

where σ is the specific plasma conductivity, $\mathbf{u}\{u,0,0\}$ is the flow rate, $\mathbf{B} = \{\mathbf{B}(0,0,B)\}$ is the magnetic field induction, Ξ is the electron mobility ($\Xi|B| = \beta$).

Taking into account the small thickness of the conducting layer, the field on the insulating walls can be written in terms of the surface wall conductivity, σ^w, and surface current density, $\mathbf{j}^w$:

$$\mathbf{E}^W = -\,\mathrm{grad}\,\phi^W,\ j_x^W = \sigma_x^W \mathbf{E}_x^W\ ,$$

$$\frac{\partial j_x^W}{\partial x} + \frac{\partial j_y^W}{\partial y} + j_z^W = 0,\ j_y^W = \sigma_y^W \mathbf{E}_y^W\ . \tag{3.14}$$

Here, $\mathbf{E}^w$ is the electric field intensity vector on the surface of the insulating channel walls, $j_z^w = j_z(x,y,b)$ is the density of current fed to the wall from the plasma. Ohm's law is written on the assumption of absence of the Hall effect in the wall ($\beta^w = 0$). However, the fact that the wall may be characterized by anisotrophic conductivity ($\sigma_x^w \neq \sigma_y^w$) has been taken into account.

The subsystem of equations analogous to (3.14) can also be written for interelectrode insulators. At the electrode surfaces, $\phi\,(x,z) = \phi_\pm = \text{const}$. The equations describing the electric field in the plasma flow and on the channel walls are interconnected by conditions at the phase separation boundaries:

$$j_n = j_n^w,\quad \phi = \phi^w + \Delta\phi\ ,$$

where j_n, j_n^w, are the normal components of the current density in the plasma and on the wall, respectively, and

$\Delta\phi = \Delta\phi(j_n)$ is the potential drop in the space charge layer that is a known function of j_n.

In addition, conditions in the channel cross sections (x_1 and x_2) limiting the region under consideration must be specified. When determining the local characteristics of a sufficiently long channel with segmented electrodes, the variation of the flow parameters per segment length, s, can be neglected and one can consider a periodic problem, specifying the boundary conditions in the form

$$j(x+s,y,z) = j(x,y,z) ,$$

$$\phi(x+s,y,z) - \phi(x,y,z) = E_x^H s = \text{const.} ,$$

where E_x^H is the transverse electric field intensity of the load.

Finally, the external electrical load of the sector of the MHD channel under consideration has to be specified in both the transverse and axial directions. Even if the problem has been separated into gasdynamic and electrodynamic parts, and the distribution of gasdynamic parameters and plasma properties are considered to be specified, the electrodynamic problem in rigid formulation is so complex that a solution, even if possible, is unacceptable for engineering practice. For everyday calculations, one must search for reasonable ways of simplifying the problem.

As a first step, the equations of the system and the boundary conditions can be averaged over each electrode segment, neglecting the variation of flow parameters along the length. For electrode segmentation that is not very coarse, such an operation apparently does not introduce large errors. As a result, the problem is reduced to a two-dimensional problem. Examples of solutions of such a problem can be found in the literature [25, 26]. However, they are available only for $\sigma_y^w \rightarrow \infty$ (this condition is realized in a frame-type Hall generator). The problem becomes more complex when the wall conductivity is finite ($1/\sigma_y^w \neq 0$). To simplify this problem, the equations of the system (3.12)-(3.14) averaged over x, can once again be averaged over z. However, before this, the distribution of flow parameters over the channel cross section must be represented by approximate relations (3.11).[9]

Second averaging makes it possible to reduce the initial system (3.12)-(3.14) to a system of two equations in terms of total derivatives describing the distribution across the channel of averaged potential in the flow $\phi^* \equiv \langle \nu\phi \rangle / \langle \nu \rangle$ and the wall potential, ϕ^w:

$$\frac{\partial}{\partial y}\left(\mu \frac{\partial \phi^*}{\partial y}\right) + u_{oo} \, Bg^* \frac{\partial \mu f}{\partial y}$$

$$- \beta E_x \frac{\partial \mu}{\partial y} = \frac{(1+\beta^2) j_z^W(y)}{\langle \nu \rangle \sigma_{oo} b} , \qquad (3.15)$$

$$\frac{\partial}{\partial y}\left(\sigma_y^W \frac{\partial \phi^W}{\partial y}\right) = j_z^W(y) , \qquad (3.16)$$

where $g^* \equiv \langle \nu g \rangle / \langle \nu \rangle$ and the symbol $\langle \ \rangle$ designates quantities averaged over the height of the channel. The sign denoting averaging with respect to x is omitted everywhere.

Unfortunately, the system (3.15), (3.16), is not closed: it contains two equations with three unknown functions ϕ^*, ϕ^w, and j_z^w. The relation between them may be determined approximately on the basis of the following considerations.

Plasma conductivity near the wall drops sharply as a result of its strong dependence on temperature. Therefore, much of the excess plasma impedance occurs in a very thin layer at the wall. Taking this into account, divide the flow cross section into two zones. It will be assumed that the conductivity varies very weakly in the first zone, while current density j_z remains constant along the thickness of the very thin ($\varepsilon \ll b$) second zone, adjacent to the insulating wall

$$\nu(z) = 1 \text{ when } 0 < |z| < b - \varepsilon ,$$

$$j_z(z) = j_z^w \text{ when } b - \varepsilon < |z| < b . \qquad (3.17)$$

[9] Calculation of the velocity profile, $f(\overline{y})$ and $g(\overline{z})$, and conductivity profiles, $\mu(\overline{y})$ and $\nu(\overline{z})$, as well as certain problems of their averaging over a channel cross section appear in the Appendix. The profiles in the Appendix were used in all calculations by these authors.

Then, integrating Ohm's law equation for the z component of the current,

$$- \frac{\partial \phi}{\partial z} = \frac{j_z}{\sigma} , \tag{3.18}$$

gives

$$\phi^w - \phi_o = \int_o^b \frac{j_z}{\sigma} dz$$

$$= - \frac{1}{\sigma_{oo}\mu} \left(\int_o^{b-\varepsilon} j_z \, dz + j_z^w \int_{b-\varepsilon}^{b} \frac{dz}{\nu} \right)$$

$$= - \frac{1}{\sigma_{oo}\mu} \left(\langle j_z \rangle \, b + j_z^w \left(\int_o^b \frac{dz}{\nu} - \int_o^{b-\varepsilon} \frac{dz}{\nu} \right) \right)$$

$$= - (b/\sigma_{oo}\mu)(\langle j_z \rangle + j_z^w (\langle 1/\nu \rangle - 1)) ,$$

or

$$- j_z^w = \frac{(\phi^w - \phi_o)(\sigma_{oo}\mu/b)}{\langle 1/\nu \rangle - 1 + \langle j_z \rangle / j_z^w} . \tag{3.19}$$

It can be shown that, in view of the similar characteristics of the distribution of ν $(\bar{z})$,

$$\phi^* \simeq \phi_o . \tag{3.20}$$

Now, the ratio $\langle j_z \rangle / j_z^w$ must be determined. Based on general considerations, it can be assumed that $\partial j_z / \partial_z = 0$ and that, overall, the current density profile $j_z(z)$ is less filled than the linear one. Then

$$0 < (\langle j_z \rangle / j_z^w) < 0.5 .$$

For a more accurate determination of $\langle j_z \rangle / j_z^w$, one can specify the profile $\Psi(z) \equiv j_z / j_z^w$ in the form of an odd, fifth order polynomial of $\bar{z}$,

$$\Psi(\bar{z}) = a_1 \bar{z} + a_3 \bar{z}^3 + a_5 \bar{z}^5 , \tag{3.21}$$

the coefficients of which are determined from the boundary conditions

$$\Psi(1) = 1,\ \Psi'(0) = 0,\ \Psi'(1) = 0\ . \qquad (3.22)$$

The last condition follows approximately from the current continuity equation and the Ohm's law:

$$\frac{\partial j_z}{\partial z} = -\frac{\partial j_y}{\partial y}\ ,\ j_y \sim \nu\ .$$

In real MHD channels, even at limitingly hot insulating walls ($\overline{T}_w = 0.8$) ν (1) << 1, so that at the wall $\partial j_z/\partial z \simeq 0$. Then, using boundary conditions (3.22), from polynomial (3.21), we obtain

$$\Psi = 2.5\bar{z}^3 - 1.5\bar{z}^5 \text{ and } \langle j_z\rangle/j_z^W = 0.375\ .$$

Substituting this value of $\langle j_z\rangle/j_z^w$ into equation (3.19),

$$j_z^W = (\phi_o - \phi^W)(\sigma_{oo}\mu/b)(\langle 1/\nu\rangle - 0.625)^{-1}\ . \qquad (3.23)$$

More accurate estimates have shown that, in the range ν (1) = 0.001-0.05 (at $\overline{T}_w$ = 0.6-0.8), formula (3.19) leads an error of 5 to 12 % in j_z^w. Using a constant value $\langle j_z\rangle/j_z^w = 0.375$, the error increases 7 to 17 %. In the same range of $\overline{T}_w$, the error in determining j_z^w, introduced by assumption (3.20), is $(\phi^* - \phi_o)/(\phi^w - \phi_o)$ = 2-5 %.

Knowing the current profile, $j_z/j_z^w = \psi(\bar{z})$, the system of equations (3.15), (3.16), can be closed without using assumptions (3.17), (3.20). Substituting the current profile $\psi(\bar{z})$ into the Ohm's law equation (3.18) and performing double integration and some simple transformations yields,

$$j_z^W = \frac{\sigma_{oo}\langle\nu\rangle\mu}{b}\left(\langle\nu\rangle\int_o^1 \frac{\psi}{\nu}\,d\bar{z}\ - \right.$$

$$\left. - \int_o^1 \nu \int_o^z \frac{\psi}{\nu}\,d\bar{z}\,d\bar{z}\right)^{-1}(\phi^* - \phi^W)\ . \qquad (3.24)$$

It can be expected that equation (3.24), obtained making fewer assumptions, should more accurately describe the current density distribution, j_z^w (y), than expression (3.23). These dependences at various $\nu_w = \nu(1)$ and different degree of filling of the conductivity profile, $\nu(\bar{z})$, differ by an insignificant amount. As was to be expected, the greatest deviation is observed in the case of a uniform plasma conductivity distribution over the channel height ($\nu(\bar{z}) = 1$). However, even in this case it does not exceed 24 %. Furthermore, formula (3.23) provides lower values of j_z^w. Taking into account the fact that, in many cases, the "$j_z = 0$" and "$E_z = 0$" models provide adequate characteristics of an MHD generator, it can be expected that the 25 % error in determining j_z will not lead to larger errors in the final results of the calculations. For profile $\nu(\bar{z})$ closer to real conditions, the difference in j_z^w calculated from equations (3.23) and (3.24) is considerably smaller. In these calculations, the simplest formula (3.23) has always been used, even though an estimate of j_z^w from equation (3.24) is also not too difficult. Closing the system of equations (3.15), (3.16), with the aid of expression (3.23), it is expedient to represent this system in dimensionless form:

$$(\mu\bar{\phi}^{*\prime})' - A_4\mu(\bar{\phi}^* - \bar{\phi}^w - \Delta\bar{\phi}_z)$$

$$= g^* (\mu f)' + \beta\bar{E}_x\mu' , \qquad (3.25)$$

$$(\bar{\phi}^w)'' - A_2\mu(\bar{\phi}^* - \bar{\phi}^w - \Delta\bar{\phi}_z) = 0 , \qquad (3.26)$$

where $\phi \equiv \phi/u_{oo}Ba$; $\bar{E}_x = E_x/u_{oo}B$, and $'$ designates differentiation with respect to $\bar{y}$

$$A_2 \equiv (a/b)^2(\bar{\sigma}_y^W R\langle\nu\rangle\langle 1/\nu\rangle)^{-1} ;$$

$$A_4 \equiv \bar{\sigma}_y^W(1 + \beta^2)\, A_2 ;$$

$$\bar{\sigma}_y^W \equiv \sigma_y^W/(\delta_{oo}\langle\nu\rangle b);\; R \equiv 1 - 0.625(\langle 1/\nu\rangle)^{-1}.$$

An external load of an MHD generator is specified by the electric field intensity, $\bar{E}_x^H = \bar{E}_x$, in equation (3.25) and cathode and anode potentials, $\bar{\phi}^+$ and $\bar{\phi}^-$, in that order,

$$\bar{\phi}^* - \Delta\bar{\phi}_y = \bar{\phi}^+, \ \bar{\phi}^W = \bar{\phi}^+ \quad \text{when } \bar{y} = -1 ;$$

$$\bar{\phi}^* = \bar{\phi}^-, \ \bar{\phi}^W = \bar{\phi}^- \quad \text{when } \bar{y} = +1 . \qquad (3.27)$$

The system of equations (3.25), (3.26), subject to boundary conditions (3.27), was solved numerically by means of successive approximations.

Having obtained distributions $\bar{\phi}^*$ $(\bar{y})$ and $\bar{\phi}^W$ $(\bar{y})$ for specified axial voltage and transverse load $\bar{E}_x^H = \bar{E}_x$, $\bar{E}_y^H \equiv \bar{V}_y \equiv (\bar{\phi}^+ - \bar{\phi}^-)/2$, it is easy to determine the load current density

$$\bar{j}_y^H = (\langle\nu\rangle/\langle 1/\mu\rangle(1+\beta^2))(\Phi(\bar{E}_x, \bar{E}_y) + \langle f\rangle\, g^*$$

$$- (\bar{E}_y^H + \Delta\bar{\phi}_y) - \beta\bar{E}_x^H) - \langle\mu\rangle\langle\nu\rangle\tilde{\sigma}_y^W\, \bar{E}_y^H ,$$

$$- \bar{j}_x^H = (\beta\langle\nu\rangle/\langle 1/\mu\rangle(1+\beta^2))(\Phi(\bar{E}_x^H, \bar{E}_y^H)$$

$$+ \langle f\rangle g^* - (\bar{E}_y^H + \Delta\bar{\phi}_y) + (G/\beta^2)\beta\bar{E}_x^H) , \qquad (3.28)$$

where

$$\bar{j} \equiv j/\sigma_{oo}u_{oo}B; \quad \bar{E} \equiv E/u_{oo}B; \quad \bar{\phi} \equiv \phi/u_{oo}Ba ;$$

$$\Delta\bar{\phi}_y \equiv \frac{\Delta\phi_y(j_y)}{2au_{oo}B} , \quad \Delta\bar{\phi}_z \equiv \frac{\Delta\phi_z(j_z)}{2au_{oo}B}$$

is the dimensionless potential drop in the spatial charge layer on electrodes, and on insulating walls,

$$\Phi(\bar{E}_x^H, \bar{E}_y^H) \equiv \frac{a}{b}\, \frac{1+\beta^2}{\langle\nu\rangle} \int_{-1}^{1} \left(\langle\frac{1}{\mu}\rangle - \frac{1}{\mu}\right) \int_{-1}^{\bar{y}} \bar{j}_z^W \, d\bar{y}\, d\bar{y}$$

$$= A_4 \int_{-1}^{1} \left(\langle\frac{1}{\mu}\rangle - \frac{1}{\mu}\right) \int_{-1}^{\bar{y}} (\bar{\phi}^* - \bar{\phi}^W - \Delta\bar{\phi}_z)\, d\bar{y}\, d\bar{y} ;$$

(3.29)

$$G \equiv \langle\mu\rangle \langle 1/\mu\rangle (1+\tilde{\sigma}_x^W)(1+\beta^2) - \beta^2 ;$$

$$\tilde{\sigma}_{x,y}^W \equiv \sigma_{x,y}^W / \sigma_o \langle\mu\rangle \langle\nu\rangle b ;$$

$$\sigma_y^W = (\sigma_y^W)_z ;$$

$$\sigma_x^W = (\sigma_x^W)_z + (\sigma_x^W)_y (b/a)(1-c/s)^{-1} ;$$

$$(3.30)$$

$(\sigma_{x.y}^W)_z$, $(\sigma_x^W)_y$ is the surface conductivity of insulating walls and interelectrode insulators, respectively, and c is the length of the electrodes.

3.3.2. Special Cases. The simplified calculation methods described makes it possible to reduce a two-dimensional problem of an electric field in the cross section of the plasma flow in an MHD channel with fairly complex boundary conditions (especially in the case of finite surface conductivity of insulating walls ($1/\sigma_y^w \neq 0$) to a solution of a system of ordinary, second order differential equations. Solution of such a system is considerably easier. However, even after this simplification, the problem is still generally very complex. At the same time, a whole series of special cases is available when the problem is simplified quite considerably. Most of these are of considerable practical interest and deserve special consideration.

3.3.2.1. Low Near-Electrode Potential Drop

When the near-cathode potential drop is many times lower than the induced voltage ($\Delta\phi \ll 2a \langle f\rangle g^* u_{oo} B$), it can be assumed that $\Delta\overline{\Phi}_y = \Delta\overline{\Phi}_z = 0$. Subject to such an assumption, the problem becomes linear and the load current density is specified by the following relations:

$$\bar{j}_y^H = A_y + A_{yx}\bar{\bar{E}}_x^H + A_{yy}\bar{\bar{E}}_y^H ,$$

$$j_x^H = A_x + A_{xx}\bar{\bar{E}}_x^H + A_{xy}\bar{\bar{E}}_y^H , \qquad (3.31)$$

and to determine the load factor it is sufficient to determine three points at different values of $\bar{\bar{E}}_x^H$ and $\bar{\bar{E}}_y^H$.

When performing calculations, it is convenient to choose points ($E_x^H = 0$; $E_y^H = 0$), ($E_x^H = 0$; $E_y^H = E_{y1}^H$) and ($E_x^H = E_{x1}^H$; $E_y^H = 0$) where E_{y1}^H and E_{x1}^H are arbitrarily chosen electric field intensities of the load that are not equal to zero.

The electric field intensities E_x^H and E_y^H can be considered to be independent variables or can be specified in terms of the "load conductivity"

$$\bar{\sigma}_y^H \equiv (R_y^H sba^{-1} \sigma_{oo} \langle\mu\rangle \langle\nu\rangle)^{-1},$$

$$\tilde{\sigma}_x^H \equiv (R_x^H s^{-1} ab\sigma_{oo} \langle\mu\rangle \langle\nu\rangle)^{-1} , \qquad (3.32)$$

where R_x^H, R_y^H are the resistances of the external electrical load for the channel section under consideration, along the axial and transverse directions, respectively.

In this case, $E_x^H = E_x^H(\tilde{\sigma}_x^H, \tilde{\sigma}_y^H)$ and $E_y^H = E_y^H(\tilde{\sigma}_x^H, \tilde{\sigma}_y^H)$ are the functions being sought that can be determined from the system (3.31), taking into account that

$$\bar{J}_y^H = \tilde{\sigma}_y^H \bar{E}_y^H, \quad \bar{J}_x^H = \tilde{\sigma}_x^H \bar{E}_x^H .$$

3.3.2.2. The "$J_z = 0$" Model

Neglecting the current flow into insulating walls (i.e., assuming $A_2 = A_4 = 0$), system (3.25), (3.26), separates into two equations that are easily integrated. The load current density in this case is specified by relations (3.28)-(3.30), in which Φ must be set equal to 0. When $j_z^H = 0$, one can easily obtain the load factor of a Faraday generator from the system of equation (3.31)

$$\bar{j}_y^H = \bar{j}_y^{s.c.}(1-K_y) ,$$

$$\bar{E}_x^H = \beta \bar{j}_y^H (1+\tilde{\sigma}_x^H)^{-1} ,$$

where

$$\bar{j}_y^{s.c.} = \langle\nu\rangle\langle\mu\rangle g^*\langle f\rangle(1+\tilde{\sigma}_x^W) G^{-1} ;$$

$$\bar{V}_y^{x.x} \equiv \bar{E}_y^{x.x} = \langle f\rangle g^*(1+\tilde{\sigma}_y^W G/(1+\tilde{\sigma}_x^W))^{-1} .$$

3.3.2.3. "$E_z = 0$" Model

Within the framework of this model, it is assumed that dc resistance of the plasma in the z direction is negligibly small and $\phi^* = \phi^w = \phi_o$. In order for $\phi^* - \phi^w = 0$, it is necessary to assume that, in formulas (3.25) and (3.26), $R = 0$. To eliminate the uncertainty, the term A_4 $(\phi^* - \overline{\Phi}^w)$ can be eliminated from formula (3.25) by expressing it in terms of $\overline{\Phi}^w$ from equation (3.26). After this, expression (3.25) is easily integrated.

In this case also, the load factor given by relations (3.28)-(3.30) remains in force and the parameter, Φ, can be represented in quadratures:

$$\Phi = -(\beta\bar{E}_x^H + \bar{E}_y^H)(\langle 1/\mu\rangle I_3/I_1 - 1)$$

$$+ g^* (\langle 1/\mu\rangle) I_2/I_1 - \langle f\rangle) , \tag{3.33}$$

where

$$I_1 \equiv \int_0^1 (\mu + \langle\mu\rangle \tilde{\sigma}_y^W (1+\beta^2))^{-1} d\bar{y} ;$$

$$I_2 \equiv \int_0^1 \mu f (\mu + \langle\mu\rangle \tilde{\sigma}_y^W (1+\beta^2))^{-1} d\bar{y} ;$$

$$I_3 \equiv 1 - \langle\mu\rangle \tilde{\sigma}_y^W (1+\beta_1^2) I .$$

In the case of a Faraday generator ($j_x^H = 0$), from expressions (3.28)-(3.30), (3.33), one can easily obtain the following equations:

$$\bar{j}_y^{s.c.} = \frac{\langle\mu\rangle\langle\nu\rangle g^* (1+\tilde{\sigma}_x^W) I_2}{\langle\mu\rangle (1+\tilde{\sigma}_x^W)(1+\beta^2) I_1 - \beta^2 I_3} ,$$

$$\bar{V}_y^{x.x.} = \frac{g^* I_2}{1 - I_3 \beta^2 \tilde{\sigma}_y^W / (1+\tilde{\sigma}_x^W)} . \tag{3.34}$$

The load characteristics (3.34) can also be obtained from the relations given elsewhere [24], provided one assumes simplified distributions of flow parameters over the channel cross section in accordance with expressions (3.11).

3.3.2.4. Infinite Surface Conductivity of Horizontal Walls of an MHD Channel ($\sigma_y^w \to \infty$)

The wall surface conductivity of an MHD channel may be quite anisotropic. Therefore, condition $\sigma_y^w \to \infty$ does not mean that the channel walls are infinitely conducting in the axial direction.

Condition $\sigma_y^w \to \infty$ is attained in a frame channel of a Hall MHD generation. Such channel is constructed from individual metal frames (with heated or unheated fire surface) located perpendicularly to its axis and well insulated from each other. The metal frames are almost perfectly conductive along the y axis ($\sigma_y^w \to \infty$) and the electrical insulation between them provides adequately low mean wall conductivity in the axial direction ($\overline{\sigma}_x^w << 1$).

When channel walls are characterized by infinite conductivity, there is no change of the wall potential in the transverse direction,

$$\frac{\partial \phi^W}{\partial y} = 0 \ .$$

In this case, equation (3.16) becomes condition ϕ^w = const. and the problem is reduced to solution of only one ordinary differential equation (3.15).

<u>3.3.3. Comparison of the Results with the Experimental Data and an Exact Solution of the Problem.</u> The results of the calculations using the simplified "$j_z = 0$, $E_z = 0$" models were compared with the experimental data acquired on the "Start" facility [10]. In performing the calculations, the cathode potential drop was neglected ($\Delta\phi_y = 0$ and $\Delta\phi_z = 0$). Therefore, when processing experimental data, the influence of this factor on the characteristics of an MHD generator was excluded by means of relations (3.10):

$$(\bar{j}_y^H)^{s.c.} \equiv j_y^H/\sigma_{oo}u_{oo}B(1-K_y-\Delta\phi_y/V_y^{x.x}) \ ,$$

$$\bar{E}_x^H \equiv E_x/u_{oo}B(1-K_y-\Delta\phi_y/V_y^{x.x}) \ .$$

The value of $\Delta\phi_y = \Delta\phi_y\ (j_y,\ T_w)$ was measured in specially conducted experiments.

When determining the no-load voltage, the cathode potential drop was neglected:

$$\bar{V}_y^{x.x} = V_y^{x.x}/2au_{oo}B \ .$$

A comparison of the experimental data with the results of the calculations has shown that the experimental points $\bar{j}_y^{s.c.}$ and $\bar{E}_x^{s.c.}$ fall close to the calculated curve "j_z = 0" (Fig. 3.15a), and points $\bar{V}_y^{x.x}$, closer to the calculated curve "E_z = 0" (Fig. 3.15b). What appears superficially, at first, to be a contradiction can be explained by

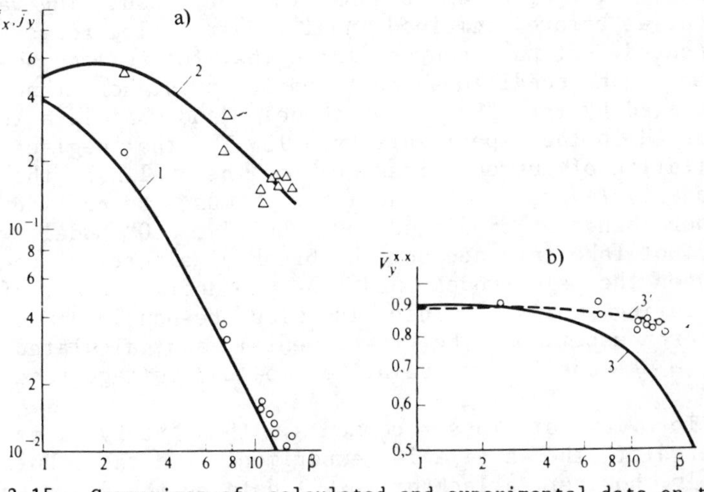

Fig. 3.15 Comparison of calculated and experimental data on the influence of the Hall effect in a nonideal MHD channel on the transverse current and axial electric fields in a short-circuit mode (a) and no-load voltage (b) ($\bar{T}_w$ = 0.57, $\bar{\sigma}^w$ = 0.01); Lines--results of the calculations; points--experimental data; 1--transverse current density; 2--axial field strength, 1-3 - "j_z = 0" model; 3' - "E_z = 0" model

characteristics of the MHD channel of the "Start" facility. As has already been noted, narrow strips with very low surface conductivity, σ^w, are located in the channel, along the edges of insulating walls. This hinders current transfer from insulating walls to electrodes and the current shorted across the walls is too low to suppress significantly the influence of the nonuniformity plasma conductivity in the boundary layers. This is confirmed by analysis that has shown that j_y^H calculated with the "E_x = 0" model, taking into account the nonconducting strips, differ very little from j_y^H calculated with the "$j_z = 0$" model.

The situation is quite different in the no-load operating mode. Electrodes operating in an open-circuit mode, insulated from horizontal walls, extract practically no current and serve only as electrostatic probes. In addition, at the high channel wall temperature in these experiments, the electrode resistance was high and that of the boundary layers was low. Therefore, the actual conditions were close to those of an ideal contact of plasma with the wall.

At a lower insulating wall temperature, their conductivity was considerably higher (see Fig. 3.5). However, the impedance of the boundary layers limiting the current flowing from plasma to the insulating walls, which, as before, remained insulated from electrodes, also increased. It has been estimated that the no-load voltage under such conditions must be even higher than that predicted by the "$E_z = 0$" model (and what is actually observed in the experiments at low T_w) that neglects the limitation of current leakage to the walls by the cold boundary layers. It definitely must be considerably higher than that predicted by the "$j_z = 0$" model, which does not take into account the break of electrical contact between the electrodes and the conducting part of the insulating walls. This is also responsible for the difference between the experimental and calculated data ("$j_z = 0$" model) for relative no-load voltage (see Fig. 3.8).

In view of these comments, the theory does not contradict the available experimental data. Unfortunately, however, a lack of exact data on the distribution of σ_y^w (y) at the junction between the insulating walls and electrode walls (near $\bar{y} = \pm 1$) in the "Start" MHD channel precludes an estimate of how accurately the model takes into account the influence of current leakages to the horizontal walls. An estimate had to be obtained by theoretical calculations.

The results of approximate calculations were compared with the exact solution of the two-dimensional problem in the (y,z) plane for a frame type ($\sigma_y^w \to \infty$), Hall channel. This case is characterized by the strongest manifestation of the effect of three-dimensional current distribution. Calculations were performed for the following conditions:

$$a/b = 3; \ \bar{\delta}_y = \bar{\delta}_z = a/3; \ \bar{T}_w = 0.6\text{–}0.8 \ ;$$

$$\beta = 1\text{–}3 \ .$$

A comparison of the results of accurate and approximate calculations of the current density in a short-circuit mode ($\bar{j}_x^{s.c.}$) and the electric field intensity in the no-load operating mode ($\bar{E}_x^{x.x}$) and is shown in Fig. 3.16. The results of the simplified calculations are represented in the form of continuous curves

$$\bar{j}_x^{s.c.}/\bar{j}_{xo}^{s.c.} \text{ and } \bar{E}_x^{x.x}/\bar{E}_{xo}^{x.x} \; (E_{xo}^{x.x} \text{ and } j_{xo}^{s.c.}$$

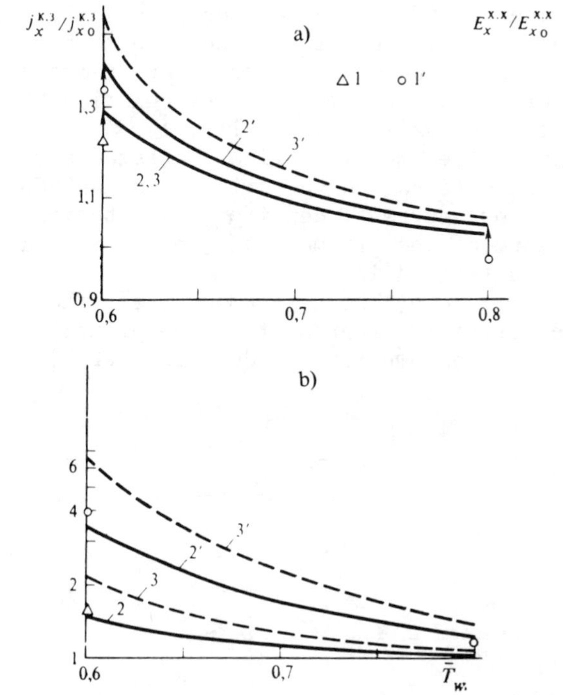

Fig. 3.16 Comparison of axial current density in a short-circuit mode (a) and transverse electric field intensity in a no-load mode (b) of a frame-type Hall MHD generator, calculated using approximate technique with the results of the solution of the problem in a rigorous formulaion; 1--1′ - exact solution; 2--2′ - "$j_z \neq 0$," "$E_z \neq 0$" model; 3, 3′ - "$E_z = 0$" model; 1-3 - $\beta = 1$; 1′ - 3′ - $\beta = 3$

(values of $E_{xo}^{x.x}$ and $j_{xo}^{s.c.}$ are calculated, using the "$j_z = 0$" model). The results of an exact solution of the two-dimensional problem are plotted as points. For comparison, Fig. 3.16 also shows the data calculated using the "$E_z = 0$" model.

The agreement between the approximate calculations and the exact solution was quite adequate through the entire range. The greatest deviation is observed in values of

$\overline{E}_x^{x.x}$ at a high degree of conductivity nonuniformity (low $\overline{T}_w$) and high β. However, in this case, the error of the approximate calculation is only 16 %. At the same $\overline{T}_w$ = 0.6 and β = 1, the difference decreases to 8 % and at $\overline{T}_w$ = 0.8 it drops to 2 % even when β = 3. In the case of $j_x^{s.c.}$, the error of the simplified calculations is the same everywhere (6 %). It should be noted that, at $\overline{T}_w$ = 0.6 and β = 3, the "E_z = 0" model leads to $\overline{E}_x^{x.x}$ values that are 1.6 times greater than the actual values, and the "j_z = 0" model leads to values that are 4 times lower than the actual values. Therefore, the results obtained appear to be quite satisfactory. There are reasons to believe that the error of the simplified calculations will be even smaller at lower wall conductivity, σ_y^w.

The simplified model provides satisfactory results not only for the external characteristics of an MHD generator, but also for current density distributions over the circumference of the frame and the potential along the channel axis (Fig. 3.17).

Hence, it can be assumed that the methodology developed correctly represents the principal processes, and that the assumptions made, considerably simplifying

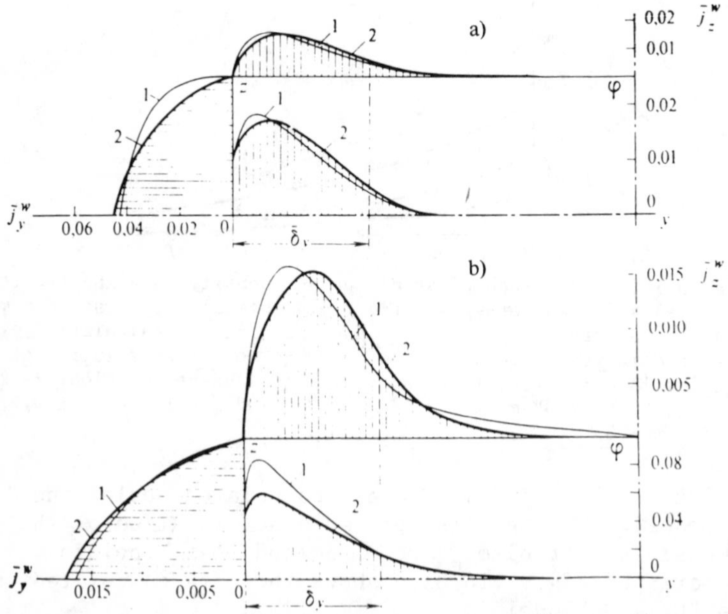

Fig. 3.17 Current distribution over the frame and axial potential in a frame channel of a Hall MHD generator with hot ($\overline{T}_w$ = 0.8) (a) and semi-hot ($\overline{T}_w$ = 0.6) (b) walls (β = 3, E_x = 0); 1--simplified model, 2--exact solution

the calculations, provide an accuracy acceptable for the solution of engineering problems. The methodology may find many different practical applications.

3.4. Nonideal Behavior of Large-Scale MHD Channels

In commercial-scale MHD generators, the dimensions of an MHD channel are so large that the significance of surface effects superposed over the background of volume processes becomes less noticeable. However, under certain conditions, the nonideal behavior of an MHD channel may exert considerable negative influence on the operation of a commercial-scale MHD generator. One must keep in mind that the requirements imposed on the commercial-scale MHD generator are completely different from those imposed on the generators of test and prototype MHD facilities. Degradation of characteristics in test facilities because of harmful effects is not important. In prototype facilities, degradation of characteristics by more than 50 % is frequently quite acceptable. However, in commercial-scale, power-generating MHD facilities, not only a few dozen percent but even a few percent drop of power density or internal efficiency is of considerable importance. This should always be kept in mind when analyzing the influence of various factors on the external characteristics of commercial-scale MHD generators.

3.4.1. The Influence of Various Factors on the Characteristics of a Commercial-Scale MHD Generator.

Experience gained on the U-02 facility [27, 29] has demonstrated that, if insulating channel walls are rationally designed, present-day technology is quite capable of providing such a low level of surface conductivity that current leakage along the walls will exert no noticeable influence on the specific characteristics of large-scale MHD generators (at least when operating on pure, i.e., ash-free, fuel). In this case, one can use the simplest variant of the technique, i.e., the "$j_z = 0$" model, to estimate local characteristics of the generator. Analysis of the influence of various factors shows that, in real commercial-scale MHD generators, a drop in the specific power will be caused primarily by plasma conductivity nonuniformity in boundary layers at the electrode walls. This effect is especially strong at the channel outlet, where the Hall parameter in commercial-scale MHD generators may reach $\beta = 4$-5. In order to illustrate the practical importance of this effect, estimates were made for a 700-MW commercial-scale MHD generator, the design of which is described in [30].

Calculations were performed for channels with hot and semi-hot walls. In the first case, it was assumed that the wall temperature varies along the length of the channel from 2300 to 2100 K and, in the second case, from 1723 to 1573 K. Results of the calculations are plotted in Fig. 3.18. In the case of hot walls, the decrease of the specific power of a commercial-scale generator is fairly small, being 5 % on the average along the channel. Such a weak influence of the conductivity nonuniformity can be attributed to the choice of the maximum temperature of hot walls at relatively low plasma temperature, used in the analysis. These quantities at the inlet and outlet

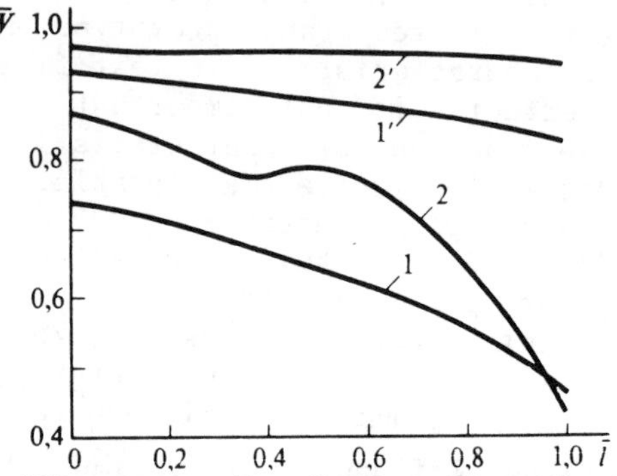

Fig. 3.18 Specific power distribution along the length of the channel of large-scale MHD generators at various degrees of plasma conductivity nonuniformity; 1, 1′ - U-25 facility; 2, 2′ - commercial-scale MHD facility [30]; 1, 2 - $\bar{T}_w$ = 0.6; 1′, 2′ - $\bar{T}_w$ = 0.8

from the channel were determined [30] on the basis of maximum efficiency considerations for the overall MHD power plant. As a result, the relative wall temperature at the channel outlet turned out to be 0.97.

When operating with semi-hot walls, the specific power at the channel outlet decreases by more than one half (by 57 %) and the decrease in power averaged over the channel is 32 %. This, of course, is not two orders of magnitude that occurred on the small-scale "Start" facility. However, in the case of a commercial-scale MHD generator, an effect that can change the volume of the channel and, consequently, the warm bore of the magnet by as much as one and a half times deserves closer scrutiny.

The influence of plasma conductivity nonuniformity in an MHD channel on the power takeoff, under conditions of the U-25 pilot plant, is qualitatively the same as that for a commercial-scale MHD generator. However, a drop in the specific power of an MHD generator on the U-25 facility is somewhat greater.

The effect of axial plasma conductivity nonuniformity on the specific power of the U-25 facility was estimated. It is interesting to note that, as a result of the large elongation of the cross section along the y axis of the MHD channel of the U-25 facility (the ratio of the sides at the channel outlet, $a/b \simeq 5$) equipped with hot walls, the axial plasma conductivity nonuniformity is responsible for a greater drop in the specific power than the transverse plasma conductivity nonuniformity. This indicates that, at a relative wall temperature of 0.8, such cross section geometry is not optimal. A more effective geometry would be that closer to a square cross section. In the presence of hot walls, the influence of the transverse plasma nonuniformity exceeds many times the influence of the axial nonuniformity and, as was shown by analysis, the ratio of the sides of the U-25 channel cross section for $\overline{T}_w = 0.6$ turned out to be close to optimal.

The negative effects of other factors, such as axial plasma conductivity nonuniformity, and velocity nonuniformity, will be considerably smaller.

3.4.2. Permissible Level of Surface Conductivity of MHD Channel Walls. When an MHD generator operates on clean fuel, the MHD channel can be adequately insulated. Therefore, the problem of permissible level of surface conductivity of walls is not important, even though the results given below still apply. This is not the case for an MHD facility with direct combustion of ash-laden fuel (coal). In this case, a conducting film of molten slag is formed on the channel walls. Unfortunately, at present we have no data on the conductivity of such film and its dependence of various factors. Therefore, as of now, we can only discuss the permissible surface conductivity of channel walls.

An analysis of the influence of surface conductivity of channel walls on the local electrical characteristics of an MHD generator, performed using a technique developed, has shown that the transverse current nonuniformity appearing in a flow as a result of current leakage to insulating walls decreases considerably the harmful effect of plasma conductivity nonuniformity in the boundary layers at the electrode walls. This effect is especially pronounced at large β. As a result, the influence of σ^w on the electrical characteristics of a generator when the flow is nonuniform is considerably less negative than in the case of a uniform flow. Under certain conditions, it may even turn out to be positive.

This can be seen from Fig. 3.19, which shows some results of analysis of the influence of the surface conductivity of channel walls on the characteristics of a Faraday MHD generator. Calculations were performed for surface wall conductivity $\tilde{\sigma}_y^w = 0.01$, $\tilde{\sigma}_x^w = 2.67\,\tilde{\sigma}_y^w$ (this corresponds to $(\sigma_x^w)_z = (\sigma_x^w)_y = \sigma_y^w$, $a/b = 3$, and $c/s = 0.8$). Transverse current density in the short-circuited operating mode, $j_{y_o}^{s.c.}$, and the transverse field strength during the no-load mode, $E_{y_o}^{x.x}$, calculated assuming ideal insulation of the MHD generator ($\sigma_x^w = \sigma_y^w = 0$), were used as reference quantities.

From the analysis, it follows that, when $\tilde{\sigma}_y^w \leq 0.01$ and $\tilde{\sigma}_x^w < 0.3$, nonideal behavior of insulation will not substantially influence the electrical characteristics of the MHD generator. The influence of surface conductivity on the internal relative efficiency, η_{oi} may turn out to be more considerable.

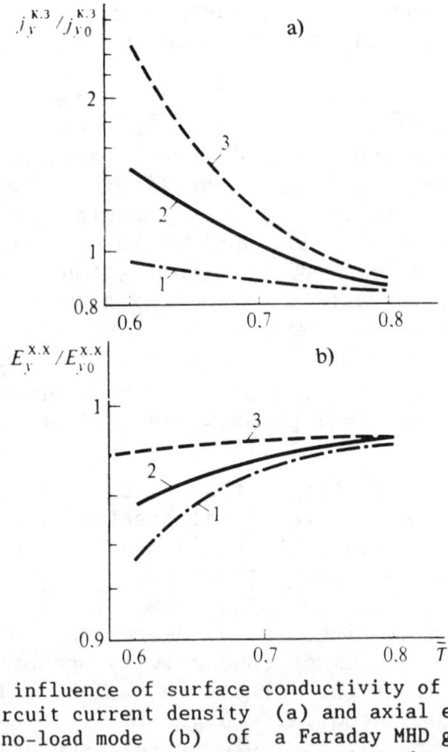

Fig. 3.19 The influence of surface conductivity of channel walls on the short-circuit current density (a) and axial electric field intensity in a no-load mode (b) of a Faraday MHD generator as a function of channel wall temperature (estimated using different models for

$\tilde{\sigma}_y^w = 0.01$, $\tilde{\sigma}_x^w = 0.027$, $\beta = 3$, $\delta = 0.33$) ;
1--"$j_z = 0$"; 2--"$j_z \neq 0$," "$E_z \neq 0$"; 3--"$E_z = 0$"

<u>3.4.3. Nonideal Insulation and Efficiency of an MHD Generator</u>. Limiting the analysis to purely electrodynamic aspects of the problem and disregarding friction and heat losses, the local efficiency of a Faraday MHD generator can be defined as

$$\eta_{oi} = j_y^H E_y^H \Big/ \int_F\!\!\int j_y uB \, dy \, dz = \bar{j}_y^H \, \bar{E}_y^H \big/ \langle\langle \bar{j}_y f g \rangle\rangle \; . \tag{3.35}$$

In an ideal MHD generator, $\eta_{oi} = K_y$. However, current leakage along the walls and reverse current flow along the boundary layer at insulating walls, caused by nonuniform flow rate distribution along the height of the channel, generate additional dissipative losses, which lower the internal relative efficiency of the MHD generator. This is especially noticeable at large electrical load factors, when the specific power of the channel is relatively low. A decrease in conductivity in the boundary layers at the insulating walls suppresses both the reverse current flow along the plasma and current leakage to the walls. However, the transverse plasma conductivity nonuniformity leads to a considerable increase of both.

In the absence of current flow to insulating walls ($j_z = 0$),

$$j_y = j_y(z), \quad E_y = E_y(y) \; . \tag{3.36}$$

In this case, the transverse current density can be taken outside the averaging sign:

$$\langle f j_y) = j_y \langle f \rangle, \quad \langle j_y/\mu \rangle = j_y \langle 1/\mu \rangle \; . \tag{3.37}$$

Then, expression (3.35) can be transformed into the form

$$\eta_{oi} = K_y \, (1-K_y) \, \frac{\langle 1/\mu \rangle (1+\beta^2)}{\langle \nu \rangle \langle f \rangle g^*} \left(\frac{\langle f \rangle g^{**}}{\langle \bar{E}_y^{x.x} \rangle \bar{j}_y^{s.c.}} - \frac{K_y}{\bar{j}_y^{s.c.}} \; \frac{1-\beta^2 \tilde{\sigma}_y^W}{1+\tilde{\sigma}_x^W} + \frac{1-K_y}{\langle \bar{E}_y^{x.x.} \rangle} \; \frac{\beta^2}{\langle \mu \rangle \langle \nu \rangle (1+\tilde{\sigma}_x^W)} \right)^{-1} , \tag{3.38}$$

where

$$g^{**} \equiv \langle \nu g^2 \rangle / \langle \nu g \rangle .$$

Strictly speaking, in the presence of current leakage from the flow to the insulating walls, conditions (3.37) are not fulfilled. However, when $\tilde{\sigma}_y^w \ll 1$ (for commercial-scale MHD generators, only this case is of interest), current leakage to the insulating walls is small and conditions (3.36) and (3.37) are fulfilled almost exactly. Apparently, equation (3.38) is also fulfilled with adequate accuracy. Using relations (3.28)-(3.30) and substituting expressions for $\bar{j}_y^{s.c.}$ and $\bar{E}_y^{x.x}$ following from these formulas into expression (3.38), gives (when $\Delta\bar{\Phi} = 0$),

$$\eta_{oi} = \frac{K_y(1-K_y)(1-\Phi_2)^{-1}(1+\Phi_1/\langle f \rangle g^*)}{(1+\tilde{\sigma}_y^W P)(1+\varepsilon Q)-K_y} , \qquad (3.39)$$

where

$$\varepsilon \equiv \frac{g^{**}}{g^*} - 1 \equiv \frac{\langle \nu g^2 \rangle \langle \nu \rangle}{\langle \nu g \rangle^2} - 1$$

is the parameter determining the influence of the nonuniformity of the flow rate and plasma conductivity distributions along the channel height on the local efficiency of the MHD generator:

$$P \equiv G/(1+\tilde{\sigma}_x^w)(1-\Phi_2) ;$$

$$Q \equiv G/(G+\beta^2)(1+\Phi_1/\langle f \rangle g^*) ;$$

and Φ_1, Φ_2 are determined from formula (3.29):

$$\Phi_1 = \Phi(\bar{E}_y = 0), \quad \Phi_2 = (\Phi(\bar{E}_y) - \Phi_1)/\bar{E}_y ;$$

when $j_z = 0$, i.e., in the absence of current leakages to the insulating walls, $\Phi_1 = \Phi_2 = 0$.

From formula (3.39) it follows that in the case of an ideal MHD channel, η_{oi} (K_y) is a nonmonotonic function (Fig. 3.20). The maximum efficiency $\eta_{oi} = \eta_{oi}^{max}$ is reached when $K_y = K_y^{opt}$. Using expression (3.39), it is easy to show that

$$\eta_{oi}^{max} = (2K_y^{opt}-1)F ,$$

where

$$K_y^{opt} = A - (A(A-1))^{1/2} ;$$

$$A \equiv (1+\tilde{\sigma}_y^W G/(1+\tilde{\sigma}_x^W)(1-\Phi_2))$$

$$\times\ (1+\varepsilon G/(G+\beta^2)(1+\Phi_1/\langle f\rangle g^*)) ;$$

$$F \equiv (1-\Phi_2)/(1+\Phi_1/\langle f\rangle g^*) .$$

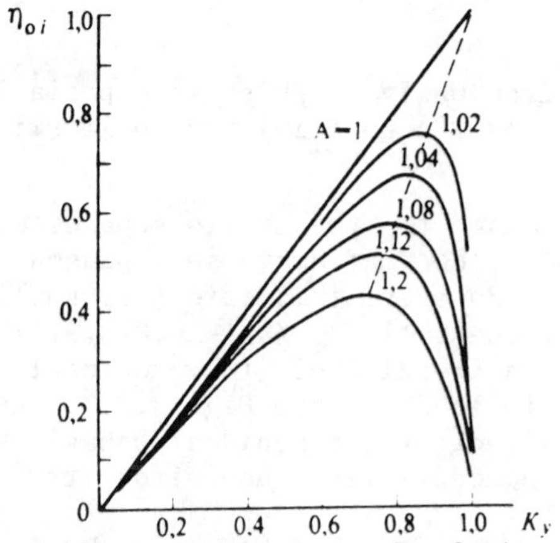

Fig. 3.20 Variation of the internal relative efficiency of a nonideal MHD generator with the electrical load factor at different degrees of nonideal behavior of an MHD channel

Parameter A characterizes the degree of nonideal behavior of an MHD channel. In the case of ideal insulation ($\tilde{\sigma}^w = 0$) and a uniform plasma flow ($\varepsilon = 0$), parameter A = 1. Figure 3.21 shows the influence of $\tilde{\sigma}_y^w$ on the maximum efficiency. Even when well insulated ($\tilde{\sigma}_y^w$ = 0.01), the maximum internal relative efficiency differs

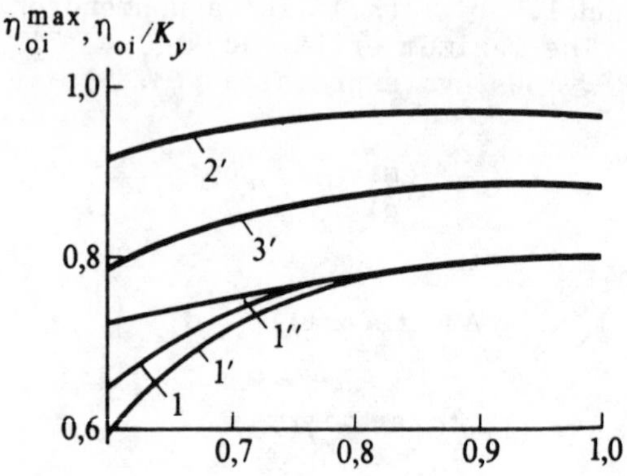

Fig. 3.21 Dependence of internal relative efficiency of a nonideal MHD generator on the relative channel wall temperature ($\tilde{\sigma}_y^w$ = 0.01); 1,1′,1′′ - maximum value of η_{oi} that can be reached, 2′,3′ - ratio of η_{oi}/K_y in the maximum power output mode and maximum efficiency mode, respectively;
1 - "$j_z \neq 0$," "$E_z \neq 0$" model; 1′2′,3′ - "$j_z = 0$" model;
1′′ - "$E_z = 0$" model

considerably from unity. This is especially true in the case of high plasma conductivity nonuniformity (relatively low $\overline{T}_w$).

Current leakage to the walls suppresses somewhat the influence of the transverse plasma conductivity nonuniformity. However, this effect is small. When $\beta = 3$ and $\tilde{\sigma}_y^w = 0.01$, the ratio $\eta_{oi}/K_y = 0.8$–0.9, i.e., a drop of efficiency as a result of the nonideal insulation is approximately 10 to 20 % (see Fig. 3.21). As β increases, the negative effect of the nonideal behavior of insulation and plasma conductivity nonuniformity on η_{oi} will increase.

A decrease of η_{oi} may be compensated by increasing K_y. However, operation of a commercial-scale MHD generator will be economically justifiable only at $\eta_{oi} = 0.75$–0.8. Under these conditions, an increase of the load factor, K_y, by only 10 % leads to a drop in the specific power (increase of the channel length) of approximately 30 %.

Hence, permissible values of $\tilde{\sigma}_y^w$ will be determined by the influence of this factor, exerted not so much on the generator current and voltage, but on the internal relative efficiency.

3.5. Certain Aspects of Optimizing Parameters of an MHD Generator

By exerting substantial influence on the characteristics of an MHD generator, nonideal behavior of an MHD channel may also affect the choice of optimal conditions of its operation. The influence of such factors as nonideal insulation and plasma conductivity nonuniformity, should be taken into account, not only during calculations of the characteristics of the MHD generator, but also during optimization of its parameters.

3.5.1. Optimal Flow Rate. Optimization of the flow rate of plasma in an MHD generator channel has been discussed in the scientific literature [8, 31, 32]. However, the analysis was conducted utilizing one-dimensional flow theory. Two-dimensional effects in the real MHD channel may result in corrections as to selection of the optimal flow rate. The gist of the problem under consideration can be reduced to the following. Specific power of an ideal Faraday MHD generator is given by,

$$W_{id} = \sigma u^2 B^2 K_y (1-K_y) \ . \tag{3.40}$$

At specific stagnation parameter, temperature T_T and pressure p_T, the increase of flow rate u leads to a decrease in the static temperature, T, and the plasma conductivity. As a result of the strong temperature dependence of the degree of ionization of the low-temperature plasma, even a small change of temperature leads to a considerable change in conductivity. This is responsible for raising the question of optimal choice of the flow rate. The dependence of conductivity on temperature, T, and pressure, p, can be represented in the form,

$$\frac{\sigma}{\sigma_T} = \left(\frac{T}{T_T}\right)^{3/4} \exp\left(-\frac{\Phi}{T_T}\left(\frac{T_T}{T} - 1\right)\right)\left(\frac{p}{p_T}\right)^{-1/2},$$

where σ_T is the plasma conductivity at stagnation parameters T_T and p_T. In the case of natural gas combustion products, plasma seeded with potassium salt, Φ = 25,150 K. Introducing gasdynamic functions

$$\tau(\lambda) \equiv 1 - \frac{k-1}{k+1}\lambda^2 \equiv \frac{T}{T_T},$$

$$\pi(\lambda) \equiv \tau^{k/(k-1)} \equiv p/p_\tau$$

(where $\lambda - u/a_{cr}$ is the flow rate factor, $a_{cr} \equiv (2k/(k+1) \times gRT_T)^{1/2}$ is the critical velocity in the flow, $k \equiv c_p/c_v$ is the isentropic index), expression (3.40) for the specific power of an ideal MHD generator may be transformed into the following form

$$W_{id} = c_1 \lambda^2 \sigma/\sigma_T = c_1 K_y(1-K_y)$$

$$\times\ (1-\tau)\ \tau^{\frac{3}{4} - \frac{k}{2(k-1)}} \exp\ (-\tau^{-1}\Phi/T_T)\ . \tag{3.41}$$

Here, c_1 depends on the stagnation parameters but is independent of λ. Neglecting the dependence of k on the temperature, the optimum flow rate can be determined from the relation

$$\frac{W'_{id}}{W_{id}} = \left(\frac{3}{4} - \frac{k}{2(k-1)}\right) \tau_{opt}^{-1}$$

$$- (1-\tau_{opt})^{-1} + \frac{\phi}{T_T} \tau_{opt}^{-2} = 0\ .$$

Here and following, ′ designates differentiation with respect to τ. From this,

$$\tau_{opt} = \frac{2(k-1)}{7-5k} \left(\frac{\Phi}{T_T} - \frac{3k}{4(k-1)} \right.$$

$$- \left(\left(\frac{\phi}{T_T} - \frac{3k}{4(k-1)}\right)^2 - 4\,\frac{\phi}{T_T}\right)^{1/2},$$

$$\lambda_{opt} = ((k+1)/(k-1)(1-\tau_{opt}))^{1/2}\ .$$

The dependence of λ_{opt} on T_T at various values of k is shown in Fig. 3.22. The value of λ_{opt} depends strongly on the isentropic index and increases somewhat with increasing stagnation temperature. In the range 2500-

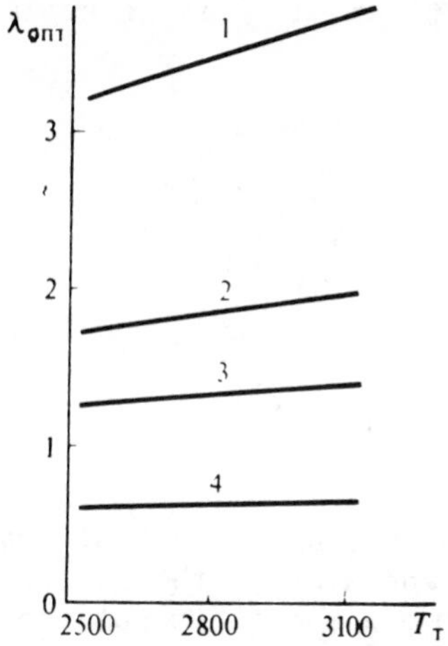

Fig. 3.22 Dependence of the optimum flow rate factor on the stagnation temperature at various values of the isotropic index (ideal MHD channel); 1 - 43/41; 2 - 23/21; 3 - 15/13; 4 - 5/3

3000 K, characteristic of the stagnation temperature in power generating open-cycle MHD generators, at k = 1.1, assuming an ideal MHD channel, λ_{opt} = 1.7-2. This result is on good agreement with the results of analysis of other authors (e.g., [8]).

As has been justifiably noted [34], it would be more correct to perform optimization at constant internal relative efficiency that when K_y = const. depends on the flow rate [8]:

$$\eta_{oi} = K_y\tau/(1-(1-\tau)K_y) \quad . \qquad (3.42)$$

Substituting K_y from formula (3.42) into expression (3.41),

$$W_{id} = c\eta_{oi}(1-\eta_{oi})\tau^{\frac{7}{4} - \frac{k}{2(k-1)}}$$

$$\times \frac{1-\tau}{(\tau(1-\eta_{oi})+\eta_{oi})^2} \quad \exp\left(-\frac{\phi}{T_T}\ \tau^{-1}\right) .$$

Under these conditions, τ_{opt} is determined from the relation

$$\frac{W'_{id}}{W_{id}} = \left(\frac{7}{4} - \frac{k}{2(k-1)}\right) \tau^{-1} - (1-\tau)^{-1}$$

$$+ \frac{\phi}{T_T} \tau^{-2} - \frac{2(1-\eta_{oi})}{\tau(1-\eta_{oi})+\eta_{oi}} = 0 .$$

Within the operating range of parameters of power generating MHD generators, the difference in λ_{opt} when K_y = const., or η_{oi} = const. is not too great:

$$(\lambda_{opt})_{\eta_{oi} = \text{const.}} = 0.97\text{–}0.98\ (\lambda_{opt})_{K_y = \text{const.}}$$

in the range 2500 K < T < 3150 K. The influence of nonideal insulation and plasma conductivity nonuniformity on λ_{opt} is considerably stronger. This can be attributed to an increase in the flow rate caused by the drop in, not only the temperature, but also the pressure. At the same time, the Hall parameter and the negative influence of plasma conductivity nonuniformity and nonideal behavior of insulation on the specific power of an MHD generator increases considerably. In a real MHD channel, specific power is given by the following relations:

$$W = \tau_{oo} u_{oo}^2 B^2 \bar{j}_y^{s.c.} \bar{E}_y^{x.x.} K_y (1-K_y)$$

$$= W_{id} (\langle f\rangle\langle g\rangle)^2 \langle\mu\rangle\langle\nu\rangle (1+\tilde{\sigma}_x)$$

$$\times\ (G(1+G\tilde{\sigma}_y^W/(1+\tilde{\sigma}_x)))^{-1} , \tag{3.43}$$

where

$$G \equiv (1\alpha)(1+\tilde{\sigma}_x)(1+\beta^2) - \beta^2 ;$$

$$\alpha = \langle\mu\rangle\langle 1/\mu\rangle - 1 .$$

The Hall parameter can be expressed in terms of $\tau(\lambda)$:

$$\beta = \beta_T (p_T/p) = \beta_T \tau^{-k/k-1)} ,$$

and the plasma conductivity nonuniformity coefficient represented in the form

$$\alpha = (\langle\mu\rangle_1 \bar{\delta} + (1-\bar{\delta}))\,(\langle 1/\mu\rangle_1 \bar{\delta}$$

$$+ (1-\bar{\delta})) - 1 = \bar{\delta}\,(X^2\bar{\delta} + Y^2(1-\bar{\delta})) ,$$

where β_T is the Hall parameter at stagnation parameters,

$$X^2 = \langle\mu\rangle_1 \langle 1/\mu\rangle_1 - 1;\; Y^2 = \langle\mu\rangle_1 + \langle 1/\mu\rangle_1 - 2 ;$$

and $\langle\mu\rangle_1$ and $\langle 1/\mu\rangle_1$ are mean values of conductivity and reverse conductivity along the thickness of the boundary layer. Both of these parameters depend on the temperature profile in the boundary layer, which depends on the relative wall temperature, $\bar{T}_w = T_w/T_{To}$, and the rate flow factor, λ, (see Appendix).

The relative thickness of the boundary layer, $\bar{\delta} = \delta/a$ also depends on λ. Even if one neglects the influence of λ on the absolute thickness of the boundary layer, δ, assuming that it depends only on the past history of the flow, $\bar{\delta}$ will still depend on the flow rate factor. This can be attributed to the fact that the mass velocity $\rho u = (\rho u)_{cr}\, q(\lambda)$, where $(\rho u)_{cr}$ is the critical mass velocity, and correspondingly the area of the cross section of the channel, F, will change with λ. At a constant ratio of the sides of the channel cross section $b \sim a \sim \sqrt{F}$. In this case, $\bar{\delta} \sim \delta\sqrt{a(\lambda)}$, or $\bar{\delta} = \bar{\delta}_{max}(q(\lambda)/q(1))^{1/2}$, where $\bar{\delta}_{max}$ is the relative thickness of the boundary layer at maximum mass velocity, which is attained when $\lambda = 1$ ($q(1) = 1$) and the reduced flow rate

$$q(\lambda) = \frac{\rho u}{\rho_{cr} u_{cr}} \equiv \left(\frac{k+1}{2}\right)^{\frac{1}{k-1}} \lambda\tau^{\frac{1}{k-1}}$$

$$= \left(\frac{k+1}{2}\right)^{\frac{1}{k-1}} \left(\frac{k+1}{k-1}\right)^{1/2} (1-\tau)^{1/2}\tau^{\frac{1}{k-1}} .$$

Actually the relative wall conductivity, $\tilde{\sigma}^w = \sigma_w/\sigma_{oo}$ $\langle\mu\rangle\langle\nu\rangle$ b, also depends on λ. However, when $\sigma^w < 0.01$, this dependence will exert only a weak influence on the final results and, as shown in the preceding, channels with $\tilde{\sigma}^w > 0.01$ are unlikely to find application. In view of these observations, the condition for an optimal flow rate may be represented in the form

$$\frac{W'}{W} = \left(\frac{W'_{id}}{W_{id}} + \frac{(\langle\mu\rangle\langle\nu\rangle)'}{\langle\mu\rangle\langle\nu\rangle}\right.$$

$$\left.- \frac{G'}{G}\left(1 + \frac{G\tilde{\sigma}^W_y}{1+\tilde{\sigma}^W_x+\tilde{\sigma}^W_y G}\right)\right)_{\tau=\tau_{opt}} = 0 \; , \tag{3.44}$$

where

$$\frac{(\langle\mu\rangle\langle\nu\rangle)'}{\langle\mu\rangle\langle\nu\rangle} = \frac{(\langle\mu\rangle_1 - 1)\delta'/\delta + \langle\mu\rangle'_1}{\delta^{-1} + \langle\mu\rangle_1 - 1}$$

$$+ \frac{(\langle\mu\rangle_1 - 1)\delta'/\delta + \langle\nu\rangle_1}{\delta^{-1} b/a + \langle\nu\rangle_1 - 1} \; ;$$

$$G' = \alpha' \; (1+\tilde{\sigma}^W_x)\left(1+\bar{\beta}^2_T \tau^{\frac{2k}{k-1}}\right)$$

$$- \; (\alpha+\tilde{\sigma}^W_x+\alpha\tilde{\sigma}^W_x)\cdot 2k/(k-1)\cdot \beta^2_T \tau^{-\frac{3k-1}{k-1}} \; ;$$

$$\alpha' = (\delta'/\delta)(2\delta(X^2-Y^2)+Y^2)\;\delta$$

$$+ \; (X^2)'\delta^2 + (Y^2)'\;\delta\;(1-\delta) \; ;$$

$$\delta'/\delta = (2\tau(k-1))^{-1} - (4(1-\tau))^{-1} .$$

Having determined τ_{opt} from condition (3.44), it is easy to determine the optimum flow rate factor:

$$\lambda_{opt} = ((k+1)/(k-1)\cdot(1-\tau_{opt}))^{1/2} \; .$$

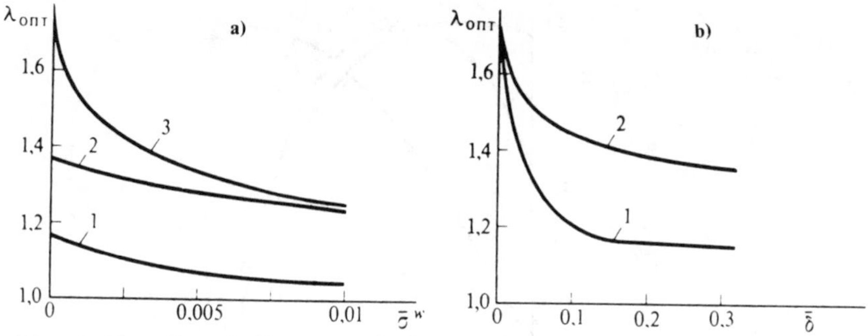

Fig. 3.23 The influence of the nonideal behavior of insulation and the thickness of the boundary layer on the optimum flow rate factor in an MHD channel, (a) and (b), respectively, at different degrees of plasma conductivity nonuniformity ($T_{T\circ}$ = 2700 K, β_T = 3) 1 - $\bar{T}_w$ = 0.6; 2 - 0.8; 3 - 1

Figure 3.23a shows the results of the calculations of λ_{opt} for a uniform flow (δ = 0) and for a flow with a fairly thick boundary layer ($\bar{\delta}$ = 0.3) for hot ($\bar{T}_w$ = 0.8) and semi-hot ($\bar{T}_w$ = 0.6) channel walls. This figure demonstrates the strong influence of the nonideal behavior of an MHD channel on λ_{opt}. The Hall parameter, β_T = 3, was chosen to be fairly large but quite realistic for power generating MHD generators with a superconducting magnet.

At such β_T, even a relatively small deviation from ideal behavior of insulation leads to a considerable drop in λ_{opt}. When $\tilde{\sigma}_y^w$ = 0.01, the optimal flow rate is approximately 1-1/2 times lower than in the ideal case. In the case of hot channel walls, plasma conductivity nonuniformity leads to approximately the same drop of λ_{opt}. However, even in the case of hot channel walls, the nonuniformity of plasma conductivity exerts considerable influence on λ_{opt} (up to 20 %), even though, in this case, the nonuniformity factor is 40 to 50 times smaller than it is for semi-hot walls, and does not exceed 0.01-0.015.

A drop in λ_{opt} because of nonideal insulation is small in the case of a nonuniform flow, and is only 10 % when σ_y^w $\simeq$ 0.01. The boundary layer thickness used in estimates may be somewhat high. However, even at considerably lower $\bar{\delta}$, the results change very little (Fig. 3.23b). A strong influence of $\bar{\delta}$ on λ_{opt} is observed only when $\bar{\delta} < 0.1$.

The conditions considered are characteristic of a flow at the channel outlet, where p_T $\simeq 10^5$ Pa. At the channel inlet, the stagnation pressure will be an order of magnitude higher and the Hall parameter, β_T, an order of magnitude lower. At such β_T, the influence exerted on λ_{opt} by the tranverse plasma conductivity nonuniformity

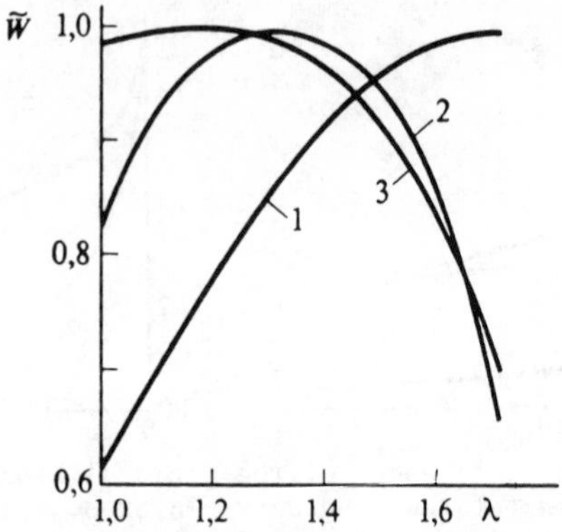

Fig. 3.24 Dependence of the specific power of an MHD generator on the flow rate factor (T_{T_o} = 2700 K, β_T = 3), 1 - $\bar{\delta}$ = 0; 2,3 - $\bar{\delta}$ = 0.1 and $\bar{T}_w$ = 0.7(2) and 0.6(3)

and nonideal insulation will be negligibly small. The dependence of the power density of an MHD generator on λ, plotted on Fig. 3.24, demonstrates a significant deviation of the flow rate from the optimum value. The conditions at the MHD channel inlet are close to being ideal, with λ_{opt} = 1.7-1.8 (transition to subsonic flows may lead to a drop in the power output by more than 30 %). The optimum flow rates at the channel outlet are considerably lower and, in terms of the specific power, operating modes that are nearly sonic do not differ significantly from the optimal mode (especially when the walls are not too hot). At the same time, operation at high flow rates close to the optimal for an ideal generator may lead to an almost one-third decrease of the specific power. Therefore, in choosing the optimum operating mode based on flow rates one cannot neglect the factors related to nonideal behavior.

3.5.2. Certain Aspects of Optimum Wall Temperature of an MHD Channel. The relative wall temperature $\bar{T}_w = T_w/T_{oo}$ is one of the principal parameters determining the degree of plasma conductivity nonuniformity in boundary layers at channel walls. As a rule, as T_w increases, the plasma conductivity nonuniformity in the boundary layer decreases. This exerts favorable influence on the characteristics of an MHD generator. However, as a result of viscous energy dissipation at high flow rates, the temperature profile in the boundary layer may be nonmonotonic and display a maximum that considerably exceeds the core flow temperature, T_{oo}. A conductivity maximum corresponds to a temperature maximum. Under these conditions, an increase of the wall temperature will lead to a further increase of the temperature maximum and, correspondingly, to an increase of the plasma conductivity nonuniformity parameter, α. Hence, it can be expected that a certain optimum wall temperature exists at supersonic flow rates. At this optimum wall temperature, the plasma conductivity nonuniformity will reach a minimum and the specific power of the MHD generator will reach a maximum. This was first noticed [33] when it was shown that, in a supersonic Hall MHD generator, the induced electric field intensity, E_x, may be greater at low wall temperature than it is at high wall temperature.

An analysis of the influence of various factors on the optimal temperature of channel walls for open-cycle MHD generators was performed to clarify this problem. An expression for the specific power that takes into account plasma conductivity nonuniformity (3.43) formed the basis of the calculation. The plasma conductivity distribution in the boundary layer was determined from the temperature profile (see Appendix).

Results of the computation of the plasma conductivity nonuniformity parameter, α, are shown in Fig. 3.25. A clearly defined minimum of α that is displaced toward

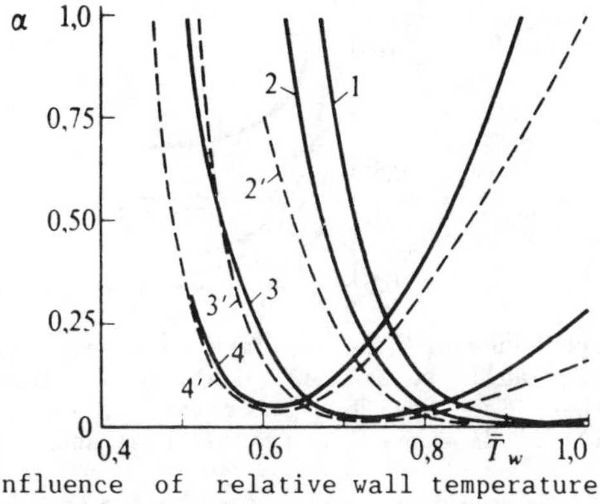

Fig. 3.25 Influence of relative wall temperature of the plasma conductivity nonuniformity factor in the boundary layer at large flow rates (k = 1.1); 1 - λ = 0; 2′ - 1.0; 3,3′ - 2.0; 4′ - 2.5; 2-4 - T_{To} = 3000 K; 2′-4′ - T_{To} = 3300 K

lower wall temperature with increasing λ is observed at supersonic flow velocities. When λ increases from 1 to 2.5, the optimum relative wall temperature, $\bar{T}_w^{opt}$ at $\bar{\delta} = 1$, decreases from 0.9 to 0.62. At a lower boundary layer thickness, $\bar{T}_w^{opt}$ will be even lower. The dependence of the specific power of an MHD generator on $\bar{T}_w$ is somewhat more complex than that for the nonuniformity parameter, α. In the case of ideal insulation, the specific power of Faraday and Hall generators is expressed by the following relations:

$$\bar{W} \equiv W/\delta_{oo} u_{oo}^2 > B^2 \sim \langle\mu\rangle/(1+\alpha(1+\beta^2)) \ ;$$

$$\bar{W} \sim \langle\mu\rangle/(1+\alpha)(1+\alpha(1+\beta^2)) \ .$$

Therefore, the optimum wall temperature, T_w^{opt}, for maximum power takeoff will be somewhat higher than the wall temperature at which parameter α reaches a minimum. The dependence $\bar{T}_w^{opt}$ (β, δ) for Faraday and Hall MHD generators is shown in Fig. 3.26. Because the mean plasma conductivity, $\langle\mu\rangle$, increases with $\bar{T}_w$, $\bar{T}_w^{opt}$, increases considerably, especially at small β. When $\beta \to \infty$, the peak on the W ($\bar{T}_w$) curve degenerates ($\bar{T}_w^{opt} \to \infty$). As β increases, the influence of $\langle\mu\rangle$ on $\bar{T}_w^{opt}$ decreases,

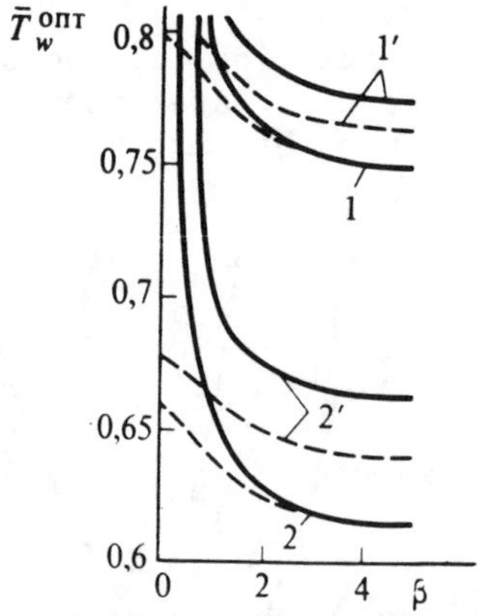

Fig. 3.26 Dependence of the optimum wall temperature of an MHD channel on the Hall parameter; continuous line--Faraday MHD generator; dashed line--Hall MHD generator; 1,2--δ = 0 and λ = 2.0 (1) and 2.5 (2); 1′,2′ - δ = 1 and λ = 2 (1′) and 2.5 (2′)

becoming negligible at β = 3. In terms of absolute temperature, the increase in $\bar{T}_w^{opt}$ will be approximately 30–50 K at high $\bar{\delta}$ and only 20 K when $\bar{\delta} \to 0$. At the present time materials, such as zirconium dioxide and magnesium oxides, capable of operating at a temperature of 1800–2000°C are available; the optimum wall temperature at high flow rates assumes not only theoretical but practical importance as well.

Excessive overheating of the wall may lead to a considerable decrease of the specific power. However, this effect is manifested strongly only when $\lambda \gtrsim 2.0$. This is already higher than flow rates optimal for power-generating MHD generators. Therefore, there is no need to worry about a substantial specific power drop because of the wall temperature exceeding optimum temperature in power-generating Faraday MHD generators. Hence, the principal practical conclusion reached from the analysis is negative. However, another important conclusion can also be made, i.e., taking into account the weak dependence of the specific power on the wall temperature near the optimum value, there is no need to attempt to reach maximum wall temperatures at high λ if this is harmful to the duration and reliable operation of an MHD channel.

3.5.3. Optimal Ratio of the Sides of an MHD Channel Cross Section. Specific power of an MHD generator of a rectangular channel cross section and with a uniform flow is independent of the ratio of the sides of a rectangle. However, plasma conductivity nonuniformity, which always occurs in MHD channels, changes this significantly, because the influence of axial conductivity nonuniformity exerted on an MHD generator is considerably weaker than that exerted by the transverse conductivity nonuniformity, even in the absence of Hall effect that amplifies this difference. It can be assumed that the thickness of the boundary layers, $\delta_{y,z}$, and the conductivity profiles in the boundary layers, $\sigma/\sigma_{oo} = \mu\ (y/\delta_y)$ and $\sigma/\sigma_{oo} = \nu\ (z/\delta_z)$, depend on the previous history of the flow and are independent of the ratio of the channel side walls, a/b. Then, by extending the cross section along the y axes, one can weaken the transverse plasma conductivity nonuniformity. However, this increases the importance of the transverse conductivity nonuniformity and, no matter how small initially, the influence of the axial plasma conductivity nonuniformity will eventually become comparable, or even greater than the influence of the transverse conductivity. The expression for the power per unit working volume of the MHD channel can be used to determine the optimum ratio of the sides of the cross section of the channel, a/b;

$$\bar{W} \equiv W/\delta_{oo}u_{oo}^2\ B^2 = K_y(1-K_y)(\langle f\rangle\langle g\rangle)^2\,\Sigma\ , \qquad (3.45)$$

where K_y is the electrical load factor, $(\langle f\rangle\ \langle g\rangle) \equiv \langle\langle u\rangle\rangle/u_{oo}$ is the flow rate averaged over the cross section, $\Sigma \equiv \sigma_{eff}/\sigma_{oo}$ is the dimensionless effective plasma conductivity. Using the approximate "$j_z = 0$" electrodynamic flow model considered above as the starting point, Σ may be represented in the form,

$$\Sigma = \langle\mu\rangle\langle\nu\rangle/(\langle\mu\rangle\langle 1/\mu\rangle(1+\beta^2)-\beta^2) \; . \tag{3.46}$$

It will be assumed that the area of the cross section of the channel, F; the core flow parameters u_{oo}, T_{oo}, σ_{oo}, p_{oo}, boundary layer thicknesses δ_y and δ_z, and conductivity profiles $\mu\ (y/\delta_y)$ and $\nu\ (z/\delta_z)$ of the channel cross section under consideration are known. For simplicity, the nonideal channel wall insulation will be neglected, even though taking $\tilde{\sigma}_x^w$ and $\tilde{\sigma}_y^w$ into account does not unduly complicate the analysis.

The dependence of average values of conductivity $\langle\mu\rangle$ and $\langle\nu\rangle$ and inverse conductivity $\langle 1/\mu\rangle$ on the ratio of the sides of the channel cross section may be represented in the form,

$$\langle\mu\rangle = 1 - (1-\langle\mu\rangle_1)\tilde{\delta}_y X \; ,$$

$$\langle\nu\rangle = 1 - (1-\langle\nu\rangle_1)\tilde{\delta}_z X^{-1} \; ,$$

$$\langle 1/\mu\rangle = 1 + (\langle 1/\mu\rangle_1 - 1)\, \tilde{\delta}_y X \; , \tag{3.47}$$

where $\langle\mu\rangle_1$, $\langle\nu\rangle_1$ and $\langle 1/\mu\rangle_1$ are conductivities and inverse conductivity averaged over the thickness of the boundary layer, $\tilde{\delta}_{y,z} = \delta_{y,z}/\sqrt{F}$, $X = \sqrt{b/a}$. Generally speaking, the thickness of boundary layers at the electrode and insulating walls may be different ($\delta_y \neq \delta_z$). This also applies to conductivity profiles in the boundary layers and, therefore, to $\langle\mu\rangle_1$, $\langle\nu\rangle_1$.

Substituting relations (3.47) into expression (3.46), we obtain

$$\Sigma = \frac{(1-mX)(1-nX^{-1})}{1+(pX-qX^2)(1+\beta^2)} \; , \tag{3.48}$$

where

$$m = (1-\langle\mu\rangle_1)\tilde{\delta}_y \; ;$$

$$n = (1-\langle\nu\rangle_1)\tilde{\delta}_z \; ;$$

$$p = (\langle\mu\rangle_1 + \langle 1/\mu\rangle_1 - 2)\tilde{\delta}_y \; ;$$

$$q = (1-\langle\mu\rangle_1)(\langle 1/\mu\rangle_1 - 1)\tilde{\delta}_y^2 \; .$$

Differentiating equation (3.45), taking into account expressions (3.46) and (3.48), and equating the derivative to zero, gives the following relation, which can be used to determine the optimum value of $X = X_{opt}$ when $K = \text{const}$.

$$\frac{\bar{W}'}{\bar{W}} = \frac{\Sigma'}{\Sigma} = \left(-\frac{m}{1-mX} + \frac{n}{X(X-n)} - \frac{(1+\beta^2)(p-2qX)}{1+(pX-qX^2)(1+\beta^2)}\right)_{X_{opt}} = 0 . \qquad (3.49)$$

Here and following, ′ designates differentiation with respect to X.

Equation (3.49) specifies the ratio of the sides of the channel cross section $(a/b)_{opt} = X_{opt}^{-2}$, that results in maximum power takeoff per unit volume of the working fluid of an MHD channel at a specified electrical load factor, $K = \text{const}$.

One should actually optimize at a specified internal relative efficiency of the MHD generator, $\eta_{oi} = \text{const}$. (instead of when $K_y = \text{const}$.). This was discussed in the section concerning optimization of flow rate in the MHD channel.

The internal relative efficiency of an MHD generator depends on the electrical load factor. With friction and heat transfer on channel walls taken into account, the relation between these two quantities is specified by the following expression [8]:

$$\eta_{oi} = K_y \tau/(1-(1-\tau)K_y + (c_f/S)(1+\bar{T}_{Tw})/(1-K_y)) , \qquad (3.50)$$

where

$$\tau(\lambda) = \frac{T}{T_T} = 1 - \frac{k-1}{k+1}\lambda^2$$

is the gasdynamic function, $S = \sigma B^2 D_{equiv}/\rho u$ is the magnetic interaction parameter, $D_{equiv} = 4F/\Omega$ is the equivalent diameter (Ω is the circumference of the cross section), c_f is the hydraulic resistance coefficient, $\bar{T}_{Tw} = T_w/T_T$ is the dimensionless stagnation temperature at the wall. Relation (3.50) is written on the assumption that friction and heat transfer processes are similar, so that the Stanton number is $St = c_f/2$. Determining the magnetic interaction parameter, S, one should use effective conductivity $\sigma_{eff} = \sigma_{oo}\Sigma$ for σ. Then, S is conveniently represented in the form,

$$S = S_o \Sigma (D_{equiv}/\sqrt{F}) = S_o \chi(X) ,$$

where

$$S_o \equiv \sigma_{oo} B^2 F^{3/2}/\dot{m} ; \quad \chi = 2\Sigma/(X+1/X) ;$$

$$\dot{m} = \langle\langle \rho u \rangle\rangle F$$

is the flow rate of the working fluid in the MHD channel. Using expression (3.49) and eliminating K from equation (3.45), obtains an expression for the specific power of an MHD generator in terms of the internal relative efficiency, η_{oi}:

$$\bar{W} = \frac{\Sigma}{2} \frac{\eta_{oi}}{C} \left(\left(1 - \frac{\eta_{oi}}{C}\right) + R + \frac{4A_o}{\chi(X)} \right) . \qquad (3.51)$$

where

$$C \equiv 1 - (1-\tau)(1-\eta_{oi}) ;$$

$$A_o = c_f (1+\bar{T}_{TW})/2S_o ;$$

$$R \equiv \left(\left(1 \frac{\eta_{oi}}{C}\right)^2 - \frac{8A_o\eta_{oi}}{C\chi(X)} \right)^{1/2} .$$

The following condition that can be used to determine X_{opt} when η_{oi} is specified follows from the relation (3.51)

$$(\bar{W}'/W)_{\eta_{oi}} = (\Sigma'/\Sigma)_{X=X_{opt}}$$

$$+ (K(1-K))'/K(1-K)|_{X=X_{opt}} = 0 , \qquad (3.52)$$

where

$$\frac{(K(1-K))'}{K(1-K)} = \frac{4A_o(1-\eta_{oi}/C)(-\chi'/\chi^2)}{(1-\eta_{oi}/C)+R+4A_o/(1-\eta_{oi}/C)\chi'} ;$$

$$-\frac{\chi'}{\chi^2} = \left(\left(1-\frac{1}{\chi^2}\right) - \frac{\Sigma'}{\Sigma}\left(\chi+\frac{1}{\chi}\right)\right) / 2\Sigma .$$

From equation (3.50), it follows that the efficiency η_{oi} increases with decreasing $\overline{T}_{Tw}$. It is known from gasdynamics (e.g., see Ref. 35) that heat transfer from the flow increases stagnation pressure, resulting in higher efficiency. However, even when $\overline{T}_{Tw} = 0$, heat transfer cannot completely compensate the stagnation pressure drop due to friction. Therefore, expansion of the circumference of the channel cross section always leads to a drop in efficiency. Therefore, when η_{oi} = const., the optimal shape of the cross section is less elongated than when T = const. Furthermore, one should keep in mind that an increase in efficiency of the energy conversion process in an MHD generator resulting from heat transfer from the MHD channel does not signify an increase in efficiency of the entire facility. Just the opposite, any heat losses during the cycle decrease the overall efficiency of the facility. Obviously, heat from cooling the MHD channel can be used in the steam power cycle. However, the use of this low-potential heat involves certain difficulties. In any case, the efficiency of using heat from cooling is always less than the efficiency of the steam power cycle. In other words, the efficiency of using heat from cooling an MHD channel in the steam power cycle will always be less than one. Taking this into account, it can be shown that, when using the heat from cooling an MHD channel in the steam power cycle, the increase in efficiency of the facility as a whole, reduced to the working volume of the MHD channel, can be given approximately by the following expression:

$$\Delta\eta/V = C_1\bar{W}(1-\bar{W}_k - (1-\xi)\bar{Q}_{cool}\eta_\pi/(1-\eta_\pi)) ,$$

where

$$C_1 = (1-\eta_\pi)\sigma_{oo}u_o^2B^2/Q_o$$

is a coefficient independent of the shape of the channel cross section, η_π is the efficiency of the steam power cycle, Q_o is the total amount of fuel used during the cycle, W is the specific power of the MHD generator, $\overline{W}_k = W_k/W$ is the fraction of the specific power of an MHD generator expended to drive the compressor, $\overline{Q}_{cool} = Q_{cool}/W$, Q_{cool} is the heat from cooling an MHD channel, ξ is the coefficient of utilizing the heat from cooling the channel in the steam power cycle.

The dependence of $\overline{Q}_{cool}$ on the shape of the channel cross section may be represented in the form

$$\overline{Q}_{cool} = \frac{2A_o(1-\overline{T}_{TW})}{(1+\overline{T}_{TW})XK_y(1-K_y)(1-\tau)} .$$

Because heat transfer from the channel will increase with expansion of the circumference, the optimum value of the ratio a/b will decrease.

Up to this point, optimization of the shape of the channel cross section has been discussed on the basis of specific indices per unit working fluid of the channel. From a practical point of view, a more important parameter is the external (gross) channel volume, V_g, because it is this parameter that determines the volume of the magnetic field and, consequently, the cost of the magnet. The ratio of the external volume of the channel with the working volume depends greatly on the skill of the designer. However, this ratio must depend considerably on the working fluid configuration. Hence, the following quantity must be optimized:

$$\Delta\eta/\overline{V}_g \sim \overline{W}(1-\overline{W}_K$$

$$- (1-\xi)\overline{Q}_{cool}\,\eta_\pi/(1-\eta_\pi))/\overline{V}_g ,$$

where

$$\overline{V}_g \equiv V_g/V = (a+\Delta_y)(b+\Delta_z)/ab$$

$$= (1+2\overline{\Delta}_y X)(1+2\overline{\Delta}_z X^{-1}) ;$$

Δ_y, Δ_z are the thicknesses of the vertical and horizontal channel walls, respectively; $\overline{\Delta}_y = \Delta_y/\sqrt{F}$, $\overline{\Delta}_z = \Delta_z/\sqrt{F}$, are the dimensionless thicknesses of the vertical and horizontal channel walls, respectively.

In the simplest case, it can be assumed that Δ_y and Δ_z are independent of the geometrical cross section of the channel. However, if one takes into account that increase of the circumference of the cross section will require strengthening the walls, an increase in the cooling water flow rate, et al., in all probability it is correct to assume that the thickness of the channel walls will depend on the dimensions of the cross section. In the first approximation, such a dependence can be assumed to be linear:

$$\Delta_y = \Delta_{yo} + \bar{\Delta}_{y1} b, \quad \Delta_z = \Delta_{zo} + \bar{\Delta}_{z1} a \; .$$

Then,

$$\bar{V}_g = (1+2\bar{\Delta}_{yo} X + \bar{\Delta}_{y1} X^{-2})$$

$$\times \; (1+2\bar{\Delta}_{zo} X^{-1} + \bar{\Delta}_{z1} X^{-2})$$

and the optimization condition assumes the form

$$\frac{(\Delta\eta/\bar{V}_g)'}{\Delta\eta/\bar{V}_g} = \Bigg(\frac{\bar{W}'}{\bar{W}} - \frac{\bar{V}'_g}{\bar{V}_g}$$

$$+ A_1 \, \frac{1-X^{-2}-(\bar{W}'/\bar{W})(X+X^{-1})}{\bar{W}/(1-\bar{W}_K)-A_1(X+X^{-1})} \Bigg)_{X_{opt}} = 0 \; , \tag{3.53}$$

where

$$\frac{\bar{V}'_g}{\bar{V}_g} = \frac{2(\bar{\Delta}_{yo}+\bar{\Delta}_{y1}X)}{1+2\bar{\Delta}_{yo}X+\bar{\Delta}_{z1}X^2} - \frac{2(\bar{\Delta}_{zo} = \bar{\Delta}_{z1}X^{-1})}{X^2+2\bar{\Delta}_{zo}\,X+\Delta_{z1}} \; ;$$

$$A_1 = A_o(1-\xi)\,\frac{\eta_\pi}{1-\eta_\pi}\;\frac{1-T_{Tw}}{(1+T_{Tw})(1-\tau)} \; .$$

Formula (3.53) expresses the most general condition for optimizing the ratio of the sides of an MHD channel

Table 3.1
Optimum Ratio of the Sides of a Cross Section of an MHD Channel

Parameters in Cross Section Considered	U-25 Facility				Industrial MHD Power Plant [30]			
	Hot Walls		Semi-hot Walls		Hot Walls		Semi-hot Walls	
	Inlet	Outlet	Inlet	Outlet	Inlet	Outlet	Inlet	Outlet
Flow rate, kg/s	50	50	50	50	750	750	750	750
Temperature, K	2873	2620	2873	2620	2650	2150	2650	2150
Pressure, kPa	200	75	200	75	427	80	427	80
Magnetic field induction, T	2.0	2.0	2.0	2.0	6.0	6.0	6.0	6.0
Channel cross sectional area, m^2	0.293	0.793	0.293	0.793	1.323	6.605	1.323	6.605
Distance from channel inlet, m	2	7	2	7	3	18	3	18
Wall temperature, K	2300	2100	1723	1573	2300	2100	1723	1573
$\overline{T}_{Tw} \equiv T_w/T_T$	0.8	0.8	0.6	0.6	0.87	0.98	0.65	0.73
$\overline{\delta} \equiv \delta/\sqrt{F}$	0.065	0.130	0.065	0.130	0.033	0.082	0.033	0.082
$\tau \equiv T/T_T$	0.950	0.950	0.950	0.950	0.945	0.945	0.945	0.945
$St \cdot 10^3$	--	--	--	--	1.97	1.85	1.97	1.85
$S_o \equiv \delta_{oo} B^2 F^{3/2}/m$	--	--	--	--	0.67	1.30	0.67	1.30
β	0.65	1.65	0.65	1.65	0.90	4.70	0.90	4.70

Table 3.1 (continued)

Optimization Condition	Optimal Ratio of the Sides of the Cross Section $(a/b/)_{opt} = X_{opt}^{-2}$							
1. $\overline{W} = \overline{W}_{max}$ when K = const.	2.0	3.3	7.8	13.7	2.24	16.4	6.1	24.0
2. Same, when η_{oi} = const.	--	--	--	--	1.20	3.4	2.7	13.6
3. Same, with cooling, Q_{cool}	--	--	--	--	1.15	3.3	2.0	11.5
4. $\Delta\eta/\overline{V}_g = (\Delta\eta/\overline{V}_g)max\|\overline{\Delta}_{y,z}$ = const.	--	--	--	--	1.02	1.4	1.32	5.9
5. $\Delta\eta/\overline{V}_g = (\Delta\eta/\overline{V}_g)max\|\overline{\Delta}^{*}_{y,z}$ = var	--	--	--	--	1.03	1.2	1.27	3.9
Gain in specific power takeoff due to optimization of a/b $(\overline{W}/\overline{V}_g)max/(\overline{W}/\overline{V}_g)(a/b)=1$ (for condition 5)					1.0002	1.0053	1.01	1.34

* At constant thickness of channel wall, $\overline{\Delta}_y=\overline{\Delta}_z=0.2(\overline{\Delta}=\Delta/\sqrt{F})$.

**At wall thickness linearly dependent on cross section dimensions $\overline{\Delta}_y=0.1+0.05\overline{X}$; $\overline{\Delta}_z=0.1+0.05\overline{X}^{-1}$.

cross section. Conditions (3.49) and (3.52) considered above are special cases of (3.53).

As an example, a channel cross section was optimized for characteristic operating conditions of an MHD generator in the U-25 facility and other commercial-scale MHD installations. Operating conditions of the commercial-scale MHD generator were taken from the literature [30]. Both the inlet and outlet cross sections of the channel were optimized at two different wall temperatures. The results of the calculations are shown in the table.

The maximum specific power per unit working fluid of the channel is achieved when the ratio a/b is sufficiently large, especially at the channel outlet, where the Hall parameter is large. However, when minimizing the external channel volume, cross sections close to quadratic are more advantageous. The optimum ratio of the sides, a/b = 3.9, applies only to outlet cross section with hot walls. In the remaining cases, $(a/b)_{opt}$ is close to unity. The last line in the table shows the ratio of the maximum specific power with the power for a square cross section of a commercial-scale MHD power plant. It turns out that, in the case of semi-hot channel walls, the gain in volume of the magnetic field through optimization of the channel cross section may be on the order of 20 % (1 % at the inlet and 34 % at the outlet of the MHD channel). Quite obviously, this is significant. For very hot walls (T_w = 2300-2100 K), the gain from optimization of the ratio of sides of the cross section of the MHD channel is very small in comparison with a square cross section. It is only 0.5 %, even at the channel outlet. Such an approach can be used for a more detailed optimization, which may include such factors as "nonisotropy" of the cost of the magnetic field volume along the length, width, and height.

3.6. Conclusions

Systematic experimental investigation of the electrodynamics of a nonideal MHD generator, performed at the Institute of High Temperatures of the Soviet Academy of Sciences throughout a broad range of conditions, made it possible to achieve complete understanding of characteristics of the MHD energy conversion process in real MHD generators, and to estimate the importance and contribution of various factors to this process. A simplified model of the phenomenon was constructed and a fairly simple method of engineering calculations of electrical characteristics of real MHD generators was developed. Although fairly simple, this method nevertheless takes into account the influence of all major factors associated with real structures, including two-dimensional and even three-dimensional effects, on the efficiency of operation of an MHD generator.

The experimental data were generalized using the model suggested. It should be noted that satisfactory agreement exists between the results of calculations and the experimental data. Results of the calculations using the method developed are in good agreement with an exact solution of the two-dimensional electrodynamic problem in the y,z plane.

The method suggested was used in analyzing the influence of various factors on the operation of commercial-scale MHD generators. This analysis has shown that plasma conductivity nonuniformity in the boundary layers at electrode walls can exert considerable influence on the characteristics of an MHD generator. This effect is especially significant at the MHD channel outlet, where the Hall parameter may reach 4-5. In the case of channel hot walls (T_w = 2100-2300 K), a drop in the specific power resulting from the transverse plasma conductivity nonuniformity is not so high. However, when the wall temperature drops to 1600-1700 K, the power loss averaged along the channel may reach 30 % or more. This is quite significant and must be taken into account in commercial-scale MHD facilities.

When the electrical load factor is constant, the specific power of an MHD generator exhibits only a weak dependence on other factors. However, according to this analysis, such factors as nonideal insulation and velocity nonuniformity in the boundary layers at insulation walls may considerably decrease the internal relative efficiency of an MHD generator. Furthermore, the negative influence of these factors on the efficiency is enhanced considerably by nonuniform transverse plasma conductivity. Just the opposite, by lowering the reverse current flow along the boundary layers at the insulating walls, as well as current leakage to these walls, the nonuniformity of transverse plasma conductivity suppresses somewhat the harmful influence of these factors on efficiency.

By affecting the parameters of the generator, the nonideal behavior of the MHD channel may also exert substantial influence on the choice of its optimum operating conditions. Thus, in commercial-scale MHD generators, the growth of the boundary layers at the channel walls leads to a decrease of the optimum flow rate along the channel length of almost 30 %. For semi-hot walls, nonideal behavior of the MHD channel may influence the optimal ratio of the sides of the channel cross section, even though, in most cases of practical importance, the optimum cross section will be nearly square. Results of the investigations were used widely in shakedown tests of the U-25 facility, and were completely justified under the operating conditions of such a large-scale facility. In particular, all of these results were used in developing the 1D MHD channel intended to attain the rated power output from the MHD generator of the U-25 facility (20 MW).

This problem was formulated toward the end of the first stage of the shakedown tests of the U-25 facility, when the principal components of the loop of the facility were brought up to the projected parameters and the power output of the MHD generator (with the 1B channel) reached 6.5 MW. Additional means had to be found to raise the power output significantly. Preliminary analysis has shown no radical ways of immediately increasing the power output 300 %. However, a whole series of under utilized capabilities were established, and it was shown that, although none of them could lead to an increase of more than 15 to 20 % of the power output, the use of all of them could solve the problem.

The complex of measures aimed at increasing the power output, as well as measures to increase reliability of the MHD channel, made it possible to design a channel with a rated power output of 20.6 MW [36]. In the first tests conducted on the MHD generator of the U-25 facility with the 1D channel, performed at less than the normal flow rate of the working fluid (50 kg/s instead of the projected flow rate of 60 kg/s), the power output reached 12.4 MW and verified to within 10 % the preliminary estimate for this flow rate (Fig. 3.27).

Tests to bring the channel to its rated parameters revealed and, thus, made it possible to eliminate certain fabrication defects. The MHD generator was then brought on line at the projected power output of 19.6 MW, which is close to the nominal value. Power output of 20.4 MW was attained during subsequent tests. Distribution of the power output along the channel length, compared with the results of preliminary calculations, is plotted in Fig. 3.27. The result achieved may be considered to indicate the adequacy of the present-day concepts of the energy conversion process in a nonideal MHD channel, as well as the possibility of developing an MHD generator with specified properties at the present level of our knowledge.

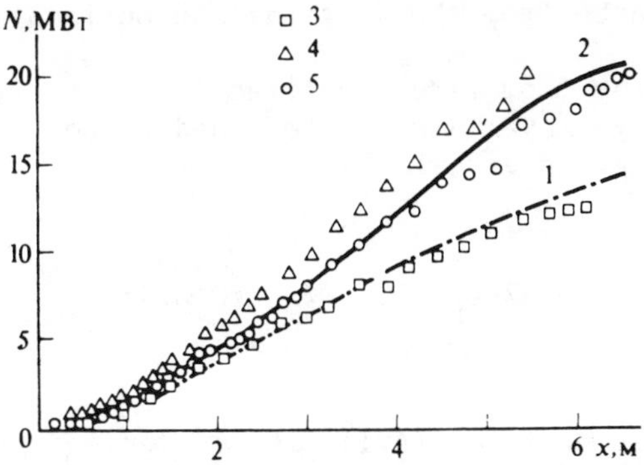

Fig. 3.27 Distribution of the power output along the length of the 1D MHD channel of the U-25 facility; 1,2--results of the calculations at flow rates of 50 and 60 kg/s, respectively; 3--run No. 47 (03/16/75), G = 50 kg/s; 4 - run No. 58 (12/27/75), G = 63 kg/s; 5 - run No. 61 (04/15/76), G = 61 kg/s

Appendix

Distribution of Flow Parameters Over the Channel Cross Section

Before applying the computational model, it is necessary to determine the profiles of the distribution of the flow parameters over the channel cross section that are assumed to be specified within the framework of the electrodynamic model. The fact that the calculated values of the external generator characteristics depend only on the averaged values of the flow of parameters simplifies the problem somewhat, making it possible to use an approximate distribution of parameters over the cross section. Certain assumptions concerning flow parameter distributions have already been made and thus, the problem is reduced to determination of f $(\overline{y})$, g $(\overline{z})$, μ, $(\overline{y})$, and ν $(\overline{z})$ in the boundary layer. Because laminar flow in the boundary layer is very unlikely under operating conditions in an MHD generator, only a turbulent boundary layer will be considered here. In cases where highly accurate computation is not required, the flow rate distribution in the turbulent boundary layer is usually specified by a power function [37]:

$$f(\tilde{y}) = \hat{y}^{1/7}, \quad g(\tilde{z}) = \tilde{z}^{1/7}, \tag{A.1}$$

where

$$\tilde{y} = \hat{y}/\delta_y, \ \hat{y} = 1 - y, \text{ and } \tilde{z} = \hat{z}/\delta_z, \ \hat{z} = 1 - z$$

is the distance from the vertical and horizontal walls, in that order.

When using flow rate profiles (A.1), the thickness of the boundary layer is determined from a well-known relation (e.g., see Ref. 37):

$$\delta_y/x = 0.37\mathrm{Re}_x^{-0.2}, \quad \delta_z/x = 0.37\mathrm{Re}_x^{-0.2}. \tag{A.2}$$

The power profile (A.1) is convenient and, in many cases, provides quite acceptable accuracy, especially for integral characteristics. However, near channel walls, it leads to flow rates that are too high and to a velocity gradient on the wall, $du/dy \to \infty$. Therefore, in most cases, when the near-wall part of the boundary layer is of importance, a two-layer model with a laminar sublayer must be used. It is preferable to reject the power profile and to use a more universal flow rate distribution instead.

The flow rate profile can be determined by integrating the equation,

$$\frac{du}{d\hat{y}} = \tau\,(\mu + A_\tau)^{-1}, \tag{A.3}$$

where μ is the dynamic coefficient of viscosity and A_τ is the turbulent coefficient of viscosity.

The distribution of the tangential voltage τ along the thickness of the boundary layer was assumed to be linear:

$$\tau = \tau_w\,(1 - \hat{y}/\delta)\ .$$

Equation (A.3) can be conveniently represented in the form,

$$\frac{d\bar{u}}{d\tilde{y}} = \frac{c_f}{2}\,\mathrm{Re}_\delta\,\frac{1-\tilde{y}}{1+\bar{\varepsilon}_\tau}, \tag{A.4}$$

where

$$c_f \equiv 2\tau_w/\rho u_{oo}^2$$

is the coefficient of friction, $\overline{\varepsilon}_\tau = \varepsilon_\tau/\nu$ (ν is the kinematic coefficient of viscosity, ε_τ is the turbulent coefficient of viscosity).

Relation suggested by Reichard [38] and tested on heat transfer problems [39] was used to determine $\overline{\varepsilon}_\tau$,

$$\overline{\varepsilon}_\tau = \Xi y^* \left(A\tilde{y}/y^* - \mathrm{th}(A\tilde{y}/y^*)\right) , \qquad (A.5)$$

where $\Xi = 0.4$, $y^* = 11$ (according to Reichard), $A = Re_\delta\sqrt{c_f/2}$. Knowing Re_x in the cross section, Re_δ can be determined from (A.2).

$$Re_\delta = 0.37 Re_x^{0.8} , \qquad (A.6)$$

and c_f can be found from a well-known formula [37]:

$$c_f = 0.576 Re_x^{-0.2} . \qquad (A.7)$$

Then

$$A = 0.0627 Re_x^{0.7} . \qquad (A.8)$$

Equations (A.4)-(A.8) complete define the velocity profile over the cross section at a specified Re_x. Generally speaking, one need not use formulas (A.6) and (A.7) to determine A. Using various A, $\overline{\varepsilon}_\tau = \overline{\varepsilon}_\tau(\tilde{y},A)$ can be found from A.5. Substituting $\overline{\varepsilon}_\tau$ into (A.4) and integrating within the limits of the boundary layer, the following relation is obtained:

$$1 = \frac{c_f}{2} Re_\delta \int_0^1 \frac{(1-\tilde{y})d\tilde{y}}{1+\varepsilon_\tau(\tilde{y},A)} = \frac{c_f}{2} Re_\delta I(A) ,$$

from which

$$Re_\delta = A^2 I(A), \quad c_f/2 = (AI(A))^{-2} .$$

At this stage, the velocity profile in the form $\overline{u} = \overline{u}(\tilde{y}, A)$ can be used in the following integral relation,

$$\frac{d}{dRe_x}\left(Re_\delta \delta^{**}(A)\right) + Re_\delta(2\delta^{**}+\delta^*)\frac{1}{u_{oo}}\frac{du_{oo}}{dRe_x} = \frac{c_f}{2} , \qquad (A.9)$$

where

$$\delta^{**} = \delta\bar{\delta}^{**} \text{ and } \delta^{*} = \delta\bar{\delta}^{*}$$

are the momentum thickness loss and expulsion thickness. Integrating expression (A.9) gives a relation between Re_x and Re_δ. It is now possible to find A in the specified cross section and to determine the flow rate profile. However, taking into account the weak dependence of the turbulent flow rate profile on Re_δ (or on A), in determining integral characteristics, it is quite adequate to limit oneself to an approximate determination of A from formula (A.8).

The conductivity profiles μ $(\bar{y})$ and ν $(\bar{z})$ are determined from the temperature profiles, T $(\bar{y})$ or T $(\bar{z})$, on the assumption that a thermodynamic equilibrium exists at each point of the flow. Knowing the flow rate profile, it is easy to obtain the temperature profile, assuming that the analogy between friction and heat transfer processes remain in force, even in the case of an MHD flow. The existence of similarity between the stagnation enthalpy and flow rate profile follows from this analogy:

$$\Theta)\tilde{y}_T) \equiv (h_T - h_W)/(h_{To} - h_W) = \bar{u}(\tilde{y}) \ ,$$

where h_T is the stagnation enthalpy, h_w is the gas enthalpy at the wall, h_{To} is the stagnation enthalpy in the core flow, $\tilde{y}_T = \hat{y}/\Delta$, Δ is the thickness of the thermal boundary layer. Strictly speaking, $\Delta \neq \delta$, and depends on a number of factors, including the Prandtl number, Pr, temperature conditions on the wall, $(\bar{T}_w\ (x))$, distribution of u_{oo} (x), etc. However, as a rule, this dependence is weak and, in approximate calculations, it can be assumed that $\Delta = \delta$. To simplify calculations, it was assumed that, in the temperature range $T_w < T < T_{oo}$, the heat capacity, c_p, and the isentropic index, k, are constant. Then, the temperature profile can be expressed in terms of the flow rate profile.

$$\bar{T} \equiv T/T_{oo}(\bar{T}_{TW} + (1 - \bar{T}_{TW})\ \bar{u}$$

$$- (1 - \tau)\bar{u}^2)\ \tau^{-1}. \qquad \text{(A.10)}$$

Here,

$$\bar{T}_{TW} \equiv T_W/T_{oo} ;$$

$$\tau \equiv 1 - \frac{k-1}{k+1} \lambda^2 \text{ and}$$

$$\lambda = \frac{u_{oo}}{a_{\kappa p}} = u_{oo} \left(\frac{2k}{k+1} gRT_{T_{oo}} \right)^{-1/2}$$

are the gasdynamic function and the flow rate coefficient, respectively.

When calculating the conductivity profile in the boundary layer, the conductivity of the combustion products of natural gas seeded with potassium salts was calculated using a method suggested in the literature [40]. The data obtained were approximated by the equation

$$\sigma = CT^{3/4} \exp(-\Phi/T)p^{-1/2}, \qquad (A.11)$$

where $\Phi = 25{,}150$ K, C is the coefficient that depends on the content of the combustion products flow and seed concentration, and p is the pressure. From relation (A.11), is obtained

$$\mu(\bar{y}) \equiv \frac{\sigma}{\sigma_{oo}}$$

$$= \left(\frac{T}{T_{oo}}\right)^{3/4} \exp\left(-\Phi\left(1 - \frac{T_{oo}}{T}\right) / T_{oo}\right),$$

where T/T_{oo} is a function of $\bar{y}$ specified by relation (A.10). An analogous method is used to determine $\nu(\bar{z})$.

When analyzing experimental data acquired on the "Start" facility, the conductivity of the argon-potassium plasma seeded with nitrogen was calculated using collision cross sections of electrons with argon and potassium atoms and nitrogen molecules [41, 42]. Coulomb electron collisions were also taken into account [43].

The results of the calculations were approximated by the following formula:

$$\sigma = 5.253 (n_e \cdot 10^{-20})(s \cdot 10^{20})^{-1} p^{-1} \sqrt{T}.$$

Here, the electron concentration

$$n_e \cdot 10^{-20} = 4190 \sqrt{X_K p T^{0.5}} \exp(-25043/T) ,$$

and the effective collision cross section

$$s \cdot 10^{20} = 3.975 \cdot 10^{-4} T + 3X_K$$

$$+ 1.7 \cdot 10^{-3} X_{N_2} \sqrt{T} + 0.1488 + 369(n_e \cdot 10^{-20})$$

$$\times (Tp)^{-1} \ln(9 \cdot 10^4 T^{3/2} (n_e \cdot 10^{-20})^{1/2}) ,$$

where X_K, X_{N2} are the potassium and nitrogen concentrations (in mass percent), p is the pressure in 10^5 Pa, T is the temperature in K.

Laws specifying the current flow in the space charge layer near the surface of the electrode are completely different from those in quasi-neutral plasma. Therefore, the concept of conductivity for this layer as a property of the medium loses its meaning. Based on these considerations, Reference 25, referring to Ref. 34, recommended exclusion of the space charge layer when determining the average inverse conductivity, $\langle 1/\mu \rangle$. Instead, the near-electrode potential drop in this layer should be taken into account individually [25]. The thickness of the space charge layer is $(1 \le N \le 10)$,

$$D = (\varepsilon_o kT/e^2 n_e)^{1/2} = 69(T/n_e)^{1/2} .$$

Taking this into account, the average inverse plasma conductivity is given by the following expression,

$$\left\langle \frac{1}{\mu} \right\rangle = \frac{1}{a} \int_{yo}^{a} \frac{1}{\mu} \, dy ,$$

where $y_o = ND$.

In the subject calculations, it was assumed N = 3. For MHD generators with hot and semi-hot walls, D is quite small. In any case, D << δ. Nevertheless, when the electrode wall temperature is low, when the concentration of electrons in the cold part of the boundary layer is low, correction $\langle 1/\mu \rangle$ prove to be substantial. At high temperature at the electrode wall, this correction is insignificant. The approach to specifying velocity and conductivity profiles described was used in processing of the experimental data and in analyzing the influence of various factors on the characteristics of an MHD generator.

REFERENCES

[1]Louis, J. F., and Brogan, T. R., "Flow Dynamics in MHD Generators," Transactions of an International Symposium on MHD Energy Conversion, Paris, July 1964 (Russian translation), Vol. 2, p. 289-301.

[2]Mattson, A. C. J., and Brogan, T. R., "Self-Excited MHD Generators," Transactions of International Symposium on MHD Electric Power Generation, Salzburg, Austria, July 1966 (Russian translation), Vol. 3, p. 3-12.

[3]Kirillin, V. A., and Sheyndlin, A. E., "Certain Results of Investigations on the Experimental U-02 Facility," Magnitohydrodynamic Method of Generating Electric Power (Magnitogidrodinamicheskiy metod polucheniya elektroenergii), 1968, Publishing House Energiya, p. 9-39.

[4]Buznikov, A. E., Vanin, V. E., Kirillov, V. V., and Sokolov, Yu. N., "Experimental 'Start' Facility for Investigating an MHD Generator on Inert Gas Plasma Seeded with Alkali Metal," Teplofizika vysokikh temperatur, Vol. 8, No. 5, 1970, p. 1064-1072.

[5]Kirillin, V. A., Sheyndlin, A. E., Shumyatskiy, B. Ya., et al., "Investigations at U-02 MHD Facility; Some Results," Fifth International Conference on MHD Electrical Power Generation, Munich, April 1971, Vol. 1, p. 353-370.

[6]Kirillin, V. A., Sheyndlin, A. E., Shumyatskiy, B. Ya., et al., "The U-02 Facility under Prolonged Continuous Operating Conditions," Teplofizika vysokikh temperatur, Vol. 9, No. 5, 1971, p. 1029-1046.

[7]Bityurin, V. A., Buznikov, A. E., Kirillov, V. V., et al., "Study of Nonideal MHD Generator Performance in the Presence of a Hall Effect," 10th Symposium on Engineering Aspects of MHD, MIT, Cambridge, Massachusetts, 1969; MIT Press, Cambridge, 1969, p. 117-119.

[8]Rosa, R. T., "Physical Principles of Magnetohydrodynamic Power Generation," Physics of Fluids, Vol. 4, No. 2, 1961, p. 182-198.

[9]Sheyndlin, A. E., Shumyatskiy, B. Ya., Kirillov, V. V., et al., "Experimental Investigation of the Influence of Boundary Layers and Certain Other Effects on Characteristics of an MHD Generator," Electricity from MHD, Proceedings of a Symposium, Warsaw, July 1968, IAEA, Vol. 4, p. 2379-2405.

[10]Sheyndlin, A. E., Buznikov, A. E., and Vanin, V. E., "Experimental Investigation of the Influence of Boundary Layers on the Characteristics of an MHD Generator Throughout a Broad Range of Hall Numbers," Teplofizika vysokikh temperatur, Vol. 9, No. 2, 1971, p. 386-394.

[11]Bityurin, V. A., Kovbasyuk, V. I., and Medin, S. A., "Certain Electrodynamic Effects in an MHD Generator Channel," AN SSSR, Izvestiya, Mekhanika zhidkosti i gaza, No. 3, 1968, p. 3-8.

[12]Hoffman, M. A., "Nonequilibrium MHD Generator Losses Due to Wall and Insulator Boundary-Layer Leakages," AIAA Journal, 1968, Vol. 6, No. 9, p. 1667-1673.

[13]McNab, I. R., Brown, R., and Hains, M. G., "Closed-Cycle MHD Experiments with Applied Electric and Magnetic Fields," Electricity from MHD, Proceedings of a Symposium, Warsaw, July 1968, Vienna; IAEA, No. 2, p. 925-934.

[14]Vanin, V. E., Kirillov, V. V., and Sokolov, Yu. N., "Investigation of Behavior of Electrical Insulation in an MHD Facility Operating on Argon-Potassium Plasma," Teplofizika vysokikh temperatur, Vol. 8, No. 6, 1970, p. 1274-1279.

[15]Beloglazov, A. A., Kirilov, V. V., Panovko, M. Ya., and Sokolov, Yu. N., "Measurement of Insulation in an MHD Generator Under Operating Conditions," Teplofizika vysokikh temperatur, Vol. 10, No. 1, 1972, p. 157-160.

[16]Hurwitz, H., Kilb, R. W., and Sutton, G. W., "Influence of Tensor Conductivity on Current Distribution in MHD Generator," Journal of Aplied Physics, Vol. 32, No. 2, 1961, p. 205-216.

[17]Celinski, Z. N., and Fisher, F. W., "Effect of Electrode Size in MHD Generators with Segmented Electrodes," AIAA Journal, Vol. 4, No. 3, 1966, p. 421-428.

[18]Oliver, D. A., and Mitchner, M., "Nonuniform Electrical Conduction in MHD Channels," AIAA Journal, Vol. 5, No. 8, 1967, p. 1424-1432.

[19]MacDonald, J. R., Mitchner, M., and Oliver, D. A., "Electrode-Size Effects in an MHD Generator with Nonuniform Electrical Conduction," AIAA Journal, Vol. 6, No. 5, 1968, p. 234-235.

[20]Zelikson, Yu. M., Ivanov, A. B., Kirillov, V. V., et al., "Near-Electrode Potential Drop and Impedance of Zirconium Dioxide Electrodes in an Ionized Gas Flow," Teplofizika vysokikh temperatur, Vol. 9, No. 3, 1971, p. 462-467.

[21]Zalkind, V. I., Zuyeva, N. V., Kirillov, V. V., et al., "Investigation of the Microarc Operating Mode of Electrodes in Open-Cycle MHD Generator," The First U.S.-U.S.S.R. Colloquium on Energy Conversion (Perviy Sovetsko-Amerikanskiy kollokvium po MGD preobrazovaniyu energii), Moscow, 1974, p. 289-304.

[22]Buznikov, A. E., Vanin, V. E., Kirillov, V. V., and Sokolov, Yu. N., "Influence of Electrode Configuration on the Characteristics of an MHD Generator at Different Hall Numbers," Magnetohydrodynamic Method of Electric Power Generation (Magnitogidrodinamicheskiy metod polucheniya elektroenergii), edited by V. A. Kirillin and A. E. Sheyndlin, Energiya Press, Moscow, 1972, p. 110-123.

[23]Vanin, V. E., Kirillov, V. V., Maslennikov, G. I., and Sokolov, Yu. N., "Influence of the Amount of Protrusion of Electrodes into the Flow on the Local Characteristics of an MHD Generator," Magnitohydrodynamic Devices (Magnitogidrodinamicheskiye ustranovki), edited by V. A. Kirillin and A. E. Sheyndlin, Nauka Press, Moscow, 1975, p. 87-93.

[24]Bityurin, V. A., and Lyubimov, G. A., "Quasi-One-Dimensional Analysis of Flow in an MHD Generator Channel," Teplofizika vysokikh temperatur, Vol. 7, No. 5, 1969, p. 974-986.

[25]Eustis, R. H., Cima, R. M., and Berry, K. E., "Current Distribution in Conducting Wall MHD Generators," 11th Symposium on Engineering Aspects of Magnetohydrodynamics, California Institute of Technology, Pasadena, California, March 1970, p. 119-127.

[26]Jett, E. S., Denzel, D. L., and Wu, Y. C. L., "Eddy Currents in an Infinitely Segmented Hall Generator," AIAA Journal, Vol. 8, No. 9, 1970, p. 1992-1997.

[27]Beloglazov, A. A., Burenkov, D. K., Vysotskiy, D. A., et al., "Investigation of Operation of Segmented Ceramic Insulating Walls of an MHD Generator Channel," Teplofizika vysokikh temperatur, Vol. 12, No. 3, 1974, p. 605-613.

[28]Sheyndlin, A. E., Beloglazov, A. A., Burenkov, D. K., et al., "Investigation of Long-Duration Operation of Insulating Walls of an Open-Cycle MHD Generator," The First U.S.-U.S.S.R. Colloquium on MHD Energy Conversion (Perviy Sovetsko-Amerikanskiy kollokvium po MGD preobrazovaniyu energii), Moscow, 1974, p. 278-288.

[29]Sheyndlin, A. E., Burenkov, D. K., Zalkind, V. I., et al., "Investigation of an Insulating Wall of a Long-Duration MHD Generator," Teplofizika vysokikh temperatur, Vol. 13, No. 1, 1975, p. 151-156.

[30]Mori, Y., "Cycle and Efficiency of Large-Scale, Open-Cycle MHD Power Plants," 5th International Conference on Magnetohydrodynamic Electric Power Generation, Munich, 1971, Vienna: IAEA, 1971, Vol. 1, p. 569-584.

[31]Rosa, R. J., Magnetohydrodynamic Energy Conversion (Russian translation), Mir, 1970.

[32]Ralph, G., "Physical Processes in an MHD Generator Channel," Russian translation from an unspecified source, 1966, p. 25-53.

[33]El'kin, A. E., and Yantovskiy, E. I., "On the Influence of Transverse Nonuniformities on the Gas Flow in a Hall Channel with a Single Load," Magnitnaya gidrodinamika, No. 1, 1971, p. 57-63.

[34]Lyubimov, G. A., "Variation of Electrical Potential Near the Channel Wall during Motion of Ionized Gas in a Magnetic Field," Zhurnal prikladnoy matematiki i technicheskoy fiziki, No. 5, 1963, p. 24-34.

[35]Abramovich, G. N., Applied Gasdynamics, Nauka Press, Moscow, 1969.

[36]Sheyndlin, A. E., Buznikov, A. E., Iserov, A. E., et al., "Experimental Investigation of the U-25 MHD Generator Facility with the 1D Channel," The Third U.S.-U.S.S.R. Colloquium on Energy Conversion, Moscow, October 1976, IHT, 1978, p. 77-89.

[37]Schlichting, G., Theory of Boundary Layer, translation from German, Nauka Press, Moscow, 1974.

[38]Reichardt, H., "Vollständige Darstellung der turbulent Geschwindigkeitsverteilung in glatten Leitungen," Zeitschriften Angewanden Math. and Mech., 1951, Bd. 31, No. 7, p. 208-219.

[39]Petukhov, B. S., and Kirillov, V. V., "On the Problem of Heat Transfer during Turbulent Flow of Fluid in Pipes," Teploenergetika, No. 4, 1958, p. 63-68.

[40]Frost, L. S., "Conductivity of Seeded Atmospheric Pressure Plasmas," Journal of Aplied Physics, Vol. 32, No. 10, 1961, p. 2029-2036.

[41]Volkov, Yu. M., and Malyuta, D. D., "Electrical Conductivity and Energy Balance in a Nonequilibrium Plasma," Electricity from MHD, Proceedings of a Symposium, Warsaw, July 1968, Vienna: IAEA, 1968, Vol. 1, p. 77-96.

[42]Veldre, V. Ya., "Collision of Slow Electrons with Atoms," Low-Temperature Plasma (Nizkotemperaturnaya plasma), Mir Press, Moscow, 1967, p 40-45.

[43]Sutton, G. W., and Sherman, A., Engineering Magnetohydrodynamics, New York, 1968 (Russian translation).

Chapter 4

Plasma Diagnostics in MHD Generators

4.1. Introduction

Successful development of MHD energy conversion depends considerably on the feasibility of producing a working fluid with optimal characteristics. In order to achieve this goal it is necessary to develop diagnostic methods for the working fluid of an MHD generator. It is especially important to be able to obtain data during operation of the facility. Combustion products flow with easily ionized alkali seed is usually used as the working fluid in an equilibrium MHD generator. Seeded combustion products in an MHD channel are a nonstationary, electrically-conducting, spatially-nonuniform gas.

The working fluid of an equilibrium MHD generator is adequately characterized by the following set of parameters: gas temperature, alkali atom seed and electron concentrations, and rate of electron collision with neighboring particles. These quantities control the electrical conductivity of the working fluid and the power output of the MHD generator. Variation of each parameter makes it possible to establish and control the optimal operating mode of the major components of the facility with an MHD generator.

Measurement of the gas temperature makes it possible to estimate the efficiency of the facility as a whole, and especially of its combustion chamber. Measurement of the concentration of atoms of easily ionized seed in the generator channel makes it possible to estimate the efficiency of seed injection into the facility. The electrode concentration and effective collision rate with the neighboring particles completely determine the electrical conductivity of the gas. These quantities respond differently to changes in the operating mode of the facility. Therefore, their measurement makes it possible to determine rational methods of increasing electrical conductivity in an MHD generator. The gas temperature and concentration of atoms of easily ionized seed are measured spectroscopically. Because the resonance lines of alkali metals are bright throughout a broad range of conditions, their emission is used in spectroscopic methods. The well-known spectral line reversal techniques used to measure the gas temperature of a spatially nonuniform emitter provide only temperature averaged along the line of sight. In order to judge correctly the operation of the facility with an MHD generator, it is necessary to determine the temperature of the core flow. Hence, it is extremely important, when developing spectroscopic diagnostic methods, to investigate the influence of the spatial gas nonuniformity on the spectral line profile of emission and the temperature being measured. In performing this investigation, it is important to be able to take into account the characteristics and adjustment of the spectral instrument, as well as the time lag of the measuring apparatus when determining the temperature of the nonstationary gas averaged over time. It is also of interest to investigate the temperature fluctuations by spectroscopic means.

No method of determining concentration of alkali metal atoms existed at the beginning of diagnostic investigations in equilibrium MHD generators. Yet, this method is so simple in practice that it could have been reliably used to investigate the working fluid. One of the problems to be solved by spectroscopic means was the development of such a technique of determining the concentration of atoms.

The following conditions must be met when conducting measurements in MHD generators. The measurements must be made continuously and the result of measurements must become available at the control panel of the facility during the experiments. In addition, no operator can be present near the object of measurement. Therefore, remote-controlled automatic temperature and concentration measurement systems were designed and fabricated for real-time diagnostics in MHD generators. Phase shift and attenuation of submillimeter laser emission in gas can be used to determine the electron concentration and rate of electron collision with neighboring particles. After choosing a laser oscillating at a certain wavelength, it is necessary to analyze conditions when the phase shift and attenuation of emission are caused by electron, and not molecular component of the working fluid of an equilibrium MHD generator. In investigations involving gasdynamics and heat transfer in MHD generators, it is necessary to know the spatial distribution of parameters of the working fluid. The probe method presents an opportunity to determine the spatial distribution of the electron temperature, which, in the case of an equilibrium MHD generator, is usually equal to the temperature of the molecules. The first problem encountered in developing the probe method is the feasibility of determining electron temperature in the region of the gas undisturbed by the probe from the voltage-current characteristic of the probe.

The major unclarified points are associated with the influence of temperature of the cold probes on the results of their readings at moderate and high pressures, when the mean-free-path of electrons is considerably shorter than the characteristic dimensions of the probe and the layer surrounding it. In order to determine conditions at which the electron current to the probe provides data on the temperature of electrons in the plasma, it is necessary to consider the physical pattern of the influence of the temperature of the cold probe on the electron probe current.

4.2. Spectroscopic Diagnostics of Spatially Nonuniform Gas (Theory)

Spectroscopic methods for MHD generator channels are based on the use of spectral lines emitted by spatially nonuniform gas. Consider the influence of the emitting nonuniform gas on the shape of the spectral line. The influence of nonuniformity of the emitter on the shape of the spectral line is usually investigated by solving the emission transfer equation. If Kirchhoff's law applies locally, the spectral distribution of intensity I_ν of an inhomogeneous emitter of length ℓ can be written in the following form:

$$I_\nu = \int_0^\ell I^0_\nu(T(x))k_\nu(x) \exp\left(-\int_x^\ell k_\nu(y)dy\right) dx \,, \tag{4.1}$$

where I°_ν (T (x)) is the intensity of monochromatic radiation of a black body at temperature T (x) near frequency ν, x is the coordinate counted from the boundary of the emitter opposite from the observer, $k_\nu(x)$ is the coefficient of absorption at frequency ν. In case of an emitter symmetric along the line of sight, with a monotonic temperature decrease from the center to the periphery, one can introduce optical coordinates that make it possible to replace expression (4.1) with the following relation [1]:

$$I_\nu = \tau_\circ \exp\left(-\frac{\tau_0}{2}\right) \int_{-1/2}^{1/2} I^\circ_\nu(\xi)\exp(\tau_\circ \xi)d\xi =$$

$$= 2\tau_o \exp\left(-\frac{\tau_o}{2}\right) \int_0^{1/2} I_\nu^o(\xi)\mathrm{ch}(\tau_o \xi)d\xi \,. \tag{4.2}$$

where

$$\tau_o = \int_0^{\ell} k_\nu(x)dx; \quad \tau = \int_{\ell/2}^{x} k_\nu(x)dx \;;$$

$$\xi = \frac{\tau}{\tau_o}, \tag{4.3}$$

where τ_o is the optical density, τ, ξ, are the absolute and relative optical coordinates.

Expressions (4.1) and (4.2) describing the influence of nonuniformity on the emission line profile in general form were theoretically investigated in a number of papers [2-4]. Expression (4.2) can be written in the following form [2]:

$$I_\nu / I_\nu^o(T(\ell/2)) = MY(\tau_o) \,, \tag{4.4}$$

where T (ℓ/2) is the temperature in the middle (along the line of sight) of the emitter;

$$M = \int_{-\frac{\tau_o}{2}}^{\frac{\tau_o}{2}} I_\nu^o(T(x))d\tau / I_\nu^o(T(\ell/2))$$

$$= \int_{-1/2}^{1/2} I_\nu^o(\xi)\, d\xi / I_\nu^o(T(\ell/2)). \tag{4.5}$$

The nonuniformity parameter, M, is independent of the optical density of the emitter, and is constant inside the line if the profile of the coefficient of absorption inside the emitter does not change.

The function Y (τ_o) depends both on the degree of nonuniformity and optical density of the emitter, τ_o:

$$Y(\tau_o) = \tau_o \exp\left(-\frac{\tau_o}{2}\right) \int_{-1/2}^{1/2} s(\xi)\exp(\tau_o \xi)d\xi$$

$$= 2\tau_o \exp\left(-\frac{\tau_o}{2}\right) \int_0^{1/2} s(\xi)\mathrm{ch}(\tau_o \xi)d\xi \,, \tag{4.6}$$

where the relative source function in optical coordinates, ξ, is given by the expression

$$s(\xi) = I^o_\nu(\xi) \Big/ \int_{-1/2}^{1/2} I^o_\nu(\xi)\,d\xi$$

$$= I^o_\nu(\xi) \,/\, I^o_\nu(T(\ell/2))\; M\,. \tag{4.7}$$

Up to this point, no approximations were used in deriving expressions for intensity. To determine I_ν/I°_ν (T (ℓ/2)), it is necessary to calculate M and Y (τ_o). These calculations have been done in a number of papers, under various assumptions about variation of the source function. Series expansion of the term under the integral sign in formula (4.6) makes it possible to represent Y (τ_o) in the form of a family of curves [2] independent of the parameter,

$$p = 12 \int_{-1/2}^{1/2} \xi^2 s(\xi)\,d\xi\,. \tag{4.8}$$

The family Y (τ_o, p) has been calculated numerically within a limited range of τ_o in the same work [2]. The theory was developed as it applies to gas discharge diagnostics [2-8]. Considerable attention was devoted to obtaining approximate, but simple relations between M and p [2, 4, 5].

Certain regularities in the behavior of the shape of the self-reversed spectral lines follow from expression (4.4) and an analysis of previously calculated functions Y (τ_o). If the assumptions of the monotonic decrease of the temperature toward the periphery and the constancy of the profile of the coefficient of absorption inside the inhomogeneous emitter are fulfilled, the ratio of intensity at the peaks of self-reversed line with the intensity of the black body at the highest temperature of the emitter, $I_{\nu max}/I^\circ_\nu$ (T (ℓ/2)), depends only on the degree of inhomogeneity of the emitter. The optical line at the peaks of the self-reversed line, i.e., frequency ν_{max} at which the intensity reaches maximum, depends on the magnitude and variation of the coefficient of absorption in the line. The dependence of the coefficient of absorption on frequency is determined by broadening, shifting, overlapping of nearby lines and continuum emission. The principal mechanisms responsible for broadening of resonance lines of alkali metal atoms in combustion products at atmospheric pressure are Doppler and collisional, with collisional broadening caused by collisions with molecules of products of combustion.

At a short distance from the center of the line, collisional broadening is described by shock theory, and the coefficient of absorption of an isolated line is given by the Voigt integral [9]. The collision cross sections of alkali atoms with nitrogen molecules and combustion products have been determined [10-15]. At longer distances from the center of the line, variation of the coefficient of absorption with frequency depends only on collisional broadening, when shock theory applies [9]. At distances even farther from the center of the line, statistical broadening theory applies [9, 16].

When investigating parameters of the working fluid of an MHD generator, one is frequently interested in conditions and distances from the center of the line, when neither the shock theory nor statistical theory provides a correct description of the coefficient of absorption. Therefore, when performing diagnostics of the working fluid of an MHD generator, it is preferable to use the results of experimental investigations of the coefficient of absorption in the line wings.

Experimental investigations of the coefficients of absorption in the wings of resonance doublets of sodium (5890/5896 Å) and potassium (7665/7699 Å) in the products of combustion of natural gas were performed [17, 18]. In these investigations, the dependence of the coefficients of absorption on the wavelength was obtained throughout a broad range of conditions. Usually, the coefficient of absorption at individual sectors of a spectral range can be represented in the form

$$k_\nu \sim n_a \Delta\nu^{-y}/\sqrt{T}, \tag{4.9}$$

where n_a is the alkali atom concentration, $\Delta\nu$ is the distance from the center of the line, y is the exponent equal to two in the collision wings of the Voigt integral and that at large distances from the center of the line was determined experimentally [17, 18]. Expression (4.9) does not describe the narrow, central part of the shape of the coefficient of absorption that is usually not used in the diagnostics of MHD generators.

Direct experimental investigation of the influence of nonuniformity of the emitter on the variation of the shape of spectral lines is of interest. The influence of the spatial nonuniformity in gas on the profile of the resonance line of sodium (5890 Å) [19, 20] emitted by combustion products of natural gas (CH_4 - 92 %) in air with an additive consisting of an aqueous solution of table salt was investigated. Combustion occurred in a Meker burner at stoichiometric fuel-to-oxidizer ratio. The gas-air mixture was fed into the combustion zone through two channels. Two flares of combustion products were formed, the outer flare surrounded the inner flare. Solution of sodium salt was fed by an atomizer into the combustion zone together with the air. The flow rates of the gas-air mixture and solution of sodium salt were controlled independently in each flare. The salt content in the solution varied. The temperature in the inner flare was constant, whereas that in the outer flare decreased linearly with increase in radius. In all experiments, the concentration of sodium atoms in the inner flare was maintained constant; in the outer flare, the concentration was varied throughout a broad range of values. The results of the experiments can be summarized as follows.

Initially, an increase of sodium atom concentration in the inner flare leads to an increase in the degree of nonuniformity of the emitter. The depth of the dip at the center of the line increases, the intensity in the maxima decreases, and the peaks move away from the center. When the seed concentration in the ring flare is sufficiently high, further increase leads only to an increase of the optical density and a very small change in the degree of nonuniformity of the emitter. In this case, the peaks move away from the center of the line and the intensity at the peaks remains practically constant. Such behavior of the experimental lines is in agreement with theoretical predictions. Practical application of the regularities describing the behavior of self-reversed contours required a method to adequately describe these regularities adequately. The most convenient description of the spectral line is in terms of the source function (4.7) and the use of relations (4.4)-(4.6).

4.2.1. On the Application of Kirchhoff's Law for a Description of Emission of Alkali Atoms in Combustion Products. Before constructing a source function adequate for an equilibrium MHD generator, one must estimate the validity of assumptions as to the local isothermal state of the combustion products at pressure close to the atmospheric pressure and as to equilibrium distribution of excited alkali atoms over the energy levels. The fulfillment of these assumptions is necessary for application of Kirchhoff's law.

The gas is locally isothermal if the characteristic energy exchange time between particles is less than the characteristic time of the process. In combustion products flow, characteristic time of the process depends on the particle flight time from nonequilibrium combustion zone to the volume under investigation, diffusion in the direction perpendicular to the flow, and flow temperature fluctuations.

With an accuracy on the order of magnitude, the exchange time of kinetic energy between molecules t_{M-M}, the transfer time of kinetic energy of electrons to molecules, t_{eM}, and the energy exchange time between electrons, t_{ee}, at atmospheric pressure and T = 2000-3000 K, are given by the following relations:

$$t_{M-M} \simeq (n_M \langle v_M \rangle \sigma_{M-M})^{-1} = 10^{-7} \text{ s} ,$$

$$t_{eM} \simeq (n_M \langle v_e \rangle \sigma_{eM} \delta_e)^{-1} = 10^{-8} \text{ s} ,$$

$$t_{ee} \simeq (n_e \langle v_e \rangle \sigma_{ee})^{-1} = 3 \cdot 10^{-7}/n_e \text{ s} ,$$

where $n_M \simeq 10^{18}$ cm^{-3} is the molecular concentration, n_e is the electron concentration in cm^{-3}, $\langle v_M \rangle \simeq 10^5$ cm/s and $\langle v_e \rangle \simeq 3 \times 10^7$ cm/s are the mean velocity of molecules and electrons, respectively; $\sigma_{M-M} \simeq 10^{-16}$ cm^2, $\sigma_{eM} \simeq 10^{-15}$ cm^2, $\sigma_{ee} \simeq 10^{-15}$ cm^2 is the interaction cross section of molecules with each other, electrons with molecules and electrons with each other; $\delta_e \simeq 4 \times 10^{-3}$ is the effective part of the energy loss during a single collision of an electron with molecules. As long as

$$n_e < 3 \times 10^{15} \text{ cm}^{-3} ,$$

time, t_{ee} is longer than time t_{eM}.

Characteristic cross sections and energy lost by electrons during collisions are taken from the literature [21, 22]. The following approximate data on the molar content of the principal molecules in the combustion products of natural gas and air are required for further estimates [23]:

	2000, K	3000, K
N_2	0.70	0.66
H_2O	0.19	0.14
CO_2	0.0090	0.029
CO	0.0030	C.058
O_2	0.0016	0.026
H_2	0.0013	0.031

An estimate of rotational relaxation time shows that it is short. For example, at atmospheric pressure, rotational relaxation of nitrogen molecules, which are the principal component of the combustion products, occurs during the time $t_\omega \simeq 10^{-9}$ s.

The vibrational relaxation of the molecules of nitrogen occurs most effectively during interaction with water molecules in translational motion. Under conditions considered, $t_{\nu N2} = 5 \times 10^{-6}$ s [24]. Finally, consider the population of resonance levels of alkali metals in the combustion products. Inelastic collisions with particles lead to the establishment of an equilibrium distribution over the energy levels, deviation from which can be caused by emission and, if plasma is nonuniform, to a loss of excited particles from the volume under consideration [25-27]. When estimating the influence of various processes determining the population of resonance levels in the combustion products, it is sufficient to compare processes leading to the loss of the excited state.

Consider the depopulation rate of the 3^2 P level of sodium (similar values are characteristic of resonance levels of other alkali metals as well). The quenching frequency of this level q_{N2}, during collisions with nitrogen molecules depends on the quenching cross section accompanied by energy transfer to the vibrational levels of molecules, $\sigma_{qN2} = 3 \times 10^{-15}$ cm^2 [24]. For a nitrogen molecule concentration $n_{N2} = 2 \times 10^{18}$ cm^{-3}, we obtain $q_{N2} = \sigma_{qN2} n_{N2} \langle v_M \rangle = 6 \times 10^8$ s^{-1}.

The quenching process by molecules turns out to be the predominant process of establishing population. Quenching by electrons is characterized by cross sections σ_{qe}, which are approximately on the same order of magnitude. Therefore,

$$\frac{q_{N_2}}{q_e} \simeq \frac{\langle v_{N_2} \rangle n_{N_2} \sigma_{qN_2}}{\langle v_e \rangle n_e \sigma_{qe}} \simeq \frac{\langle v_{N_2} \rangle n_{N_2}}{\langle v_e \rangle n_e} .$$

From this it follows that the quenching rate of electrons becomes comparable to the quenching rate of nitrogen molecules only at a sufficiently high degree of ionization.

The quantity inverse of the lifetime of sodium atoms at the 3^2 P level is equal to 5×10^7 s^{-1} [9] and is less than 0.1 of the quenching rate of nitrogen molecules. Actually, this relation is even smaller, because the characteristic emission time usually increases because of radiation capture and because nitrogen molecules, as well as other molecules of the combustion products components, participate in quenching. The time required to establish equilibrium distribution over the resonance levels of alkali metal atoms, $t_p = q_{N2}^{-1} = 2 \times 10^{-9}$ s, is on the order of magnitude of the quenching time or time required to excite this level during most intense collisions with molecules. This is true if an equilibrium distribution of molecules over the rotational states is attained. The time required to establish rotational equilibrium of nitrogen is considerably greater ($t_{\nu N2} = 5 \times 10^{-6}$ s). It also exceeds considerably other characteristic times, i.e., time required to establish kinetic energy of electrons and molecules, rotational relaxation. Therefore, $t_{\nu M}$ determines the time lag in establishing equilibrium distribution over the levels. If some temperature disturbance appears, equilibrium distribution will be reestablished in not less than 5×10^{-6} s.

Let us determine the spatial locality of the equilibrium state. As a result of diffusion, the molecules are displaced a distance $\ell_D \simeq \sqrt{D_M t_{\nu M}} = 5 \times 10^{-3}$ cm during the time $t_{\nu M}$. In this formula, $D_M = 5\text{–}10$ cm^2/s is the coefficient of diffusion in the combustion products at atmospheric pressure. If the characteristic dimensions of the nonuniformity are smaller than ℓ_D, as a result of diffusion the nonuniformities will smooth out during the time of vibrational relaxation. The gas inside such regions can be assumed to be spatially homogeneous and all distributions with respect to energy are characterized by a single temperature.

Finally, the characteristic distance required to establish equilibrium in the combustion products flow leaving the nonequilibrium combustion zone will be estimated. At a flow rate $v_\pi = 10^5$ cm/s, characteristic of an MHD generator, this distance $v_\pi t_{\nu M} = 0.5$ cm. In laboratory burners, where

$$v_\pi < 10^4 \text{ cm/s} ,$$

the distance required to establish an equilibrium is less than 0.05 cm. It should be noted here that we have not estimated the time required to establish the equilibrium content of seed atoms for different methods of seed injection into the combustion zone. We have considered only establishment of the distribution over the excited states at a specified number of metal atoms.

The characteristic dimensions and the process times in plasma of MHD generators is considerably longer than those estimated above. Therefore, it will be assumed that the combustion products are locally isothermic and that equilibrium distribution over the energy levels has been attained. This means that one can apply Kirchhoff's law locally to volumes of combustion products characterized by a single temperature.

4.2.2. Emission Source Function of Resonance Lines of Alkali Atoms in MHD Generators. When Kirchhoff's law is fulfilled, determination of the source function from formula (4.7) requires the knowledge of the temperature distribution in the emitter. The flow mode in MHD generator channels is usually turbulent and the temperature distribution in the boundary layer and its thickness may be found from known relations [28]. The principal problem is transformation from a geometrical coordinate to the optical coordinate with the aid of relation (4.3).

Consider the source function of an emitter with a homogeneous central part, when the coefficient of absorption is given by formula (4.9). In accordance with this relation, the shape of the contour of the coefficient of absorption is independent of location inside the emitter. The relative optical coordinate, ξ, characterizes emission at any wavelength within the spectral line. Let δ be the thickness of a nonuniform boundary layer. Then, in accordance with formulas (4.3),

$$\xi = 0.5\,\frac{\ell/2-\delta+Q(\ell-\delta,x)}{\ell/2-\delta+Q(\ell-\delta,\ell)}\,, \tag{4.10}$$

where

$$Q(b_1 b_2) \equiv \int_{b_1}^{b_2} \frac{n_a(x)}{n_a(\ell/2)} \left(\frac{T(\ell/2)}{T(x)}\right)^{1/2} dx\,.$$

At the boundary of the uniform part of the emitter, the relative optical coordinate is given by the following relation:

$$\xi_b = \frac{0.5(\ell/2-\delta)}{\ell/2-\delta+Q(\ell-\delta,\ell)} \; . \qquad (4.11)$$

The known temperature distribution, which also specifies the equilibrium distribution of the concentration of atoms, must be used to transform from the geometrical coordinate, x, to the optical coordinate, ξ, using relations (4.10), (4.11). It is calculated at a specified total flow rate of the emitting substance. It is possible to introduce an optical coordinate that characterizes the relative optical coordinate, z, only inside the boundary layer and is independent of its thickness. Using expressions (4.10) and (4.11), obtains

$$z \equiv \frac{(\xi/\xi_b-1)(\ell/2\delta-1)}{Q(\ell/\delta-1,\ell/\delta)} = \frac{Q(\ell/\delta-1,y)}{Q(\ell/\delta-1,\ell/\delta)} \; . \qquad (4.12)$$

When $\ell/2\delta = 1$ and $\xi_b = 0$, instead of relation (4.12), we have

$$z = \xi/0.5 = Q(\ell/2,\, y)/Q\,(\ell/2,\, \ell) \; . \qquad (4.13)$$

Relations (4.10)-(4.13) make it possible to change from geometrical coordinates to optical coordinates. Calculation of the source function in relative coordinates in the boundary layer of a turbulent combustion products flow with alkali seed [29, 30] shows that the source function $i^\circ(z) \equiv I^\circ(z)/I^\circ(T(\ell/2))$, depends considerably on the temperature of the core flow, $T(\ell/2)$ and channel wall temperature, T_w, and is characterized by a sharp drop along the boundary layer sectors adjacent to the wall (Fig. 4.1). Calculations performed for alkali metals (sodium, potassium, rubidium, and cesium) in the range of concentrations $n_a(\ell/2) = 8 \times 10^{13} - 1.5 \times 10^{16}$ cm^{-3} result in very similar curves. The data on equilibrium content of alkali metal atoms at various temperatures were used in the calculations.

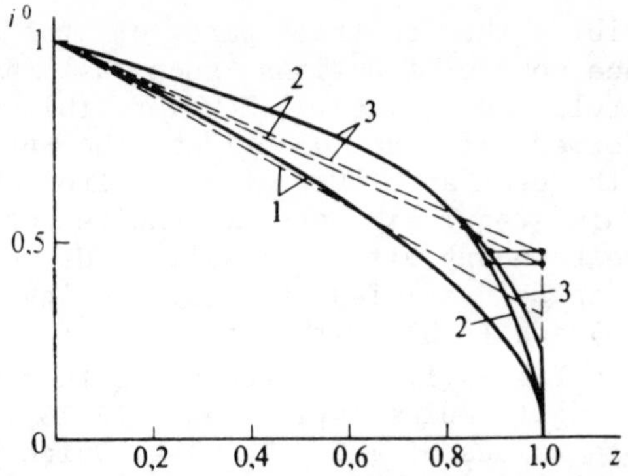

Fig. 4.1 Relative distribution of the source function in the boundary layer of a turbulent flow at various core flow and wall temperatures (K). Continuous lines - results of calculations; dashed lines - results of approximations (modeling) 1 - $T(\ell/2) = 2800$, $T_w = 1000$; 2 - $T(\ell/2) = 2800$, $T_w = 2000$; 3 - $T(\ell/2) = 2200$, $T_w = 1500$

Source functions analogous to those shown in Fig. 4.1 at a specific ratio δ/ℓ is adequate to construct the shape of a spectral line.

From a practical point of view, it is more convenient to represent the line shape in terms of elementary functions, using a simplified source function. In particular, true source functions such as the one shown in Fig. 4.1 can be replaced by simple functions characterized by a linear decrease in the boundary layer:

when $$0 \le \xi \le \xi_b$$

$$i^\circ(\xi) = 1 \ ;$$

$$\xi_b \le \xi \le 0.5$$

when
$$i^\circ(\xi) = 1 - (1 - i^\circ(T_{min}))\,\frac{(\xi - \xi_b)}{(0.5 - \xi_b)} \ . \tag{4.14}$$

Then, from formulas (4.5) and (4.8),

$$M = 1 - (1 - i^\circ(T_{min}))(0.5 - \xi_b) \ , \tag{4.15}$$

$$p = M^{-1}\,(1 + (-0.375 + \xi_b - 2\xi_b^4)$$

$$\times\ (1 - M)/(0.5 - \xi_b)^2) \ . \tag{4.16}$$

In describing the central part of the profile, one cannot replace source functions such as those shown in Fig. 4.1 by relation (4.14). However, this substitution can be performed if one describes the profile in the vicinity of the peak and even farther from the center of the line. This requires correct choice of value of i° (T_{min}) at the boundary with the wall. The correctness of this choice can be verified by calculating M and p from formulas (4.5) and (4.8) with the aid of curves such as those shown in Fig. 4.1, then comparing them with M and p calculated using formulas (4.15) and (4.16). The choice of approximate values of i° (T_{min}) at which parameters M and p are equal, calculated on the basis of both theories, verified the feasibility of such an approximation. This method of choosing parameters is based on the fact that two emitters with the same parameters M and p provide approximately the same characteristics of spectral lines near the peak [2].

Computations have led to the conclusion that true distributions of the source function in a turbulent flow can be replaced by a model (4.14), provided the following relation is fulfilled:

$$i^\circ\,(T_{min} \simeq i^\circ\,(\ell\text{-}0,\ 1\delta)\ .$$

This indicates that, when using model (4.14), the wall temperature must be replaced with the temperature at a distance from the wall equal to 0.1 δ (corresponding source functions are shown in Fig. 4.1 by dashed lines).

To apply model (4.14), it is necessary to connect the ratio δ/ℓ, determined using the usual formulas of gasdynamics [28] with ξ_b. Calculations performed for various turbulent flow conditions have shown that

$$\xi_b \simeq 0.5\,(1-1.6\delta/\ell)\ .$$

In all cases considered, the coefficient does not deviate more than 10 % from the value 1.6. Substituting the simplified source function given by formula (4.14) into expression (4.7), the contour of the spectral line determined from formula (4.6) is given by the following expression:

$$I_\nu\, I_\nu^0\,(T(\ell/2)) = Y\,(\tau_0)M = i^\circ\,(T_{min})(1-e^{-\tau_0}) +$$

$$+ (1-i^\circ (T_{min})) (1+e^{-\tau_o} - e^{(\xi_b-0.5)\tau_o}$$

$$- e^{-(\xi_b+0.5)\tau_o})/(0.5-\xi_b)\tau_o \ . \quad (4.17)$$

When $i^\circ (T_{min}) \to 0$, this expression becomes the far field source function for a trapezoid [4]. Representation in terms of elementary functions makes it possible to obtain easily the shape of the spectral line when parameters ξ_b and i° (T_{min}) (or δ/ℓ and T_w/T $(\ell/2)$) are known. In particular, using expression (4.17), one can find the density τ_{omax}, and intensity, $I_{\nu max}/I^\circ_\nu$ $(T (\ell/2))$, at the peaks of the self-reversed lines. Results of the calculations of $\tau_{omax} = f (i^\circ (T_{min}))$ at certain values of ξ_2 (Fig. 2) and i° (T_{min}) do not change considerably. The values of τ_{omax} obtained may be used in plasma diagnostics.

4.2.3. Determination of the Concentration of Atoms with the Aid of Self-Reversed Spectral Lines. If the dependence of the coefficient of absorption on frequency is known and the distance between the center and the peak of the self-reversed line, $\Delta\nu_{max}$, has been measured, the calculated value of τ_{omax} can be used to determine the mean concentration of atoms along the line of sight.

$$\langle n_a\rangle = \tau_{omax}/\ell\langle k_{1\,\Delta\nu max}\rangle \ , \quad (4.18)$$

where $\langle k_{1\,\Delta\nu max}\rangle$

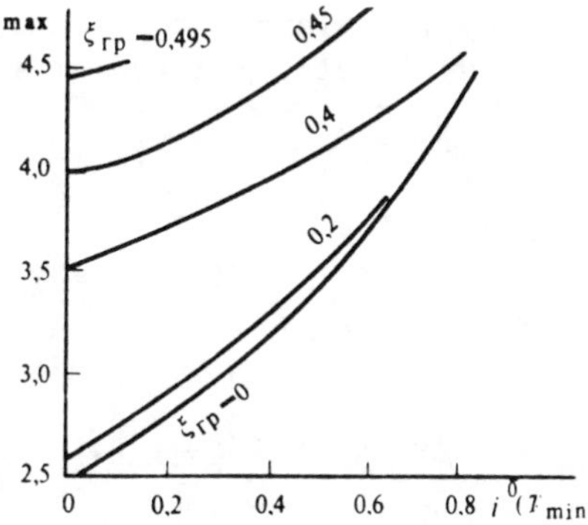

Fig. 4.2 Optical density at the maxima of the self-reversed contour of a line (calculated using formula (4.17) for various values of ξ_b

is the atomic (i.e., computed per single atom) coefficient of absorption along the distance measured from the center of the line, $\Delta\nu_{max}$. averaged along the line of sight. If $\Delta\nu_{max}$ is located within the region of collisional wing of the Voigt integral, the use of known expressions for the atomic coefficient of absorption [9] leads to the following expression [31]:

$$\langle n_a \rangle = \frac{\pi^{3/2}}{2^{5/2}} \frac{m_e c_0 \sqrt{k}}{e^2} \times \frac{\langle T \rangle^{1/2} \Delta\nu^2_{max}}{\sqrt{m_a^{-1} + m_{\pi.c}^{-1}}\, f\sigma P} \frac{\tau_{0\,max}}{\ell}, \tag{4.19}$$

where e, m_e are the charge and mass of the electron, respectively; c_o is the speed of light; k is the Boltzmann constant; m_a and $m_{\pi.c.}$ are the mass of the emitting atom and molecule of the combustion products; f is the oscillator strength, σ is the collisional cross section of atoms with neighboring molecules of the combustion products. P is the pressure (in the CGS system of units).

In order to determine $\langle n_a \rangle$ from relation (4.19), it is necessary to determine $\Delta\nu_{max}$ from the experimental shape of the line and to measure the mean temperature, $\langle T \rangle$. However, one can utilize temperature determined by the self-reversal method instead. In addition, it is necessary to know the pressure, P.

A large Δν, if only the dependence of the relative coefficients of absorption in the wings on the frequency is known, the expression for the optical density requires the knowledge of the frequencies at which it becomes the Voigt integral. Let us use an approximation for the coefficient of absorption along individual sectors in the form of relation (4.9). Let y = 2 (Voigt integral) when

$$\Delta\nu < \Delta\nu_1, \ y = y_1$$

when

$$\Delta\nu_1 < \Delta\nu < \Delta\nu_2, \ y = y_2$$

when

$$\Delta\nu_2 < \Delta\nu < \Delta\nu_3, \text{ etc.}$$

Then, when

$$\Delta\nu_2 < \Delta\nu < \Delta\nu_3$$

we have

$$k_{\Delta\nu} = k_{\Delta\nu 1}\left(\frac{\Delta\nu_2}{\Delta\nu_1}\right)^{-y_1}\left(\frac{\Delta\nu}{\Delta\nu_2}\right)^{-y_2}. \tag{4.20}$$

Here, $K_{\Delta\nu 1}$ depends on the Voigt integral and connects $k_{\Delta\nu}$ with the concentration of atoms. When

$$\Delta\nu_1 < \Delta\nu < \Delta\nu_2,\ y_2$$

in formula (4.20) must be set equal to y_1.

Analogous to the expression (4.19),

$$\langle n_a\rangle = \frac{\pi^{3/2}}{2^{5/2}}\ \frac{m_e c_0\sqrt{k}}{e^2}\ \frac{\langle T\rangle^{1/2}\Delta\nu_1^2}{\sqrt{m_a^{-1}+,m_{\pi.c}^{-1}}\ f\sigma P}$$

$$\times\ \frac{\tau_{\circ\max}}{\ell}\left(\frac{\Delta\nu_2}{\Delta\nu_1}\right)^{-y_1}\left(\frac{\Delta\nu_{\max}}{\Delta\nu_2}\right)^{-y_2}. \tag{4.21}$$

The same quantities $\Delta\nu_{max}$, P and $\langle T\rangle$ must be measured in the experiments. The values of $\Delta\nu_1$, $\Delta\nu_2$, y_1, y_2, are predetermined from the experimental dependence of the coefficients of absorption on frequency.

When the absolute value of the coefficient of absorption, $k_{1\Delta\nu}$, is known at specified values of P and $\langle T\rangle$, it is not necessary to tie-in the coefficient of absorption to the center of the line by means of the Voigt integral. In this case, after determining $\Delta\nu_{max}$ from the experimental contour, it is sufficient to determine the corresponding value of $k_{1\Delta\nu max}$ and to use formula (4.18). The optical density of the emission source with a uniform central sector can be written in the form

$$\tau_\circ = k_\nu\ (\ell/2)\ (\ell-2\delta) + 2\ {}_0\!\int^{\delta} k_\nu(x)\ dx$$

$$= k_\nu\ (\ell/2)\ (\ell-2\delta) + 2\ \delta\ \langle k\rangle_\delta\ . \tag{4.22}$$

If the relative temperature and concentration of emitting atoms in the boundary layer are known, expression (4.22) makes it possible to change from the mean concentration of atoms,

$$\langle n_1\rangle$$

(4.21), to the concentration in the central part of the emitter, n_a (ℓ/2). When the dependence of the coefficient absorption on n_a and T is given by expression (4.9), we have

$$n_a\ (\ell/2) \simeq \langle n_a \rangle\ (1-(1-\delta^{-1}Q(0,\delta))2\delta/\ell)^{-1}\ .$$

4.2.4. Measurement of the Temperature of a Spatially Nonuniform Gas. The most widely used methods of measuring the temperature of the equilibrium products of combustion are the visual and generalized self-reversal methods [32-34]. The visual reversal method is simple and is characterized by a low instrumental error (1-1.5 %). The generalized method is easier to automate.

Consider the difference between the temperature measured by the self-reversal methods and the temperature of the hottest central part of a spatially nonuniform gas [35]. Measuring temperature differs from measuring concentration of the atoms in that the shape of the emission line is not always resolved. This is associated with the fact that spectral lines used for temperature measurements are relatively narrow and the requirements imposed on the accuracy of measurements make it necessary to obtain relatively high emission intensities. When using results of investigations of the spectral line contours in temperature measurements, it is necessary to take into account the parameters and the tuning of the spectral instrument.

The generalized method is based on the determination of the ratio of emissivity of gas with its absorptivity in the range of some spectral line. The emissivity of a body is characterized by the emission flux passing through the spectral instrument into the exit slit. We shall write out the expression for the emission flux, omitting the factor that includes the transmission coefficient of optics in the spectral instrument, the solid angle through which emission enters the instrument and the dimensions of the inlet slit. This factor is the same for all fluxes compared in the reversal methods.

Let $A(\nu - \nu_1)$ be the function of the spectral instrument, satisfying the condition

$$\int_{-\infty}^{\infty} A(\nu-\nu_1)\ d\nu = 1\ .$$

The flux characterizing emissivity of the gas passing into the outlet slit of the instrument of spectral width S'_{out} near frequency ν^* can be written in the form

$$\Phi_{em}(\nu^*) = \int_{\nu^* - \frac{S'_{out}}{2}}^{\nu^* + \frac{S'_{out}}{2}} d\nu_1 \int_{-\infty}^{\infty} I_\nu A(\nu - \nu_1)\, d\nu \,. \tag{4.23}$$

The absorptivity is characterized by the ratio of the light flux absorbed in the object. The flux from the reference source at the outlet of the spectral instrument can be written in the following manner:

$$\Phi_s(\nu^*) = I^o_{\nu^*}(T_s)\, S'_{out} \,, \tag{4.24}$$

where T_s is the brightness temperature of the reference source calibrated so as to take into account transmission of the optics between the gas and the reference source. In formula (4.24), it is assumed that the intensity, $I^o_{\nu^*}(T_s)$, in the spectral range, S'_{out}, is practically constant.

The light absorbed in the gas is characterized by the following quantity:

$$\Phi_{ab}(\nu^*) = I^o_{\nu^*}(T_s) \int_{\nu^* - \frac{S'_{out}}{2}}^{\nu^* + \frac{S'_{out}}{2}} d\nu_1$$

$$\times \int_{-\infty}^{\infty} (1 - e^{-\tau_0})\, A(\nu - \nu_1)\, d\nu \,, \tag{4.25}$$

where $(1 - e^{-\tau_0})$ is the monochromatic absorptivity at frequency ν. The flux absorbed in the object can be represented in the form of a sum:

$$\Phi_{ab}(\nu^*) = \Phi_s(\nu^*) + \Phi_{em}(\nu^*) - \Phi_\Sigma(\nu^*) \,, \tag{4.26}$$

where Φ_Σ is the emission flux from the reference source transmitted through the gas.

From relations (4.23), (4.24), and (4.26), and Wien law we obtain

$$T(\ell/2) = T_s \left(1 - \frac{kT_s}{h\nu^*} \ln \frac{\phi_{em}(\nu^*)}{\beta\Phi_{ab}(\nu^*)} \right)^{-1} , \tag{4.27}$$

where

$$\mathcal{B} = \frac{\int_{\nu^* - \frac{S'_{out}}{2}}^{\nu^* + \frac{S'_{out}}{2}} d\nu_1 \int_{-\infty}^{\infty} I_\nu A(\nu - \nu_1)\, d\nu}{i^o_{\nu^*}(T(\frac{\ell}{2})) \int_{\nu^* - \frac{S'_{out}}{2}}^{\nu^* + \frac{S'_{out}}{2}} d\nu_1 \int_{-\infty}^{\infty} (1 - e^{-\tau_0}) A(\nu - \nu_1)\, d\nu} \, ; \tag{4.28}$$

h is the Planck constant.

Parameter $\mathcal{B}$ is the ratio of two light fluxes: flux emitted by a nonuniform real emitter and flux from a spatially uniform emitter at temperature T (ℓ/2) and the same absorptivity as in the object being investigated. In a homogeneous object, $\mathcal{B}$ = 1 and expression (4.27) assumes the usual form of a formula for the generalized reversal temperature.

When using the visual reversal method, the following intensities become equal:

$$\Phi_\Sigma(\nu^*) = \Phi_s(\nu^* + \Delta\nu) \; . \tag{4.29}$$

Here it was taken into account that Φ_s is usually observed not inside the line ($\nu = \nu^*$), but in its neighborhood. In this case, expression (4.27) reduces to the following:

$$T(\frac{\ell}{2}) = T_r\left(1 - \frac{kT_r}{h\nu^*} \ln \mathcal{B}\right), \tag{4.30}$$

where T_r is the reversal temperature at which equality (4.29) is fulfilled. When the gas is uniform, instead of (4.30), $T = T_r$ is obtained.

In the case of a nonuniform plasma, one can use formula (4.27) to calculate the difference between temperature T (ℓ/2) and the temperature averaged over the volume, neglecting the nonuniformity in this determination. When, as usual, T (ℓ/2) is close to T_s,

$$\frac{T(\ell/2) - \langle T \rangle}{T(\ell/2)} = -\frac{kT_s}{h\nu^*} \ln \mathcal{B} \; . \tag{4.31}$$

Formula (4.31) with $\langle T \rangle = T_r$ is also valid in the visual reversal method.

In the case of a spatially nonuniform gas, $\mathcal{B}$ and, consequently, $(T(\ell/2) - \langle T \rangle)/T(\ell/2)$ depends not only on the characteristics of the emitter, I_ν and $(1 - e^{-\tau_0})$, but also on the characteristics of calibration of the spectral instrument: the instrumental functions $A(\nu - \nu_1)$, width of the exit slit, S'_{out}, and frequency, ν^*, for which the instrument is calibrated. For a uniform gas, instrument calibration affects only the measurement accuracy.

The absorptivity of a spatially nonuniform gas in a spectral integral, S'_{out} near ν^*, is expressed in terms of experimentally determined fluxes, in a manner similar to that for a uniform gas:

$$\frac{1}{S'_{out}} \int_{\nu^* - \frac{S'_{out}}{2}}^{\nu^* + \frac{S'_{out}}{2}} d\nu_1 \int_{-\infty}^{\infty} (1 - e^{-\tau_0}) A(\nu - \nu_1) d\nu$$

$$= \frac{\Phi_{ab}(\nu^*)}{\Phi_s(\nu)} .$$

This expression differs somewhat from that obtained in Ref. 36, where the shape of the spectral line in the instrumental function is assumed to be rectangular. Consider a few special cases of determining $\mathcal{B}$. Let the width of the instrumental function be so small in comparison with the width of the line that the instrument does not introduce distortion into the record of the line. Then,

$$\mathcal{B} = \frac{\int_{\nu^* - \frac{S'_{out}}{2}}^{\nu^* + \frac{S'_{out}}{2}} I_{\nu 1} d\nu_1}{I^0_{\nu^*}(T(\frac{\ell}{2})) \int_{\nu^* - \frac{S'_{out}}{2}}^{\nu^* + \frac{S'_{out}}{2}} (1 - e^{-\tau_0}) d\nu_1} . \tag{4.32}$$

In this formula, $\mathcal{B}$ depends only on two characteristics of the spectral instrument: ν^* and S_{out}.

An analogous expression for correction coefficient for temperature in the case $(1 - e^{-\tau_o}) = 1$ appears in Ref. 37. If, in addition, the width of the exit slit is small in comparison with the width of the line, then

$$\mathscr{B} = I_{\nu^*}/I^o_{\nu^*}(T(\ell/2))\,(1-e^{-\tau_o(\nu^*)})\ . \tag{4.33}$$

In this case, the spectral instrument does not introduce distortions and the value of $\mathscr{B}$ depends only on the frequency, ν^*, inside the line at which the signal, Φ, is recorded. The farther away from the center of the line is the frequency ν^*, the closer is $\mathscr{B}$ to unity and the smaller is the correction provided by formula (4.31).

Another special case is that when the width of the instrumental function is narrow, i.e., equality (4.32) is fulfilled; however, frequency ν^* is such and the exit slit is so wide that practically all of the integrated line intensity passes into it. Then, instead of expression (4.32), we have

$$\mathscr{B} = \frac{\int_{-\infty}^{\infty} I_\nu d\nu}{I^o_{\nu^*}(T(\ell/2))\int_{-\infty}^{\infty}(1-e^{-\tau_o})\,d\nu}\ . \tag{4.34}$$

The quantity $\mathscr{B}$ is the ratio of integrated intensity of the line with the integrated intensity of a homogeneous emitter at temperature T $(\ell/2)$ and with the same absorptivity as in the object being investigated. In this case, there is no explicit dependence on the characteristics of the spectral apparatus.

In practice, the width of the function $A(\nu - \nu_1)$ (or spectral interval corresponding to the image of the inlet slit, S'_{in}) is close to the width of the line. The signals utilized are comparable to each other, and the influence of gas on the passage of light from the reference source is considerable, decreasing the overall measurement error. Let the inlet slit, S'_{in}, be sufficiently large. If in a limited spectral interval, $\Delta\nu$, the intensity of the spectral line differs from zero, the spectral width of the image of the line at the exit of the spectral apparatus depends on the quantity $\Delta\nu + S'_{in}$. If $S'_{in} > \Delta\nu + S'_{in}$, the absorption line and emission line completely pass through the inlet slit. Then, $\mathscr{B}$ is described by formula (4.34). When the width of the instrumental function is much larger than the width of the emission line and the width of the absorption line, $\mathscr{B}$ is also given by formula (4.34).

It should be emphasized that, when using the general formula (4.28) to calculate $\mathcal{B}$, one should have considerably more data on the emitted line in the spectral instrument system than in the special cases described by formulas (4.32), (4.34). Therefore, it makes sense to adjust the width of the inlet and outlet slits so that one of these formulas is valid. The relation (4.31) together with formulas for $\mathcal{B}$ has been verified experimentally [19, 20, 38, 39]. Investigations were performed, using a laboratory burner and on the U-02 facility. Visual reversal method, because it was more accurate, was used in both cases. The core flow temperature $T\ (\ell/2)$, was measured by injecting emitting substance only into the core. The temperature averaged along the line of sight, $\langle T \rangle$, was determined both in the case when only a narrow spectral sector inside the line was identified (most frequently the maximum of the line) and in the case when the spectral sector comparable to the width of the line was used. Various amounts of the emitting substance were introduced into all parts of the flow cross section.

The results of measurements are clearly correlated with the behavior of the spectral line contours. Varying the content of alkali metal changed the degree of nonuniformity and, as a result, intensity at the maximum of the line and the difference, $T\ (\ell/2) - \langle T \rangle$. When the degree of nonuniformity and the intensity in the maxima remained constant even when the amount of metal was varied, the difference $T\ (\ell/2) - \langle T \rangle$ also remained constant. Using the U-02 facility, it was shown experimentally (and in agreement with formulas (4.28), (4.32)-(4.34)) that, when the flow contains traces of sodium (n_{Na} on the order of 10^{11} cm^{-3}), reversal temperature of the maximum $\langle T \rangle$ is equal to the temperature of the core flow, $T\ (\ell/2)$. As the concentration of sodium atoms increases, $\langle T \rangle$ first decreases and, when sufficiently high concentration ($n_{Na}\ (\ell/2) > 10^{13}$ cm^{-3}) is achieved, becomes almost independent of it. The degree of nonuniformity in the channel remains practically constant. Under the usual operating conditions of an MHD generator, $n_{Na} > 10^{13}$ cm^{-3}. This makes it possible to calculate reliability the difference, $T\ (\ell/2) - \langle T \rangle$, from relation (4.31).

<u>4.2.5. Temperature Fluctuations in the Gas</u>. The fluctuations of emission intensity in spectral lines observed in equilibrium MHD generators may be caused by temperature fluctuations or fluctuations of the number of emitting atoms. These fluctuations must be taken into

account when performing measurements by means of the generalized reversal method. Spectral emission fluctuations may be used to obtain an idea of the nature of temperature pulsations.

The generalized reversal method formula (4.27) was derived on the assumption that all three signals, Φ_{em}, Φ_s, and Φ_Σ, are measured during a time interval when neither the gas temperature nor the degree of blackness can change. In reality, this condition is frequently not fulfilled. Let the characteristic time of change of the temperature or the degree of blackness, τ_{em}, be considerably longer than the time required for all particles in the gas to reach the same temperature. Depending on the relationship between τ_{em} and the time, which depends on the time lag of the measuring apparatus, the results of the calculation of temperature using formula (4.27) may turn out to be different.

In practice, there are two different versions of the generalized reversal method, differing by the number of emission detectors. When the number of detectors is equal to the number of emission fluxes required for calculation, all signals are measured simultaneously and continuously [10]. In measuring systems equipped with a single photodetector to record all emission fluxes, the signals are separated in time. When the two emission fluxes are separated in time, the duration of the impulse U_{K1}, characterizes the time during which the flux, Φ_{em}, shines on the exit; U_{k2} is the duration of the flux ϕ_Σ (see Fig. 4.3).

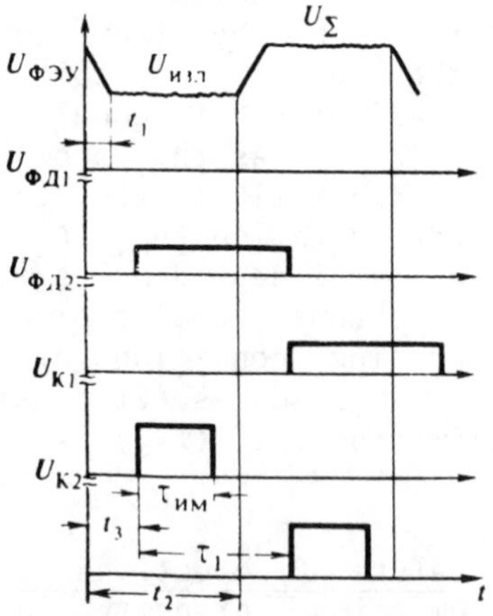

Fig. 4.3 Onset of signals at the load of the photo-multiplier and switching of the load to one of the output channels of the device

Let τ_{im} be the duration of each impulse; τ_1, the time between onsets of the two adjacent impulses; $u\tau_1$, time between identical signals, where u is two or three, depending on whether all three signals are fed into the system during the experiments or one of them has been measured previously. Designate the time during which the signals being recorded are averaged by τ_{reg} and time during which the calculated temperature is averaged by τ_{av}. When discrete signals are recorded during time τ_{reg}, the number of impulses entering the system $n = \tau_{reg}/u\tau_1$, and the total arrival time of signals $\tau_s = \tau_{im}n = \tau_{im}\tau_{reg}/u\tau_1$. If $\tau_{em} >> \tau_{reg}$, all time lag properties of the apparatus exert no influence on the results of measurements and formula (4.27) gives the true temperature. Consider some representative cases where it is necessary to take into account the nonstationary plasma behavior. If $\tau_{em} << \tau_{reg}$ and, during registration of discrete signals, $\tau_{em} << \tau_t$, time averaged values $\langle\Phi_{em}\rangle_t$ and $\langle\Phi_\Sigma\rangle_t$ are substituted into (4.27).

Mean values of the flux $\langle\Phi_{em}\rangle_t$ differ from the flux $\langle\Phi_{em}\rangle(T_o)$, which depends on the temperature averaged over time, T_o (e.g., see Refs. 32, 40, 41). This difference is reflected directly on the results of temperature measurements. Absolute and relative intensity techniques are known to have been used to investigate the influence of intensity fluctuations on the temperature measurements of axially symmetric emitters [42, 43].

We now attempt to determine the difference between the temperature measured by the reversal methods and the temperature averaged over time [44, 45]. In this case, in order to simplify the problem, it is assumed that the gas along the line of sight is homogeneous at each instant of time. Then, instead of formula (4.27), we have

$$T_{eff1} = T_S \left(1 - \frac{kT_S}{h\nu^*} \ln \frac{\langle\Phi_{em}(\nu^*)\rangle_t}{\langle\Phi_{ab}(\nu^*)\rangle_t}\right)^{-1} , \tag{4.35}$$

where

$$\langle\Phi_{ab}(\nu^*)\rangle_t = \langle\Phi_S(\nu^*)\rangle_t$$

$$+ \langle\Phi_{em}(\nu^*)\rangle_t - \langle\Phi_\Sigma(\nu^*)\rangle_t .$$

In order to determine the relation between the temperature T_{eff1} given by formula (4.35) and the gas temperature averaged over time, the expressions for the emission fluxes (4.23), (4.26), are represented in a more convenient form:

$$\Phi_{em}(t) = I^0_{\nu*}(T(t))F(t) \,, \tag{4.36}$$

$$\Phi_\Sigma(t) = \Phi_s + \Phi_{em}(t) - \Phi_s F(t) \,, \tag{4.37}$$

where

$$F(t) = \int_{\nu* - \frac{S'_{out}}{2}}^{\nu* + \frac{S'_{out}}{2}} d\nu_1 \int_{-\infty}^{\infty} (1 - e^{-\tau_0(t)}) A(\nu - \nu_1) d\nu \,.$$

The time variation of F (t) is caused primarily by fluctuations of the number of emitting atoms. Let $T(t) = T_o + \Delta T(t)$, $F(t) = F_o + \Delta F(t)$, where F_o and T_o are time independent. If $h\nu\Delta T/kT_o^2$ is not very large, then expanding the exponent in expression (4.36), where $I_\nu^\circ = 2\pi h\nu^3 c_o^{-2} \times \exp(-h\nu/kT)$, gives the following expression, accurate to quadratic terms:

$$\Phi_{em}(t) \simeq \frac{2\pi(\nu*)^3 h}{c_o^2} \exp\left(-\frac{h\nu*}{kT_o}\right) F_o\left(1 + \frac{\Delta F(t)}{F_o}\right)$$

$$\times \left(1 + \frac{h\nu*}{kT_o}\frac{\Delta T(t)}{T_o} + \frac{h\nu*}{kT_o}\left(0.5\frac{h\nu*}{kT_o} - 1\right) \times \left(\frac{\Delta T(t)}{T_o}\right)^2\right) . \tag{4.38}$$

When $h\nu*\Delta T(t)/kT_o^2 = 1$ and $h\nu*/kT_o = 8$, such a representation differs from the exact expression (4.36) by 6 %. Expression (4.38) is convenient to use in averaging with accuracy up to the second order terms.

$$\langle\Phi_{em}(\nu*)\rangle_t \simeq \frac{2\pi(\nu*)^3}{c_o^2} \exp\left(-\frac{h\nu*}{kT_o}\right)$$

$$\times F_o \left(1 + \frac{h\nu*}{kT_o}\left\langle\frac{\Delta T(t)}{T_o}\frac{\Delta F(t)}{F_o}\right\rangle_t\right.$$

$$\left. + \frac{h\nu*}{kT_o}\left(0.5\frac{h\nu*}{kT_o} - 1\right)\left\langle\left(\frac{\Delta T(t)}{T_o}\right)^2\right\rangle_t\right) ,$$

$$\langle\Phi_{ab}(\nu*)\rangle_t = \frac{2\pi h(\nu*)^3}{c_o^2} \exp\left(-\frac{h\nu*}{kT_s}\right) F_o \,. \tag{4.39}$$

In deriving these relations, it was assumed that $\langle \Delta T(t) \rangle_t = \langle \Delta F(t) \rangle_t = 0$. Substituting expression (4.39) into formula (4.35) and expanding the logarithm into a series, with an accuracy of up to quadratic terms in $\Delta T(t)$ and $\Delta F(t)$, obtains the following expression for the case when the characteristic fluctuation time is considerably shorter than the averaging time:

$$T_{eff1} = T_o \left(1 + \left\langle \frac{\Delta T(t)\Delta F(t)}{T_o F_o} \right\rangle_t + \left(0.5\,\frac{h\nu^*}{kT_o} - 1\right) \left\langle \left(\frac{\Delta T(t)}{T_o}\right)^2 \right\rangle_t \right) . \quad (4.40)$$

If no correlation exists between changes in temperature and absorptivity and line width, or if $\Delta F(t) = 0$, one term in relation (4.40) drops out. Furthermore, in the special case of harmonic oscillations of temperature,

$$\Delta T(t) = \Delta T \cos\omega t \ ,$$

$$\langle (\Delta T(t))^2 \rangle_t = 0.5\ (\Delta T)^2 \ .$$

When $\Delta T/T_o = 0.125$ and $h\nu^*/kT_o = 8$, then $T_{eff1} = 1.02\ T_o$. Such a difference between the temperature being measured and the average temperature must be taken into account during measurements in an MHD generator. Another case is also of practical interest. It deals with measurements using discrete signals when τ_{em} is comparable to τ_1, $\tau_{im} < \tau_{em}$, and the temperature is calculated from single impulses. Signals $\Phi_{em}(t)$ and $\Phi_\Sigma(t)$ are emitted by gas at different temperatures; however, the temperature does not change during the time required to record one signal. In this case, if signal Φ_Σ follows signal Φ_{em}, the expression under the logarithm sign in formula (4.27) must be replaced with

$$\Phi_{em}(t)/(\Phi_{em}(t) + \Phi_s - \Phi_\Sigma(t+\tau_1)) \ ,$$

if the signed Φ_Σ follows the signal Φ_{em}. Using expressions (4.36), (4.37) and the approximate representation (4.38), one can obtain the temperature with an accuracy of up to quadratic terms:

$$T_{eff2} = T_o \left(\left(1+ \frac{\Delta T(t)}{T_o}\right)\left(1-\exp\left(\frac{h\nu^*}{kT_s}\left(1-\frac{T_s}{T_o}\right)\right)\frac{\Delta T(t)-\Delta T(t+\tau_1)}{T_o}\right)\right.$$

$$- \exp\left(\frac{h\nu^*}{kT_s}\left(1-\frac{T_s}{T_o}\right)\right)\left(0.5\,\frac{h\nu^*}{kT_o} - 1\right)\frac{(\Delta T(t))^2-(\Delta T(t+\tau_1))^2}{T_o^2}$$

$$\left. + \exp\left(2\,\frac{h\nu^*}{kT_s}\left(1-\frac{T_s}{T_o}\right)\right)\left(0.5\,\frac{h\nu^*}{kT_o} + 1\right)\frac{(\Delta T(t)-\Delta T(t+\tau_1))^2}{T_o^2}\right) .$$

(4.41)

When $\tau_1 \to 0$, according to (4.41) T_{eff2} is the true temperature of the gas. If the temperature of the gas is a harmonically oscillating function,

$$\Delta T(t+\tau_1) = \Delta T\cos\omega(t+\tau_1) ,$$

when $\omega = 2\pi/\tau_{em}$, relation (4.41) can be replaced with the following expression

$$T_{eff2} = T_o\ ((1+\Delta T\cos\omega t/T_o)\ (1$$

$$- \exp\ ((1-T_s/T_o)\ h\nu^*/kT_s)\ 2\Delta T\ \sin\ \omega\ (t+\tau_1/2)$$

$$\times \sin\ (\omega\tau_1/2)/T_o) + (\Delta T/T_o)^2\ ((1-0.5\ h\nu^*/kT_o)$$

$$\times \exp\ ((1-T_s/T_o)\ h\nu^*/kT_s)\ \sin\ (2\omega(t+\tau_1/2))$$

$$\times \sin\ \omega\tau_1 + 4\ (1+0.5\ h\nu^*/kT_o)\ \exp\ (2(1-T_sT_o)$$

$$\times h\nu^*/kT_s)\ \sin^2\ (\omega(t+\tau_1/2))\ \sin^2\ (\omega\tau_1/2)))\ . \qquad (4.42)$$

The temperature calculated using (4.41) or (4.42) may be averaged in the instrument, where $\tau_{av} \gg \tau_{em}$.

In the case of harmonic oscillation of the gas temperature, the result of averaging can be represented in the form

$$\langle T_{eff2}\rangle_t = T_o\ (1-(\Delta T/T_o)^2\ \exp\ ((1-T_s/T_o)$$

$$\times\ h\nu^*/kT_s)\ (\sin^2(\omega\tau_1/2)\ -\ 2\ (1+0.5\ h\nu^*/kT_o)$$

$$\times\ \exp\ (2(1-T_sT_o\ h\nu^*/kT_s)\ \sin^2\ (\omega\tau_1/2)))\ .$$

When $\tau_1/\tau_{em} = m$, where m is a whole number, $\langle T_{eff2}\rangle_t = T_o$, while when $\tau_1/\tau_{em} = (2m + 1)/2$, the difference between $\langle T_{eff2}\rangle_t$ and T_o is a maximum. In the case considered above ($h\nu^*/kT_o = 8$, $\Delta T/T_o = 0.125$, and $T_s = T_o$, $T_{eff2} = 1.14$, i.e., the temperature is considerably too high. If averaging is performed over a not very large number of temperature values, it is expedient to use summation formulas for trigonometric functions.

Thus, sufficiently simple relations make it possible to estimate the difference between the time-averaged temperature and the temperature measured. In addition to the temporal characteristics of the measuring apparatus, correct choice of formula for the effective temperature requires some understanding of the characteristic time, τ_{em}, and amplitude pulsations.

Now consider some aspects of determination of the nature of temperature fluctuation from emission intensity fluctuations. An expression for the pulsating intensity of the gas emission will be written in the following form:

$$\Phi_{em}(t) = \langle\Phi_{em}(t)\rangle_t + \Delta\Phi_{em}(t)\ ,$$

where $\langle\Phi_{em}(t)\rangle_t$ is a constant and $\Delta\Phi_{em}(t)$ is the variable component of the signal. Using the approximate expressions (4.38) and (4.39), one can easily obtain the temporal variation of emission intensity with an accuracy of up to second order terms in $\Delta F(t)$ and $\Delta T(t)$:

$$\frac{\Delta\Phi_{em}(t)}{\langle\Phi_{em}(t)\rangle_t} = \frac{\Delta F(t)}{F_o} + \frac{h\nu^*}{kT_o}\ \frac{\Delta T(t)}{T_o}$$

$$+ \frac{h\nu^*}{kT_o}\left(\frac{\Delta T(t)}{T_o}\ \frac{\Delta F(t)}{F_o} - \left\langle\frac{\Delta T(t)}{T_o}\ \frac{\Delta F(t)}{F_o}\right\rangle_t\right)$$

$$+ \frac{h\nu^*}{kT_o}\left(0.5\ \frac{h\nu}{kT_o} - 1\right)\left(\left(\frac{\Delta T(t)}{T_o}\right)^2 - \left\langle\left(\frac{\Delta T(t)}{T_o}\right)^2\right\rangle_t\right)\ .$$

The variation of intensity depends on the fluctuations of the degree of blackness and temperature. In interesting cases, fluctuations of the degree of blackness may be insignificant. Then, the expression for the temporal variation of emission intensity is simplified:

$$\frac{\Delta\Phi_{em}(t)}{\langle\Phi_{em}(t)\rangle_t} = \frac{h\nu^*}{kT_o}\,\frac{\Delta T(t)}{T_o}$$

$$+\frac{h\nu^*}{kT_o}\left(0.5\,\frac{h\nu^*}{kT_o} - 1\right)\left(\left(\frac{\Delta T(t)}{T_o}\right)^2\right.$$

$$\left. - \left\langle\left(\frac{\Delta T(t)}{T_o}\right)^2\right\rangle_t\right) . \qquad (4.43)$$

In order to determine temperature fluctuations from the experimentally observed emission fluctuations, $\Delta\Phi_{em}(t)/\langle\Phi_{em}(t)\rangle_t$, relation (4.43) will be written in the form:

$$\frac{\Delta T(t)}{T_o} = \frac{\Delta\phi_{em}(t)}{\langle\phi_{em}(t)\rangle_t}\,\frac{kT_o}{h\nu^*} + \left(1-0.5\,\frac{h\nu^*}{kT_o}\right)$$

$$\times\left(\left(\frac{\Delta T(t)}{T_o}\right)^2 - \left\langle\left(\frac{\Delta T(t)}{T_o}\right)^2\right\rangle_t\right) .$$

When the temperature changes are small, this equation may be solved by successive approximations. In the first approximation,

$$\frac{\Delta_1 T}{T_o} + \frac{\Delta\phi_{em}(t)}{\langle\phi_{em}(t)\rangle_t}\,\frac{kT_o}{h\nu^*} . \qquad (4.44)$$

In this case, emission fluctuations completely mimic temperature oscillations; however, relative temperature fluctuations are $h\nu^*/kT_o$ times smaller than the relative emission fluctuations. In addition, it follows from formula (4.44) that frequency spectra of temperature oscillations and plasma emission are the same.

In the second approximation, we obtain

$$\frac{\Delta_2 T}{T_o} = \frac{\Delta\phi_{em}(t)}{\langle\phi_{em}(t)\rangle_t}\,\frac{kT_o}{h\nu^*}$$

$$+ \left(1-0.5\,\frac{h\nu^*}{kT_o}\right)\left(\left(\frac{\Delta\Phi_{em}(t)}{\langle\Phi_{em}(t)\rangle_t}\right)^2\right.$$

$$\left. - \left\langle\left(\frac{\Delta\phi_{em}(t)}{\langle\phi_{em}(t)\rangle_t}\right)^2\right\rangle_t\right)\left(\frac{kT_o}{h\nu^*}\right)^2 . \qquad (4.45)$$

When $|\Delta\Phi(t)/\langle\Phi_{em}(t)\rangle_t| \leq 0.2$ and $kT_o/h\nu^* = 0.125$, the second approximation correction does not exceed 4 %. When $|\Delta\Phi(t)/\langle\Phi_{em}(t)\rangle_t| = 1$, second order approximation correction reaches 20 %.

For second approximations and higher, the frequency spectrum of temperature fluctuations does not mimic the frequency spectrum of emission fluctuations. However, the frequency spectrum of temperature fluctuations can be obtained by use of Fourier series expansion and expressions such as (4.45).

Here, a simplified model has been used to investigate the influence of the nonuniformity of the emitter on the spectral line profile and the temperature of gas being measured by reversal methods. The results of the investigations make it possible to:

- calculate the difference between the temperature measured by means of the reversal methods and the temperature of the central part of a uniform emitter,
- take into account the influence of the spectral instrument on the gas temperature measurements, and
- determine the atom concentration from the distance between the center and the maximum of the self-reversed profile of the spectral line.

In addition, as a result of the analysis of the influence of fluctuations in plasma on emission, we have been able to obtain relations between the temperature measured by means of the generalized reversal method and plasma temperature averaged over time, and to investigate temperature fluctuations by means of intensity fluctuations in lines.

4.3. Spectroscopic Gas Temperature and Atom Concentration Measurements in Equilibrium MHD Generators

Measurements of the principal parameters of the working fluid of an MHD generator must satisfy the following conditions. The results of measurements must be available during the experiments and the measuring systems must be automatic and remote-controlled. Existing systems used for automatic gas temperature measurements by means of reversal methods either require processing of primary data after the experiments are completed [32, 34, 46-52], or are applicable only within narrow temperature range [53, 54]. In designing devices for temperature and atom concentration measurements, it is necessary to take into account the maximum acceptable measurement errors. It is different in these two cases, because the electron concentration given by Saha equation depends very strongly on the temperature of the gas and weakly on the concentration of alkali atoms. Conductivity and, consequently, power output of an MHD generator depend strongly on concentration of electrons. For a maximum acceptable error of measuring electron concentration of 10 to 15 %, the temperature must be known to within 2-3 % and the concentration of alkali metal atoms must be determined with an error of several dozen percent.

4.3.1. Devices for Measuring Temperature by the Generalized Self-Reversal Method.

Two models of an automatic temperature measuring system based on the generalized reversal method have been developed [55, 56]. It can be seen from formula (4.27) that three beams (Φ_{em}, Φ_s, Φ_Σ) must be measured to determine the temperature. A single photomultiplier in each system is used as a detector of all light beams. In the first, a two-channel model of the automatic device, the flux Φ_s, is determined beforehand and is input into the computer with the remaining quantities (k, h, ν^*, $\mathcal{B}$, T_s) required to perform calculations using formula (4.27). This model of the system is deficient in that, when the sensitivity of the photomultiplier is changed during the experiment, the value of Φ_s input into the computer may not correspond to the actual emission of the reference source. The sensitivity of a photomultiplier can be affected by a change in the magnetic field. This requires correction of measurements during the experiments. All three light beams are measured in the second, three-channel model of the system, in which a change of sensitivity of a photomultiplier for any reason is automatically taken into account. Such a device is more complex.

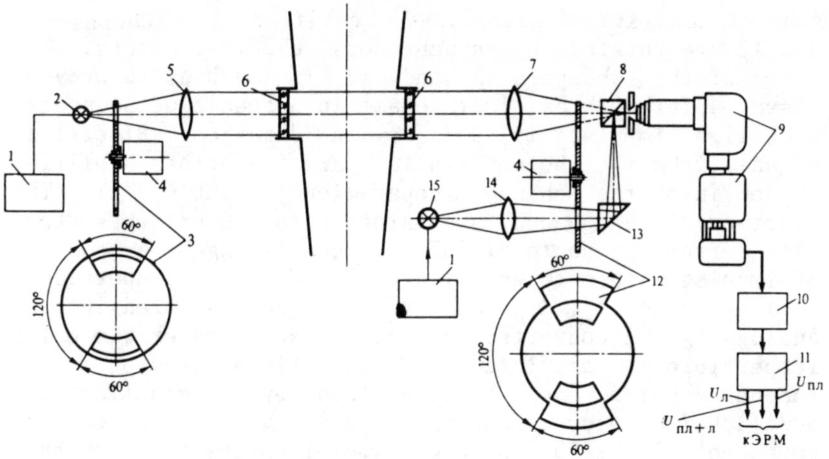

Fig. 4.4 Schematic block diagram of the three-channel system, 1--power supply, 2--main reference light source, 3,12--choppers, 4--synchronous motor, 5,7,14--lens, 6--protective window, 8--semitrans-parent beam splitter, 9--spectrograph with a photoelectric attachment, 10--amplifier, 11--channel separation assembly, 13--rotating prism, 15--additional reference light source

Figure 4.4 is a schematic block diagram of the three-channel system. Emission from the MHD generator passes through the optical tubes with protective windows 6 and lens 7 and is incident on the inlet slit of the spectrograph, 9. The spectrograph separates a bright emission line, usually D-line of sodium, directing a light beam, Φ_{em}, onto the photomultiplier of the attachment to the spectrograph. Emission from the reference source, 2, consisting of a tungsten ribbon lamp, is collimated by lens 5 across protective windows, 6, into the MHD generator and onto the spectrograph. Thus, a beam, Φ_{Σ}, reaches its outlet. During measurements, it is quite complicated to obtain a light beam, Φ_s, from the main reference source, 2, located behind the MHD generator channel. Therefore, the system is equipped with an additional reference source, 15, the light from which is collimated by lens 14, passes rotatable prism 13 and a prism cube with a semitransparent diagonal, 8, and is incident on the inlet slit of the spectrograph, 9. In the absence of plasma and a magnetic field, light beams from the main and additional reference sources are equalized.

A change of the sensitivity of the photomultiplier affects all three signals to almost the same degree. The photomultiplier is placed into a magnetic shield. However, this is insufficient to weaken the influence of the magnetic field. The power supply of the reference sources consists of stabilized rectifiers 1. Choppers 3 and 12 are rotated by synchronous electric motors. The shape of the choppers 3 and 12 is such as to provide passage of the three light beams in a specified sequence. Each light impulse is 3.3 ms in duration. Electrical signals from the photomultiplier are fed across amplifier 10 and into the channel separation assembly, 11. The function of the channel separator is to convert the output signal from the photomultiplier, consisting of a sequence of impulses, into three voltages U_s, U_{em}, U_Σ, proportional to light beams Φ_s, Φ_{em}, and Φ_Σ. Signals are fed into an analog-digital converter and into a computer where the temperature is calculated. A schematic diagram of a two-channel system can be obtained by eliminating the additional reference source, second chopper, and optical components 8, 12-15 in Fig. 4.4 from the three-channel system.

One of the principal assemblies of the temperature recording system is the channel separation assembly, 11. An electrical schematic diagram of the two-channel model is shown in Fig. 4.5 and the onsets of the principal signals in Fig. 4.3.

The signals from the photomultiplier load are fed into the transistor switches, one of which (VT7, or VT8) is opened only by the signal U_Σ, and another (VT9, or VT10)

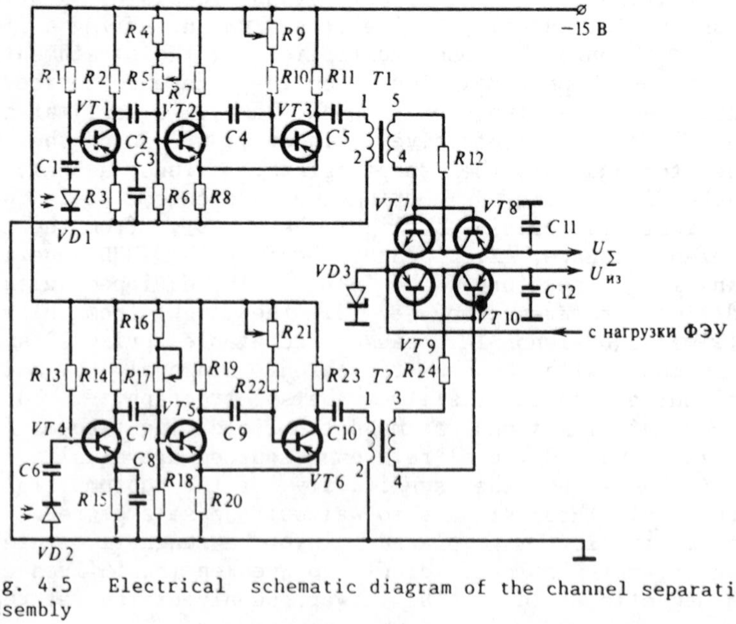

Fig. 4.5 Electrical schematic diagram of the channel separation assembly

only by the signal, U_{em}. Proper opening and closing of the switches is provided by photodiodes VD1 and VD2. The photodiodes are located directly behind chopper 3 (see Fig. 4.4) and are connected so that one of them generates a signal, U_{PhD2}, when maximum light from reference source is transmitted by the chopper, and the other generates signal, U_{PhD1}, in the absence of light from the reference source (see Fig. 4.3). The signals from the photodiodes are fed into transistorized amplifiers VT1 and VT4, then to the delay multivibrators VT2, VT3, and VT5, VT6. Impulses U_{K1} and U_{K2} alternately open each of switches, VT7, VT8, and VT9, VT10. Resistors R9 and R21 control duration of passage.

From the output of the switches, the signals are fed into capacitors C11 and C12, which are charged to voltages proportional to light beams Φ_{em} and Φ_{Σ}. The following conditions must be fulfilled for the voltages at the output of the channel separation assembly, U_{em}, U_s, U_{Σ}, to be proportional to light beams Φ_{em}, Φ_s, and Φ_{Σ}: the light response characteristic of the photomultiplier and the response of amplifier 10 (see Fig. 4.4) must be linear, and the voltage transfer coefficient in channel separation switches must be independent of the signal.

Tests have shown that the light response characteristic of the FEU-17A photomultiplier used in the system is linear with an error not exceeding 3 % at anode load voltage up to 25 V [55]. Usually, the entire operation range of the photomultiplier is not used. Therefore, deviation from linearity is considerably lower (it can be assumed to be 1 to 2 %). In our experiments, deviation from linearity of the amplifier response did not exceed 0.5 % and the voltage transfer error did not exceed 0.1 %. Finally, one must properly choose the time constants of the photomultiplier load circuit from which the signal is fed to the switches VT7 and VT8, and VT9 and VT10, and of the charge and discharge circuits of capacitors C11, C12 (see Fig. 4.5), the voltage from which is fed into the computer.

The time constant of the load circuit of the photomultiplier will now be determined. An equivalent schematic diagram for switching on the photomultiplier can be represented as consisting of a current source connected in a series with a load resistor, R, shunted by capacitor C. Assume that the plasma parameters do not change during one light signal, the duration of which is 5 ms in a two-channel system. Let the light flux be intersected by the edge of the opening of the chopper during time t_1 and completely transmitted through the opening during time t_2-t_1 (see Fig. 4.3). In estimating, one can assume that the light beam intersected by the edge of the opening varies linearly with time.

In this case, we can write:

when $0 \le t \le t_1$,

$$RC \frac{dU_C}{dt} + U_C = U_o + \frac{U_{max} - U_0}{t_1} t , \qquad (4.46)$$

when $t_1 \le t \le t_2$

$$RC \frac{dU_C}{dt} + U_C = U_{max} , \qquad (4.47)$$

where U_c is the capacitor voltage, U_{max} is the stationary value of the signal corresponding to the maximum light beam divergence, and U_0 is the stationary value of the signal in the preceding impulse.

From equations (4.46) and (4.47):

when $0 \le t \le t_1$

$$U_C = U_o + \frac{U_{max} - U_o}{t_1} RC \left(\frac{t}{RC} - 1 + e^{-\frac{t}{RC}} \right) , \qquad (4.48)$$

when $t_1 \le t \le t_2$,

$$U_C = \frac{U_{max} - U_o}{t_1} RC \left(1 - e^{\frac{t_1}{RC}} \right) e^{-\frac{t}{RC}} + U_{max} .$$

The voltage, U_c, during the time interval between t_3 and $t_3 + \tau_{im}$ must be fed into the amplifier. Assume that it is necessary to fulfill the condition $U_c(t_3 + \tau_{im}) \ge 0.995\, U_{max}$. In addition, assume that $U_o = 0.5\, U_{max}$. Determining t_1 from the design data and assigning specific values to t_3 and τ_{im}, one can use (4.48) to determine the required value of RC. The capacitance, C, which depends on the capacitance of the mounting, is 70 pF and, therefore, R = 1 MΩ.

Consider capacitor charge and discharge circuits C11, C12. Because the signal from the capacitor is fed into the computer continuously, rather than as impulses, both circuits must be considered. The capacitors are charged by an impulse of duration τ_{im}. They must be charged as rapidly as possible and discharged as rapidly as possible between impulses. They are charged across direct impedance of switch R_d connected in series with it and discharged across reverse impedance of switch R_r, and the impedance of the analog-to-digital converter at the input into the computer, R_H. The impedance of the amplifier that serves as a signal source is insignificant. Then, the charge time constant $\tau_{ch} = R_d C_{11}$ and the discharge time constant $\tau_{dis} = C_{11} R_r R_H / (R_r + R_H)$.

In the subject experiments, $R_H \simeq 1.0$ MΩ, $R_r \simeq 0.5$ MΩ. If the capacitance of the condensers, $C_{11} = C_{12} = 10.0$ µF, then discharge time $\tau_{dis} = 3.33$ s. Let the duration of signal $\tau_{im} = 2$ ms. Then the time between signals is 8 ms. At the end of this time, the capacitor is discharged 0.34 % of its initial voltage.

The direct impedance of transistors with maximum emitter-collector voltage is 40-50 Ω. Therefore, τ_{ch} is 0.5 ms. In this case, at the end of the signal, $U_{max} - U_C = 0.02\,(U_{max} - U_o)$. Usually, the properties of plasma do not change significantly during the time between impulses. In obtaining an estimate assume that the intensity of the light beam during this time changed 10 %. Then, $U_{max} - U_C = 2 \times 10^{-3}\, U_{max}$.

The total error resulting from conversion of each light signal into electric voltage to the computer does not exceed 3 %. The maximum error of determining the temperature corresponding to such a voltage error of the three signals is approximately 0.9 % (at a gas temperature of 2000-3000 K). Half of this error is the result of nonlinearity of the light response of the photomultiplier. Because identical systematic deviation of all three signals does not result in a temperature error (see formula (4.27), apparently the temperature measurement error is smaller.

The foregoing signal conversion errors can be attributed to the influence of constant parameters of the system, i.e., systematic errors. It should be noted that the errors associated with charge or discharge time constants of capacitors C11, C12 can change sign, depending on whether alternating signals fed to C11 and C12 capacitors increase or decrease. Therefore, systematic error of measuring the temperature does not exceed 1 %.

Now, replace Φ in formula (4.27) with proper voltages. In addition, assume for simplicity that $\mathcal{B} = 1$ and T $(\ell/2)$ = T. Then the expression for a random error of temperature measurement by means of the generalized reversal technique assumes the form:

$$\left(\frac{\Delta T}{T}\right)^2 = \left(\frac{k}{h\nu}\right)^2 T^2 \left(\frac{\Delta a}{a}\right)^2 + \Delta T_S^2 \frac{T^2}{T_s^4}, \qquad (4.49)$$

where

$$a = U_{em}/(U_{em}+U_S-U_\Sigma) \; ; \qquad (4.50)$$

$$(\Delta a/a)^2 = ((U_S+U_\Sigma)/U_{em})^2 \; \Delta U_{em}^2$$

$$+ \Delta U_S^2 + \Delta U_\Sigma^2)/(U_{em} + U_S - U_\Sigma)^2 . \qquad (4.51)$$

Random signal noises, ΔU, are caused by noise in the photomultiplier load circuit and by amplifier instability. The principal source of noise in the photomultiplier load circuit is the shot effect in the photomultiplier [57]. The signal-to-noise ratio in the case of the shot effect increases with increasing current, load impedance, and capacitance in the photomultiplier load circuit. On the other hand, the load impedance and capacitance must be limited because of their influence on the systematic error, and the photomultiplier current must be limited because of the fatigue effect.

In addition, the signal-noise ratio depends on the voltage of the photomultiplier power supply. The optimum voltage was determined by shining a constant light flux on the photomultiplier of the system, varying the photomultiplier power supply voltage, and measuring signal at the load and noise by means of a dc voltmeter and an ac voltmeter with a broad passband.

The optimum operating voltage of 800 V and the load parameters R = 1 MΩ, C = 70 pF, at a useful load voltage on the order of a few volts results in the ratio $\Delta a/a$ due to photomultiplier noise not exceeding 10 %. Because in $kT/h\nu \simeq 0.1$ in formula (4.49), random errors in determining voltage due to photomultiplier noise result in temperature errors of nearly 1 %.

The random error can be determined more accurately

after choosing a relation between signals U_s and U_{em}. Let us find an optimum relation between these signals for a case that is simple, but, of practical importance. Let the spectral line used to perform measurements be absolutely black in the frequency range of the spectral device. Then, $U_{em} = U_\Sigma$, $\Delta U^2_{em} = \Delta U^2_\Sigma$. Noticing that, in the case of the shot effect, $\Delta U^2 = cU$, where c is a constant, we have the following expression instead of (4.51):

$$\left(\frac{\Delta a}{a}\right)^2 = \frac{cU_{em}}{U_s^2}\left(\left(\frac{U_s}{U_{em}} - 1\right)^2 + \frac{U_s}{U_{em}} + 1\right) .$$

Let us determine the extremum $(\Delta a/a)^2$ when T_s varies:

$$\frac{\partial}{\partial U_s}\left(\frac{\Delta a}{a}\right)^2 = \frac{cU_{em}}{U_s^3}\left(\frac{U_s}{U_{em}} - 4\right) = 0 .$$

The following condition follows from the above equality:

$$U_s/U_{em} = \Phi_s/\Phi_{em} = 4 .$$

The second derivative is positive:

$$\frac{\partial^2}{\partial U_s^2}\left(\frac{\Delta a}{a}\right)^2 = \frac{2c}{U_s^3}\left(-1 + 6\frac{U_{em}}{U_s}\right) > 0 ,$$

i.e., the error resulting from the shot effect is minimal when the brightness of the reference source is four times larger than the brightness of the plasma.

A decrease of the mean square error with increasing reference source temperature was experimentally observed on a burner. In experiments, one usually must work at a nonoptimal signal ratio ($\Phi_s < \Phi_{em}$). However, because the dependence $(\Delta a/a)^2$ on U_s is sufficiently smooth, this does not lead to a considerably higher error.

The noise in the amplifiers and other components of the circuit of the temperature recording system is almost insignificant compared with the photomultiplier noise. This is verified directly by observing noise on the screen of an oscillograph connected to, versus disconnected from, the photomultiplier. For a 10-15-V signal, the amplifier drift does not exceed 12 mV, i.e., is 0.1 %. This error is significantly lower than that of the photomultiplier.

Error in determining the reference source temperature depends on the calibration error of the source and the accuracy of controlling and measuring the current in the lamp. The reference source is usually calibrated for brightness temperature to accuracy of 0.4 %. The error of determining current consists of an error of the precision resistor (0.04 %) and an error of measuring the voltage drop across this resistor with a digital voltmeter (0.01 %). At a current stabilization accuracy of 0.05 %, the total current measurement error is 0.1 %. Error in temperature measurement determined by means of a calibrated curve, is 0.1 %. From this, the total error $|\Delta T_s|/T_s \simeq 0.5$ %. Hence, the total square random error depends on the shot effect of the photomultiplier and the error of the temperature reference source, and is on the order of 1.2 %.

Two- and three-channel temperature measuring systems were tested on a laboratory burner and on the U-02 facility. The results were compared with the results of measurements performed visually by means of the reversal method. These experiments have verified the usefulness of the systems. Special experiments to determine the influence of the fringe fields of the magnetic system on the operation of the three-channel system were performed on the U-25 facility. The plasma temperature being recorded remains practically constant when the magnetic field is switched on, even though the photomultiplier signals vary considerably (Fig. 4.6). When the nominal value of the current of the magnetic system was reached, observed fluctuations did not exceed 2.5 %. However, this had a noticeable effect on the U_{em}, U_s, and U_Σ signals, i.e., the sensitivity of the photomultiplier changes. At the same time, fluctuations of the temperature being recorded are insignificant (50 K), are not connected with variation of signals at the photomultiplier output, and apparently depend on the instability of the thermal mode of the loop of the device.

The automatic temperature measuring system has been in constant operation on the U-25 facility from the first runs, initiated in 1971. The temperature measurements are performed in MHD generator channels and on a dummy channel (gas duct) during technical tests of the facility. The optical part of the system and, also, the photoelectric attachment to the spectrograph and the amplifier are located in the immediate vicinity of the magnetic system, on both sides of the channel. The distance between the main reference source and the inlet slit of the spectrograph is approximately 12 m. Therefore,

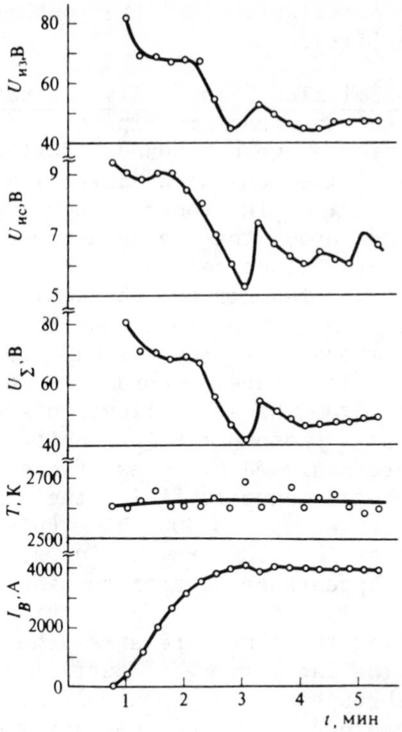

Fig. 4.6 Temporal variation of signals at the output of the three-channel system (U_{em}, U_s, U_Σ) and of the temperature and current in the magnetic system of the U-25 facility (T and I_v, respectively)

considerable attention must be given to the optical circuit. Most of the electronics of the system and power supply assemblies are located at a distance of 80 m from the channel. The choppers, reference source currents, and photomultiplier power source voltage are controlled remotely. Proper operation of individual assemblies of the systems is ascertained from oscillograph records of signals at the output of the amplifier, 10, and the channel separation assembly, 11, (Fig. 4.4). The relative difference between temperature measured on the U-25 facility and the core flow temperature are estimated from the foregoing relations (see 4.2.4.) and are 1-2 % in typical operating modes.

In accordance with the U.S.-U.S.S.R. cooperative program on MHD energy conversion, Soviet plasma diagnostics equipment, including a three-channel temperature measuring system, were supplied to the Avco-Everett firm. This system was successfully used in joint

U.S.-U.S.S.R. investigations of the combustion products on the MARK-VI facility.

4.3.2. Emission Intensity Fluctuations. The spectrograph and the photomultiplier of the temperature measuring system were used during experimental investigations of the emission intensity fluctuations on the U-25 facility. In those cases, when it was not necessary to measure, the temperature reference sources and choppers were disconnected. The signal from the photomultiplier was fed across the amplifier and into the cable of the system, then to a tape recorder registering the constant component of the signal U_{em} by a recording potentiometer. The tape recorder recorded frequencies between 50 and 16,000 Hz at nonlinear distortion of 2-3 %. The records of photomultiplier noise during control experiments were analyzed by means of a spectrum analyzer. In the absence of a magnetic field, the highest pulsations appear in the range of 50-300 Hz. Pulsations are small above frequencies of 500 Hz. Pulsations increase and their rate of appearance increases in an operating MHD generator.

In conducting these measurements the amount of sodium injected into the facility was sufficient for the degree of blackness throughout the spectral interval to be equal to one. Therefore, it can be assumed that emission pulsations are the result of temperature pulsations that can be attributed to phenomena such as nonuniform fuel, oxidizer and seed injection, vibration combustion mode, and turbulent properties of flow in the channel. Relative pulsation intensity, $\Delta\Phi_{em}/\langle\Phi_{em}(t)\rangle_t$, usually does not exceed 15-20 %. Formula (4.44) can be used to determine the temperature variation from the emission intensity changes. This means that the fluctuation intensity spectrum adequately describes the spectrum of temperature fluctuations and that $\Delta T/T_o$ = 1.3-2 %. In cases when the system was used to measure the temperature, emission pulsations were observed on an oscillograph connected to the amplifier at the output of the photomultiplier.

Oscillograph records acquired during various runs of the U-25 facility display signals U_Σ, U_{em}, U_s (Fig. 4.7). The signal from the reference source, U_s, is smaller and is characterized by pulsations that are small in comparison with other signals. Because the degree of blackness of gas in these experiments in the spectral range under consideration is close to unity, signals U_{em}, U_Σ, are practically identical. Sudden deviations (downward before the start of the signal U_s and upward

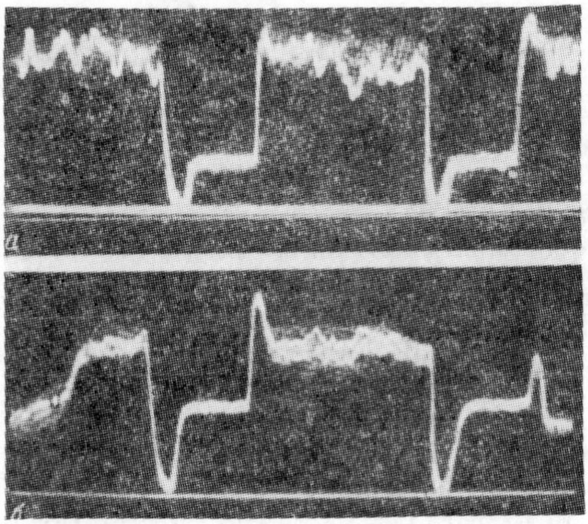

Fig. 4.7 Records of signals U_s, U_Σ, U_{em} at the output of the amplifier of the photomultiplier during different operating modes of the U-25 facility, a -run No. 28, T = 2800 K; b - run No. 30, T = 2620 K

before the start of the signal U_Σ) are the result of inaccurate mutual positions of the choppers when their synchronization system was not operational. Controlling duration of the signal transmission across the switches of the channel separation assembly made it possible to cut off sudden variations, which, therefore, were not passed to the output capacitors of the system and into the computer.

In conclusion, we will estimate the difference between the measured and time-averaged temperature. The components of fluctuation spectra at different frequencies may exert different effects on the temperature being measured (see 4.2.5.). To estimate the upper limit, assume that the relative temperature pulsations at all frequencies, $\Delta T(t)/T_o = 0.02$ and that $h\nu/kT = 8$. Components of fluctuations considerably higher than 300 Hz must be averaged over the impulse when capacitors C11, C12, of the channel separation assembly are being charged. In this case, the difference between the temperature being measured and the mean temperature is given by formula (4.40) and is 0.1 %, which is considerably less than the measurement error.

The lowest fluctuation frequency, 50 Hz (τ_{em} = 0.02 s), may result in an error, because signals U_{em} and U_Σ arrive at times characterized by different temperatures. The temperature calculated from these signals can then be averaged on a computer. The relative difference between the temperature being measured and the mean temperature is not more than 0.02 %.

In the usual operating modes of the U-25 facility, and in the absence of anomalously large pulsations, the temperature being measured is practically equal to that averaged over time.

4.3.3. Measurement of Concentration of Alkali Metal Atoms. Measurements of the concentration of alkali metals using the technique considered in 4.2.4. were conducted in MHD generators on the U-02 and U-25 facilities. Photographic recording of line contours with subsequent processing to determine the distance from the center of the maximum of the self-reversed line were used on the U-02 facility. This distance was then used to calculate the concentration.

Photoelectric recording of spectral line contours was used on the U-25 facility. The system for measuring the atom concentration is simpler, in many respects, than that for measuring the temperature. This can be attributed to the method of finding concentration from the position of the maximum that does not require knowledge of the absolute emission intensity. All that is required is the relative distribution of intensity over frequencies. This eliminates the use of a reference source during measurements. It is only necessary to obtain, during preparation of experiments, the dependence of the photomultiplier sensitivity on the frequency in the frequency range of the line being used. Variation of sensitivity of the photomultiplier, resulting from magnetic field variation or other causes, is not very significant. It is only important that no noticeable changes of sensitivity should occur during photographing of line profiles, and that sensitivity should not be too low. A spectrograph with a photoelectric attachment is located the same way as spectral equipment during temperature measurements. However, in this case it is sufficient to provide magnetic shielding only for the photomultiplier. Emission reaches the spectrograph through optical tubes.

The photoelectric attachment consists of a base with guide bars. A standard spectral slit with an attached photomultiplier is moved along the guide bars by means of a remotely controlled motor with a micrometer screw. The housing to which the base is attached is made of steel covered with permalloy strip wound around the housing that also serves as a magnetic shield. An induction position sensor is used for remote control of the displacement of the slit along the line. Maximum linear displacement of the spectral slit is 25 mm and is controlled by end switches mounted on the base of the photo-attachment. An additional electric motor can be used to rotate the dispersion system so as to position the required spectral sector onto the slit of the photo-attachment.

The output signal from the photomultiplier is fed across the amplifier into a cable 100 m long and into a room that controls operation of all of the diagnostic and recording equipment. The lines are recorded by means of a quick response potentiometer and a computer. As an example, Fig. 4.8 shows the contour of potassium doublet acquired on the U-25 facility by means of the concentration measurement system.

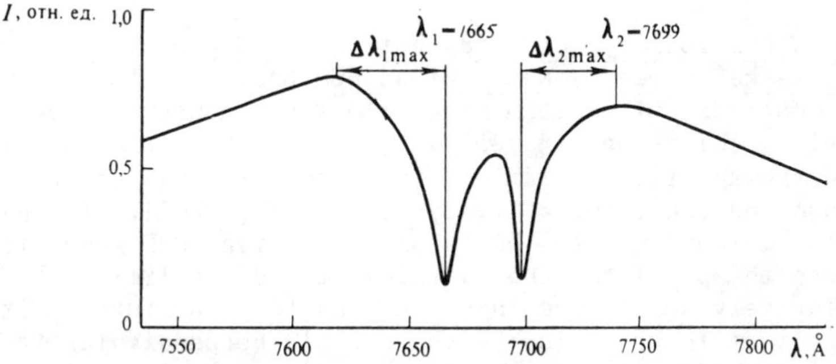

Fig. 4.8 Self-reversed contour of a potassium doublet acquired in a channel of the U-25 facility

Determination of the concentration of atoms from the line contours may be automated. Automation is achieved with the aid of a computer that determines the necessary distances, $\Delta\nu_{max}$, and calculates the concentration. The temperature and pressure in the channel, measured in the same operating mode, are automatically introduced into the calculations.

Error in determining the concentration on the U-25 facility (see formulas (4.19) and (4.21)) depends almost completely on the errors in the coefficients of absorption and the optical density, τ_{omax}. Recording of the dependence of the relative intensity on frequency, determination of the maximum, as well as calculations are performed with a higher degree of accuracy, i.e., instrumental errors are not too significant. For the case under consideration, the dependence of the coefficient of absorption on frequency is known to within an error of 10 % [17, 18]. Similar error applies to determination of τ_{omax} (see 4.2.3.). This error can be attributed primarily to using insufficiently accurate value of the ratio, δ/ℓ in a specific experiment. Error in determining the temperature and position of the line maxima contributes several percentage points to the total error. Thus, the total error does not exceed 25 %.

The automatic, remotely-controlled temperature measuring systems based on the generalized reversal method, and atom concentration measurement systems utilizing self-reversed spectral lines make it possible to conduct measurements with acceptable accuracy during operation of an MHD generator. The data on the parameters being measured can be used directly during the experiments. These systems have been designed, fabricated, and used during operation of equilibrium MHD generators.

4.4. Feasibility of Diagnostics of Combustion Products with Alkali Seed by Means of an HCN Laser

Methods of determining electron concentration and their collision rates by probing plasma with electromagnetic radiation from an external source are known and are being actively pursued [58-64]. If one of these methods is to be applied in real MHD generators, where the ports to the working fluid are limited by the relatively small openings and narrow apertures, it is expedient to probe the working fluid perpendicular to its surface and to make observations along the line of sight. It would be expedient to use methods that provide data on the working fluid of the MHD generator both in a magnetic field and in its absence. In this case, the most suitable technique is measurement of the phase shift and attenuation of plane-polarized electromagnetic waves.

A plane transverse electromagnetic wave is characterized by a complex wave number

$$k' = (\mu - i\Xi)\ \omega_o c_o \equiv \beta - i\alpha.$$

Here, ω_o is the angular frequency of the wave; μ and Ξ are the refraction and attentuation indices, in that order; β is the phase constant and α is the attenuation constant. Both electron and heavy particles affect the wave number. In weakly ionized gas, it can be assumed that the attenuation constant and refraction, $\mu - 1$, of different components are additive [60]. From basic theory of interaction of an electromagnetic wave with free electrons [65], when $\nu_{eff}^2 \ll \omega_o^2$ and $\omega_p^2 \ll \omega_o^2$, the following expressions are obtained for the refraction and attenuation indices:

$$\mu_e = 1 - 0.5\omega_p^2/\omega_o^2\ , \tag{4.52}$$

$$\Xi_e \equiv \alpha_e \frac{c_o}{\omega_o} = 0.5 \frac{\nu_{eff}}{\omega_o} \frac{\omega_p^2}{\omega_o^2} , \qquad (4.53)$$

where ν_{eff} is the effective collision rate of electrons with neighboring particles and ω_p is the plasma frequency.

In the opposite case, when $\nu_{eff}^2 >> \omega_o^2$, basic theory provides an expression for the low-frequency electronic electrical conductivity in the form

$$\sigma_e = n_e e^2/m_e \nu_{eff} \equiv n_e b_e e , \qquad (4.54)$$

where b_e is the electron mobility. It is this low-frequency electrical conductivity that must be measured in MHD generators. One of the major limitations of the basic theory is that the collision rate of electrons with neighboring particles cannot be a function of velocity of electrons. Only in such a case are frequencies ν_{eff} in expressions (4.53) and (4.54) the same. In other cases, these frequencies may differ considerably [66]. This aspect of the problem will be considered further in the discussion of experimental results. The refraction and attenuation indices that depend on the neutral gas component are described by well-known dispersion equations [67]. Calculations using these equations for combustion products seeded with alkali metals are practically impossible, because of lack of data on parameters of the molecules. The following expression for refraction [60, 68] can be used to calculate approximate indices,

$$\mu_M - 1 = 2\pi\bar{\alpha}_M n_M , \qquad (4.55)$$

where $\bar{\alpha}_M$ is the static polarization of molecules, numerical values of which can be found elsewhere [69-72]. When performing calculations, the content of the products of combustion must be taken into account.

The range of wavelengths of the probing radiation that is convenient for investigations of the working fluid of an MHD generator will next be considered. The plasma in an MHD generator may be characterized by the following parameter values: $n_M = 10^{18}-10^{19}$ cm^{-3}; $n_e = 10^{13}-10^{15}$ cm^{-3}; $\nu_{eff} = 10^{11}-10^{12}$ s^{-1}. When using even shortwave microwave radiation at a wave length of 2 mm ($\omega_o = 9.4 \times 10^{11}$ s^{-1}), the critical electron concentration ($n_{cr} = 2.7 \times 10^{14}$ s^{-3}) is lower than that in MHD generators. In addition, as a result of the large ratio ν_{eff}/ω_o, the electromagnetic wave is rapidly attenuated even at $n_e <$

n_{cr} (hence, when $n_e = 10^{14}$ cm^{-3} and $\nu_{eff} = 3 \times 10^{11}$ s^{-1}, the attenuation constant is $\alpha_e = 2.2$ cm^{-1}). When the width of the MHD generator channel is about 1 m, attenuation makes it very difficult to use microwave radiation.

Attenuation and refraction by electrons in the visible and near-infrared ranges are small. For example, at a wavelength of 0.63 μm and a change of n_e from 10^{13} cm^{-3} to 10^{15} cm^{-3}, attenuation and refraction are characterized by the following parameters: $\alpha_e = 10^{-8}$-10^{-6} cm^{-1}; $1 - \mu_e = 10^{-8}$-10^{-6} (or phase shift of 10^{-4}-10^{-2} rad/cm). At a wavelength of 10.6 μm, $\alpha_e = 10^{-6}$-10^{-4} cm^{-1} and $1 - \mu_e = 10^{-6}$-10^{-4} (phase shift of 10^{-3} - 10^{-1} rad/cm). This indicates that measurements can hardly be made at an electron concentration of 10^{13} cm^{-3}. The difficulties become even greater as a result of the molecules beginning to exert considerable influence on refraction. Calculations performed using formula (4.55), taking plasma content into account [23], lead to a value of refraction that depends only on the amount of molecules of nitrogen and water ($\sum_M (\mu_M - 1) = (1 - 2) \times 10^{-4}$), which is considerably greater than refraction due to electrons at $n_e = 10^{13}$ cm^{-3}.

Estimates given show that it is expedient to use an intermediate submillimeter range of probing waves with wavelength of 0.1 - 1 mm. An active development of submillimeter lasers encourages the use of this range. Electric discharge lasers operating on vapors of three-atom molecules [73-75] have been developed. The NCN laser oscillating at the wavelength of 337 μm is the most advanced of such lasers. The choice of the probing wavelength of a laser for the determination of electron concentration and effective collision rate in an MHD generator channel requires preliminary determination of how much of the phase shift and attenuation in combustion products plasma at this wavelength is caused by electrons and not by the heavy component of plasma.

The available data on propagation of electromagnetic radiation at wavelengths of 337 μm is reduced to the following.

Experimental determination of the index of refraction of air and water vapors at normal conditions [76, 77] has shown that the influence exerted by water on propagation of radiation is much greater than is indicated by calculations performed using formula (4.55). This can be attributed to the large number of absorption bands of water in the submillimeter spectral range that, according to the dispersion equation, affects the index of refraction. Recalculation of data in Ref. 76, to conditions existing in combustion products, provides an expression for refraction, $\mu_{H_2O} - 1 = (4\text{-}5) \times 10^{-4}$.

Measurements of absorption in water vapors and air in the submillimeter spectral range [75, 78] indicate that, under conditions of combustion products, one can expect $\alpha = 3 \times 10^{-4}$ cm^{-1}. These estimates are inadequate, because they take into account only molecules of water and nitrogen and neglect the complex content of combustion products, particularly molecular compounds of alkali atoms. In this connection, it became necessary to perform experiments for the determination of the feasibility of using radiation of a wavelength of 337 μm in combustion products plasma with alkali seed [79, 80]. Spectroscopic methods were used primarily as a check.

4.4.1. Experimental System and Measuring Methods. The object of the investigation was plasma of combustion products of propane in air enriched with oxygen. A burner with a guard flare was used. The combustion products flow rate in the burner was 0.6-0.7 g/s and the flow rate was approximately 10 m/s. The diameter of the inner homogeneous flare was 2 cm and the thickness of the guard flare was 1 cm. The seed usually consisted of alcohol solutions of acetates of potassium, rubidium, and cesium. The seed was injected only into the inner flare of the burner. The measurement zone was located approximately 5 cm from the zone of the primary reactions. The time of flight of gas from the primary reaction zone was approximately 5 ms. When this level was displaced upward or downward by 1-1.5 cm, the emission brightness in resonance lines of alkali metals did not change. This indicates that the temperature and atom concentration near the measurement zone is constant. The investigation of the dependence of temperature and concentration of atoms of alkali metals on height above the burner indicate the presence of a fairly flat maximum of both quantities in the range of measurements.

Correct interpretation of the experimental results for the determination of feasibility of using emission at 337 μm for plasma diagnostics of combustion products plasma with alkali seed requires, not only equality of the temperature of molecules and charged particles, but also equilibrium population of energy levels of alkali atoms and the presence of ionization equilibrium between alkali atoms and electrons. It will be assumed that the characteristic time for establishing ionization equilibrium is approximately equal to the recombination time. The data in Ref. 81 that dealt with recombination of electrons and alkali ions accompanied with ionization energy transfer to vibrational levels of nitrogen and water will be used. Characteristic recombination time can be estimated from the relation

$$t_R = (\alpha_R n_e)^{-1} , \tag{4.56}$$

where α_R is the recombination coefficient.

In a mixture of nitrogen and water (proportions are close to those in combustion products) at atmospheric pressure and T = 2500 K, $\alpha_R \simeq 2 \times 10^{-9}$ cm³/s. If n_e is equal to 10^{13} cm^{-3}, from formula (4.56) it follows that $t_R = 5 \times 10^{-5}$ s.

At a flow rate of 10 m/s and recombination time of $t_R = 5 \times 10^{-5}$ s, equilibrium is characteristically established at a distance of approximately 0.5 mm. Characteristic dimensions of the regions dealt with in these experiments are on the order of centimeters. Therefore, one can easily assume that ionization equilibrium in the regions under consideration has time to become established. A cw HCN laser operating on a mixture of methane and nitrogen was used as the emission source. The principal parameters of the laser are as follows: single-mode emission power output at a wavelength of 337 µm is approximately 1 mW, beam divergence is 3.5-4°, power fluctuations do not exceed 5 %, and power output loss from detuning of the resonator is approximately 5 %/h. Duration of operation of the laser is 200-300 h, after which it is disassembled, the cathode and tube with the anodes is washed with water, and the resonator mirror is spray-coated.

A Michelson interferometer was used. In some experiments, laser emission at the inlet into the interferometer was mechanically modulated by a chopper, the use of which, jointly with a narrow-beam amplifier at the inlet into the detector, enhances noise-immunity of the system. The phase shift of the carrier frequency was measured in this technique. In these experiments, phase shifts in plasma were less than interference bands. In such cases, the phase shift was determined from the mirror displacement in one of the interferometer arms, required for compensation of the interferometer imbalance caused by plasma [82-87].

Phase modulation in the reference channel of the interferometer was used in other experiments. Submillimeter radiation was phase-modulated using different techniques, i.e., using a cylindrical diffraction grating or a mirror rotating according to a sinusoidal dependence [88-91].

In order to increase the measurement accuracy and to attain continuous phase shift display, it is desirable to perform phase modulation so as to make it possible to transpose phase shifts from the submillimeter wavelength range into the microwave range and, thus, to measure phase shift by means of a radio phase meter. However, the foregoing phase modulation methods do not offer such a possibility.

Measurements obtained using amplitude modulation (Fig. 4.9) will next be discussed. Crystalline quartz lens, 6, with a focal length of 0.5 m focuses the beam of the laser, 7, in plasma, 2. Laser emission is modulated at a frequency of 32 Hz by chopper 5. Beam splitter 4, made of lavson film 50 μm thick, splits part of the beam into the reference channel and the other part, into the measurement channel containing plasma. Aluminized optical glass was used for interferometer mirrors. A spherical mirror, 3, with a radius of curvature of 1 m provides additional focusing of radiation in plasma. A flat mirror, 1, can be displaced by means of a micrometer screw. All components of the interferometer are mounted on guides of a standard optical bench located on a heavy table surface. Emission from the reference and measurement channels is incident on detector, 10, consisting of a crystalline InSb point contact detector [92]. Signal at the modulation frequency is fed to the amplifier, 9, and is recorded by oscillograph, 8.

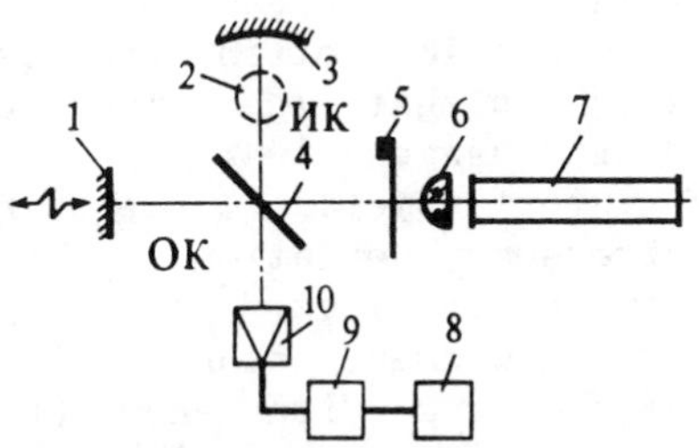

Fig. 4.9 Block diagram of an interferometer with amplitude modulation. RC, MC - reference and measurement channels; 1,3--plane and spherical mirros, respectively 2--plasma; 4--beam splitter; 5--chopper; 6--lens; 7--laser; 8--oscillograph; 9--amplifier; 10--detector

Before the appearance of plasma, the interferometer is tuned to an intensity minimum at the detector by means of movable mirror, 1. In the presence of plasma, phase shift, ϕ, is determined from the displacement d of a movable mirror, 1, required to compensate the interferometer imbalance. We have $\phi = 4\pi d/\lambda$, where λ is the wavelength.

When measuring attenuation of emission in plasma, the reference beam is overlapped. Attenuation is a function of voltage amplitudes at the output of the amplifier connected to the output of a square detector,

$$2\alpha\ell = 0.5 \ln(U_o/U_p) , \tag{4.57}$$

where ℓ is the plasma thickness and U_o and U_p are the voltage amplitudes at the output of the amplifier, in the absence of plasma, and in the presence of plasma, in that order.

Consider the errors associated with phase shift measurements using amplitude modulation. The displacement error of a movable mirror $\Delta d = 0.005$ mm. This determines the error of measuring the phase shift caused by the recording device:

$$\Delta\phi = 4\pi\Delta d/\lambda = 0.06\pi \text{ rad} \simeq 11^0 .$$

The error associated with the instrument recording the signal peak during phase shift measurements can be neglected.

In order to estimate the measurement error of the phase shift in the submillimeter channel of the interferometer, one must take into account the slow loss of power caused by detuning of the laser cavity (5 %/h). In addition to this, it is necessary to take into account "detuning" of the location of interferometer components. Observations have shown that the signal at the output of the interferometer changes 10 % per hour because of detuning of the interferometer. Detuning of the laser cavity leads to frequency deviation, and "detuning" of the interferometer components, i.e., to changes in length of the reference and measurement channels.

The change of the laser frequency can be estimated by assuming that the line contour is Doppler with a width of 10^7 Hz [93]. A 0.5 % change of laser emission power during the experiment that lasts several minutes may be caused by frequency changes not exceeding 4×10^5 Hz. A difference in length between the reference and measurement interferometer channels, X = 0.1 m, causes a phase shift, $\Delta\phi = X\Delta\omega_o/c_o \leq 3 \times 10^{-4}\pi$ rad = 0.5 degrees. A 1 % detuning of the interferometer estimated from the power loss during the several minutes required for the experiment, leads to a phase shift of approximately 2^0.

The total error of measuring the phase in an interferometer with amplitude modulation that depends on the error of recording the shift (11^0) and the error in the submillimeter channel (2^0-3^0) does not exceed 14°. The attenuation error is caused by laser power instability. In accordance with formula (4.57), in the case of random power fluctuations of 2-5 %, $\Delta(2\alpha)$ = 0.02-0.04. Because the signal-to-noise ratio is less than 100, the noise of the detector of submillimeter radiation and the detector-amplifier channel can be neglected.

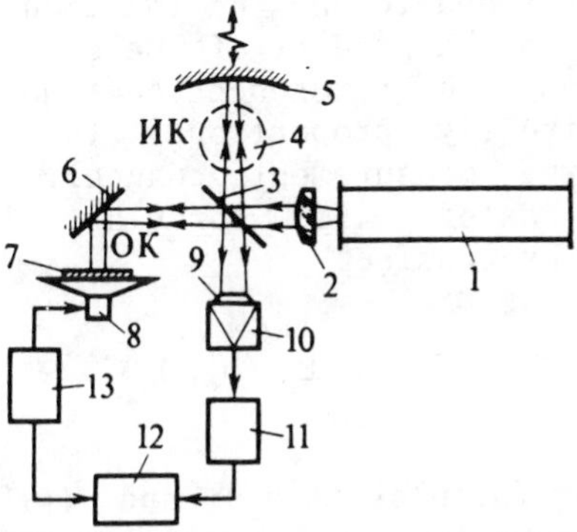

Fig. 4.10 Block diagram of an interferometer with phase modulation. 1--laser; 2,9--lens; 3--beam splitter; 4--plasma; 5-7--mirrors (spherical, plane rotatable mirror, and oscillating mirror, in that order); 8--electrodynamic system; 10--radiation detector; 11--amplifier; 12--phase meter; 13--modulator of the control of the electrodynamic system

Consider, next, an interferometer with phase modulation in the reference channel (Fig. 4.10). Laser 1, lens 2, splitter 3, mirror 5, and detector 10, are identical to the same components in the block diagram of the interferometer with amplitude modulation. The path difference between the reference and measurement channel beams is modulated by means of an oscillating mirror, 7. Chopper 13 controls the electrodynamic system, 8, that displaces mirror 7. A mirror, 6, that can be turned directs radiation to mirror 7, located horizontally on the electrodynamic system. The mirror is modulated in accordance with a two-tonal law with the modulating voltage generated by superposition of two harmonic components with even frequencies.

Amplitude and phase can be chosen in such a way that transposition of the carrier frequency phase to one of the modulating frequencies is attained almost without distortion [94]. Two frequencies, at 57 Hz and 114 Hz, are generated in the interferometer modulator, 13. The phase is transposed to a frequency of 114 Hz. After the signal is amplified in amplifier 11, the phase shift in plasma 4 is measured by means of a phase meter, 12. In the absence of plasma, the phase difference at the detector is reduced to zero by displacing the movable mirror in the measurement channel. In the presence of plasma in the measurement channel, the phase shift is measured on the phase meter scale.

When measuring signal attenuation in plasma by means of an interferometer with phase modulation, the reference channel is not overlapped. The voltage at the outlet of the amplifier tuned to one of the modulation harmonics is proportional to the product of the electric field strength of waves from the reference and measurement channels, i.e., is directly proportional to the electric field strength in the measurement channel. On the basis of this, and by analogy with formula (4.57), attenuation is given by the expression

$$2\alpha\ell = \ln (U_o/U_p) \ . \tag{4.58}$$

Comparing formulas (4.57) and (4.58) shows that, for the same attenuation in plasma, the change in the voltage at the amplifier output is smaller for a system with phase modulation.

Let us now estimate measurement errors in an interferometer with phase modulation. First, look at the errors associated with determination of the phase shift.

The errors of the submillimeter channel remain the same as during amplitude modulation. The recording error drops sharply and depends on the error of the phase meter (approximately 1^0). In addition, another error associated with phase transposition into the radio frequency range is added; however, it does not exceed 5^0 [94], and may be decreased considerably by improving the chopper. The total error does not exceed $5\text{-}8^0$. In some experiments with rigid control of phase transposition, it is $1\text{-}3^0$.

The error associated with determination of attenuation during phase modulation is twice as large as during amplitude modulation. In accordance with formula (4.58), random fluctuations of the laser power output, 2-5 % result in $\Delta (2\alpha\ell) = 0.03\text{-}0.07$.

Phase shift and attenuation measurements were accompanied by spectroscopic measurements of the temperature and concentration of atoms of alkali metals. This was performed with an ISP-51 spectrograph. The reference source consisted of an SI-10-300 lamp with an SIP-30 power supply. The temperature was measured with an accuracy of 1 % by means of the visual reversal method. The integral intensity of potassium doublet (4044/4074 Å), rubidium doublet (4202/4216 Å), and cesium line (4555 Å) were determined, in order to measure the concentration of alkali metal atoms. Concentration of atoms from the integral intensity was calculated using the growth curve with a Voigt contour [95]. Known oscillator strengths [9, 96] and collisional cross sections with combustion product molecules were used [11-15].

The method of integral intensity was chosen because of its experimental simplicity. Let us determine the error of measuring the concentration of atoms. Under the experimental conditions, the growth curve depends on collisional broadening, and the integral intensity of the line can be expressed as follows [9, 95]:

$$J \sim \exp(-h\nu/kT)\,(f(n_\alpha T^{-1/2}\sigma)^{1/2} .$$

The integral intensity was measured a number of times so that the random square error caused by fluctuations in plasma, as well as the instrumental error of the digital voltmeter, could be neglected. Photography of the flare of the combustion products was used to determine thickness of the plasma. An interference filter with a transmission band in the range of resonance lines of alkali metals was used. Photometric evaluation of the film was performed. The error of determining thickness of the emitter did not exceed 5 %. Then, taking into account the errors associated with f values and σ, and determining temperature by means of the visual reversal method, the error of determining the concentration of atoms is estimated to be

$$\Delta n_\alpha / n_\alpha \simeq 40^0/c .$$

It should be noted that the doublet lines may partially overlap and that this is not taken into account in growth curves [95]. Estimates show that the concentration of potassium atoms is higher than it should be, but not by more than 40 %.

Hence, the object of investigation is an equilibrium flow of combustion products with alkali seed. An interferometer with an NCN laser makes it possible to measure phase shifts and attenuation of emission at the wavelength of 337 μm. Spectroscopic equipment is used to determine temperature and concentration of alkali atoms in plasma [79, 97].

4.4.2. Experimental Verification of the Feasibility of Using an HCN-Laser in Combustion Products with Alkali Seed. Preliminary estimates and the results of measurements show that the inequalities

$$\omega_{eff}^2 << \omega_o^2 \quad \text{and} \quad \omega_p^2 << \omega_o^2 \ ,$$

are fulfilled under the experimental conditions. These inequalities must be fulfilled in order for formulas (4.52) and (4.53) to be valid. From these formulas, it follows that

$$n_e = 2n_{cr}(1-\mu_e) \ ,$$

$$\nu_{eff} = \omega_o \Xi_e/(1-\mu_e) \ ,$$

where

$$n_{cr} = m_e \omega_o^2/4\pi e^2 = 9.84 \cdot 10^{15} \ cm^{-3} \ .$$

Electron concentration and the effective collision rate expressed in terms of the phase shift, ϕ (rad), and attenuation, $2\alpha\ell$, in plasma observed in the Michelson interferometer are as follows:

$$n_e = n_{cr}\phi/2\pi \cdot \lambda/\ell \ , \tag{4.59}$$

$$\nu_{eff} = \omega_o 2\alpha\ell/\phi \ . \tag{4.60}$$

From formula (4.59) and estimated errors, it follows that the minimal $n_e\ell$ measured in an interferometer with amplitude modulation is 10^{13} cm^{-2}, and that in an interferometer with phase modulation is less than 4×10^{12} cm^{-2}. Within the limits of the measurement error, no phase shift was observed in the interferometer with amplitude modulation in the absence of seed in the flare. Phase shifts between 1^0 and 2^0 were observed when a phase meter was used. The phase shift was opposite of that when seed was injected. In the presence of seed, phase shifts were observed in both interferometers (in different experiments they varied between 15^0 and 150^0).

The electron concentration was determined from the phase shifts using formula (4.59). The concentration of alkali atoms and the temperature were measured spectroscopically, to assure that the phase shifts observed are associated with the electron component rather than alkali metal compounds that appear in the combustion products in the presence of alkali seed.

Fig. 4.11 shows the dependence of electron concentration measured on the concentration of potassium, rubidium, and cesium atoms. Each experimental point is a result averaged over several measurements. The number of measurements of electron and atom concentrations were chosen so that the mean square measurement error resulting from plasma parameter pulsations was smaller than the instrumental error.

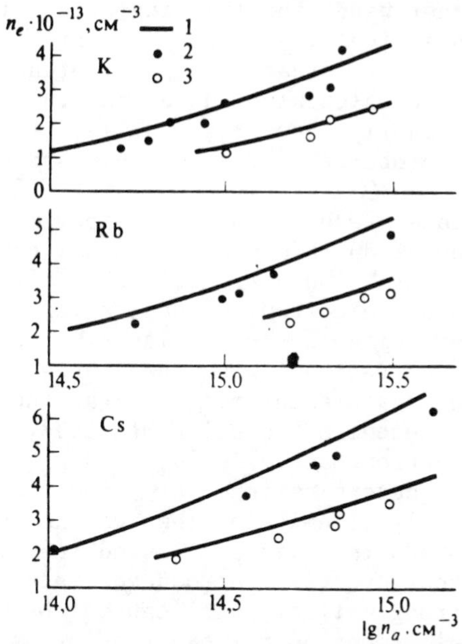

Fig. 4.11 Variation of the concentration of electrons with the concentration of alkali metal atoms. 1--calculated using Saha equation; 2,3--experimental results at T = 2500 and 2380 K, respectively

Saha equation was used to determine the electron concentration (neglecting trapping) from the measured concentration of atoms and the temperature (see Fig. 4.11). Calculated curves are in satisfactory agreement with the experimental data. Apparently, two factors may be responsible for the small difference between the calculated and the experimental data, i.e., either trapping of electrons by the OH hydroxyl decreases the real electron concentration in comparison with the calculated concentration, or the phase shift observed in the submillimeter interferometer is affected by the alkali metal compounds in the combustion products when seed is injected [23, 98].

Hydroxide is the predominant alkali metal compound in the combustion products. It is known that variation of temperature exerts a strong influence on the concentration of alkali metal hydroxide. In the case of cesium, for example, it decreases 1-1/2 to 2 times as the temperature increases from 2380 to 2500 K. If the relatively high values of the experimental data were associated with refraction by metal hydroxide, the deviation should decrease with increasing temperature (refraction by electrons and molecules has opposite signs).

On the other hand, the trapping effect in the range of interest is practically independent of temperature. In view of the fact that the difference between the experimental and calculated data in this range does not change significantly (see Fig. 4.11), it can be assumed that the most probable cause of this systematic shift is trapping.

In this case, results of the experiments showed that the combustion products and alkali metal compounds at $n_e > 10^{13}$ cm^{-3} exert no significant influence on the determination of electron concentration by means of an interferometer with an HCN laser. Obviously, this statement is correct within the range of errors that depends on the measurement methods (at the lower limit of the electron concentration being measured, $n_e \simeq 1 \times 10^{13}$, the error is approximately 20 %).

Consider determination of the collision rate. Attenuation of laser emission that was not observed in the combustion products without seed became immediately noticeable when seed was introduced into the flare. An especially strong effect was caused by cesium. Other kinds of alkali seeds do not exert such a strong influence because of the smaller concentration of electrons. Formula (4.60) and the phase shift and attenuation measured were used to determine the effective collision rate of electrons that is practically independent of the concentration of atoms (Fig. 4.12a). The error in measuring the effective collision rate was estimated from the relation

$$\Delta\nu_{eff}/\nu_{eff} = \Delta\alpha/\alpha + \Delta\phi/\phi + \Delta\ell/\ell$$

and varied when the concentration of cesium atoms changed from 20 to 40 %.

It should be emphasized that the effective collision rate in formula (4.53) describing attenuation of a high-frequency wave was determined in the submillimeter interferometer. The collisional cross section of the combustion products plasma at a temperature of 2500 K changes by two orders of magnitude when the energy of electrons increases from 0.1 to 1.0 eV [99]. Such a dependence of the cross section on energy results in a difference between effective rates in formulas (4.53) and (4.54) of not more than 10-12 % [66].

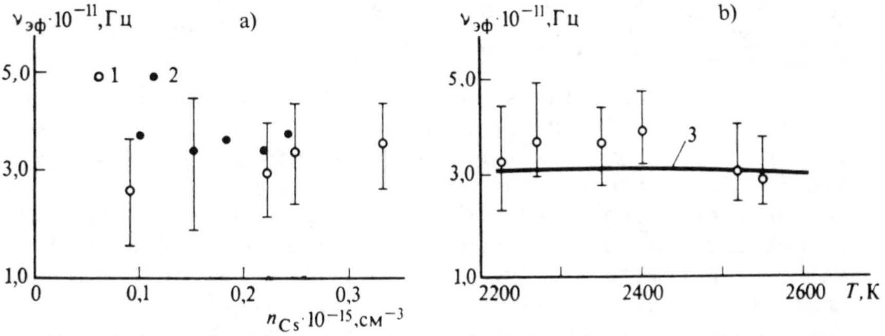

Fig. 4.12 Variation of the effective electron collision rate with concentration of cesium atoms (a) and combustion products temperature (b). 1,2--experimental measurements of the attenuation of the submillimeter signal and electrical conductivity, respectively; 3--results of calculations

The electron collision rate was determined not only from the attenuation and phase shift of the submillimeter laser radiation, but also by means of the following method: The electrical conductivity of plasma was measured using the Q-factor technique [100] and the electron concentration was measured from the phase shift in the submillimeter interferometer. These quantities were used to calculate the collision rate (see Fig. 4.12a). Because low-frequency electrical conductivity is measured by the Q-factor technique, the effective collision rate also falls into the low-frequency range.

The results obtained using different methods are in good agreement with each other, even when the difference between effective collision rates determined by these techniques is taken into account. The dependence of the effective electron collision rate on temperature determined from the phase shift and attenuation of submillimeter emission is in good agreement with the theoretically calculated data [30] (see Fig. 4.12b).

Thus, it was shown experimentally that refraction of emission at a wavelength of 337 μm and its attenuation at electron concentration $n_e \geq 2 \times 10^{13}$ cm^{-3} in combustion products plasma with alkali additives, atmospheric pressure, is caused by the electronic component of plasma. Therefore, both attenuation and the phase shift of emission from an HCN laser can be used to determine the electron concentration and effective electron collision rate with the neighboring particles in the working fluid of an equilibrium MHD generator when the degree of ionization is greater than 10^{-5}.

4.5. Local Measurement of the Electron Temperature with Cold Electric Probes.

Contact-free methods of measuring parameters of the working fluid of an equilibrium MHD generator do not require insertion of a foreign body into the gas and are, hence, quite accurate and convenient. The results of such measurements are used to determine spatially averaged parameters. However, even determination of the maximum value of parameters represents an additional problem, and that of obtaining spatial distributions requires much more complex processing of the experimental data.

The present chapter deals with feasibility of measuring the electron temperature by means of relatively cold electric probes, when the mean free path of charged particles is considerably shorter than the characteristic dimension of the probe. Investigations were initiated in response to a need of determining local values of the main parameter, i.e., temperature in the working fluid of an equilibrium MHD generator, where the electron temperature, T_e, is equal to temperature of molecules, T_M.

For many reasons, inserting the probe into an electrically conducting gas results in a disturbance. The difference between the probe potential and the potential of the surrounding space results in an electrical disturbance. The loss of charged particles to the non-emitting probe decreases their number in the vicinity of the probe. A "thermal" disturbance occurs when the temperature probe differs from the temperature of the gas. Flow characteristics are changed when the probe is inserted into the moving gas. All of this can be complicated by emission from the probe. By choosing the size, shape, location, temperature, and potential of the probe, it is possible to decrease, or practically eliminate, the influence of some of these disturbances. A theory being actively developed at the present time takes into account electrical disturbances of the plasma by the probe under conditions when the following inequalities apply:

$$L_e \ll r_3, \quad L_p \ll r_3 ,$$

where L_e, L_p, are the mean free path of an electron and ion, in that order, and r_3 is the characteristic dimension of the probe. The principal equations describing these effects are the continuity equations for each type of charged particles and expressions for the particle flux and Poisson equation (e.g., Refs. 101-104). In certain papers, it was assumed that $L < R_D$ and $L_p < R_D$, where R_D is the Debye radius. In the investigations, particle interactions were taken into account in different ways and the boundary conditions were written differently. Probes of different geometry were considered. Mathematical methods used also differed. However, in all of these investigations, it was assumed that the electron temperature inside the plasma region disturbed by the probe does not vary. The following result, significant for subsequent calculations, is obtained from an analysis of the electric current to the probe, independently of the specific problem formulation, and a method for its solution. At sufficiently high negative potentials of the probe, close to the floating potential, the dependence of the electric current on the probe potential becomes exponential. The constant factor in the exponential depends on the electron temperature. The physical meaning of this result is that the electron flux to the non-emitting surface of the probe at sufficiently high negative potentials is negligibly small, compared with the random flux, and that, independent of specific conditions, the electrons are uniformly distributed throughout the electric field of the probe.

This means that determination of the temperature at sufficiently high negative potentials may be performed using the same techniques as at low pressures, i.e., from the slope of the semilogarithmic dependence of the electric current on the potential, or with the aid of a double probe. The probe theory in gas flows is being actively developed, with particular attention devoted to calculations of the ionic current to the probe [104].

Note one of the results of investigations that must be taken into account in diagnostics of the working fluid of an MHD generator. An increase in the flow rate may lead to an increase of ionic current that, in turn, leads to a decrease of the absolute value of the floating potential. And, even though the gas flow affects the electron current considerably less than ionic current, the slope of the voltage-current characteristic in the range of low floating potential may not be characteristic of the electron temperature. If this is neglected, and the temperature is determined by the usual technique, it will be higher than the actual temperature. Hence, theoretical voltage-current characteristics cannot be used to determine the electron temperature in the flow, because they require additional, usually unavailable, data.

Taking into account the thermal disturbance introduced into the gas by the probe [105-112] leads to the following conclusions. In those cases where it is assumed that the temperature of the relatively cold probe affects only the temperature of heavy particles (molecules) in its vicinity, and that the electron temperature remains the same as in undisturbed plasma, the probe temperature has only a very weak influence on the measurement of the electron temperature [105-107]. When it is assumed that the probe affects both the temperature of heavy particles and the electron temperature and that $T_e = T_M$, in the vicinity of the probe, the results of numerical calculations of various special cases indicate that the temperature of the probe influences both the distribution of parameters in its vicinity [108] and the voltage-current characteristics [107].

When it is assumed that the temperature of heavy particles near the probe does not change and that the electric field of the probe influences the electron temperature, the variation of T_e is reflected in the results of numerical calculations of the voltage-current characteristics [111, 112]. Analysis of the influence of the temperature of the cold probe on the probe characteristics through formation of negative ions in the cold region near the probe (which may occur in certain flames and electronegative gases) has shown that neglecting these processes leads to a high estimate of the temperature [109, 110, 113].

Experimental investigations dealing with the determination of electron temperature by means of probes at average and high pressures provide contradictory results [114-122]. Let us consider in more detail the physical meaning of the influence of the probe temperature on the electron current at average and high pressures. As a result of energy losses by probes through emission and thermal conductivity, the probe temperature, T_3, is usually lower than the temperature of heavy particles in gas. This leads to formation of a hot boundary layer (with characteristic length x_T) near the probe. In this layer, the temperature of heavy particles, T_M, varies between the temperature of undisturbed plasma, T_0, and the temperature of the probe, T_3.

Variation of T_M exerts an influence on T_e inside the thermal layer. This can be estimated by comparing the time required to equalize the temperature of electrons and heavy particles,

$$\tau_{rel} \simeq L_e \langle\nu_e\rangle^{-1} \delta_e^{-1} , \qquad (4.61)$$

with the time required to exchange kinetic energy between electrons through electronic thermal conductivity inside the thermal layer. The order of magnitude of this time depends on the diffusion coefficients of electrons, D_e:

$$\tau_{e.t.} \simeq x_t^2/D_e . \qquad (4.62)$$

When the inequality,

$$\tau_{rel}/\tau_{e.t.} \gg 1 , \qquad (4.63)$$

applies, the electron temperature inside the thermal layer does not change and, apparently, the probe temperature should not exert much influence on the results of measurements. When reverse inequality,

$$\tau_{rel}/\tau_{e.t.} \ll 1 ,$$

applies, one can assume $T_e = T_M$ inside the thermal layer. In this case, one can expect an increase in the influence of the probe temperature on the electronic current. This effect must depend on the ratio of characteristic dimensions of the thermal (x_T) and electric (x_{el}) layers [123]. The electric layer is an arbitrary concept.

Assume that most of the voltage applied to the probe is concentrated in this layer. The ratio, x_T/x_{el}, can frequently be estimated from the ratio, r_3/R_D. If the electric layer lies inside the thermal layer, it is to be expected that, because the electrons inside the electric layer become colder as the layer decreases, the smaller x_{el}, the greater the influence of the temperature of the probe exerted on electronic current. Consider the ratios x_{el}/x_T and $\tau_{rel}/\tau_{e.t.}$ at which investigation of the influence of temperature relative to the cold probes on the measurement of electron temperature is of practical interest. If $x_{el}/x_T \ll 1$ and $\tau_{rel}/\tau_{e.t.} \ll 1$, the probe response curve may include data only on gas in the vicinity of the probe, deep inside the thermal layer. Usually such data are of no interest.

When $\tau_{rel}/\tau_{e.t.} \gg 1$, probe temperature should not exert noticeable influence on its readings. Conditions such as $\tau_{rel}/\tau_{e.t.}$ high, but not so high that the temperature of the electrons inside the thermal layer remains constant, and $x_{el}/x_T \ll 1$ frequently occur in products of combustion and in arcs at pressure near average and near atmospheric pressure. In this case, one must have ratios of the electron temperature near the surface of the probe with the temperature in the undisturbed plasma. In this case, it can be assumed that the electron temperature inside the electric layer does not change, and one can calculate electron motion in the same manner as in the absence of thermal disturbance of the gas by the probe.

The case where $\tau_{rel}/\tau_{e.t.} \ll 1$, i.e., when the electron temperature is equal to the temperature of the gas and x_{el} and x_T are comparable, may take place in combustion products with alkali seed at pressures close to atmospheric.

4.5.1. The Influence of the Temperature of the Probe on the Temperature Being Measured in the Case of High Electronic Thermal Conductivity. Let us now obtain relationships that can be used to connect the temperature of electrons in the region of the electric layer, i.e., near the surface of the cold probe, with the temperature of electrons in the undisturbed plasma in the presence of considerable influence of electronic thermal conductivity on the establishment of the temperature of electrons inside the thermal layer [124]. This problem will be solved using the electron energy balance equation that must be solved simultaneously with charged particles and energy of molecules balance equations.

Keeping in mind that it is necessary to obtain approximate solutions, the problem can be simplified by specifying the change of the electron concentration and the temperature of molecules inside the thermal layer. Assume that the thermal layer near the probe is plane and that the plasma is weakly ionized, i.e., the collision rate of electrons with neutrals is much higher than the collision rate with ions. In addition, it will be assumed that the temperature is measured close to the floating potential and that neither gas flow nor emission of electrons from the probe exert much influence on its value.

Two limiting cases of the electron concentration distribution in the thermal layer will be considered. In the first case, assume that $n_e(x)$ is determined by diffusion to the probe. Then

$$n_e(x) = n_e(0)\ (1+sx/x_T)\ , \qquad (4.64)$$

where

$$s = (n_e(x_T) - n_e(0))/n_e(0)\ ;$$

the x coordinate is counted from the probe to the boundary of the thermal layer with plasma. Because concentration and temperature of electrons is fairly low, these conditions are realized if electrons neither disappear nor are generated in the layer. The validity of assumptions (4.64) can be estimated by comparing the diffusion length, ℓ_D, with the size of the thermal layer. Formula (4.64) will be applicable, provided the inequality, $\ell_D > x_T$, is valid. The electron concentration at the boundary with the wall, $n_e\ (x_T)$, may differ from zero. Because the electric layer is small in comparison with the thermal layer, this boundary can be assumed to coincide with the external boundary of the electric layer.

In the second case, assuming that only particles of the same type are ionized, also assume the existence of a local ionization equilibrium at each point in the layer, i.e.,

$$n_e(T_e = \left(n_M \frac{g_e g_p}{g_M} \frac{(2\pi m_e k T_e)^{3/2}}{h^3}\right)^{1/2}$$

$$\times \exp\left(-\frac{eV_i}{2kT_e}\right)\ , \qquad (4.65)$$

where g_e, g_p, g_M, are statistical weights of electrons, ions, and neutrals; h is the Planck constant, and V_i is the ionization potential of the particle. The relation (4.65) may apply at sufficiently high electron concentration, when $\ell_D < x_T$.

Independently of assumptions (4.64) or (4.65), it is necessary that the condition $n_e^{-1/3} \ll x_T$ is fulfilled. Otherwise, one cannot use the concept of electron concentration inside the thermal layer.

The electron energy balance equation inside the thermal layer will be written in the following form:

$$\sigma_e E^2 - W + \frac{d}{dx}\left(\Xi'_e \frac{dT_e}{dx}\right) = 0 , \qquad (4.66)$$

where σ_e is the electrical conductivity given by formula (4.54), $\Xi'_e = 2/3 n_e k < v_e > L_e$ is the electron thermal conductivity coefficient of weakly ionized plasma, $\sigma_e E^2$ is the energy transferred to the electrons by the electric field of the arc (zero when considering collision-free plasma, e.g., combustion products plasma),

$$W = \frac{8}{3} k\,(T_e - T_M) n_e \sum_M \nu_{Meff} \gamma_M \frac{m_e}{m_M}$$

$$= \frac{8}{3} k\,(T_e - T_M) n_e \gamma \frac{m_e}{\tilde{m}_M} \nu_{eff} , \qquad (4.67)$$

are the bulk energy losses during elastic and inelastic interactions of electrons with heavy particles of various types in a weakly ionized plasma; $\tilde{m}_M$ is the mean mass of heavy particles, and γ is the coefficient of inelastic losses in a mixture of gases. After introduction of $\tilde{m}$ and γ, formula (4.67) makes it possible to use the same expression, W, for the mixture of gases as for one type of heavy particles when $\tilde{m}_M = m_M$ and $\gamma = \gamma_M$.

The coefficients of inelastic losses, γ_M and γ, are arbitrary. Assume that γ does not take into account processes associated with energy transfer from one part of the thermal layer into another. An increase of gas density in the thermal layer leads to an increase in energy losses of electrons. Usually, it can be assumed that $n_M \sim T_M^{-1}$. The last term in the left part of equation (4.66) is the energy acquired by electrons as a result of electron thermal conductivity ($W_{e.t.}$).

The energy balance equation (4.66) does not include three other possible sources of electron energy: enthalpy transfer to the probe by the electron flux, cooling of electrons by the electric field of the probe, heating of electrons in a certain region of the layer through collisions of the second kind with the excited particles from the outside. If excitation occurs in the same region, it will be taken into account by γ in the expression for W. Let us show that all of these sources of energy are small in comparison with the remaining sources in the balance equation. In obtaining an estimate, it will be assumed that the electrons collide with one type of heavy particles.

Consider heating of electrons as a result of enthalpy transfer. The energy source per unit volume can be represented in the form

$$W_\pi = \frac{d}{dx}\left(n_e u_e\left(\frac{5}{2}kT_e + eV_i\right)\right)$$

$$= n_e u_e \frac{d}{dx}\left(\frac{5}{2}kT_e\right) + \left(\frac{5}{2}kT_e + eV_i\right)\frac{d}{dx}(n_e u_e) ,$$

where u_e is the drift rate of electrons.

The first term in the righthand side of W_π describes an increase in the thermal energy electrons caused by the electron flux into the region under consideration from areas where the thermal energy of electrons is higher. The second term arises in connection with the appearance and disappearance of electrons, and characterizes the energy of the electron gas, provided the electrons produce ionization and recombination occurs with participation of an electron as the third particle that receives the recombination energy.

If the probe potential is close to the floating potential, the electron flux depends on ambipolar diffusion, and the following expression is valid:

$$W_\pi = -D_a \frac{dn_e}{dx}\frac{d}{dx}\left(\frac{5}{2}kT_e\right)$$

$$-\left(\frac{5}{2}kT_e + V_i\right)\frac{d}{dx}\left(D_a\frac{dn_e}{dx}\right) ,$$

where

$$D_a = 1/3L_p \langle\nu_p\rangle (1+T_e T_M) .$$

When condition (4.64) applies (when ionization and recombination in the thermal layer are insignificant) one can write

$$\frac{W_\pi}{W_{e.t.}} \simeq D_a \frac{dn_e}{dx} \frac{d}{dx}\left(\frac{5}{2} kT_e\right)\left(\Xi'_e \frac{d^2T_e}{dx^2}\right)^{-1}$$

$$\simeq \frac{1}{2} \frac{\sigma_{eM}}{\sigma_{pM}} \frac{\langle\nu_M\rangle}{\langle\nu_e\rangle}\left(1 + \frac{T_e}{T_M}\right) \ll 1 ,$$

where σ_{eM}, σ_{pM}, are the average cross section of momentum transfer of electrons and ions.

When condition (4.65) is fulfilled,

$$\frac{W_\pi}{W_{e.t.}} = eV_i \frac{d}{dx}\left(D_a \frac{dn_e}{dx}\right)\left(\Xi'_e \frac{d^2T_e}{dx^2}\right)^{-1}$$

$$\simeq \frac{1}{2} \frac{\sigma_{eM}}{\sigma_{pM}} \frac{\langle\nu_M\rangle}{\langle\nu_e\rangle}\left(\frac{eV_i}{2kT_e}\right)^2\left(1+ \frac{T_e}{T_M}\right) .$$

Usually, even in this case, $W_\pi/W_{e.t.} \ll 1$. This inequality indicates that the electron thermal conductivity exerts considerably greater influence on the electron temperature than the enthalpy transfer, even in the case when particles are generated and recombined.

Electron cooling by the electric field of the probe exerts the same influence as enthalpy transfer, i.e., the work performed by the electric field of the probe at the floating potential can be written in the form,

$$W_E = j_e E = - kT_e D_a n_e^{-1}\left(\frac{dn_e}{dx}\right)^2 .$$

For diffusion-type distribution of the concentration (see condition [4.64]), we obtain

$$\frac{W_E}{W_{e.t.}} \simeq - 2 \frac{\sigma_{em}}{\sigma_{pM}} \frac{\langle\nu_M\rangle}{\langle\nu_e\rangle}\left(1+ \frac{T_e}{T_M}\right) \ll 1 ,$$

and for distribution of concentration according to condition (4.65)

$$\frac{W_E}{W_{e.t.}} \simeq \frac{\sigma_{eM}}{\sigma_{pM}} \frac{\langle \nu_M \rangle}{\langle \nu_e \rangle} \left(1+ \frac{T_e}{T_M}\right) \frac{eV_i}{2kT_e} \ll 1 \ .$$

The upper limit of the influence of electron heating resulting from collisions of the second kind can be estimated by representing the energy transferred during these collisions to electrons in the form,

$$W_B = n_e n_B \sigma_{qe} \langle \nu_e \rangle V_B e \ ,$$

where n_B is the number of excited particles that enter the elementary volume under consideration from the outside, V_B is the excitation potential, σ_{qe} is the quenching cross section of an excited particle by an electron. Such heating is possible if the particle is not quenched or does not lose energy as a result of emission, with radiation leaving the boundaries of the layer during the diffusion time of the excited particle in the volume under consideration. In practice, this may occur in the case of motion of metastables or during diffusion of resonance radiation. Compare W_B with W.

Assuming that $n_B(x) \leq n_B(0)$, we can write

$$\frac{W}{W_B} \geq \frac{8}{3} \frac{kT_e}{eV_B} \frac{m_e}{m_M} \gamma_M \frac{n_M}{n_B(0)} \frac{\sigma_{eM}}{\sigma_{qe}} \geq$$

$$\geq \frac{8}{3} \frac{kT_e}{eV_B} \frac{m_e}{m_M} \gamma_M \frac{g_o}{g_B} \exp\left(\frac{eV_B}{kT_e}\right) \frac{\sigma_{eM}}{\sigma_{qe}} \ .$$

Usually, it can be assumed that $W/W_B \gg 1$ (even though it is necessary to check this inequality under specific conditions). However, if the heating of electrons during collisions of the second kind is negligible, T_e in the thermal layer may increase. Let us transform the energy balance equation (4.66) into a dimensionless form, introducing dimensionless variables,

$$\chi = x/x_T, \quad \Theta_e(\chi) = T_e(\chi)/T_e(0) \ ,$$

$$\Theta_M(\chi) = T_M(\chi)/T_M(0), \quad \zeta(\chi) = n_e(\chi)/n_e(0) \ , \qquad (4.68)$$

and dimensionless coefficients

$$\varepsilon = T_e(0)/T_M(0), \quad \eta_i = eV_i/2kT_e(0) ,$$

$$\psi = \sigma_e(0)\, E^2/W\,(0) ;$$

$$\Lambda = \Xi'(0)\, T\,(0)\, (\varepsilon-1)/x^2 W\,(0) . \quad (4.69)$$

The coefficient $\psi \geq \varepsilon - 1$ characterizes the relation between energy transferred to the electrons by the heating field in a certain volume of the undisturbed plasma with the energy transferred by them to the heavy component in the same volume. The quality $\psi = \varepsilon - 1$ corresponds to the case when all losses are volume losses in undisturbed plasma and the energy transfer is of no importance. In the absence of a heating field, $\psi = 0$.

The coefficient Λ characterizes the relative importance of electron thermal conductivity and volume interaction of electrons with the heavy component of plasma in the thermal layer. At a low degree of ionization and a predominance of collisions with particles of one type,

$$\Lambda_e = 0.25\ \varepsilon/\gamma_M \cdot m_M/m_e \cdot (L_e/x_T)^2 .$$

Within order of magnitude, this coefficient coincides with the ratio $\tau_{rel}/\tau_{e.t.}$. Conditions (4.64) and (4.65) now assume the following form,

$$\zeta(\chi) = 1 = s\chi , \quad (4.70)$$

$$\zeta(\chi) = \theta_M^{-1/2}\theta_e^{3/4} \exp\,(-\eta_i/\theta_e) . \quad (4.71)$$

The cross section of electron interaction with a heavy particle depends on the electron temperature. In a small range of variation of T_e, it can be assumed that this dependence is an exponential function ($\sigma_{eM} \sim T_e^m$). The collision rate is proportional, $T_e^{m-1/2}$. Similar exponential dependence can be assumed for ν_{eff} in a gas mixture (e.g., see Ref. 98). Taking this into account, let us transform equation (4.66) into the following form:

$$\Lambda \frac{d}{d\chi} \left(\zeta \Theta_M \Theta_e^{1/2-m} \frac{d\Theta_e}{d\chi} \right)$$

$$= \left(\left(\frac{\Theta_\varepsilon}{\Theta_M} \varepsilon - 1 \right) \Theta_e^{m+1/2} - \psi \Theta_M \Theta_e^{-m-1/2} \right) \zeta \; . \tag{4.72}$$

We will specify the following boundary conditions:

$$\Theta_e = 1 \text{ when } \chi = 0; \quad \frac{d\Theta_e}{d\chi} = 0 \text{ when } \chi = 1 \; . \tag{4.73}$$

The boundary condition for $\chi = 1$ (on the wall) means the following. A flux of thermal energy of electrons onto a solid wall, resulting from electron thermal conductivity is usually negligible as a result of a weak energy exchange between electrons and the wall during elastic reflection of low energy electrons from the wall. By effectively mixing fast and slow electrons from adjacent regions, electron thermal conductivity may play a predominant role in determining the nature of the temperature distribution of electrons in plasma. However, because, in this case, no energy is transferred to the wall, in practice it is of no importance to the overall balance of the electron energy in the volume of plasma bound by solid walls. However, an important contribution to the overall energy balance is made by the enthalpy transfer to the walls by an electron flux, where recombination energy is released and transferred either to the lattice of a solid or to the neutral gas molecules.

The boundary condition $\Theta_e(0) = 1$ indicates that the electron temperature at the external boundary of the thermal layer is not disturbed. If the gas temperature inside the thermal layer at its boundary drops sharply, and the importance of the electron thermal conductivity is considerable, electron temperature may be disturbed beyond the limits of the thermal layer. This occurs as a result of the appearance of a flux of thermal energy of electrons into the below-normal temperature region of the gas. Usually, the temperature of heavy particles near the external boundary of the thermal layer varies slowly. Therefore, this boundary condition is fulfilled. However, even for sharp variation of T_M in the thermal layer, the boundary condition is fulfilled in two cases: when the electron temperature follows the gas temperature ($\chi \ll 1$) and when, just the opposite, the importance of the electron conductivity is so great ($\chi \gg 1$) that, as a result of the absence of a flux of "electron" heat to the wall, the electron temperature inside the thermal layer changes very little.

Equation (4.72) is nonlinear. It will be solved by successive approximations for the case when the electron thermal conductivity plays the most important role in establishing the temperature distribution in the layer.

The equation for the first approximation ($\Lambda^{-1} = 0$) has the form

$$\frac{d}{d\chi}\left(\theta_{e1}^{1/2-m}(\chi)\,\zeta(\chi)\,\theta_M(\chi)\,\frac{d\theta_{e1}(\chi)}{d\chi}\right) = 0 \,.$$

Integrating twice and taking boundary conditions (4.73) into account, we obtain

$$\theta_{e1}(\chi) = 1 \,.$$

This indicates that, in the first approximation, the electron thermal conductivity equalizes the electron temperature inside the boundary layer.

The equation for the second approximation is obtained by substituting the following first approximation into the right side of equation (4.72)

$$\frac{d}{d\chi}\left(\zeta\theta_M\theta_{e2}^{1/2-m}\,\frac{d\theta_{e2}}{d\chi}\right)$$

$$= \Lambda^{-1}\left(\frac{\varepsilon}{\theta_M} - 1 - \psi\theta_M\right)\zeta \,. \qquad (4.74)$$

In integrating this equation, it will be assumed that the temperature of heavy particles in the layer varies linearly between the temperature of the unperturbed plasma, T_M (0), and the temperature of the probe, T_B, i.e.,

$$\theta_M(\chi) = 1 + \mathcal{J}\,\chi \,, \qquad (4.75)$$

where

$$\mathcal{T} = (T_3 - T_M(0))/T_M(0) \; .$$

In the subject problem, assuming such a temperature distribution for heavy particles is an adequate approximation, formula (4.75) describes accurately the distribution of T_M when it depends only on the thermal conductivity of gas.

Let us first find the second approximation, assuming a diffuse mode, when (4.70) is valid. In this case, equation (4.74) assumes the form

$$\Lambda \frac{d}{d\chi} \left((1+s\chi)\ (1+ \mathcal{T}\chi)\theta_{e2}^{1/2-m}(\chi)\ \frac{d\theta_{e2}(\chi)}{d\chi}\right.$$

$$= (\varepsilon/(1+ \mathcal{T}\chi) - 1 - \psi\,(1+ \mathcal{T}\chi))\ (1+s\chi) \; .$$

Double integration, taking into account boundary conditions (4.73) leads to the following expression:

$$\Lambda\,(\theta_{e2}^{3/2-m}(\chi) - 1)/(3/2-m) = f(\chi) \; , \qquad (4.76)$$

where m = 3/2

$$\Lambda \ln_{e2}(\chi) = f(\chi) \; . \qquad (4.77)$$

In formulas (4.76) and (4.77), the right side is given by the following expression,

$$f(\chi) = -\ \psi\chi^2/6\ -\ \chi(3s(1-\varepsilon) + 3\ \mathcal{T}\ \varepsilon$$

$$+\ \psi\,(s+ \mathcal{T}\,))/6s\ \mathcal{T}\ +\ (\psi(\ \mathcal{T}\ -\ 3s\ -\ s^2\ (2s\ \mathcal{T}$$

$$+\ 3s\ +\ 3\ \mathcal{T}\ +\ 6))/6\ +\ 0.5\ (1+s)((\varepsilon-1)\ s\ (1+s)$$

$$+\ \mathcal{T}\ \varepsilon\ (1-s)))\ \ln\,(1+s\chi)/(\ \mathcal{T}\ -s)\ s^2$$

$$+ (\psi (-s+3 \mathcal{T} + \mathcal{T}^2(2s \mathcal{T} + 3(s+ \mathcal{T}) + 6))/6$$

$$+ 0.5 (1+ \mathcal{T})((\varepsilon-1) (s (1- \mathcal{T}) -2 \mathcal{T})$$

$$- \mathcal{T} \varepsilon (1- \mathcal{T}))) \ln (1+ \mathcal{T} \chi)/(\mathcal{T} -s) \mathcal{T}^2. \tag{4.78}$$

Second integration was performed approximately. The term $(1 + \mathcal{T} \chi)$ under the integral sign was expanded in series, and only the first two terms were kept. Even when the probe is five times colder than the plasma gas ($\mathcal{T} = -0.8$), this approximation leads to a 20 % error. The error is smaller when the probes are warmer. If the electron concentration inside the thermal layer does not change ($s = 0$), relation (4.78) can be replaced by the following relation:

$$F(\chi) = - 1/4 (\psi+\varepsilon) \chi^2 + \chi (-\psi-2+3\varepsilon)/2 \mathcal{T}$$

$$+ (\psi (1/2 + 1/2 \mathcal{T}^{-2}+ \mathcal{T}^{-1})$$

$$- (1+ \mathcal{T}) (3\varepsilon-2- \mathcal{T} \varepsilon)/2 \mathcal{T}^2) \ln (1+ \mathcal{T} \chi)$$

In the case of a local ionization equilibrium inside the layer, when expression (4.71) is valid, the equation for the second approximation (4.74) assumes the form:

$$\Lambda \frac{d}{d\chi} \left((1+ \mathcal{T} \chi)^{1/2} \theta_{e2}^{5/4-m} \exp \left(-\frac{\eta_i}{\theta_{e2}}\right) \frac{d\theta_{e2}}{d\chi}\right)$$

$$= ((\varepsilon/(1+ \mathcal{T} \chi) - 1) (1+ \mathcal{T} \chi)^{1/2}$$

$$- \psi (1+ \mathcal{T} \chi)^{1/2}) \exp (1-\eta_i) . \tag{4.79}$$

After integration of equation (4.79), subject to the boundary conditions (4.73), we obtain

$$\exp\,(-\eta_i(\theta_{e2}^{-1} - 1))\ \theta_{e2}^{13/4-m}(1-(13/4-m)\ \theta_{e2}/\eta_i)$$

$$= 1 - (13/4-m)/\eta_i - 2\eta_i\ (\varepsilon \ln (1+ \mathcal{F}\ \chi)$$

$$+ \ \mathcal{F}\ \chi + \psi\ ((1+ \mathcal{F}\ \chi)^2-1)/6$$

$$- 2\ ((1+ \quad \chi)^{1/2}-1)\ (\varepsilon/(1+ \quad)^{1/2}$$

$$+ (1+ \quad)^{1/2} + \psi\ (1 + \quad)^{3/2}/3))/\Lambda^{\ 2}\ . \tag{4.80}$$

Integration was performed approximately, with an accuracy of up to η_i^{-2} terms.

Expressions (4.76) and (4.80) can be used to find the electron temperature in the second approximation (θ_{e2}), near the probe when $\chi = 1$, provided it is first established whether diffuse or equilibrium distribution of electron concentration in the layer applies.

Electron temperature can be estimated from relations (4.76) and (4.80), if the dimensions of the thermal layer and the temperature of the probe are known, which can be estimated or determined experimentally. When applying relations (4.76), it is necessary to use parameter s, which can be determined from the experimental ion current density to the probe (j_p) at a floating potential:

$$s = \frac{d\zeta}{d\chi} = - j_p x_T/eD_a n_e(0)\ .$$

Various factors influence the electron temperature variation inside the thermal layer differently. A decrease in the temperature of gas in the layer is one of the factors responsible for deviation of $\theta_{e2}\ (\chi)$ from unity. Therefore, when $\mathcal{F} \to 0$ and $\psi = \varepsilon - 1$, in accordance with expressions (4.76) and (4.80) the second approximation does not differ from the first, i.e., $\theta_{e2} \to 1$.[1] Taking into account the decrease of electron

[1] When $\psi > \varepsilon - 1$, it may turn out that $\theta_{e2} > 1$, because introduction of the wall that does not cool the gas, but only prevents energy loss from plasma through electron electrical conductivity, may only increase the electron temperature.

concentration in the thermal layer leads to a drop in the value of the coefficient of electron thermal conductivity and energy losses during interactions with gas. These changes lead to an increase of the electron temperature in the layer, as compared with the case when the electron temperature is constant. The influence of n_e on T_e through volume losses is more substantial than the influence of n_e on T_e through Ξ_e'. Therefore, T_e is higher than when $\zeta = 1$ in the layer.

An increase of Λ leads to an increase in the electron temperature in the layer. Heating by current ($\psi \neq 0$) also raises the temperature. Consider two examples. Experimental conditions chosen are as follows: argon arc, $P = 6.7 \times 10^3$ Pa, $T_e(0) = 12{,}000$ K, $T_M(0) = 1600$ K, $T_3 = 800$ K, $n_e(0) = 5 \times 10^{12}$ cm^{-3}, $j_p = 2 \times 10^{-4}$ A/cm^2, $\gamma = 5$, $x_T = 0.5$ cm, $m = 1$, $L_e = 8.7 \times 10^{-3}$ cm, $D_a = 1.4 \times 10^2$ cm^2/s. Under these conditions, $\varepsilon = 7.5$, $\psi = 6.5$, $\Lambda = 8.4$, and $\mathcal{G} = -0.5$.

The diffusion length is

$$\ell_D = (D_a/\alpha_R n_e)^{1/2} = 0.53 \text{ cm} .$$

Here, $\alpha_R = 10^{-10}$ cm^3/s is the coefficient of volume recombination. It is assumed that relations (4.76) and (4.78) can be used. Then, from the experimental conditions, $s = -0.9$. The calculations show that the temperature of electrons near the probe, calculated in the second approximation, differs by 20 % from the electron temperature in the undisturbed plasma.

Calculation of the same example on the assumption that $s = 0$ (i.e., n_e inside the thermal layer does not change) shows that, as was to be expected, the temperature is lower than in the previous case, when $n_e(x_T) = 0.1\, n_e(0)$. However, the difference is not too great (by a factor of one and a half).

As a second example, consider a probe in the flow of combustion products seeded with potassium at $P = 3.1 \times 10^4$ Pa, $T_M(0) = T_e(0) = 2800$ K, $T_3 = 1400$ K, $L_e = 1.2 \times 10^{-3}$ cm, $x_T = 10^{-2}$ cm, $\gamma = 25$. Under these conditions, $\Lambda = 8.4$, $\psi = 0$, $\eta = 9$, $\varepsilon = 1$, $m = 1.25$ and $\mathcal{G} = 0.5$. Relation (4.80) will be used.

Results of the calculations show that, for the same $\mathcal{G}$ and Λ as in the previous example, but at a considerably stronger temperature dependence of the electron concentration (in accordance with the Saha equation), the variation of the electron temperature in the thermal layer is not as great. However, it is still 10 % lower in comparison with T_e (0).

Thus, the algebraic relations obtained make it possible to estimate the decrease of electron temperature in the thermal layer with respect to a cold probe when the predominant influence is exerted by electron thermal conductivity. When the electric layer is considerably smaller than the thermal layer, this decrease makes it possible to obtain a relation between the temperature being measured and the electron temperature in the undisturbed plasma.

4.5.2. The Influence of the Temperature of the Probe on the Temperature Being Measured When the Electron Temperature is Equal to That of Heavy Particles. Let us next consider the feasibility of determining electron temperature when a thermal layer comparable to the electric layer exists near the probe [125, 126]. The relation between the electron mean free path and characteristic dimensions of the layer depends on the condition that the electron temperature is equal to that of the heavy particles. This condition is fulfilled provided $\tau_{rel}/\tau_{e.t.} \ll 1$, which in accordance with formulas (4.61) and (4.62), leads to the relation

$$L_e(m_M/m_e\gamma)^{1/2} \ll x_T. \qquad (4.81)$$

Assume that no ionization, recombination, and attachment of electrons to heavy particles takes place inside the region of considerable electrical disturbance of the plasma by the probe. Then, excluding the region adjacent to the surface of the probe, with dimensions equal to L_e, the constant density of the electron current inside the plane electric layer is given by the expression,

$$- j_e = - eD_e \frac{dn_e}{dx} + en_e b_e \frac{dV}{dx}$$

$$- eD_e K_T \frac{dT_e}{dx} \frac{n_e}{T_e}, \qquad (4.82)$$

where K_T is the thermal diffusion ratio, V is the potential, and the x coordinate is counted from the surface of the probe. In the case of the thermionically emitting probe, the electron concentration at its surface may be represented in the following form [127]:

$$n_e(L_e) = -2(j_e - 2j_a)/e\nu_e(L_e) , \qquad (4.83)$$

where $n_e(L_e)$, $\nu_e(L_e)$, are the concentration and mean velocity of electrons at a distance equal to the mean free path from the probe, j_{em} is the current density emitted from the probe. The potential drop at a distance equal to the mean free path from the probe will be neglected.

Solving equation (4.82), gives n_e and, taking into account relation $b_e/D_e = e/kT_e$ from (4.83), we obtain the expression

$$j_e = \frac{-1/2\, e n_e(x_{el})\nu_e(L_e)\theta^{-K_T(\bar{T})}\mathcal{E}(0,x_{el}) - 2j_a}{1 + \frac{3}{2}\int_0^{x_{el}} \frac{1}{L_e}\mathcal{E}(0,\xi)d\xi}, \qquad (4.84)$$

where

$$\mathcal{E}(a,b) \equiv \exp\left(-\int_a^b \frac{e}{kT_e(x)}\frac{dV}{dx}dx\right) ;$$

$$\theta \equiv T_3/T(x_{el}); \quad \bar{T} = T_3(\theta-1)^{-1}\ln\theta .$$

At a constant temperature in the neighborhood of the probe and in the absence of emission, expression (4.84) is analogous to that obtained in Ref. 128. Consider in more detail the dependence of j_e on the probe potential. If one deals with probes with a potential close to the floating potential, the emission current density, j_{em}, may be, first of all, characterized by thermal emission, and be independent of the potential. An example will clarify the importance of the numerator in expression (4.84) for the dependence of current on the potential.

Assume that the variation of potential and the temperature inside the electric layer is:

$$V(x) = V_3\,(1 - x/x_{el}) ;$$

$$T(x) = T_3\,(1 + (\theta^{-1} - 1)x/x_{el}) .$$

In this case, the exponential in the numerator in expression (4.84) assumes the form

$$\mathcal{E}(0,x_{el}) = \exp\left(\frac{eV_3}{kT_e(x_{el})}\ \frac{\ln\theta}{1-\theta}\right) .$$

Using inequality (4.81), the numerator in expression (4.84) can be represented approximately in the form of a sum,

$$1 + 3/2x_{el}/L_e \cdot kT_3/eV_3 .$$

At high pressure, one can neglect 1 in comparison with the second term. In the example under consideration, the dependence of the denominator in expression (4.84) on the potential at sufficiently high $|V_3|$ shows a much weaker dependence than that of the numerator on the potential.

A weak dependence of the numerator in (4.84) on the potential at sufficiently high negative values occurs in practically all of the interesting cases of distributions of V(x) and T(x) near the probe. As the probe potential approaches potential of the plasma, the influence of the numerator in expression (4.84) on the dependence of the electric current on potential $j_e(V_3)$ increases.

At a sufficiently high negative potential of the probe, formula (4.84) can be rewritten in the form

$$j_e = A \exp(eV_3/kT_{eff}) - B , \qquad (4.85)$$

where A, B, are coefficients that are almost independent of the probe potential.

$$T_{eff} = eV_3/-k \int_0^{x_{el}} \frac{e}{kT_e(x)}\ \frac{dV}{dx}\, dx . \qquad (4.86)$$

In the absence of a thermal layer, $T_{eff} = T$, in plasma. When a thermal layer is present, T_{eff} depends on $T(x_{el})$ and on variation of T inside x_{el}. It should be noted that the temperature of electrons arriving at the probe is practically equal to the probe temperature. However, their number depends considerably on the diffusion coefficient and mobility at the outer boundary and inside the electrical layer. Therefore, the temperature at the outer boundary of the layer can also be estimated from the electron current.

The total current density to the probe may be represented by the following expression:

$$j_3 = j_e + j_a + j_p = j_+ + A \exp (eV_3/kT_{eff}) \, , \tag{4.87}$$

where j_p is the density of ion current to the probe, $j_+ = j_a + j_p - B$.

If the dependence of j_+ and A on the potential is not too significant, one can use the usual voltage-current probe characteristic to determine T_{eff}. This occurs when the emission current is the thermal emission current and the ion current to the probe and thermal emission current do not substantially exceed ion current in plasma at rest. Only in this case the floating potential, at which electron current is still noticeable and convenient two-probe characteristics can be obtained, is sufficiently different from the plasma potential.

Thus, simple analysis shows that, when electron temperature and temperature of heavy particles inside the thermal layer are equal, one can use the probe characteristic to determine T_{eff} that depends on the temperature at the outer boundary of the electric layer. When the dimension of the electric layer changes, T_{eff} is given by the ratio x_{el}/x_T.

4.5.3. Experimental Investigation of the Influence of the Temperature of the Probe on the Temperature Measurements. Experimental investigations of the influence of the temperature of the probe on the temperature being measured were conducted in combustion products of natural gas in air at atmospheric pressure [45, 129] and in a Meker burner (see 4.2 and 4.4).

The probes consisted of two cylindrical, stainless steel horizontal electrodes placed into alundum ceramic tubes so that only 20 mm of the probe was in contact with combustion products. The diameter of the probes varied between 0.5 and 5 mm. The working sector of the probe was located at a distance of 12 mm from the orifice of the burner in the inner flare. Both probes had identical dimensions and were exposed to identical conditions. The distance between them was chosen so that thermal and electric layers did not overlap. As with other probe measurements, the distance between probes, x_{1-2}, cannot be too long, because the potential drop inside the layer of the spatial charge, V_s, should be much larger than in plasma between the probes, V_p. Because at an adequately high negative potential of the probe the current in the region of the spatial charge is caused by ion emission and the conductivity inside the plasma is electronic, the following condition is obtained:

$$V_s/V_p \simeq b_e x_{el} b/b_p x_{1-2} >> 1 .$$

Because the distance between the probes was very short (x_{1-2} = 6 mm), this inequality was satisfied. The plasma temperature in the region undisturbed by the probe, measured by the reversal method, remained constant during all experiments (T_o = 2050 ± 20 K). The temperature of the probes was varied between 300 and 1650 K by changing the diameter of the probes and also by using air- or water-cooled probes in the form of tubes. The water-cooling system of each probe was designed to prevent current leakage from the probe circuit. The temperature of the probes was measured by a pyrometer or Chromel-Alumel thermocouple built into the external surface of the probe, at a distance of 2 mm from the working zone. Air cooling made it possible to decrease the temperature to 770 K and water cooling to 300 K. The seed, in the form of an aqueous solution of nitrogen chloride, was injected only into the inner flare with constant temperature over the cross section. The sodium atom concentration in the inner flare, measured from the width of the spectral (line determined by means of a Fabri-Perot interferometer) varied between 3 x 10^{10} and 1.0 x 10^{14} cm^{-3}.

Figure 4.13, from a photograph taken from the end from which insulation was removed for photography, shows the flow around one of the probes. The MZ-2 film used is sensitive in the visible part of the spectrum, where combustion products do not radiate. Therefore, one can see only the central flare colored by sodium salt. A laminar thermal layer that appears dark on the photograph surrounds the bright tube probe. Such photographs were used to determine the temperature distribution in the boundary layer of the probe. The distribution of the emission intensity determined from the darkening of the film was used for that purpose. In these experiments, the sodium atom concentration in the undisturbed plasma (n_{Na0})

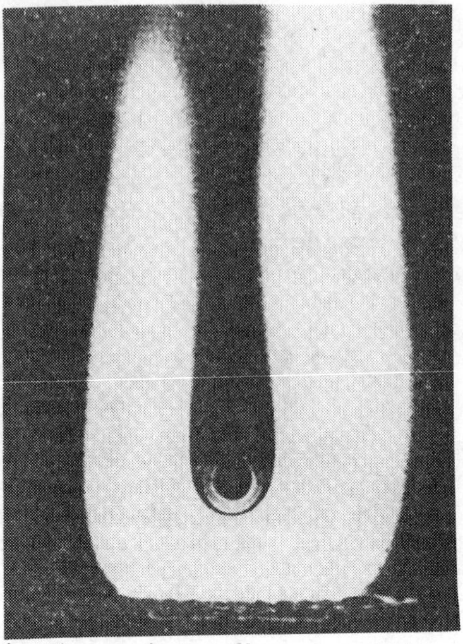

Fig. 4.13 Combustion products flow around the probe (T_3 = 1140 K, d_3 = 6 m, n_{Na0} = 2 x 10^{13} cm^{-3})

was 2 x 10^{13} cm^{-3}. The recorded intensity of emission depended almost completely on the integral intensity of the 5890/5896 Å sodium resonance doublet. The distribution of the integral intensity of the line at adequate optical thickness depends on the distribution of the concentration of emitting atoms and temperature in accordance with the growth curves for lines with a Voigt contour [95] in the following manner:

$$\frac{J(x)}{J_o} = \left(\frac{n(x)T(x)}{n_o T_o} \right)^{1/2} \exp \left(\frac{h\nu}{kT_o} \left(1 - \frac{T_o}{T(x)}\right)\right) . \tag{4.88}$$

Subscript $_o$ refers to the undisturbed plasma, and the coordinate, x, characterizes the position inside the boundary layer. Because the dependence of intensity on the temperature is considerably stronger than on the concentration of atoms, the temperature distribution was computed from the measured intensity distributions, using successive approximations. First, it was assumed that the concentration of sodium atoms in the boundary layer remains constant, i.e., in expression (4.88), $n_{Na}(x)$ =

n_{Nao}. The first approximation to the temperature distribution was then found from the measured intensity distribution $J(x)/J_o$. Then, the temperature distribution was used to determine the distribution of concentration of sodium atoms in the boundary layer, with chemical reactions taken into account. The distribution $n_{Na}(x)/n_{Nao}$ obtained in such a manner was substituted into expression (4.88), which then specified the temperature distribution in the second approximation. The temperature distributions determined in the first and second approximations are close to each other. However, deviations increase with decreasing distance from the probes, but do not exceed 50 K. The next approximation is practically identical to the second approximation.

Optical measurements made it possible to construct a temperature distribution in the thermal layer at distances from the surface of the probe where, because of low temperature, luminescence of seed was too weak to expose the film (Fig. 4.14). As the diameter of the probe decreased, the thickness of the thermal layer decreased also.

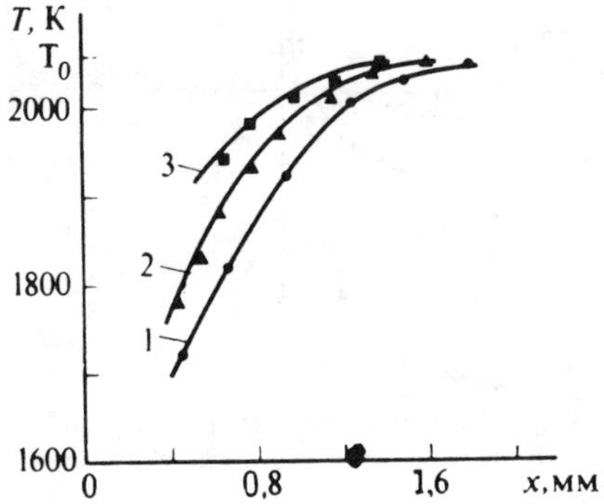

Fig. 4.14 Temperature distribution in the boundary layer near the probe at a frontal point. Points--probe measurements, lines--optical measurements; 1 - T_3 = 1140 K, d_3 = 6 mm, $x_T \simeq 1.8$ mm, 2 - T_3 = 1260 K, d_3 = 4 mm, $x_T \simeq 1.6$ mm, 3 - T_3 = 1330 K, d_3 = 2 mm, $x_T \simeq 1.4$ mm

The voltage-current characteristics of a double probe were recorded at various temperatures, probe diameters, and sodium atom concentrations. The probe characteristics in the temperature range 750-1300 K were obtained at 50-60 K intervals; when $T_3 > 1300$ K, every 80-100 K and when water cooling was used ($T_3 < 750$ K), the probe characteristics were obtained only at 300 and 400 K. The maximum temperature of the probe without forced cooling, at which voltage-current characteristics were obtained at probe diameters of 6, 4, 2, and 0.5 mm, were 1140, 1260, 1330, and 1650 K, in the same order. Figure 4.15 shows the voltage-current characteristics of a double probe at various temperatures of the probe and at constant concentration of sodium atoms. The voltage-current characteristics obtained in combustion products at atmospheric pressure (Fig. 4.15) have the same form as at low pressure.

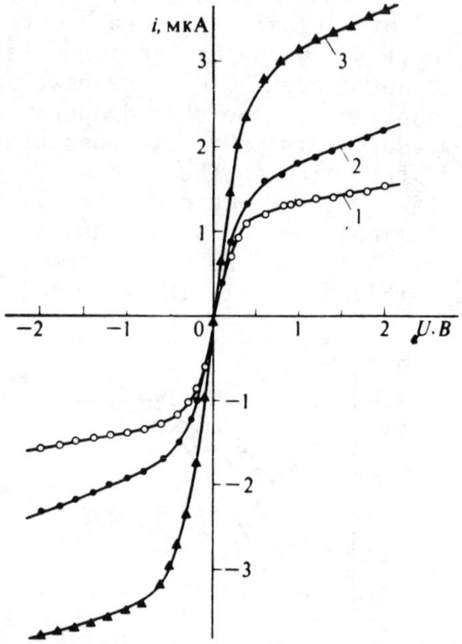

Fig. 4.15 Voltage-current characteristics of a double probe: 1 - T_3 = 750 K, d_3 = 6 mm; 2 - T_3 = 970 K, d_3 = 6 mm; 3 - T_3 = 1260 K, d_3 = 4^3 mm

Temperature determination using such characteristics was carried out using the following formula [101, 130, 131]:

$$\frac{kT_e}{e} = \frac{i_{1+}\, i_{2+}}{i_{1+} + i_{2+}} \frac{di}{dU}_{i=o} , \qquad (4.89)$$

where i_{1+}, i_{2+} are saturation currents to the probes, $(d_i/dU)_{i=o}$ is the slope of the probe characteristic at a floating potential, U is the voltage between the probes.

Formula (4.89) was derived for the low-pressure case, i.e., when $L_e \gg r_3$, $L_e \gg x_{el}$ [130, 131]. Derivation of formula (4.89) is based on the assumption of an exponential dependence of the electron current on the potential and a weak dependence on the current potential directed opposite to that of the electron current. From relation (4.87), it follows that these assumptions can be fulfilled even when T_e in formula (4.89) is replaced with T_{eff}. The error of determining T_{eff} because of measurement error i_{1+}, $i_{2+}(di/dU)_{i=0}$ during the experiments was 3 %.

At constant seed concentration, both currents i_{1+} and i_{2+}, as well as the derivative at zero (see 4.15), increase with increasing probe temperature. The influence of changes of currents i_{1+} and i_{2+} is more substantial, leading to a temperature increase given by formula (4.89).

A probe of 6 mm diameter was used during the investigation of the dependence of T_{eff} on the temperature of the probe in the range of 300–1140 K probe temperature. Thinner probes were used to obtain higher values of T_3. This, of course, was accompanied by a decrease in the thermal layer near the probe.

The temperature being measured increases with increasing temperature of the probe (slowly at first, then more rapidly), reaching the temperature of plasma in the undisturbed region. Further increase of probe temperature had no effect (Fig. 4.16). The lower the concentration of sodium atoms, the lower the probe temperature at which the temperature being measured becomes equal to the temperature of the undisturbed plasma.

The dependence of the temperature being measured by a double probe on the concentration of easily ionized sodium

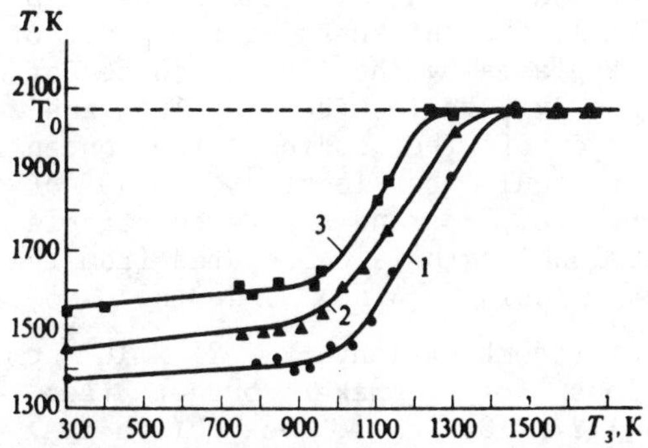

Fig. 4.16 Variation of the effect of temperature being measured by a double probe on the temperature of the probe at various concentrations of sodium atoms in the undisturbed plasma. 1 - 3 - 10^{14} cm^{-3}, 2 - 10^{13} cm^{-3}, 3 - 10^{12} cm^{-3}

at a constant temperature of the probe varied between 3 x 10^{10} and 6 x 10^{13} cm^{-3}. The nature of the dependence of the temperature being measured on the concentration of sodium atoms remains the same at all probe diameters. When the concentration of atoms is small, the temperature being measured is equal to the temperature of the undisturbed plasma. As the number of sodium atoms increases, the temperature being measured decreases (Fig. 4.17). A decrease of the probe diameter and an increase of its temperature result in the temperature being measured being equal to plasma temperature at large values of n_{Nao}.

To understand the results properly, it is necessary to determine ratios x_{el}/x_T, $\tau_{rel}/\tau_{e.t.}$, under experimental conditions. The electrical resistance of a cold thermal layer depends strongly on its thickness. It can be assumed that the probe current recorded passes primarily through the thinnest frontal part of the thermal layer. The experimentally determined thickness of the thermal layer in the frontal region of the probe was found to be 1-2 x 10^{-1} cm (see Fig. 4.15).

The thickness of the electric layer can be estimated by calculating n_e in the undisturbed plasma. Sodium atoms are practically the only ones ionized in the combustion products plasma. Calculating the electron concentration from Saha equation (4.65), using measured temperature and concentration of sodium atoms, shows that, when n_{Nao} changes from 3 x 10^{10} to 10^{14} cm^{-3} the electron concentration at T = 2050 K varies between 2 x 10^9 and 1.2 x 10^{11} cm^{-3}. This electron concentration leads to a change in the Debye radius, which varies between 9 x 10^{-4} and 7 x 10^{-3} cm. On the basis of many papers dealing with computation of ion current to the probe at high pressure (e.g., Ref. 132), the thickness of the layer of electrical disturbance of plasma by the probe can be estimated to be 10-20 R_D. Thus, x_{el} varies between 10^{-1} and 2 x 10^{-2} cm. During a change of the sodium atom concentration, the electric layer could be located inside or outside the thermal layer, i.e., x_{el} and x_T are comparable.

The diffusion length is calculated from the expression $\ell_D = (t_R D_a)^{1/2}$. Using (4.56) and the same value of the coefficient of recombination, α_R = 2 x 10^{-9} cm^2 x s^{-1}, as in calculations for a Meker burner (see 4.4.1.) and setting n_e = 1.2 x 10^{11} cm^{-3}, we obtain t_R > 4 x 10^{-3} s. When D_a is equal to 5 cm^2/s, ℓ_D > 0.45 cm. The diffusion length exceeds x_{el}. One can approximately assume that neither ionization nor recombination of electrons takes place.

Let us estimate $\tau_{rel}/\tau_{e.t.}$, or what's the same, let us verify the validity of equality (4.81). In accordance with foregoing estimates, we obtain

$$L_e = (n_M \sigma_{eM})^{-1} \simeq 10^{-4} \text{ cm} ,$$

$$m_M/m_e = 2 \cdot 10^4 , \quad \gamma = 10^2 ,$$

then

$$L_e (m_M/m_e \gamma)^{1/2} \ll x_T .$$

From this, it follows that $\tau_{rel}/\tau_{e.t.} < 1$ and the electron temperature in the thermal and electric layers is equal to the temperature of heavy particles. This indicates that the experimental results can be explained on the basis of an analysis of mutual location of the electric and thermal layers that control the temperature being measured, T_{eff}.

The change of the sodium atom concentration leads to a change of x_{el}. As the concentration of atoms decreases the layer grows, leading to a higher temperature at the outer boundary layer and higher average temperature in the layer. The temperature being measured also increases, but remains below the temperature of the undisturbed plasma. Further increase in concentration of atoms may result in the electric layer leaving the boundaries of the thermal layer. Then, the temperature being measured becomes equal to the temperature of the undisturbed plasma. Furthermore, additional expansion of the electric layer does not change this pattern. All of these factors clarify the dependence of T_{eff} on the concentration of atoms (see Fig. 4.17).

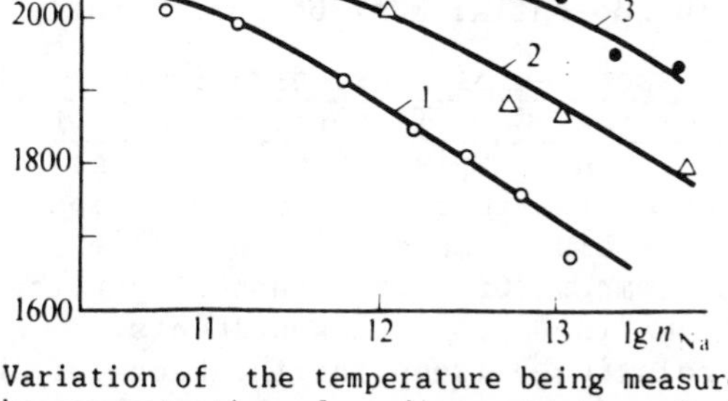

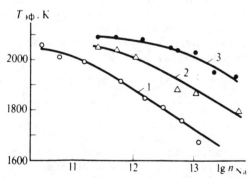

Fig. 4.17 Variation of the temperature being measured by double probe with the concentration of sodium atoms. 1 - T = 1140 K, d_3 = 6 mm; 2 - T = 1260 K, d_3 = 4 mm; 3 - T = 1330 K, d_3 = 2 mm

If the temperature and the probe dimensions are changed while the sodium atom concentration remains the same, the thickness of the electric layer remains practically constant. Only the temperature of the outer boundary of the electric layer changes. The higher the temperature of the probe and the smaller its diameter, the higher the temperature at the outer boundary of the electric layer. This explains the dependence of T_{eff} on the probe temperature (see Fig. 4.16).

The association between the temperature, T_{eff}, being measured and the temperature at the outer boundary of the electric layer can be determined by using formula (4.86). This requires that the dependence of the potential on the distance to the probe V(x) be specified. Assume that this dependence is linear; then, using the experimental distribution $T_e(x)$ and $x_{el} = x_T$, we have $T_{eff} = 0.87\, T(x_{el})$. The influence of the temperature drop is even weaker in the case of a more completely filled profile of the potential distribution inside the layer. Therefore, it can almost be assumed that $T_{eff} \simeq T(x_{el})$.

The results of the temperature measurements indicate that the floating potential of the probes was adequately high, i.e., neither the gas flow at low flow rate nor the electron emission that could appear at the maximum probe temperature had any effect on the floating potential. Otherwise, the probes would give higher temperature readings.

The effects of negative ions formation [109, 110] under conditions of the experiment are insignificant. This is verified, not only by the results of measurements, but also by the very low concentration of negative ions in the vicinity of the probes in comparison with the electron concentration [23].

Therefore, two conditions must be fulfilled in order to be able to use cold probes in the plasma flow of an MHD generator. The electric layer must lie beyond the boundaries of the thermal layer and the absolute magnitude of the floating potential must be sufficiently high.

4.5.4. Experimental Investigation of the Relative Distribution of the Gas Temperature Near the Surface of the Probe. An explanation of the experimental investigations of the influence of the temperature of the probe on its readings makes it possible to attempt to construct the temperature distribution near the surface of the probe from the probe measurements. This requires obtaining a relation between the thickness of the electric layer and the concentration of the easily ionized sodium atoms [126, 133].

As has been noted in foregoing text, characteristic thickness of the electric layer is specified by the Debye radius. This indicates that x_{el} is proportional to $n_e^{-1/2}$. In addition, one can assume that the electron concentration in the region of a spatial charge is proportional to the electron concentration in the undisturbed gas (n_e), i.e., $x_{el} \sim n_{eo}^{-1/2}$. Apparently, this assumption is justified when the diffusion length exceeds the thickness of the thermal boundary layer, i.e., when ionization and particle recombination in the thermal layer can be neglected.

The concentration, n_{eo}, outside the thermal layer when seed concentration is changed is given by the Saha equation, i.e.,

$$n_{eo} \sim n_{Nao}^{1/4} .$$

Hence, it can be assumed that

$$x_{el} \sim n_{Nao}^{-1/4} .$$

The thickness of the electric layer can be affected by the change of the temperature inside the layer, which may lead to two opposite effects: a decrease of the layer in accordance with the relation $R_D \sim T_e^{1/2}$ and, apparently, expansion of the layer as a result of electrons leaving the probe because of thermal diffusion. Under the experimental conditions characterized by a weak temperature variation in the region of investigations, this effect should be negligible, and will be neglected.

Having determined the concentration, n_{Na1}, at which the electron temperature measured by the probe becomes equal to the temperature in the undisturbed plasma, i.e., when the electric layer is equal to the thermal layer, one can find the relative thickness of the electric layer:

$$x_{el}/x_T = (n_{Na1}/n_{Nao})^{1/4} . \qquad (4.90)$$

Using this relation and scaling the dimension along the abscissa axis, it is possible to change from the dependence, $T_{eff}(n_{Nao})$, (see Fig. 4.17) to the temperature distribution in the thermal layer. The quantity n_{Na1} is determined from curves in Fig. 4.17. This is the same concentration at which T_{eff} ceases to change with further decrease n_{Nao}. When conducting the experiments, the thickness of the thermal layer, x_T, is known from optical measurements. Therefore, it was possible to construct the temperature distribution on absolute scale (see Fig. 4.14). Even though relation (4.90) is quite approximate and it assumed that all of the current passes near the frontal point of the zone, satisfactory agreement exists between this distribution and that determined optically. Accordingly, it is possible to take into account the influence of the temperature of the relatively cold probe on the temperature measurements in important cases of practical importance:

- when the electric layer is considerably thinner than the thermal layer and the electron thermal conductivity exerts predominant influence on the establishment of the electron temperature in the thermal layer, and
- when the electric layer is comparable to the thermal layer and the electron temperature is equal to the temperature of the gas in the thermal layer.

In the first case, the approximate relations obtained make it possible to derive a relation between the temperature of the electrons in the electric layer and the temperature of electrons in the undisturbed plasma.

In the second case, the temperature determined from probe characteristics at sufficiently large negative probe potentials characterizes the temperature at the external boundary of the electric layer. Therefore, correct measurements of temperature in the undisturbed gas required that the electric layer extend outside the thermal layer. In addition, by varying thickness of the electric layer, e.g., by changing the concentration of easily ionized atoms in gas, it is possible to determine the temperature distribution in the thermal boundary layer when the electron temperature is equal to the gas temperature in the thermal layer.

REFERENCES

[1]Zwicker, G., "Determination of Parameters of Optically Thick Plasma," unspecified translation from English, 1971, p. 169-194.

[2]Bartels, H., "Uber Linienemission aus inhoneogener Schict," Zeitschrift der Physik, Bd. 125, T. 1, 1949, p. 597-614; Bd. 126, T. 2, 1949, p. 108-140.

[3]Cowan, R. D., Dicke, G. H., "Self-absorption of Spectrum Lines," Reviews of Modern Physics, Vol. 20, No. 2, 1948, p. 418-455.

[4]Peobrazhenskiy, N. G., Spectroscopy of Optically Thick Plasma (Spektroskopiya opticheski plotnoy plazmy), Novosibirsk, Publishing House Nauka, 1971.

[5]Fishman, I. S., Nurmatov, E. M., "Intensity in the Maximum of Self-Reversed Lines," Teplozika vysokikh temperatur, Vol. 11, No. 4, 1973, p. 946-951.

[6]Kolobova, G. A., "Determination of Temperature from Maxima of Self-Reversed Lines," Zhurnal prikladnoy spektroskopii, Vol. 14, No. 2, 1971, p. 246-249.

[7]DeGroot, J. J., Jack, A. G., "Plasma Temperature Measurements Using Self-absorbed Spectral Lines: A Discussion of the Method Due to Bartels and Kruithof," JQSRT, Vol. 13, No. 7, 1973, p. 615-626.

[8]Wesselink, G., deMooy, D., van Gemert, M. Y. C., "Temperature Determination of High Pressure Optically Thick Gas Discharges by a Modified Bartels Method," Journal of Physics, D: Applied Physics, Vol. 6, No. 4, 1973, p. 27-30.

[9]Frish, S. E., Optical Spectra of Atoms (Opticheskiye spektry atomov), Moscow, Publishing House Fizmatgiz, 1963.

[10]Sobolev, N. N., "Optical Methods of Measuring Plasma Temperature," Trudy FIAN (Transactions of FIAN), Vol. 7, 1956, p. 159-229.

[11]Chen, Sh., Takeo, M., "Broadening and Shifting of Spectral Lines Produced by Other Gases," Uspekhi fizicheskikh nauk, Vol. 66, No. 3, 1958, p. 391-470.

[12]Hinnov, F., Kohn, H., "Optical Cross-Sections from Intensity-Density Measurements," Journal of the Optical Society of America, Vol. 47, No. 2, 1957, p. 156-162.

[13]Hofman, F. W., Kohn, H., "Optical Cross Sections of Resonance Lines Emitted by Flames under Conditions of Partial Thermal Ionization," Journal of the Optical Society of America, Vol. 51, No. 5, 1961, p. 512-521.

[14]Behmenburg, W., Kohn, H., "An Acetylene--Oxygen Flame Using Various Diluents for Study of Broadening and Shift of Spectral Lines," JQSRT, Vol. 4, No. 1, 1964, p. 163-176.

[15]Behmenburg, W., "Broadening and Shift of the Sodium Line by Various Perturbing Gases under Flame Conditions," JQSRT, Vol. 4, No. 1, 1964, p. 177-190.

[16]Sobel'man, I. I., Introduction to Theory of Atomic Spectra (Vvedenie v teoriyu atomnykh spektrov), Moscow, Publishing House Nauka, 1977.

[17]Vasil'yeva, I. A., Deputatova, L. V., Nefedov, A. P., "Experimental Investigation of the Coefficient of Absorption in the Wings of Resonance Lines of Sodium and Potassium Doublets," Optika i spektroskopiya, Vol. 39, No. 1, 1975, p. 15-20.

[18]Vasil'yeva, I. A., Deputatova, L. V., Nefedov, A. P., "Investigation of the Far Wings of Resonance Lines of Alkaline Metals in Combustion-Products Plasma," Second U.S.-U.S.S.R. Colloquium on Magnetohydrodynamic Electric Power Generation, Washington, D.C., June 5-6, 1975, Washington, 1975, p. 231-249.

[19]Vasil'yeva, I. A., Deputatova, L. V., Kirillov, V. V., Nefedov, A. P., "Influence of Plasma Nonuniformity on the Spectral Line Profile and Reversal Temperature," Optika i spektroskopiya, Vol. 33, No. 5, 1972, p. 825-831.

[20]Vasil'yeva, I. A., Deputatova, L. V., Nefedov, A. P., "Investigating the Flame with the Aid of Self-Reversed Contours of Spectral Lines," Combustion and Flame, Vol. 23, 1974, p. 305-311.

[21]Massey, H. S. W., Burhop, E. H. S., Electronic and Ionic Impact Phenomena, Oxford University Press, N.J., 1958.

[22]Brown, S., Elementary Processes in Arc Discharge Plasma, Russian translation from an unspecified source, 1961.

[23]Yungman, V. S., Gurvich, L. V., Ptishcheva, N. P., "Composition and Thermodynamic Properties of Product of Combustion of Methane with Ionizing Seed," Teplozika vysokikh temperatur, Vol. 4, No. 4, 1966, p. 507-512.

[24]Mnatsakanyan, A. Kh., "Kinetics of Elementary Processes in Plasma of Inert Gases, Molecules, and Vapors of Alkali Metals," Teplozika vysokikh temperatur, Vol. 12, No. 4, 1974, p. 858-875.

[25]Biberman, L. M., Vorob'yev, V. S., Yakubov, I. T., "Kinetics of Impact-Radiation Ionization and Recombination," Uspekhi fizicheskikh nauk, Vol. 107, No. 3, 1972, p. 353-387.

[26]Lagar'kov, A. N., "On Conditions of Aplying Local Thermodynamic Equilibrium," Teplozika vysokikh temperatur, Vol. 4, No. 3, 1966, p. 305-313.

[27]Ivanov, V. V., Radiation Transfer and Spectra of Celestial Bodies (Perenos izlucheniya i spektra nebesnykh tel), Publishing House Nauka, Moscow, 1969.

[28]Kutateladze, S. S., Borishanskiy, V. M., Handbook of Heat Transfer (Spravochnik po teploperedache), Publishing House Gosenergoizdat, Moscow, 1959.

[29]Vasil'yeva, I. A., "Source Function in Combustion Products Flow with Alkali Additive," Zhurnal prikladnoy spektrosckopii, Vol. 26, No. 2, 1977, p. 235-242.

[30]Vasil'yeva, I. A., Gurvich, L. V., Munvez, S. S., Nefedov, A. P., Shumyatskiy, B. Ya., Yundev, D. N., "Development of

Spectroscopic and Laser Investigations of Parameters of Working Fluids of Open-Cycle MHD Generators with Various Alkali Seeds," The Third U.S.-U.S.S.R. Colloquium on MHD Energy Conversion, Moscow, 1976; Publishing House IVTAN, Moscow, 1978, p. 339-348.

[31]Vasil'yeva, I. A., Deputatova, L. V., Nefedov, A. P., "Experimental Investigation of the Concentration of Atoms of Seed in an MHD Channel of the U-02 Facility"; Magnetohydrodynamic Method of Generating Electricity (Magnitogidrodinamicheskiy metod polucheniya elektroenergii), Publishing House Energiya, Moscow, 1972, p. 134-142.

[32]Kadyshevich, A. E., Measurement of Flame Temperature (Izmereniye temperatury plameni), Publishing House Metallurgizdat, Moscow, 1961.

[33]Sobolev, N. N. (ed.), "Optical Pyrometry of Plasma," Collection of non-Soviet papers, 1960.

[34]Heyden, A., Flame Spectroscopy, Russian translation, 1959.

[35]Vasil'yeva, I. A., "Determination of Temperature of Nonuniform Plasma by Generalized Reversal Method," Zhurnal prikladnoy spektroskopii, Vol. 22, No. 2, 1975, p. 203-209.

[36]Fayzulov, F. S., "Pyrometric Investigation of the State of Air, Nitrogen and Argon Behind a Shock Wave," Trudy FIAN (Transactions of FIAN), Vol. 18, 1962, p. 105-158.

[37]Litski, Ya., Sukever, Sh., "Analysis of the Influence of Self-Reversal of Spectral Lines on the Measurement of the Plasma Temperature," Teplozika vysokikh temperatur, Vol. 8, No. 3, 1970, p. 500-507.

[38]Vasil'yeva, I. A., Gaponov, I. M., Kirillov, V. V., et al., "Diagnostics of Plasma of Combustion Products in an MHD Generator Channel," Fifth International Conference on MHD Electrical Power Generation, Vol. 1, Munich, April 19-23, 1971, p. 27-42.

[39]Vasil'yeva, I. A., "Methods of Investigating Equilibrium Plasma in MHD Generator Channels," Magnetohydrodynamic Facilities (magnitogidrodinamicheskiye ustanovki), Publishing House Nauka, Moscow, 1975, p. 38-55.

[40]Vulis, L. A., "On the Influence of Temperature Pulsation on the Rate of Turbulent Combustion," AN Kazakh SSR, Izvestiya, Seriya energetika, No. 1, issue 1(15), 1959, p. 66-79.

[41]Musayev, D., Nusupbekova, D. A., Sakipov, Z. B., "Influence of Temperature Pulsations on the Emission Intensity of High Temperature Gases," AN Kazakh SSR, Vestnik, No. 11, 1973, p. 55-58.

[42]Zabotina, E. A., Tukhvatullin, R. S., "Influence of Emission Intensity Pulsations on the Measurement Accuracy of Plasma Jet Temperature by the Method of Absolute Intensity of the Spectral Line," Zhurnal prikladnoy spektroskopii, Vol. 19, No. 5, 1973, p. 796-799.

[43]Pikalov, V. V., Preobrazhenskiy, N. G., "Optical Diagnostics of Arc Plasma in the Presence of Temperature Fluctuations," AN Sibirskiy otdel SSR, Izvestiya, Seriya tekhnicheskikh nauk, No. 3, No. 1, 1976, p. 47-50.

[44]Vasil'yeva, I. A., "Determination of Temperature of Nonstationary Plasma by Generalized Reversal Method," Zhurnal prikladnoy spektroskopii, Vol. 22, No. 6, 1975, p. 981-984.

[45]Vasil'yeva, I. A., "Generalized Reversal Methods in Nonuniform, Nonstationary Plasma Flow in MHD Generator," Second U.S.-U.S.S.R. Colloquium on Magnetohydrodynamic Electrical Power Generation, Washington, D.C., June 5-6, 1975, Washington, D.C., 1975, p. 207-230.

[46]Kays, G. P., Methods and Instruments to Measure Parameters of Nonstationary Thermal Processes (Metody i pribory izmereniya parametrov nestatsionarynykh teplovykh protsessov), Publishing House Mashgiz, Moscow, 1959.

[47]Stupochenko, E. V., Losev, S. A., Osipov, A. I., Relaxation Processes in Nonstationary Thermal Processes (Relaksatsionnyye processy v udarnykh volnakh) Publishing House Nauka, Moscow, 1965.

[48]Nesterekhin, Yu. E., Soloukhin, R. I., High-Rate Measurements in Gasdynamics and Physics of Plasma (Metody skorostnykh izmereniy v gazodinamike i fizike plazmy), Publishing House Nauka, Moscow, 1967.

[49]Adams, J. M., "The Spectral Comparison Method for Temperature Measurement in Two-Phase Flames in Temperature, its Measurement and Control in Science and Industry," edited by Harmon H. Plumb, Pittsburgh, Vol. 4, No. 1, 1972, p. 627-643.

[50]Liddon, D. T., "An Automatic Remotely Operated Sodium D Line Reversal Temperature Measuring Technique," Combustion and Flame, Vol. 12, No. 6, 1968, p. 569-575.

[51]Jeanmarie, P., "Aplication of Line Reversal Method to Measurement of Shock Flow Electron Temperature," Physics of Fluids, Vol. 17, No. 2, 1974, p. 353-359.

[52]Maleyev, E. N., Ermakov, D. S., "Single-Beam, Two-Frequency Pyrometer for Plasma Temperature Measurements," Optika i spektroskopiya, Vol. 13, No. 3, 1962, p. 598-605.

[53]Huster, R., Kisch, D., Mahnig, M., "Opticah electronisches Verfahren zur Messung Hoher Gastemperaturen T. 1: Absorptionsverfahren," Ber. Ges. Phys. Chem., Bd. 73, N 1, 1969, p. 49-55.

[54]Hentschel, B., Huster, R., Kisch, D., Mahnig, M., "Optisch--electronisches Verfahren zur Messung hoher Gastemperaturen. T. 2. Emissionsverfahren," Ber. Ges. Phys. Chem., Bd. 73, N 1, 1969, p. 55-58.

[55]Vasil'yeva, I. A., Kirillov, V. V., Maksimov, I. A., et al., "Measurement of Plasma Temperature by a Spectroscopic Method with Continuous Automatic Recording," Teplozika vysokikh temperatur, Vol. 11, No. 4, 1973, p. 838-845.

[56]Belashov, N. A., Vasil'yeva, I. A., Gaponov, I. M., et al., "Plasma Diagnostics on the U-25 Facility" (Diagnostika plazmy na ustanovke U-25), Teplozika vysokikh temperatur, Vol. 12, No. 2, 1974, p. 417-424.

[57]Chechik, N. O., Faynshteyn, S. M., Lifshits, G. M., Electronic Photomultipliers (Elektronnyye umnozhitely), Publishing House Gostekhteorizdat, Moscow, 1957.

[58]Golant, V. E., Superhigh Frequency Methods of Plasma Investigations (Sverkhvysokochastotnyye metody issledovaniya plasmy), Publishing House Nauka, Moscow, 1964.

[59]Hild, M., Worton, S., Microwave Plasma Diagnostics, translated from English, 1968.

[60]Haddlestone, R., Leonard, S., Plasma Diagnostics, translated from English, 1967.

[61]Dushin, L. A., Pavlichenko, O. S., Plasma Investigations with Lasers (Issledovaniye plasmy s pomoshchyu lazerov), Publishing House Atomizdat, Moscow, 1968.

[62]Bartoli, U., Badyali, M., DeMarco, P., Sounding of Plasma by Electromagnetic Waves, translated from Italian, 1973.

[63]Kuznetsov, E. I., Shcheglov, D. A., Diagnostic Methods of High-Temperature Plasma (Metody diagnostiki vysokotemperaturnoy plazmy), Publishing House Atomizdat, Moscow, 1974.

[64]Pyatnitskiy, L. N., Laser Plasma Diagnostics (Lazernaya diagnostika plasmy), Publishing House Atomizdat, Moscow, 1976.

[65]Ginzburg, V. L., Propagation of Electromagnetic Waves in Plasma (Rasprostraneniye elektromagnitnykh voln v plazme), Publishing House Nauka, Moscow, 1967.

[66]Shkarovskiy, I., Johnson, T., Bachinskiy, M., Plasma Particle Kinetics, translated from English, 1969.

[67]Landsberg, G. S., Optics (Optika), Publishing House Gostekhizdat, Moscow, 1949.

[68]Tamm, I. E., Basics of Theory of Electricity (Osnovy teorii elektrichestva), Publishing House Gostekteorizdat, Moscow, 1949.

[69]Hirshfeld, G., Curtis, Ch., Bird, R., Molecular Theory of Gases and Fluids, 1961, translated from English.

[70]Allen, K., Astrophysical Quantities, 1960, translated from English.

[71]Smirnov, B. M., Atomic Collisions and Elementary Processes in Plasma (Atomnyye stolknoveniya i elementarnyye protsessy v plazme), Publishing House Atomizdat, Moscow, 1968.

[72]Sternheimer, R. M., "Electronic Polarizabilities of the Alkali Atoms," Physical Review, Vol. 183, No. 1, 1968, p. 112-122.

[73]Basharinov, A. E. (ed.), SHF Emission of Low-Temperature Plasma (SVCh izlucheniye nizkotemperaturnoy plazmy), Publishing House SAvetskoye Radio, Moscow, 1974.

[74]Mozer, I. F., Shtefen, G., Kneybel, R., "Submillimeter Wave Techniques," Uspekhi fizicheskikh nauk, Vol. 99, No. 3, 1969, p. 469-502.

[75]Valitov, R. A., Dyubko, S. F., Kamyshan, V. V., Submillimeter Wave Techniques (Tekhnika submillimetrovykh voln), Publishing House Sovetskoye Radio, Moscow, 1969.

[76]Chamberlain, J., Findlay, F. D., Gebbie, H. A., "Refractive Index of Air at 0.337 mm Wavelength," Nature, Vol. 206, No. 4986, 1965, p. 886-887.

[77]Kutovoy, V. D., Petrov, G. D., Samarskiy, P. A., et al., "Submillimeter Plasma Inteferometry on Two Wavelengths," Fizika plasmy, Vol. 1, No. 5, 1975, p. 857-860.

[78]Birch, J. R., Burroughs, W. J., Emery, J., "Observation of Atmospheric Absorption Using Submillimeter Maser Sources," Infrared Physics, Vol. 2, 1969, p. 75-83.

[79]Vasil'yeva, I. A., Shumyatskiy, B. Ya., Yundev, D. N., "Possibility of Measuring Electron Concentration and Electron Collision Rates in Combustion Products Flow by Means of a 337 μm Laser Interferometer," Teplozika vysokikh temperatur, Vol. 13, No. 6, 1975, p. 1242-1247.

[80]Vasil'yeva, I. A., Golubkova, A. S., Shumyatskiy, B. Ya., Yundev, D. N., "Investigation of Concentration and Electron Mobility in Combustion Products Flow Using an HCN Laser," Teplozika vysokikh temperatur, Vol. 14, No. 5, 1976, p. 1055-1060.

[81]Mnatsakanyan, A. Kh., Naydis, G. V., "Ionization and Recombination of Atomic Ions and Electrons in a Non-equilibrium Atomic-Molecular Plasma," Preprint of Institute of High Temperatures, No. 1-43, Moscow, 1979.

[82]Parkinson, G. J., Dangor, A. E., Chamberlain, J., "The Resolved Measurements of Electron Number Density and Electron Temperature Using Laser Interferometry at 337 μm Wavelength," Aplied Physics Letters, Vol. 13, No. 7, 1968, p. 233-235.

[83]Petrov, G. D., Petryakov, A. I., Samarskiy, P. A., "Submillimeter Laser Interferometry of a Coal Arc," Teplozika vysokikh temperatur, Vol. 10, No. 1, 1972, p. 181-182.

[84]Kolerov, A. N., Petrov, G. D., "Determination of Electron Concentration in Plasma with a Three-Mirror Submillimeter Laser Interferometer," Teplozika vysokikh temperatur, Vol. 11, No. 5, 1973, p. 1107-1108.

[85]Kolerov, A. N., Petrov, G. D., "Plasma Diagnostics in the Far Infrared (Submillimeter) Wavelength Range with a Mach-Z'ehnder Interferometer," Optika i spektroskopiya, Vol. 37, No. 3, 1974, p. 604-606.

[86]Kolerov, A. N., Petrov, G. D., "Submillimeter Laser Interferometer," Radiotekhnika i elektronika, Vol. 19, No. 6, 1974, p. 1282-1286.

[87]Berezhnaya, V. I., Petrov, G. D., Petryakov, A. I., "Submillimeter Polarimetry of Plasma in a Transverse Magnetic Field," Teplozika vysokikh temperatur, Vol. 12, No. 3, 1975, p. 684-686.

[88]Peterson, R. W., Jachoda, F. C., "A Far-Infrared Coupled Cavity Interferometer," Applied Physics Letters, Vol. 18, No. 10, 1971, p. 440-442.

[89]Véron, D., "High Sensitivity HCN Laser Interferometer for Plasma Electron Density Measurements," Optical Communications, Vol. 10, No. 1, 1974, p. 95-97.

[90]Etievant, C., "Les méthodes de mesure pour l'étude du plasma du Tokomak TFR," Entropie, Vol. 11, No. 61, 1975, p. 55-69.

[91]Throop, A. L., "A Pulsed Submillimeter Interferometer for Transient Plasma Diagnostics," IEEE, Transactions, Nuclear Science, Vol. 19, No. 1, 1972, p. 761-769.

[92]Dyubko, S. F., Efimenko, M. N., "Detecting Properties of Metal-InSb Point Contact at a Wavelength of 337 μm at T = 300 K," Pis'ma v zhurnal eksperimental'noy i teoreticheskoy fiziki, Vol. 13, 1971, p. 533-535.

[93]Heckenberg, N. R., Tait, G. D., Whitbourn, L. B., "A 337-μm HCN-Laser Interferometer for Plasma Diagnostics," Journal of Applied Physics, Vol. 44, No. 10, 1973, p. 4523-4533.

[94]Shchelokov, V. A., Yundev, D. N., "Submillimeter Laser Phase Meter for Plasma Diagnostics," Pribory i tekhnika eksperimenta, No. 6, 1975, p. 145-147.

[95]Penner, S. S., Quantitative Molecular Spectroscopy and Emissivity of Gases, translated from English, 1963.

[96]Frish, G. E. (ed.), Spectroscopy of Gas-Discharge Plasma (Spektroskopiya gasorazryadnoy plazmy), Publishing House Nauka, Leningrad, 1970.

[97]Ashin, M. I., Vasil'yeva, V. I., Kosov, V. F., et al., "Development of Methods for Determining the Electron Density in an MHD Generator Plasma," Sixth International Conference in

Magnetohydrodynamic Electrical Power Generation, Washington, D.C., June 9-13, 1975, Vol. 3, 1975, p. 185-199.

[98]Frost, L. S., "Conductivity of Seeded Atmospheric Pressure Plasmas," Journal of Applied Physics, Vol. 32, No. 10, 1961, p. 2029-2036.

[99]Atrazhev, V. M., Yakubov, I. T., "Electron Mobility in Dense Gases and Fluids," Teplozika vysokikh temperatur, Vol. 18, No. 6, 1980, p. 1292-1311.

[100]Gaponov, I. M., Poberezhsky, L. P., Chernov, Yu. G., "Study of the Electric Conductivity of Plasma of Combustion Products with Seedings in U-02 MHD Generator Channel and on a Laboratory Installation," Combustion and Flame, Vol. 23, No. 1, 1974, p. 29-35.

[101]Chen, F., "Electrical Probes," Russian translation from an unspecified source, 1967, p. 94-164.

[102]Shott, L., "Electrical Probes," Russian translation from an unspecified source, 1971, p. 459-505.

[103]Kozlov, O. V., Electrical Probe in Plasma (Elektricheskiy zond v plazme), Publishing House Atomizdat, Moscow, 1969.

[104]Smy, P. R., "The Use of Langmuir Probes in the Study of High Pressure Plasmas," Advances in Physics, Vol. 25, No. 5, 1976, p. 517-553.

[105]Grey, J., Jacobs, F. F., "Cooled Electrostatic Probe," AIAA Journal, Vol. 5, No. 1, 1967, p. 84-91.

[106]Lyddon, D. T., "Continuum Theory of Cooled Spherical Electrostatic Probes," Physics of Fluids, Vol. 12, No. 2, 1969, p. 356-360.

[107]Chapkis, R. L., Baum, E., "Theory of Cooled Spherical Electrostatic Probe in Continuous Gaseous Medium," Aeronautics and Cosmonautics, Vol. 9, No. 10, 1974, p. 92-98.

[108]Genkin, A. L., Lebedev, A. D., "Diffusion of Charged Particles in the Near-Electrode Boundary Layer," Magnitnaya gidrodinamika, No. 2, 1969, p 38-48.

[109]Tverdokhlebov, V. I., "Influence of Traping Process on the Probe Measurements," Zhurnal teckhnicheskoy fiziki, Vol. 38, No. 3, 1968, p. 465-468.

[110]Bailey, P. B., "Continuum Electrostatic Probes in the Presence of Negative Ions: A Numerical Solution," AIAA Journal, Vol. 11, No. 9, 1973, p. 1225-1226.

[111]Wen-Heul, Jou, Siu-I-Cheng, "Nonisothermal Theory of an Electrostatic Probe in a Weakly Ionized Gas," Physics of Fluids, Vol. 14, No. 10, 1971, p. 2144-2151.

[112]Barad, M., Cohen, I. M., "Continuum Theory of Spherical Electrostatic Probe in a Stationary Moderately Ionized Plasma," Physics of Fluids, Vol. 17, No. 4, 1974, p. 724-734.

[113]Sahni, O., "On Probe Measurements of Electron Temperature in Flame Plasmas," Journal of Physics, Vol. D2, No. 3, 1969, p. 471-473, Vol. D3, No. 2, 1970, p. 176-183.

[114]Gorelov, V. A., "Probe Measurements Behind a Front of a Strong Shock Wave in Air," Zhurnal tekhnicheskoy fiziki, Vol. 40, No. 1, 1970, p. 198-205.

[115]Sobolenko, K. M., "Measurement of Temperature and Electron Concentration in Flame by a Probe Method," Kharkov Politekhnical Institute, Vestnik, No. 2(50), 1965, p. 100-104.

[116]Ivanov, Yu. A.,Ovsyannikov, A. A., Oliver, D. Kh., Polak, L. S., "Probe Diagnostics of a Plasma Jet at Atmospheric Pressure," Experimental and Theoretical Investigation of Nonequilibrium Physical-chemical Processes (Eksperimental'noye i teoreticheskoye issledovaniye neravnovesnykh fiziko-khimicheskikh protsessov), Publishing House INKhS AN SSR, Moscow, Vol. 3, 1973, p. 496-505.

[117]Ferdinand, J., "Note on the Probe Technique in the Region of Intermediate Pressures," Physics Letters, Vol. A32, No. 6, 1970, p. 400-401.

[118]Ferdinand, J., "The Comparison of the Probe and Spectroscopic Diagnostic Methods on the H_2-O_2 Flame Plasma," Czechoslovakian Journal of Physics, Vol. B20, No. 7, 1970, p. 832-839.

[119]Cozens, J. R., "New Evidence for Excessive Electron Temperature in Torch Flames," International Symposium on Properties and Aplication of Ion-Temperature Plasma, International Congress on Theoretical and Aplied Chemistry, Publishing House Sovetskoye Radio, Moscow, 1965, p. 1-6.

[120]Rohatgi, V. K., "Study of Dense Plasmas by Double Floating Probe Method," Journal of Physics, Vol. D1, No. 4, 1968, p. 485-489.

[121]Bradley, D., Ibrahim Said, M. A., "Electrostatic Probe Theories and Measurements in Flame Plasmas," Journal of Physics D, Applied Physics, Vol. 6, No. 4, 1973, p. 465-478.

[122]Wolf, R. I., "Electrode Effects in Seeded Plasma," Aeronautics and Cosmonautics, Vol. 6, No. 12, 1966, p. 99-123.

[123]Yanagi, Tetsui, "New Technique of Electrostatic Single-Probe in Combustion Products," Japanese Journal of Applied Physics, No. 6, 1968, p. 605-611.

[124]Vasil'yeva, I. A., "Influence of Temperature of a Double Nonemitting Probe on Temperature Being Measured at Moderate and High Pressures," Teplozika vysokikh temperatur, Vol. 7, No. 6, 1969, p. 1053-1060.

[125]Ashin, M. I., Vasil'yeva, I. A., Nefedov, A. P., "Investigation of the Influence of the Temperature of the Probe on the Temperature of Electrons," Voprosy fiziki nizkotemperaturnoy plazmy (Problems of Low-Temperature Plasma Physics), Publishing House Nauka i tekhnika, Moscow, 1970, p. 33-36.

[126]Ashin, M. I., Vasil'yeva, I. A., Nefedov, A. P., "Experimental Investigations of the Relative Temperature Distribution Near Conducting Surfaces with the Aid of Voltage-Current Characteristics of the Gas Gap," Teplozika vysokikh temperatur, Vol. 8, No. 5, 1970, p. 944-950.

[127]Lyubimov, G. A., Mikhaylov, V. N., "Analysis of the Region of Disturbed Plasma Near an Electrode," AN SSSR, Izvestiya, Mekhanika zhidkosty i gaza, No. 3, 1968, p. 9-17.

[128]Kagan, Yu. M., "Probe Measurements at Intermediate and High Pressures," MHD Generators, Third International Symposium on Electrical Power Generation by Means of MHD Generators, Salzburg, Austria, July, 1966, Publishing House VINITI, Moscow, Vol. 1, 1967, p. 165-172.

[129]Ashin, M. I., Vasil'yeva, I. A., Nefedov, A. P., "Investigation of the Influence of the Temperature of a Double Cooled Probe on the Electron Temperature Being Measured in Plasma at Atmospheric Pressure," Teplozika vysokikh temperatur, Vol. 7, No. 4, 1969, p. 626-632.

[130]Biberman, L. M., 'Panin, B. A., "Measurement of Parameters of a High-Frequency, Electrode-Free Discharge with Two Probes," Zhurnal tekhnicheskoy fiziki, Vol. 21, No. 1, 1951, p. 12-17.

[131]Kagan, Yu. M., Perel', V. I., "Probe Measurements of Plasma," Uspekhi fizicheskikh nauk, Vol. 81, No. 3, 1963, p. 409-452.

[132]Kiel, R. E., "Continuum Electrostatic Probe Theory for Large Sheets on Spheres and Cylinders," Journal of Applied Physics, Vol. 40, No. 9, 1969, p. 3668-3673.

[133]Vasil'yeva, I. A., Nefedov, A. P., "The Effect of Thermal Boundary Layer Near the Probe upon Measuring the Temperature of Plasma Electrons of Medium and High Pressures," Proceedings of the International Conference on Gas Discharges, London, September, 1970, p. 162-166.

Chapter 5

Phenomena in the Near-Electrode Region of a Constricted Discharge

5.1. Introduction

The most important aspect of designing high-power output, stationary MHD generators is the development of erosion-resistant electrode walls of long-duration channels exposed to high heat fluxes. Therefore, much attention is focused on cooled metal electrodes. Their advantages (convenient fabrication, high electric and thermal conductivity, relative thermal stability, etc.) are frequently discussed in the literature. The results of tests in MHD facilities indicate the feasibility of long-duration operation (hundreds of hours). Longer operation of metal electrodes depends on ability to predict their erosion characteristics in the arc discharge mode, at relatively high mean current densities ($j \gtrsim 1$ A/cm^2) over the entire surface of the electrode.

According to the available experimental data [1-13], electric contact at the plasma-metal electrode boundary occurs in the form of individual constricted discharges, referred to as arc discharges. High heat fluxes released at the locations of such discharges destroy the electrodes, reducing their operating lifetime. The degree of damage at a specified heat flux depends on the thermophysical parameters of the electrode material and the average area exposed to the heat flux.

In other words, erosion in the presence of arc discharges depends considerably on the size of the area (or current density at a specified current to the spot) and its temperature. In general, determination of these parameters requires careful analysis of all processes occurring in the narrow near-electrode zone of the arc discharge and the governing transfer mechanism of electric charges in the plasma-electrode contact region. Accordingly, calculation of arc spot parameters under specific conditions must be based on a model of the near-electrode region. Experience shows that analysis of electrode spots generated by arc discharges in various media can be based on modifications of models used to investigate a vacuum discharge.

Analysis of parameters of a constricted discharge under conditions of an MHD generator channel [14, 15] also leads to the conclusion that description of this type of discharge is identical to that of an arc burning in the vapors of electrode material. Furthermore, processes in spots in the channel of an MHD facility are in many respects similar to processes on film electrodes during a discharge in vacuum (material vaporization, formation of a neutral medium, electron emission, charge particle generation in plasma, etc.) [16].

Therefore, in investigating the cathode region, the authors will analyze parameters and models of arc spots burning on both solid and film cathodes. The results obtained will be used to construct a model and to investigate parameters of an arc discharge under conditions existing in an MHD generator channel.

5.2. Present State of the Problem

Numerous observations and investigations of the cathode region over a period of many years [17-28] show that the arc discharge at the cathode is constricted into very small, bright formations, referred to as cathode spots, which can be stationary or can be displaced at different rates. As the arc current increases, the number of cathode spots also increases. Furthermore, they can appear and reappear as a result of decay of old spots, the lifetime of which is determined by their internal processes.

Cathode spots refer to a region containing an emitting sector of a cathode surface, a region of cathode potential drop, where the spatial charge is concentrated and a bright part of the arc (referred to as negative luminescence, i.e., an assumed site of intensified ionization) occurs [24, 25]. The exceedingly small dimensions, high mobility, high particle concentration, and short lifetime of a cathode spot makes the experimental investigation very complex and difficult. Reliable investigation of smaller regions of the cathode spots is frequently a completely unsolvable problem [25]. Therefore, most of the experimental data pertain to the cathode region as a whole.

A cathode spot is usually characterized by the following parameters: I, total current to the spot, u_k, cathode potential drop; G, cathode material erosion rate, or the mass of the metal vapor removed from the cathode region per unit time; ν, displacement rate; j, the mean current density; j_e, its electron components, T_k, the temperature of the cathode surface in the spot; $s = j_e/j$, electron component of the total current density of the spots; T_e and T, temperature of electrons and heavy particles, respectively, in the near-cathode region; α, degree of ionization, or the length of the characteristic regions of the cathode spot; h, regions of spatial charge; ℓ, ionization region; n_i and n_a, concentration of charged and neutral particles, respectively, in the near-cathode region; E, the electric field strength at the cathode. The data on these parameters are scattered throughout the literature and may differ considerably. No data are available for some of the parameters.

For example, the results of measurements of n_a, n_i, T_e, T, s, α, E, condensation coefficients, particle neutralization, energy accommodation, as well as the size of individual regions of the cathode spot, are practically nonexistent. Reference 29 describes an attempt to determine the ion component of current at the cathode in a magnetic field under conditions in a plasma accelerator, by comparing the excess force at the cathode, resulting from ion current, with an analogous force at the anode [29]. It was established that, for various electrode materials and gas in the vicinity of the electrode, the ion current is 10 % of the total current.

This value was obtained under very special conditions, assuming equality of pulses of neutrals at the cathode and the anode, which generally do not have to be fulfilled because of the difference in corresponding concentrations of neutrals. The data on the ion concentration are given only for rapidly moving spots, where $n_i \simeq 5 \times 10^{17}$ cm^{-3} and $T_e \simeq 2$ eV [30]. In experiments conducted at the present time, the data recorded include current to the spot, current density, cathode potential drop, cathode material erosion rate, spot displacement rate, heat flux into the electrode. These data are used in various models in order to calculate unavailable parameters of the spot.

Consider some of the principal results of the experimental investigations of an arc in a vacuum under conditions in an MHD channel required for further analysis. The most reliable and accurate characteristic of the arc, the cathode potential drop, u_k, has been determined with an error not exceeding 1 % for solid and film cathodes made of many different materials [25, 31-33]. It was determined that the voltage at the arc undergoes temporal oscillations. However, its minimum value remains at a constant level, which is assumed to be the cathode potential drop, u_k. It depends weakly on the current on solid cathodes and is completely independent of current on film cathodes, where the cathode drop is a function of the thickness of the film. The cathode potential drop is somewhat lower for a copper film on a copper substratum ($u_k \simeq 11$ V) than for solid copper cathode ($u_k \simeq 15$ V) [25, 31-33]. Only the effective near-electrode potential drop was measured under conditions in MHD generators in the presence of constricted discharges. Depending on conditions in an MHD generator channel, the effective near-electrode potential drop, consisting of a sum of potential drops in the space charge zone, ionization zone (analog of u_k) and the spreadout zone may reach several hundred volts [2, 4, 10, 12, 13, 34-37]. No direct measurements of u_k are available under these conditions. Indirect, calorimetric data on water cooling of electrodes indicate that the potential drop decreases somewhat with increasing current and is on the order of 17 V [13, 37].

Experimental data on erosion rates of various metal cathodes (parameter G) for arc discharges in vacuum and in air were acquired in the range of currents of $10\text{-}10^3$ A [25, 27, 30, 38-46]. The principal result of these measurements is the determination of two characteristic regions as a function of erosion by the discharge current. At low discharge current ($I_o < 100$ A), the cathode material erosion rate at room temperature depends linearly on the current and is relatively low. According to various data, the electron transfer coefficient lies within the range of $10^{-4}\text{-}10^{-6}$ g/C. At high currents (between 10^2 and 10^3 A) the rate of cathode material erosion in vacuum increases exponentially, and is on the order of $10^{-3}\text{-}10^{-1}$ g/s, or 10^{-4} g/C.

The published results [46] of experiments that provide the most complete data for copper electrodes acquired in vacuum and in air at atmospheric pressure will next be discussed here. As the current in air decreases to $I_o \simeq 10$ A, the cathode material erosion rate is 10^{-5} g/C. At a discharge current of $5 \times 10^2\text{-}10^4$ A, the erosion rate varies between 10^{-4} and 10^{-2} g/C. In vacuum, at a current of up to 10^3 A, the cathode erosion rate remains constant at 10^{-4} g/C. At $I_o > 10^4$ A in vacuum, the erosion rate increases, reaching a value on the order of 10^{-2} g/C.

Hence, for currents of $10^3\text{-}10^4$ A, the cathode erosion rate in air is considerably higher than in vacuum. At a current $I_o \simeq 3 \times 10^2\text{-}10^3$ A, parameter G is approximately the same under these and other conditions; whereas at $I_o < 10^2$ A, the erosion rate measured in air is considerably less than in vacuum.

Erosion of electrodes operating in a constricted discharge mode, under conditions in an MHD facility, has been investigated [1, 11, 13, 47-49]. It should be noted that the erosion rate of cooled copper electrodes of an MHD facility ($T_o \simeq 500$ K) is considerably lower than in vacuum (or in air), varying between 10^{-7}-10^{-6} g/C at a current $I_o = 5$-10^2 A. As the electrode temperature increases to $T_o \simeq 1000$ K, the cathode erosion rate may increase to 10^{-4} g/C.

Results of the experiments show that determination of the erosion rate, G, is closely associated with investigation of the dynamics of cathode spot development in time, as well as variation of the spot current density. The difficulties of determining the current density, j, in the cathode spot was dealt with in many references (e.g., Refs. 25, 27). Two principally different methods are used in the experimental investigations of this parameter [39, 40].

One of these is the method of autographs [50-56], based on the relationship between the diameter of the current-conducting area and the width of the trace caused by thermal interaction of the spot with the cathode surface. The current density determined by using this method for various metals lies between 10^4-10^8 A/cm^2 and reportedly [56] may even be as high as 10^{11} A/cm^2.

In practice, however, it is difficult to determine the relation between dimensions of the faulted zone of the cathode and the spot [25, 27, 39, 40]. An attempt to establish such a relation leads to nonunique results, characterized by a significant scatter of the experimental data even under the same conditions.

At present, the expediency of the method of autographs of the determination of current density in arc spots has not been proved.

Another method of determining j, frequently referred to as the optical method, involves high speed photography of the luminescent region of the cathode spot and identification of its dimensions with the current conducting channel [57-59]. This method makes it possible to investigate simultaneously the dynamics of variation of the external form of the cathode spot, and to determine the displacement rate along the surface of a solid cathode [39, 40].

The principal deficiency of the optical method is the dependence of data about the size of the luminescent region on light sensitivity of the equipment [30]. The use of this method to determine parameters of the spot with adequate spatial and temporal resolution requires high sensitivity, i.e., ability to obtain an image of the spot during a short exposure [40].

If the sensitivity is high enough, the measurement error of the mean current density may be associated with failure to take into account the real distribution of luminescence and, consequently, current along the spot. In general, the distribution may differ greatly from the equilibrium distribution. Therefore, simply recording a luminescent region is not sufficient to estimate the current density. Furthermore, in using the optical method to analyze spots in extraneous gas, difficulties are encountered in separating the luminescence of the initial sector of the column from the spot. The results of investigations of the dynamics of development of cathode spots on various metals throughout broad pressure and current ranges by means of high-speed photography with adequate light sensitivity and high temporal (on the order of 10^{-7} s) and spatial resolution, make it possible to classify cathode spots primarily on the basis of their displacement rates and lifetimes [30, 39, 40, 60].

Rapidly moving spots ($\nu \simeq 10^3$-10^4 cm/s) with relatively short lifetime and high current density appear during the initial period of the discharge in practically all metals. Slowly moving spots ($\nu \simeq 10^2$ cm/s) are formed a short time (on the order of 10^{-4} s) after the discharge is initiated, and appear more frequently at above-normal pressure on metals, characterized by low values of thermophysical coefficients. The mean current density of rapidly moving spots on copper is $(1$-$4) \times 10^4$ A/cm^2. The spots consist of fragments with j up to 10^5 A/cm^2.

At high current ($I_o \simeq 600$-1000 A), cathode spots usually exist in the form of separate, low mobility groups. For example, in the case of copper in vacuum at $I_o \simeq 600$ A, the current to the group is approximately 200 A. Furthermore, the thermal fields of individual spots comprising the group overlap. For silver, the current to the group is 100-200 A and, for tungsten, it is 200-300 A. For example, the mean current density to a group of spots on copper varies between 2×10^4 and 2×10^5 A/cm^2 [39, 40, 61].

In the case of rapidly moving spots, the current to the spot is between 1 and 20 A. The threshold current at which a spot is no longer sustained has been investigated in detail and reported in monograph [25]. The nature of destruction of the cathode surface depends on the type of the spot, and the near-electrode processes and the type of spot depend on thermal processes governing the erosion rate of electrode material [39, 61].

The results of experimental investigations of the dynamics of development of the parameters of cathode spots appearing on electrodes under conditions existing in an MHD channel will now be discussed in more detail. First, it should be noted that, at present, such data are extremely limited and the most complete data appear in only a few works. The first investigations where arc discharges with spots on the cathode of an MHD generator channel were recorded were reported in Refs. 62, 63.

Further experimental investigations of the characteristic parameters of arc spots under MHD conditions [1, 3-10, 13, 48, 64-67] were performed primarily by means of high-speed photography. The following parameters were determined as a function of the discharge current, I_o, magnetic field induction, B, cathode material and parameters of the combustion products flow: lifetime, t; number of spots, N; their dimensions (area F or radius r); current density, j; current to a single spot, I; its displacement rate reduced to the total cathode current, heat flux into the electrode Q_T.

The data in Table 5.1 show that the experimental parameters of the spot vary throughout a sufficiently broad range of values and, as a rule, consists of data on individual characteristics of discharge on the cathode.

The most systematic data were acquired in experiments performed on the U-02 and "Temp" facilities at the Institute of High Temperatures of the Soviet Academy of Sciences, as well as on the K-1 MHD facility at the Institute of the Electrodynamics of the Ukrainian Academy of Sciences [6, 7, 12, 48, 65, 67].

The plasma flow, in the form of combustion products of natural gas enriched with oxygen and seeded with potassium (aqueous solution of potash), in the channel of the U-02 MHD facility was characterized by a core flow rate of 300-500 m/s at a channel pressure of 70-90 kPA and plasma temperature of 2500-2700 K, see 3.2.2. Heat fluxes into the electrodes were measured calorimetrically. A high speed SFR-L photo-recorder, operating at a rate of 1.2 x 10^5 frames per second and equipped with an additional optical attachment to attain the necessary spatial resolution was used in the investigation of the cathode spots. The plasma flow parameters of the "Temp" and K-1 facilities were somewhat different from those of the U-02 facility. Investigation of spots was conducted by means of an SKS-1M photo-recorder (at a rate of up to 10^4 frames per second). The experiments were conducted with the use of cooled, hemispherical copper and steel electrodes protruding into the flow and, also, electrodes flush with the inner surface of the channel. The principal result of these investigations was the determination of the dependence of cathode spots on the gasdynamics of plasma flow around the body of the electrode and on its temperature (degree of cooling).

Table 5.1
RESULTS OF THE EXPERIMENTAL INVESTIGATIONS OF CATHODE SPOT PARAMETERS ON THE CONDITIONS EXISTING IN AN MHD GENERATOR CHANNEL

$\nu \cdot 10^{-2}$, cm/s	I, A	$t \cdot 10^3$ s	N	I_o, A	$j \cdot 10^{-4}$ A/cm^2	Q_T, W/A	T_o, ^{0}C	$F \cdot 10^3$, cm^2	$r \cdot 10^2$ cm	Ref.
					Copper					
2-3	10	1	1-10	$\leq$ 50	–	20	150-700	–	–	[1]
0,1-2		0.3-10	1-10	–	–	–	–	–	–	[2,10]
–		–	2	65-75	–	–	300-600	–	3	[4]
0,1-2	10-15	0.1-1	2-10	15-100	1-3	15-30	150,140	0.2-2	–	[7]
–	2	–	–	–	–	13	–	–	1	[9,11]
–	5.5	–	–	–	10^2-10	–	–	–	0.3-0.7	[12]
0.05-5	5-15	1-10	1-12	5-200	–	17	100-600	–	–	[13,67]
					Copper, steel					
2	2-5	1-10	1-2	2-10	1-0.05	20	100,90	2-10	–	[6,48]
–	1-15	0.1-1	2-30	$\leq$ 60	1-10	–	200-1300	0.1-10	–	[66]
					Molybdenum					
–	–	–	–	–	1.5	–	200-1700	–	3	[8]

It was determined that primarily spots (type I) displaced at a relatively high rate (on the order of 10^2 cm/s) appear in the zone exposed to the flow. Low-mobility spots (type II), the area and lifetime of which are somewhat higher, are concentrated in the separation zone of the flow. The number of spots on relatively cold electrodes is limited (depending on the discharge current) and the current to a single spot is 5-15 A. As the electrode temperature increases, the number of spots and the area increase, and the current to the spot decreases. Overall, the magnetic field induction in the range of interest (up to 2 T) has a negligible effect on discharge parameters. The lifetime of cathode spots is comparable to the characteristic duration of thermal processes in a metal, and is on the order of 10^{-4}-10^{-3} s.

The erosion of electrodes is caused primarily by localized type II spots and depends to a considerable degree on the electrode temperature, T_o. One of the photographs of erosion traces left by cathode spots of various types on a copper electrode in an MHD generator channel obtained during the experiments is shown in Fig. 5.1 [13, 67].

Comparison of the results of experimental investigations of a vacuum arc discharge and a discharge burning in the channel of an MHD generator leads to the following conclusions. Long-duration, slow cathode spots under

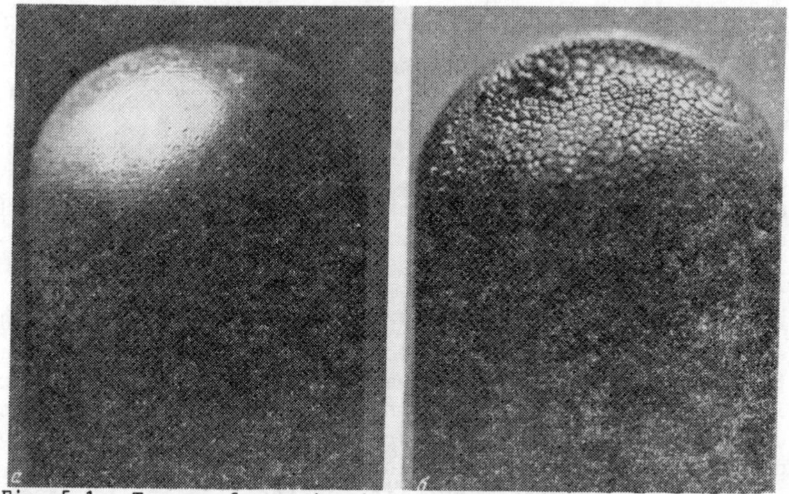

Fig. 5.1 Traces of erosion interactions in the zones exposed to the flow (a) and separation zone (b) of the flow in the case of a strongly constricted discharge (I = 200 A) in the channel of an MHD generator (copper electrode, test duration on the order of 10^2 h) [67].

conditions in an MHD generator and in a vacuum differ with respect to the current to the spot and the resulting electrode erosion. Rapid spots with relatively low current to the spot ($I \simeq 10$ A) appearing on electrodes of an MHD generator also differ considerably from analogous spots in a vacuum, with respect to their displacement rate and lifetime.

The spot formation mechanism is apparently also different. Whereas, in a vacuum or in air, arc discharge is formed as a result of electrical breakdown of the nonconducting medium, arc spots under conditions existing in an MHD generator appear as a result of electrical breakdown of the near-cathode plasma of relatively high conductivity (on the order of 10 S/m).

Phenomena associated with electrical breakdown in an MHD generator channel have been investigated [68-75]. Arc discharge may form between electrodes of a segmented channel wall (Hall breakdown) as well as in the transverse direction of the channel, between electrodes and the plasma. The data acquired as a result of analysis of the breakdown conditions and its electrical characteristics as a function of the parameters of the facility are important in developing theory of a constricted discharge in a combustion products flow of natural gas seeded with alkali metals.

Arc discharge spots on cathodes under various conditions (including in MHD generators) have common characteristics. These include the high energy concentration in the spot (10^5-10^6/W/cm^2) and the occurrence of spots moving individually at different rates (fast and slow), the number of which depends on the total current to the electrode, etc.

One should note especially the establishment of general regularities of electrode erosion in the case of concentrated spots in a vacuum and slow spots in an MHD generator. Experiments have shown that, in both cases, electrode erosion occurs as a result of the thermal effect of energy released in the near-cathode region.

Relatively high current densities in spots of an MHD generator make it possible to assume a high concentration of charged particles, comparable to that found under different conditions, e.g., in vacuum. The corresponding level of heavy particle concentration may be caused by additional vaporization of the cathode material, or the film of seed adsorbed on its surface. In this case, correct understanding of processes occurring in an arc with a cathode undergoing vaporization is extremely important for description of discharge under conditions of an MHD generator.

5.3. Near-Cathode Region of Arc Discharge in Electrode Material Vapors

Consider reports, available in the literature, of past attempts to investigate theoretically the phenomena in the near-cathode region. We will describe, here, a mathematically complete model of processes occurring on the cathode affected by an arc spot that makes it possible to perform theoretical analysis and to calculate its parameters. This model differs from the earlier ones in that it incorporates processes in the near-cathode plasma. It is based on analysis of hydrodynamic equations for a multi-component, partially-ionized mixture with different temperatures of components. This approach has made it possible to apply, sequentially, conclusions of the theory of elementary processes to interaction of charged and neutral particles formed under conditions in the cathode region of an arc discharge, to separate characteristic equilibrium and nonequilibrium regions of the near-electrode plasma, and to obtain the required additional relations that are lacking in existing models.

Numerical investigations have been performed that make it possible to reach important conclusions on the self-sustaining mechanism of discharge on the cathode, types of spots, and electron emission on the relatively large component of current in heat spots and other parameters of the near-cathode region of the arc with a cathode.

5.3.1. Analysis of the Theory. At present, investigations of near-cathode processes are performed using, primarily, equations for a solid cathode and a layer of volume charge near its surface [24, 76-78]. Such models postulate a cathode spot of a vacuum arc sustained by vaporization of cathode material and emission of electrons. These processes are caused by heating of the cathode and the effect of volume charge concentrated near the cathode surface. The volume charge and the energy flux to the cathode depend primarily on the concentration of ions outside the layer of volume charge that remained a free parameter in other models [24, 76-78]. Variation of these processes is a smooth and continuous function of time and a coordinate (discontinuities or sharp local variation of parameters do not occur in the region of the spot). In the future, this model will be referred to as "continuous-thermal" or as "thermal." As has been already noted, the occurrence of primarily two types of cathode spots of a vacuum arc can be assumed to be an experimentally established fact [27, 30, 39, 40, 60, 61]. The two types of cathode spots are rapidly moving ($\nu \simeq 10^3$-10^4 cm/s) and short-duration ($t \lesssim 10^{-5}$ s) and slowly moving ($\nu \simeq 10$ cm/s) and long-duration ($t \gtrsim 10^{-4}$ s) spots.

The mean cathode temperature and the electric field intensity in fast spots ($j \simeq 10^4$ A/cm^2) do not provide the required electron emission and vapor concentration [79] required to sustain the spot and, consequently, processes inside spots are not described by concepts that form the basis of stationary models [24, 76-78]. An exposive model of fast spots was suggested as one possibility [80-82]. For slow and long-duration spots, a model in which continuous vaporization of the cathode material in the region of the spot is the principal process [24, 76-78] is noncontradictory. This model was further developed [83-93].

In view of the broad application of its various modifications, the thermal model will be considered in more detail. The system of equations usually used for the near-cathode region consists of the following four principal equations [68, 83-93]:

1) for electron current density, j_e, (electron emission law), depending on the emission loss assumed, the form of this equation may be different;

2) for the electric field intensity, E, (as a rule, McKeown's relation [77], the solution of which depends substantially on the ionic current, is usually used for this equation);

3) for total current, in the form $I = \pi r^2 j$, where r is the effective spot radius, and

4) for the mean temperature of the spot.

In describing stationary spots, an expression obtained by solving thermal conduction equations for a half-space with a circular heat source at a specified heat flux, Q_T, is usually used for the last equation. In this case, $T_k \simeq Q_T/\lambda r$, where λ is the thermal conductivity coefficient of the cathode.

Thus, four always-identical equations are actually used to determine the five parameters of the cathode spot (T_k,r,s,j, and E). Furthermore, the cathode potential drop, u_k, current to the spot, and physical constants of the cathode material are given. From this, it follows that in principle the existing models differ by only one equation (or, if the equation contains additional parameters, by a system of equations). As a rule, various semi-empirical relations [89, 90] or limiting conditions [7, 8, 83-85] making it possible to determine s are used as the additional equation. In certain cases, it is assumed to be arbitrary [25, 94]. In some reports [87,

88], equations describing the state of the near-cathode plasma[1] are added to the system of equations [1-4].

Computation of the current density and other spot parameters depends substantially on s. In particular, a slightly higher value of s results in a substantial increase of j [79, 88]. Therefore, using a different additional equation with the system of equations 1-4 may lead to substantially different values of current density and minimum current to the spot. For example, results of [89-91] differ quite considerably (the current density values for copper are 10^6-10^8 A/cm^2). On the other hand, the possibility of occurrence of a spot on a copper cathode with current density of $j \simeq 10^5$-10^6 A/cm^2 has been pointed out [78]. From Ref. 88, it follows that the current density for copper varies between 3×10^4-10^5 A/cm^2 and depends on the total current to the spot (calculations are given for I - 100-500 A). An analogous scatter of values is also characteristic of minimum current to the spot.

Minimum current to the spot is defined as an experimental value below which the spot is "extinguished" (when the current is measured in the presence of a single spot). The term "minimum current to the spot" frequently refers to two different physical concepts. Either it is the current at which arc discharge is still sustained (with a single spot at the electrode) [25], or "cut-off" current of the arc during commutation of alternating current in vacuum [78]. Theoretically, this concept is sometimes identified with the minimum current at which solution of a system of equations for the near-cathode region still exists [78, 83-85], even though this is not always valid.

In view of the difficulty of developing a closed (no free parameters) model of cathode processes, a method of constructing a possible region of existence (in T_k, j) of solutions of a system of equations for stationary cathode spot or a method of diagrams [83-85] are of interest. In view of a lack of accurate data on physical processes and several parameters of the near-cathode region, it is suggested that the exact conservation laws in the method of diagrams be replaced with certain limiting relations that, in one sense or another, limit the region of possible variation of current density and temperature in the spot [96].

These limiting relations define lines in the T_k, j plane that form the upper or lower limits of the range of

[1] An analogous approach is also used [92, 93]; however, the system of equations is written incorrectly [95].

possible values. The limiting relations are obtained from equations 1,2, and one of the following conditions.

$$j_i/j = u_k/(u_k+u_i), \quad j_i = (1-s)\, j\ ; \tag{5.1}$$

$$j_i = eW\,(T_k); \tag{5.2}$$

$$j((u_k+u_i)\, j_i\,(j,\, T_k)/j - (\phi - Q_j))$$

$$= \lambda_v W(T_k) + \lambda T_k(\pi j/I)^{1/2}\ . \quad 5.3)$$

Here, u_k is the cathode potential drop, u_i is the ionization potential, $W(T_k)$ is the vaporization rate of cathode material at a spot temperature T_k, ϕ is the work function of a metal cathode, Q_j are heat losses to Joule heating within the electrode, λ_v is the heat of vaporization.

Relation (5.1) expresses the fact that the ion flux from the near-cathode plasma depends on the ion concentration formed in the plasma if all of the energy of emitted electrons ($j_e u_k = sju_k$) is spent on ionization (all ions formed return to the cathode). Obviously, the ion current density calculated from relation (5.1) will be too high. Further, at a given j, the temperature in the spot is limited from below. Because formula (5.2) expresses the fact that the ion flux is equal to the flux of the vaporized atoms, it, therefore, serves as the upper limit of j_i.

Expression (5.3) represents an approximate energy balance at the cathode surface. It provides the upper limit of the range of possible values of cathode temperature [83-85]. The approximate nature of expression (5.3) can be attributed to the use of equations 3 and 4 in its derivation. Because equation 4 specifies temperature at the center of the spot that is above a certain average temperature, T_k, equations 3 and 4 provide values of temperature that are too high. The coefficient of accommodation of energy of ions is set equal to unity.

The method of constructing regions of existence was analyzed in [96]. In view of certain published reports [85, 97, 98], consider the highlights of this method following conclusions obtained in Ref. 96.

The limits of the region of existence for copper and current I = 200 A (current to one group spot), (region A in Fig. 5.2) are as follows: $j \simeq 10^5$-10^6 A/cm^2, $T_k \simeq$ 3600-4000 K (approximately the same limits as for I = 100 A [85]. The region decreases sharply with decreasing total current, being reduced to a point at 13 A [83-85].

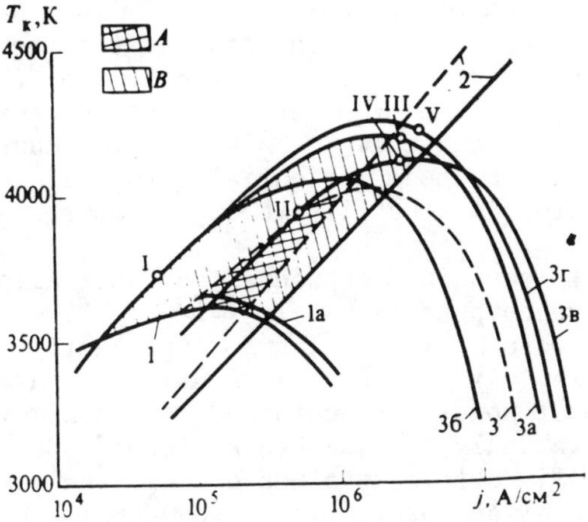

Fig. 5.2 Regions of existence of possible solutions of the system of equations of the near-cathode region of an arc discharge on copper (I = 200 A);

A - calculations reported [83-85]; B - calculations performed in this paper; 1,2 - solutions of equations 1,2, together with conditions (5.1) or (5.2), respectively; 1a - constructed using the results of [78]; 3a-3d - solution of equations 1,2, together with equations (5.3) in different form: 3a - (5.4), 3b - (5.4) when $\phi_{eff} = \phi$ (neglecting the Schottky effect), 3c - (5.4) when G = W(T), $\phi_{eff} = \phi - (e^3E)^{1/2}$, 3d - (5.4) taking into account plasma electrons [87]; I-V - exact solutions of the system of equations [87, 88].

Minimum current required to sustain a stationary cathode spot on copper is assumed to be I_{min} = 13 A [85].

In addition to the region shown in Fig. 5.2, there is another region of existence with fairly extreme parameters ($j \simeq 10^9$ A/cm^2, $T_k \simeq 6 \times 10^3$ K), the physical meaning of which is not yet clear.

Designation of curves in Fig. 5.2 corresponds to the numeration of the limiting relations given in the foregoing. All continuous curves (except 1a) were calculated by the authors. Curve 1 coincides approximately with an analogous curve in Ref. 85. The difference between continuous and dashed curves 2 [83-85] can be attributed to the use of somewhat different emission laws [96].

Hence, even without improving the position of curve 3,

one can see that changes in calculations of curve 2 lead to an expansion of the region of existence. Because the energy balance equation used in Ref. 78 is analogous to condition (5.3), the minimum current required to sustain the spot (corresponding to the current at which curve 3 passes through the intersection point of curves 1 and 2) for the region of occurrence will be 2-10 A [78].

The location of curves 1 and 2 in the T_k, j plane depends only on the emission law and the assumed value of u_k. It is easy to see from condition (5.1) that the change of u_k by 1-2 V exerts an insignificant effect on the location of curve 1. This fact is important because the measured cathode potential drop consists of the potential drop in the fly-through layer and the plasma column.

Present measuring techniques do not allow separation of these two components. In view of the fact that the potential drop in plasma apparently cannot be considerable, it is of interest to investigate the dependence of solutions obtained on small variations of u_k. Such an analysis has been performed [88]. When using experimentally obtained values of u_k, the weak dependence of the location of curve 1 on small changes of u_k indicates that the region of existence bounded by curves 1 and 2 in the T_k, j, plane is determined fairly definitely. This is verified by calculations that have been performed [85].

A comparison of continuous curves 1 and 2 with curve 1a [78] and a dashed curve 2 that coincides with a similar curve [78], shows that a small change in emission laws, resulting from different approximation of the penetration coefficient of the potential barrier, results in a change of the lower boundary of the region of existence, which is insignificant, however, and decreases near the intersection points of curves 1 and 2.

Let us now discuss relation (5.3) and location of curve 3. Elsewhere [88], the energy balance on the cathode surface was described by the following relation:

$$j((u_k+u_i)\ j_i\ (j,T_k)/j - \phi_{eff})$$

$$= \lambda_v Gj/I + 2\lambda\ (T_k-300)(j/I)^{1/2}\ , \qquad (5.4)$$

where

$$\phi_{eff} \equiv \phi - (e^3E)^{1/2}\ ,$$

and G is the total electrode material erosion rate that was determined experimentally.

A comparison of relations (5.3) and (5.4) shows that the first one [83, 84] does not take into account the influence of the Schottky effect on cooling, associated with the work function of the electrons and the specific (per unit area) electrode material erosion rate assumed to be equal to the vaporization rate $W(T_k)$.[2]

Comparing curves 3a-3c in Fig. 5.2 demonstrates the significant influence of the above-noted factors. First of all, it follows that calculations taking into account the energy balance in the form (5.4) (curve 3a) leads to a significant expansion of the region of existence (region B). Its extent along the axis of the current density corresponds to almost two orders of magnitude of variation of this quantity. A comparison of curves 3a and 3c demonstrates the influence of different assumptions of electrode material erosion rate, comparison of curves 3a and 3b demonstrates the influence of the Schottky effect. The difference between curves 3a and 3b can be attributed to the fact that Joule heat release, Q_j, was neglected in relation (5.4) and to the condition $\phi = \phi_{eff}$. Note that the use of expression for Q_j in the energy balance at the cathode [99] apparently is important when $j \gtrsim 10^7$ A/cm^2.

A comparison of curves 3a-3c for differently determined energy balance at the cathode shows not only that they can differ considerably, but also that in general, one cannot determine the variation of the curve from general physical considerations when some physical process is neglected.

In terms of limiting relations, this means that it is impossible to indicate whether a curve, corresponding to the assumed approximate energy balance equation, forms the upper or lower limit of the region of existence. A comparison of curves 3a and 3c illustrates this conclusion.

Assuming that the specific material erosion rate is equal to the vaporization rate, the total material erosion from the spot depends on its area and, consequently, on current density. When current density for a copper electrode is small and I = 200 A, the condition $G \ll \pi r^2 W(T_k)$ applies (e.g., see Ref. 88). Therefore, curve 3a lies above curve 3c. However, the situation changes at high current densities, because the small area of the spot at the same temperature cannot provide experimentally measured material erosion rate ($G > \pi r^2 W(T_k)$).

Because the balance equation, (5.3) or (5.4), includes specific, rather than total energy fluxes, it is

2 The small difference in the coefficient of the last few terms of equations (5.3) and (5.4) is associated with differently specified distribution of the external energy inflow.

impossible to predetermine (without analyzing the proper relations) the changes of curves 3 resulting from a simplification of the balance equation. Furthermore, our concepts of energy balance of the cathode surface may change as we learn more about the cathode processes. An example of this is the difference between relations (5.3) and (5.4).

Analysis of processes in the near-cathode plasma led to the following "improvement" of the balance equation: at high electron temperature in the near-cathode plasma, the energy influx from plasma caused by the energy of electrons transmitted across the potential barrier [87, 88] must be taken into account in equation (5.4). (The curve corresponding to this energy balance equation is shown in Fig. 5.2 (curve 3d)). Analogous changes may be introduced after measuring the coefficients of accommodation.

Hence, the use of the energy balance equation at the cathode as a limiting relation to limit the possible region of existence of solution may lead to unjustifiably narrow regions. However, when the energy balance equation is written out exactly, the solution has to lie along line 3. An attempt to write out this equation in the most exact form contradicts the concept of the method of the regions of existence [83-85].

Because the region of existence is fairly definitely bound from below by curves 1 and 2, by writing out the exact energy balance equation one can refer to the interval of solutions consisting of a sector of line 3 between curves 1 and 2. Let us once again emphasize that the possible temperature and current density ranges depend considerably on the assumed form of the energy balance equation, and that analysis of reliability of various approximate solutions based on the use of the energy balance equation becomes impossible. Obviously, the minimum current required to sustain a spot determined by use of the method of limiting relations will also depend on the assumed form of the energy balance equation, even though the current density and temperature at $I = I_{min}$ is fixed by the intersection of curves 1 and 2. For example, for equation (5.4), I_{min} = 8 A. At the same time, the form of the energy balance equation will depend on the characteristics of the specific type of the spot [96].

For equation (5.4) describing stationary, low-mobility spots with relatively large cathode material erosion rate, $I_{min} = 10^2$ A. Deviation from this value may be caused by inaccuracy of measuring the experimental values of u_k and G. The value I_{min} = 2-10 A, determined with the aid of limiting condition [78], coincided with the measured arc cut-off current during commutation. However, in view of the foregoing discussion, this coincidence cannot be considered important, because no stationary spots exist at a current $I \simeq 10$ A, and because I_{min} [78] was obtained using the energy balance equation (5.3) that may lead to inaccurate results [96].

In addition, it should be noted that the experimentally determined cut-off current depends on the commutation conditions [100, 101]. No current cutoff is observed when large currents are switched off. In view of the absence of stationary spots at currents close to the cut-off current or minimum arc current [25], these quantities cannot be compared with the minimum arc current required to maintain a spot, determined from some theory for stationary spots based on continuous-thermal model.

If one rejects the energy balance equations on the assumptions that our knowledge is insufficient for its accurate determination, the region of existence can be assumed to be part of the T_k, j, plane bounded below by curves 1 and 2. However, such a region turns out to be too broad, even though it apparently may serve, in some cases, to verify the validity of assumptions used in approximate solutions.

On the other hand, large scatter of parameter values obtained for different predeterminations of the system of equations 1-4 may not lead to a contradiction from the point of view of limiting relations (5.1) and (5.2), because of the wide region of possible existence of the solution. Hence, analysis leads to the conclusion that the existing systems of equations for the near-cathode arc region are mostly of the same type, do not take into account processes in the near-cathode plasma, and include free parameters that, in certain cases, are determined from the limiting relations. For cases of high current to the spot ($I \gtrsim 100$ A), characteristic of heat spots on copper, the method of diagrams based on the use of such relations leads to regions of existence that are too broad (along the current density axes).

At low current, the method of diagrams makes it possible to obtain narrow regions of existence. However, current densities characteristic of these regions (e.g., in the case of copper) exceed considerably the experimentally recorded current densities at low current in nonstationary spots. In addition, the spots may be sustained at a current that is below the minimum, determined by the method of diagrams using stationary relations. This conclusion cannot be explained when the method of constructing regions of existence is based on the use of limiting relations.

By themselves the facts considered indicate close coupling and strong dependence of the spot parameters on the physical processes in the near-cathode plasma. Therefore, to obtain solutions corresponding to parameters of real spots, it is necessary to add relations that take into account real physical processes in the near-cathode plasma to equations 1-4 [87, 88]. Exact solutions for I = 200 A [88], denoted by points I-V in Fig. 5.2, depend on the assumed energy balance equation. As can be seen from Fig. 5.2, exact solutions of a system of equations for the near-cathode have two values.

<u>5.3.2. Model and the System of Equations</u>. Estimates and systematic calculations [87, 88, 96] show that high concentration of neutral, n_a, and charged, n_i, particles ($n_a + n_i \simeq 10^{19}$-10^{20} cm^{-3}) occurs in the near-cathode region, at sufficiently high ionization ($\alpha \gtrsim 10^{-2}$) and cathode temperature $T_k \simeq$ 3000-4000 K. These quantities may also be roughly estimated by simple analysis. With the experimental values of current density (10^4-10^5 A/cm^2) taken into account, the cathode energy balance equation provides the temperature required for substantial vaporization only at adequately high ion component of current ($s \simeq 0.5$). In other words, ion current must be comparable to the total current. Then, the charged particle concentration determined from the relation for ion flux exceeds 10^{18} cm^{-3}.

The estimates given were obtained under the assumption of the predominant importance of ions appearing near the cathode in thermal and emission processes. A model of the near-cathode processes based on this assumption can be reduced to the following [79, 86-88].

In the range of parameters under consideration, the near-cathode plasma at relatively low electric field (intensity $E_\pi \simeq 10^3$ V/cm) is separated from the cathode by a collision-free layer where most of the cathode potential drop occurs. Plasma ions being accelerated in a collision-free layer are incident on the cathode, heating it through the transfer of kinetic and potential energy. In addition, the cathode can be heated by high-energy plasma electrons from the "tail" of the Maxwellian distribution. Cathode cooling occurs through thermal conductivity, electron emission, vaporization, and radiation.

The potential and electric field distributions in a collision-free layer depend on the positive volume ion charges from plasma. The electric field in this layer makes easier emission of electrons from the cathode, with a Fermi-Dirac energy distribution that provides energy to charged particles moving in the layer.

The emitted electrons, accelerated in the region of positive volume charge, transfer to the near-cathode plasma energy that is expended on maintaining a certain degree of ionization, heating plasma components, and transferring energy from plasma by high-energy electrons from the tail of the Maxwellian distribution.

Ionization is attained by means of both impact ionization by fast electrons of the beam and thermal ionization by plasma electrons. The ions formed are recombined in plasma, and diffuse to the surface of the cathode. The diffusion rate depends on the resonant charge transfer.

In the model under consideration, the following quantities are assumed: total current to the spot, I, cathode potential drop, u_k, erosion rate, and physical constant of cathode material. The quantities to be determined are: current density, j; electron component of the total current, $s = j_e/j$, where $j_e = j_{em} - j_{eT}$ (j_{em} is the emission current density, j_{eT} is the current density returned to the cathode by the plasma electrons), the cathode temperature, T_k, heavy particle temperature, T, electron temperature, T_e, degree of ionization, α, microscopic rate of electrode material transfer from the near-cathode region, $\nu_o = Gj \div (Im\ (n_a + n_i)$, specified by continuity equations for a mixture of gases, plasma particle concentration, electric field intensity near the surface of the cathode E.

Consider the complete system of equations for the near-cathode region of a vacuum arc discharge and the basic physical processes they describe [87, 88].

First, let us write out equations describing the near-cathode plasma: equation for the ion flux

$$\Gamma_i = -(1+\Theta)\, D_{ia} \frac{dn_i}{dx} + n_i \nu_o + \alpha D_{ia} \Psi , \tag{5.5}$$

where

$$\Psi = \frac{1}{kT} \left(\frac{dp}{dx} - n_i k \frac{dT_e}{dx} - (n_a + n_i)\, k \frac{dT}{dx} \right) ,$$

$$D_{ia} = 0.4 \nu_T/(n_a + n_i)\ \sigma_{ia} ,$$

$$\alpha = n_i/(n_a + n_i), \qquad \Theta = T_e/T ;$$

continuity equation for ions

$$\frac{d\Gamma_i}{dx} = \sigma_i n_a N_{eo} \exp\left(- {}_o\int^x (n_a \sigma_i + 2n_i \sigma_C)\, dx\right)$$

$$+ \beta n_o^2 n_i (1 - n_i^2/n_o^2); \qquad (5.6)$$

boundary conditions

$$(1+\Theta) D_{ia} \left.\frac{dn_i}{dx}\right|_{x=o} = \frac{n_w \nu_T}{2}, \quad n_i|_{x=\infty} = n_o;$$

energy balance equation for electrons in plasma

$$s\,(eu_k + 2kT_k) - (1 - s + \alpha\zeta) = 3kT_e\,(1 + \alpha\zeta)$$

$$+ (1-s)\, 2k\, (T_e + T_k)\, (mT_e/m_e T)^{1/2}$$

$$x \exp(-eu_k/kT_e)\,. \qquad (5.7)$$

Here, D_{ia} is the coefficient of diffusion of ions determined from analysis of the transfer process of a three-component, partially-ionized mixture with different temperature of its components; k is the Boltzmann constant; P is the equilibrium pressure; $n_a = (P/kT) - (1 + \Theta)\, n_i$; ν_T is the mean ion thermal velocity, σ_{ia} is the resonant charge transfer cross section; σ_i is the ionization cross section; N_{eo} is the electron flux across a unit surface on the wall ($x = 0$); σ_C is the Coulomb cross section for electrons emitted; β is the recombination coefficient; n_o is the equilibrium concentration of charged particles, specified by Saha's equation; n_w is the charged particle concentration on the wall; $\zeta = eG/mI$; u_i is the ionization potential; m, m_e, are the ion and electron masses, in that order; e is the electron charge.

Equation (5.6) takes into account that both impact ionization by beam electrons, N_{eo}, emitted by the cathode and thermal ionization by plasma electrons take place. The ions are removed by triple recombination. The electrons leave the beam as a result of collisions with neutral atoms, leading to ionization and as a result of elastic collisions with plasma electrons and ions. Because scattering of electrons by atoms of metals is essentially anisotropic in the energy range under consideration, variation of the momentum and energy of electrons of the beam during their elastic collisions with atoms can be neglected. In other words, the probability of beam electrons being scattered by atoms at a large angle is considerably less than the probability of ionization of the same atom.

Equation (5.7) takes into account that plasma electrons are heated by an energetic beam of electrons accelerated in the region of the cathode drop to energy eu_k and are cooled as a result of ionization and reverse "flux of plasma electrons" from the "tail" of a Maxwellian distribution. It is assumed that the temperature of heavy particles, T, is equal to the cathode temperature, T_k. Therefore, the energy equation for the heavy component [87, 88] was not considered.

Let us now proceed to analysis of equations describing processes in a solid body and on its surface. Equations that make it possible to determine the electron current in collision-free regions directly near the cathode will be written in the form:

$$sj = \frac{4\pi m_e kT_k}{h^3} \int_{-\infty}^{\infty} \frac{\ln(1+\exp(-\varepsilon/kT_k))d\varepsilon}{\exp(6.85\cdot 10^7(\phi-\varepsilon)^{3/2}\Theta(\bar{y})/E)}$$

$$- \frac{en_i}{4} \nu_{eT} e^{-\frac{eu_k}{kT_e}} , \qquad (5.8)$$

where ν_{eT} is the thermal velocity of electrons, ϕ is the work function of cathode material electrons.

The first term on the right-hand side of equation (5.8) specifies the emission current density during interaction of the field of a volume positive charge with electrons, the energy distribution, ε, of which in the conduction zone at the proper cathode temperature is described by the Fermi-Dirac statistics.

The second term on the right-hand side specifies the plasma electron current density, j_{eT}, with energy exceeding eu_k. The following expression for the Nordheim function was used in actual calculations [102]:

$$\Theta(\bar{y}) = 1 - \bar{y}^2 \, (1+0.85 \sin((1-\bar{y})/2)) \, .$$

The system of equations will be solved for type II group spots, for which elementary gasdynamic processes are stationary and nonstationary behavior of discharge burning is caused only by the establishment of thermal processes in the electrode.

In this case (assuming that the cathode material erosion rate, G, does not change), the energy balance equation at the cathode has the form

$$(1-s)u_{eff}j+1/2en_i kT_e \nu_{eT} \exp(-eu_k/kT_e)$$

$$= 2\lambda \left(\frac{j}{I}\right)^{1/2} (T_k-300) \left(\frac{2}{\pi}\arctan\left(\frac{4\pi a j t}{I}\right)\right)^{-1}$$

$$+ \left(\lambda_V + \frac{2kT_k}{m}\right) G \frac{j}{I} + 1.4\cdot 10^{-12} T_k^4 , \qquad (5.9)$$

where

$$(1-s)u_{eff} = (1-s)(u_k+u_i) - \phi + (e^3E)^{1/2} ; \qquad (5.10)$$

λ_v is the heat of vaporization, a is the coefficient of thermal conductivity.

Equations (5.9) and (5.10) take into account that the cathode is heated by transfer of kinetic and potential energy of ions and high-energy electrons from the "tail" of a Maxwellian distribution bombarding the cathode. Heat flux into the electrode is determined from the condition that the current distribution in the cathode spot is Gaussian and that, in view of the effective dimensions of the heat source, the electrode can be treated as a semi-infinite body. The cathode is cooled by vaporization and material transfer from the cathode region determined from the experimental data [27, 38], as well as radiation losses in accordance with Stefan-Boltzmann law. In addition, electron emission also contributes to cooling. The energy loss through electron emission is assumed to be equal to the work function, ϕ, with the Schottky effect taken into account ($T_k \simeq 4000$ K) [88]. The electric field intensity at the cathode, E, is usually determined from McKeown equation [77], which was obtained on the assumption that the parameters in the electric layer are a function of motion of ions from plasma and electrons emitted by cathodes, neglecting the influence of plasma electrons and the initial velocities of charged particles. In addition, the electric field intensity, E_o, at the boundary between the volume charge layer and the quasi-neutral plasma was assumed to be zero and the relation between the ion current density, j_i, ion concentration, n_i, and ion velocity, ν_i, depending on the potential distribution in the layer, was given in the form $j_i = en_i \nu_i$.

Subject to the assumptions made, the distribution, n_i, in the region of volume charge is determined everywhere except at the external boundary of the region, where its value is n_{io}. Determination of n_{io} from the relation for the ion current density is required for the determination of solution a matching conditions at the boundary of the electric layer and the region of the quasi-neutral plasma, where $n_{io} = n_{eo}$. In order to determine this condition and to obtain data on the influence of plasma electrons on parameters of the layer, it is necessary to solve the problem under consideration, with the initial velocity of charged particles taken into account.

Results of the analysis of the region of the near-spatial charge at the cathode [85, 103, 104] shows that the potential distribution near the outer boundary (when $T_e > T$) with the plasma electron flux taken into account is monotonic, provided it is subject to the condition

$$(\nu_i \geq \sqrt{(2kT_e/m)} \,.$$

If this condition is not fulfilled, solution near the boundary has a singularity. However, the singularity is eliminated, even if the condition above is not fulfilled, if production of particles or rare collisions occur in the layer [103, 105]. Estimates for a cathode spot show that particle collisions in the layer of volume charge can be neglected, that the electric field at the boundary is weak ($e\ell_i E_\pi << kT$), and that ion production does not occur. Under these assumptions, analysis of the influence of parameters of the quasi-neutral plasma on the solution of the problem in the electric layer may be performed, provided n_{io} is expressed in terms of a relation for ion fluxes from plasma into the layer.

For a thermal spot, the density of ion current from plasma is $k'en_{io}\nu_{iT}$ [87, 88]. The coefficient, k', depends on the type of velocity distribution function of ions and, in the case of Maxwellian function, $k' = 1/4$.

In general, a thin transition layer on the order of

the Debye layer (kT_e) may appear at the boundary of the region of the spatial charge with the quasi-neutral plasma, where the ion distribution function changes from equilibrium and isotropic to one shifted by velocity corresponding to the thermal energy ($\nu_i = 2kT_e/m$). The particle flux density into the electric layer is then given by the expression,

$$j_i/e = 1/4 n_i \nu_{iT} \exp(-T_e/T) + 1/2 n_i \nu_i (1+\Phi(T_e/T)) ,$$

from which it follows that, when $T_e \gg T$, Bohm condition is automatically fulfilled.

For the conditions under consideration, an expression for the electric field may be obtained from the solution of Poisson equation in the form,

$$E^2 = 7.57\cdot 10^5 ((1-s) m m_e^{-1} ((F_i+1)^{1/2} - F_i^{1/2} - R(1-\exp(-3eu_k/2w_e))) - (s+j_{eT}/j)((F_k+1)^{1/2} - F_k^{1/2} - R(1-\exp(-3eu_k/2w_e))/(1+F_k^{-1})^{1/2})) \times ju_k^{1/2}/k' + E_o^2 , \quad (5.11)$$

where

$$R \simeq 0.362 w_e (w_i eu_k)^{-1/2} ;$$

$$F_i = 0.85 w_i/eu_k ; \quad F_k = 0.85 w_k/eu_k ;$$

$$j_{eT} = 1/4 n_{io} \nu_{eT} \exp(-3eu_k/2w_e) ;$$

$$3/2 kT_{e,i,k} = w_{e,i,k} ; \quad s = j_e/j ; \quad j_e = j_{eo} - j_{eT} .$$

Here, j_{eo}, j_{eT} are the current density of electrons emitted from the cathode and plasma electrons returning to the cathode, ν_{eT} and m_e, are the thermal velocity and mass of plasma electrons, in that order.

Equation (5.11) takes into account that quasi-neutrality condition applies at the boundary of the layer and that the initial velocity of all ions is v_{iT}.

Subject to these assumptions, the potential distribution in the layer will be monotonic, provided $w_e < 8w/\pi$. This condition is satisfied by the first solution of the (5.5)-(5.11) system of equations (see 5.3.3.1.), for which $w_e \simeq 1$ eV. In other words, for the first solution the deviation of the ion distribution function from the isotropic half-Maxwellian proves to be insignificant.

For the second solution, $w_e \gtrsim 2$ eV, and the above condition is not satisfied, i.e., it is necessary to take into account the variation of ion concentration at the boundary of the layer, associated with a shift of the distribution function. However, numerical analysis of equation (5.11) shows that, for a thermal, spot terms describing the plasma electron flux, associated with the initial velocity of particles, exert a negligible effect on E, influencing only the potential distribution near the external boundary of the spatial charge layer (nonmonotonic region, in view of the fact that $eu_k/kT_e \gg 1$ is very small). It turns out that, when small terms in (5.11) are neglected, this relation differs from an analogous expression given [77] by a factor of 1/k′.

Therefore, it will be assumed, here, that the ion velocity distribution function is Maxwellian and, in analyzing parameters of the spot, $E = E_M$ is assumed to depend on the McKeown relation [77] or $E = 2\ E_M$, which utilizes expression (5.11). The second assumption makes it possible to determine the influence of deviation of E from that obtained using the results of Ref. 77.

The (5.5)-(5.11) system of equations was obtained under the assumption that the velocity, v_o, and pressure, p, in a narrow, near-cathode region vary insignificantly. This assumption, as well as the assumption that the parameters in the region of the spot are constant, have to be verified by additional experimental data and the results of the solution of the problem of expansion of the vapor jet.

Usually, an arc discharge on metal electrodes in vacuum is accompanied by formation of intense plasma fluxes, comprised of the products of erosion of electrodes (predominantly in vapor phase). The jet velocity of the cathode, measured far away from the electrodes, $v_\infty \simeq 10^6$ cm/s, and the potential drop across the jet, u_π, is on the order several volts [24-28, 42].

To analyze plasma parameters directly in the jet, equations (5.5)-(5.11) must be written out taking this expansion into account. Predominantly, this refers to the mass, energy, and momentum conservation equations. Such an investigation was performed on the assumption that the geometry of the jet is a truncated cone, with the flare angle assumed to be a parameter [106, 107]. The following integral energy and momentum conservation laws for a jet were obtained assuming that the pressure and concentration at the outer end of the jet are equal to zero [106, 107]:

$$G\nu_{\infty}^{2} = 2Iu_{\pi} ; \tag{5.12}$$

$$Gv_{0}j = Ip , \tag{5.13}$$

where $p = p_T + p_e$ is the sum of heavy particle pressure and electron pressure.

Relations (5.12), (5.13) are equations additional to the (5.5)-(5.11) system for the initial sector of the jet. In other words, when, in addition to the above given quantities, the total arc voltage, u, is also known from the experimental data, simultaneous solution of the (5.5)-(5.13) system makes it possible to determine parameters of the jet, ν_{∞} and u_{π}. However, under the experimental conditions used for copper [42], the value of ν_{∞} determined in this manner is an order of magnitude higher than the experimentally determined value. The disagreement of the experimental data and the results of the calculations will next be analyzed. Such analysis is required in order to justify assumptions that form the basis of the model used in obtaining the system of equations (5.5)-(5.11) describing parameters of the spot.

An important parameter determining processes near the cathode is the metal vapor pressure in the spot. In the model under consideration, it is assumed to be equilibrium pressure. The basis of such an assumption is the fact that the velocity of the gas, ν_o, near the base of the jet determined from the mass conservation relation using experimental values of I, G, and j, is considerably lower than the thermal velocity of particles, ν_T. In addition, another question that frequently arises concerns the applicability of the relation describing the dependence of the pressure on the temperature (e.g., see Ref, 108). Therefore, let us extend the analysis of this relation.

Because the heat flux associated with the flow of products of erosion is determined experimentally, the cathode temperature depends primarily on the energy released in the spot and thermal conductivity of the electrode. At a specified body temperature, the saturated vapor tension is given by the Clausius-Clapeyron equation. Assuming an ideal gas, integration of this equation leads to an expression in the form

$$\lg p_T = A - B/T \, , \tag{5.14}$$

where constants A and B can be calculated if the enthalpy and the free energy of the system are known.

Average values of constants A and B characterizing the variation of vapor pressure in the temperature range between room temperature, T_b, and boiling temperature, T_b, are given elsewhere [108]. For example, if $T_{cr} = T_b$, the pressure of saturated vapors is equal to atmospheric, and, from the relationship $p_T = nkT$, we get $n \simeq 2 \times 10^{18}$ cm^{-3} for copper. Further use of constants from Ref. 108, when $T > T_b$ leads to an increase of p and n. However, the accuracy of the values obtained depends on the variation of the enthalpy and its dependence on temperature, and the degree of deviation of gas under investigation from an ideal gas.

The variation of these parameters is especially strong near the critical temperature T_{cr} [109] (for the ionized gas under consideration, the parameter characterizing deviation from the ideal is also small). Therefore, it is to be expected that, away from T_{cr}, relation (5.14) with A and B taken into account [108] will not lead to considerable error. The distribution function of heavy particles under conditions being considered is assumed to be Maxwellian. Unfortunately, the data on the critical parameters, p_{cr} and T_{cr}, are not available for copper. However, such data are available for alkali metals [109]. Within the limits of experimental accuracy, estimates for these metals show that, at a specified T_{cr}, the experimental value of p_{cr} coincides with that calculated using relation (5.14) and tabled values of A and B [108]. When estimating T_{cr} of metals for which no experimental data are available, one can use the following relation [109]:

$$T_b/T_{cr} = C - \text{const.}$$

Using the experimental data on T_{cr} for alkali metals that have been given [109] $C \simeq 0.45$–0.55. For most other metals, $C \simeq 0.6$ [109]. Taking into account the range of variation of C for copper given above, $T_{cr} \simeq 5000$–6000 K.

At this temperature, $n \geq 10^{21}$ cm^{-3}. Under conditions considered here, $T_{cr} \simeq 4000$ K and n 10^{20} cm^{-3} and, therefore, relation (5.14) provides correct order of magnitude for p_T and n. Thus, we are interested only in the analysis of small deviation of constants in relation (5.14) in connection with their temperature dependence.

Before proceeding to analysis of the calculated parameters of the jet, we will discuss, separately, relations (5.12) and (5.13) as they apply to the experimental results [42], according to which, when I = 300 A, for copper, G = 1.56×10^{-2} g/s, $v_\infty = 9.8 \times 10^5$ cm/s, u = 20 V. For these parameters, relations (5.12) and (5.13) are satisfied identically, provided $p_T = 2.6 \times 10^2 (1 + \alpha\Theta)$ kPa, where $\Theta = T_e/T$. Then, $n = 5.4 \times 10^{18}/(1+\alpha\Theta)$ cm^{-3} corresponds to this pressure. Estimates were made for $j = 5 \times 10^4$ A/cm^2 and T = 3500 K, corresponding to the first solution of the system of equations (see 5.3.3.1.).

However, the particle concentration required for proper ion current of $n \simeq 5.5 \times 10^{18}/\alpha$ cm^{-3}, i.e., is greater than the total concentration of heavy particles. Therefore, in general, experimental data in Ref. 42 cannot be described by (5.12) and (5.13), for any dependence of the vapor pressure on the cathode temperature, if the ion current is equal to saturation current. For this purpose, ion current must be increased not less than 10 times.

On the other hand, from relation (5.13) it follows that the contradiction above can be eliminated by increasing j. In this case, the second solution of the system of equations (see 5.3.3.1.) is of interest. For example, when I = 300 A, G = 3.3×10^{-2} g/s. When E = $2E_M$, calculations show that for a spot $j \simeq 10^7$ A/cm^2, $T_k \sim 3950$ K and for a jet $v_\infty \simeq 5 \times 10^5$ cm/s, $u_\pi \simeq 3$ V. Analogous results may be given for other initial parameters. In other words, in a general case, parameters of the second solution may correspond to the experimental data on jets [42]. However, the value of j in these solutions is still (even when $E = 2E_M$) too high (10^6-10^7 A/cm^2). Even at present, no experiments were performed that would verify these values for $I \simeq 300$ A for a single spot.

Consequently, according to estimates, it can be assumed that difficulties of theoretical description of experimental parameters of the jet may be associated with an error in determining the initial experimental data used in the calculations.

Analysis of the results of the calculations of the jet velocity, v_∞, and the potential drop across the jet, u_π, based on the (5.5)-(5.14) system of equations for various initial parameters for u_k = 15 V (Table 5.2), shows that the values of ν_∞ and u_π may vary considerably, depending on the accuracy of specifying I, ϕ, G, and constants of the cathode material that determine its vapor tension. Furthermore, a tendency for the results of the calculations to approach the experimental data is stronger at E = $2E_M$ than at E = E_M. From Table 5.2, it follows that parameters ν_∞ and u_π drop considerably with increasing I, ϕ, and decreasing G.

An especially strong influence is exerted by variation of the work function of electrons. Even at ϕ = 3.4 eV, when I = 200 A and G = 2.11 x 10^{-2} g/s, the calculated value of ν_∞ is practically equal to the experimental value. For parameter G, this is attained when it changes by a factor of 6, when I = 300 A, and by a factor of 3 when I = 200 A. A decrease of current to 100 A also leads to $\nu_\infty \simeq 10^6$ cm/s.

In addition to the results of the calculations for copper, given in Table 5.2, computations were performed for aluminum at a current of 100 A. The experimental value, ν_∞ = 8.5 x 10^5 cm/s [42], was attained at G = 1.8 x 10^{-2} g/s, u_k = 18 V, and ϕ = 3.5 eV. The value of G used in these experiments corresponds to a current of I = 300 A [42], at which the calculated value of ν_∞ considerably exceeds the experimental value.

A decrease of the cathode potential drop and the lifetime of the spot also results in a drop of ν_∞ and u_π. For example, when u_k varies between 19 and 11 V, ν_∞ decreases three times and u_π decreases by an order of magnitude. All calculations were performed for a steady-state. However, a change of a single parameter, t, to 10^{-3} s leads to $\nu_\infty \simeq 3$ x 10^6 cm/s and, for t = 10^{-4} s, we get $\nu_\infty \simeq 2$ x 10^6 cm/s when I = 200 A and G = 2.1 x 10^{-2} g/s. It should be noted that, in all calculations where ν_∞ is close to the experimental value, n $\simeq 10^{19}$ cm^{-3}, $\nu_0 \lesssim 10^4$ cm/s and the electron energy lies within 1.2-2 eV (excluding calculations for ϕ = 3.4 eV, where $W_e \simeq$ 3.5 eV). Electron pressure becomes comparable with the pressure of heavy particles at $W_e \simeq$ 1.5-2 eV, when the degree of ionization increases. Further, electron pressure may lead to an increase in the velocity of the jet by a factor of almost 2-3 or more. This can be seen from Table 5.2, which in addition to ν_∞, also shows ν_∞^T corresponding to calculations performed at p = p_T. Calculations demonstrate the influence of the variation of constants in relation (5.14), where the value of A was decreased from 11.96 [108] to 11.2.

Table 5.2
CALCULATION OF PARAMETERS OF A CATHODE JET

A	I,A	$G\cdot10^2$, g/s	ϕ,eV	$\nu_\infty^T\cdot10^{-6}$, cm/s	$\nu_\infty\cdot10^{-6}$, cm/s	u_π,V	$j\cdot10^{-4}$, A/cm	w_e,eV
Calculations based on the assumption that $E = 2E_M$								
11.96	300	1.56	4.5	8.2	11.6	350	2.5	1.27
	300	1.56	4.0	2.2	6	98	2	1.48
	300	3.3	4.5	3.6	5.1	142	2.9	1.2
	300	10.9	4.5	0.5	1	18	7.1	1.51
	200	2.11	4.5	2.4	4.2	93	4.5	1.42
	200	2.11	4.0	0.46	2.6	36	6.6	2.1
	200	6.0	4.5	0.4	1	14.7	12	1.7
	100	1.4	4.5	0.56	2.1	30	16	2
11.2	300	1.56	4.5	1.1	5.4	76	4.5	2
Calculations based on the assumption that $E = E_M$								
11.96	300	1.56	4.5	17.8	20.9	1140	3.1	1.2
	200	2.11	4.5	5.9	7.55	301	5.3	1.3
	200	2.11	4.0	1.9	3.7	72	4.6	1.5
	200	2.11	3.8	1	3	47	4.9	1.7
	200	2.11	3.65	0.6	2.6	37	5.1	1.8
	200	2.11	3.4	0.19	1.6	13	33.7	3.5
	100	1.4	4.5	2.3	3.6	89	14.2	1.6
11.2	300	1.56	4.5	2.1	4.8	60	3.9	1.53

Thus, variation of initial parameters determined from the experimental data within reasonable limits makes it possible to obtain an overall agreement between the measured jet velocity and the calculated velocity. However, as a rule, this result is achieved at almost the minimum parameter values, which makes it difficult to perform optimum calculations. In addition, the values of u_π obtained from relation (5.12) are still too high in comparison with the experimental data. Apparently, this is associated with additional energy losses in the jet that are not taken into account in expression (5.12). One way to improve the calculations is to take into account energy losses from radiation. Assuming that, as a result of a high concentration of particles near the electrode, the jet is a black body at a temperature T_e, it is easy to obtain relations for the energy flux, $N = 2.04 \times 10^4 W_e^4 F_r$, W, where F_r is the area of the emitting surface. For the variant, $\phi = 3.4$ eV (see Table 5.2), existing estimates lead to a drop of u_π from 13 to 3.4 V already at F_r equal to the area of the spot. For other variants, where $u_\pi \simeq 10^2$ V and $\nu_\infty \simeq 2 \times 10^6$ cm/s, a drop of u_π to 1 V occurs at a characteristic jet dimension that exceeds only a few times the radius of the spot. Taking into account that the jet expands and that its area with a high concentration of particles reaches 10^{-1} cm or more, the influence of radiation on the energy balance of the jet prove to be quite substantial. Results of specific calculations will depend on the temperature distribution and the length of the emitting sector of the jet.

Hence, a model with constant parameters in the spot does not contradict experiments with vacuum jets, and, in principle, it may be possible to describe their integral characteristics by the (5.5)-(5.13) system of equations. This requires a detailed analysis of the influence of quantities estimated from the experimental data on the results of the solution of such a problem, taking into account processes in the region of spatial charge at the cathode.

If one is not interested in characteristics of the jet of vapor, and analyses parameters directly in the spot (current density, spot temperature, etc.), it is possible to use a relatively simple system of equations (5.5)-(5.11). According to calculations, deviation of the constants of cathode material and quantities determined from the experimental data throughout broad ranges exerts negligible effect on the variation of parameters of the spot (with accuracy of the experimental data) [88].

5.3.3. Investigation of the Solution of the System of Equations. Depending on the lifetime, arc current, condition of the electrode surface, etc., processes in the near-cathode region as a function of time play different roles in the mechanisms sustaining the discharge. Therefore, favorable conditions arise for the appearance of various types of cathode spots.

From the point of view of the external behavior, different cathode spots (experimentally observed) may exist because the occurrence of identical processes and, vice versa, spots of similar nature in terms of burning are described by different near-electrode mechanisms. The spot type characterizes the degree of heating of the cathode and its destruction. Therefore, as a rule, experimentally determined characteristics of the discharge require theoretical understanding of the importance of these or other processes occurring in the electrode and also in the plasma layer adjacent to it, and determination of the principal characteristics, in order to develop simple engineering methods for calculations of spot parameters.

5.3.3.1. Solid Cathode

The (5.5)-(5.11) system of equations was solved for an arc discharge burning on solid tungsten, copper, silver, and nickel cathodes. The current to the spot at various stages of its existence varied across 10-500 A. Calculations were performed assuming $E = E_M$. For comparison, the same analysis was performed for $E = 2E_M$.

A characteristic of this system of equations is the existence of two solutions for copper at a given u_k and current to the spot and a lack of solution for tungsten [96]. In order to represent more completely the variation of parameters of the spot described by equations (5.5)-(5.11), it is convenient to represent their solution in the form of two intersecting dependences (Fig. 5.3). The first dependence is obtained from solution of equations of the electric intensity, energy balance at the cathode, and electron energy balance in the near cathode plasma simultaneously with relations for the ion flux. The second relation is obtained from the same equations solved simultaneously with the electron emission equation.

Here and in other figures, the dashed curves that become continuous at low j are calculated by the use of energy balance of plasma electrons in the form (5.7). All continuous curves were calculated using an analogous balance equation; however, energy transferred out by the plasma electrons was assumed to be equal to $2k(T_e + T_k) +$

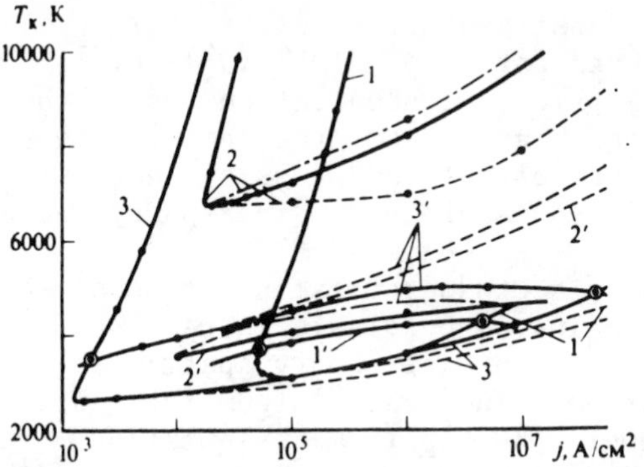

Fig. 5.3 Diagram illustrating the existence of solutions of a system of equations for the near-cathode region of the arc (I = 200 A). 1-3 and 1′-3′ - neglecting and taking into account electron emission equations, respectively; 1,1′ - copper; 2,2′ - tungsten; 3,3′ - nickel.

$2u_k$, i.e., was increased by a factor of $2u_k$ in comparison with an analogous term in equation (5.7). Dot-dashed curves near continuous curves were calculated neglecting the influence of high-energy plasma electrons.

Such an analysis makes it possible to determine the importance of these electrons in calculating parameters of the spot. Curves for silver are located close to those for copper, and are not plotted in Fig. 5.3. The existence of two solutions (marked by dots in the figure) can be attributed to curve 1 having two sets of values. Analysis of curves 1-3 and 1′-3′ shows that the existence and behavior of the second solution of the system of equations may depend on the effects associated with high electron temperature that were taken into account in energy emission and energy balance equations, and also on the electric field strength at the cathode.

The dependences of T_k on j for tungsten (see Fig. 5.3, curves 2, 2′) clearly demonstrate the absence of a general solution of the system of equations. These curves intersect only in the region $j \simeq 10^{10}$ A/cm^2, $T_k \simeq 2 \times 10^4$ K [96] (similar calculations were also performed as a function of time). The physical meaning of this is that for thermophysical and emission constants of tungsten (bonding energy of tungsten atoms, equal to 7.6 eV, considerably exceeds the work function of electrons equal to 4.5 eV) at any cathode temperature, the concentration of atoms calculated assuming that the pressure of the near-cathode vapor is an equilibrium pressure is always insufficient to generate ion current j_i (T_k) to the cathode at emission current of electrons j_{eo} (T_k).

Current density of 10^5 A/cm^2 [27] requires a heavy particle concentration of 10^{19}–10^{20} cm^{-3}. At the same time, equilibrium concentration of the above-noted particles, calculated from the temperature and satisfying emission equation at the same current density, is 10^{14} cm^{-3}. An analogous discrepancy occurs at other values of j.

The analysis conducted makes it possible to reach the following conclusions. First, in real spots on a tungsten cathode, the pressure of heavy particles may differ from the equilibrium value calculated for pure metal at the temperature of the spot. Calculations show that the difference between concentrations of heavy particles is quite substantial. The tungsten vapor tension may change because of the presence of impurities and oxidation of the surface of the cathode. For example, it is known that the oxide WO_3 is more volatile than pure tungsten. Theoretical analysis of discharge under such conditions requires additional experimental investigations to obtain complex data on the properties of the surface of the cathode and the discharge.

A description of cathode spots on pure tungsten requires the determination of conditions of generation of a neutral medium (particle flux from the anodes, influence of the anode wall on the expansion of the cathode vapor [96, 110], pressure of extraneous gas, etc.).

The electron current near refractory cathodes may differ considerably from the emission current. Investigation of this phenomenon requires vigorous analysis of processes in the region of the space charge at the cathode, taking into account reverse current of plasma electrons.

Consider characteristic dependences of parameters of spots on copper, silver, and nickel electrodes. Calculations show that, in general, the dependence of current density on the cathode potential drop for these three metals (Fig. 5.4) is characterized by two sets of values and a certain minimum value, u_k^{min} (10 V for copper and silver and 8 V for nickel), that does not coincide with the experimental value of u_k, which is 15 V, 13 V, and 18 V for copper, silver, and nickel, in the same order [25, 31, 33].

Dependences $j(u_k)$ for copper and silver are sufficiently close, but differ significantly from the analogous dependence for nickel. The current density at a specified u_k in the first solution (lower branches of the curve in Fig. 5.4) for nickel are more than an order of magnitude lower than those for copper and silver. This

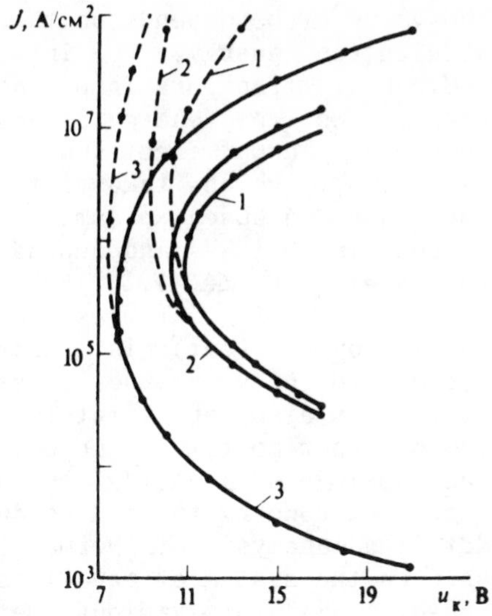

Fig. 5.4 Dependence of the current density on the cathode potential drop (I = 200 A). 1 - copper; 2 - silver; 3 - nickel.

can be attributed to the differences between thermophysical properties of materials under consideration and, in particular, to nickel vapor tension being considerably higher than that of copper and silver.

In general, u_k^{min} differs from the ionization potential (even though, for nickel, these values are close to each other) and, at a specified value of j, is a function of the relation between energy expended in the near-cathode plasma and in the solid body. The connection between u_k and parameters of the spot and, therefore, u_k and j, can be obtained easily from equations for the energy balance in the above-noted discharge regions using expressions (5.7), (5.9), (5.10):

$$eu_k = \phi - (e^3E)^{1/2} + 2\,(1+\alpha\zeta)w_e - 4sw_k/\zeta$$

$$+ 56.6\,(1-s)\,w_k\,(A_iw_e/w_k)^{1/2}\,\exp\,(-3u_k/2w_e)$$

$$+ ((\lambda_{ev}+2kT_k/m)\,G + Q_T + \alpha\zeta u_iI/e)\,e/I\ .$$

It should be kept in mind that the relation obtained is not an additional equation for the (5.5)-(5.11) system of equations, and serves only to estimate the importance of individual terms of the balance at certain experimentally determined parameters (G, u_k, I, etc.). It can be seen that the value of u_k depends on the electric field at the cathode, electron energy, w_e, in plasma, material erosion rate G, heat transfer into the cathode Q_T, ionization energy, and the energy transferred to the cathode by high-energy plasma electrons. For copper, for example, the term $\alpha\zeta u_i$, which takes into account energy losses in the near-cathode plasma associated with ionization of atoms, is 0.1 V and depends weakly on the current (approximately as $\zeta(I)$). Energy expended on electron emission and heating of plasma electrons (which also depends weakly on current) is relatively high and approximately equal to 6 V. Energy expended through thermal conductivity is also approximately the same (8 V).

Note that, in experiments conducted by the authors [111, 112], an analogous quantity characterizing heat transfer into the electrode by thermal conductivity was 6-7 V. A more detailed analysis of thermal fluxes into the electrodes, based on the system of equations for the near-cathode region [87, 88] for various metals, has been conducted [113]. This analysis demonstrated good agreement between the experimental data and the results of calculations.

Hence, a combination of processes that are characterized by emissivity of the cathode, its thermal conductivity, and cathode material erosion rate, exert considerable influence on the formation of the cathode potential drop. Thus, because the importance of processes noted above varies for different types of spots [88], the value of u_k may apparently vary, depending on the mechanism of sustaining a specific type of spots. Measurements of u_k available at present [25, 31, 33] do not permit determination of this dependence, because they were conducted at relatively low currents ($I_o \lesssim 30$ A) and correspond to a specific type of spot. According to the experimental data, only rapid spots exist at low currents to the cathode [30, 40]. If the level of thermal losses in the energy balance of such spots is insignificant, u_k for fast spots may depend on the energy losses in plasma [79] that depend very little on the current to the spot (at small current to the spot the coefficient ζ is practically constant [30, 46]). Therefore, the value of u_k will also change very little. This was observed in experiments [25, 31, 33]. Generally speaking, in the case of slowly displaced spots, where (current dependent) thermal processes at the cathode are the predominant processes, u_k may not coincide with the measured value [25, 31, 33]. Even though this difference may be small, its experimental determination is important to explaining the mechanism of sustaining various types of spots.

It should be noted that a report [114] provides some data on experiments dealing with measurements of the cathode potential drop of vacuum arcs, and observes that, when arc current is switched stepwise from an average current (10-30 A) to high current (400-700 A), the voltage increases by 3-5 V. It is also reported that the cathode potential drop depends on the cathode temperature, work function, and the arc burning regime. However, no rigorous experimental analysis of this problem exists at the present time.

Let us now consider the dependence of the current density on current (Fig. 5.5a), work function, and the lifetime of the spot (Fig. 5.5b, c) obtained using

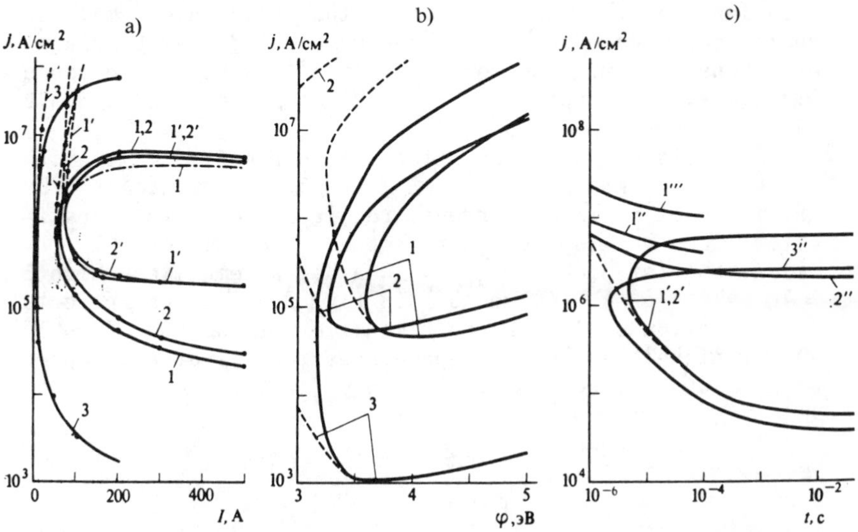

Fig. 5.5 Dependence of current density in a cathode spot on current (a), work function of the cathode (b), and the lifetime of the spot (c) for electrodes made of various materials.

1,1′, 1′′, 1′′′ - copper; 2,2′ 2′′ - silver; 3,3′′ - nickel; 1-3 - steady-state regime, 1′, 2′, 1′′-3′′, 1′′′, nonsteady-state regime; 1-3, 1′, 2′ - I = 200 A, G = 2 x 10^{-2} g/s; 1′′, 1′′′ - I = 10 A for G = 10^{-4} and 5 x 10^{-4} g/s, respectively; 2′′, 3′′ - I = 50 A for G = 10^{-4} and 5 x 10^{-3} g/s, respectively.

experimental values of u_k provided in the literature [25, 31, 33]. Analysis of the parameters of the spot as a function of lifetime was performed, using relations (5.9), (5.10), where the heat flux into the cathode was assumed to be constant and was determined from the solution of the problem. As an illustration, parameters of the spots in a steady state mode are given in Table 5.3. All results were obtained at $T = T_k$. Calculations at $T = T_e$ differ very little from the calculations above. For nickel, when

I = 15 A, solutions 2 and 2* coincide. In addition, it follows from Table 5.3 that, as the current decreases, the calculated parameters for nickel become close to parameters of group spots on copper and silver.

From Fig. 5.5 it can be seen that the curves for silver are located close to the curves for copper and the corresponding second solutions (continuous curves) are practically identical. As in Fig. 5.4, the curve for nickel differs sharply from curves for copper and silver. From Fig. 5.5b, it follows that a minimum work function, ϕ_{min}, exists for all three metals (analogous character of the dependence j (ϕ) for copper was already noted [88]) at which the (5.5)-(5.11) system of equations still has solutions and, consequently, the discharge may be sustained. Depending on the energy carried away by plasma electrons, the minimum work function varies in the range 3.3-3.6 eV for copper, 2.9-3.2 eV for silver, and 2.6-3.2 eV for nickel.

Analysis of the variation of current density with current to the spot (see Fig. 5.5a) indicates the existence of minimum current required to sustain quasi-established hot spots (for copper and silver, $I_{min} \simeq 10^2$ A, and for nickel, $I_{min} \simeq 10$ A, at a relatively high G $\simeq 10^{-2}$-10^{-3} g/s). Curve 3'' in Fig. 5.5c, which reflects the variation of j with the lifetime of the spot (I = 15 A) for nickel (analogous dependences for silver and copper at I $\simeq$ 10 A may be constructed only for

$$G \lesssim 10^{-4} \text{ g/s} \quad \text{and} \quad j \gtrsim 10^{7} \text{ A/cm}^2$$

(see curves 1'', 2'', 1''' in Fig. 5.5c)), shows that spots with low current (close to I_{min}), appearing at the cathode made of this metal, are described by a system of equations, (5.5)-(5.11), when their lifetimes are relatively short ($t \simeq 10^{-6}$ s) and current densities are low ($j < 10^6$ A/cm^2). See Table 5.3.

Numerical values of I_{min} and t_{min} for nickel obtained from these calculations coincide with the experimental data for copper for rapidly displaced spots [30, 40]. Unfortunately, no experimental investigation of the dynamics of cathode spots in vacuum are available for nickel. Therefore, one cannot determine at present which type of spots burning on cathodes made of nickel is characterized by I_{min} and t_{min} determined in the foregoing. However, the results of calculations for nickel with low current to the spot makes it possible to assume that short-lived spots, if they are ever observed

Table 5.3
CALCULATION OF PARAMETERS OF THE SPOT FOR A SOLID CATHODE MADE OF VARIOUS MATERIALS ($E = E_M$)

Cathode Material	I, A	Solution	α	s	T_K, K	w_e, eV	$E \cdot 10^{-6}$ V/cm	$f \cdot 10^{-4}$ A/cm^2	$n_i \cdot 10^{-18}$, cm^{-3}	$(n_i+n_a) \cdot 10^{-19}$, cm^{-3}	$v_0 \cdot 10^{-3}$ cm/s
Cu	200	1	0.10	0.450	3723	1.31	5.4	5.3	6.5	6.5	0.84
		2	0.86	0.977	5110	5.21	57.5	16000	710	82	200
		2*	0.753	0.857	4322	3.26	29.5	618	180.5	24	26
Ag	200	1	0.067	0.487	3530	1.32	6.97	7.8	11.8	17.6	0.38
		2	0.733	0.99	4990	5.0	72	58800	1400	190	270
		2*	0.55	0.864	3921	2.81	30.7	575.3	220.2	40.2	12.4
		2**	0.87	0.829	3690	3.96	30	420	220	20	14
Ni	200	1	0.017	0.356	3475	0.86	11	0.18	0.25	1.46	0.074
		2**	0.975	0.877	4532	6.76	39	1170	270	28	25
		2	0.86	0.943	4877	4.66	51	4550	480	56	49
	15	1	0.083	0.401	3890	1.23	4.9	3.8	4.7	5.71	2.35
		1***	0.787	0.51	4050	2.8	19	70	70	8.9	0.55
		2	0.72	0.746	4347	2.92	27	269	135	19	50

Comment. Solutions 1 and 2 on the upper and lower branches of the curve, respectively, expressing the dependence of the current density on current (see Fig. 5.5a); 2 - using plasma electron energy equation in the form (5.7); 2* - energy removed by plasma electrons, increased by a factor of 2 u_k in comparison with equation (5.7); 2** - neglecting "inverse electrons"; 1*** - at t - 10^{-6} s and G = 10^{-4} g/s.

experimentally, may be described by the continuous-thermal model. It should be kept in mind that the minimum current, I_{min}, is the current at which solution of a closed system of equations still exists.

Numerical investigations of equations (5.5)-(5.11) make it possible to analyze the mechanism responsible for sustaining various types of spots appearing on cathodes of different materials.

Preliminary estimates of the heat flux have shown that the thermal model [87, 88] is not applicable to fast spots (with current density $j \simeq 10^4$ A/cm^2) on copper [79]. Calculations performed here with a system of equations (see curves 1 and 2 in Fig. 5.4 and 5.5a and curves 1′, 2′, 1′′, 2′′, and 1′′′ in Fig. 5.5c) verify this conclusion and make it possible to extend this conclusion to silver. Taking into account the results for nickel, one can assume that spots with a short lifetime ($t \simeq 10^6$ s) on various metals may be described by different physical models. Fast spots on film cathodes may be described by a thermal model [16] (see 5.3.3.2.). On the other hand, analysis of phenomena on a tungsten cathode [96] shows that, even though in terms of classification according to type, spots on this material are close to the group spots on copper and silver, nevertheless cannot be described by the (5.5)-(5.11) system of equations. Therefore, cathode spots of a specific type (classification based on the experimental data according to external characteristics, such as displacement rate, lifetime, and current to the spot and the assumed type of emission) can be sustained by various processes.

Therefore, in analysis of various types of spots, it is apparently necessary to take into account not only the method of generating charged particles (i.e., type of emission and ionization), but also the methods of generating a neutral medium, as well as the interconnection of processes occurring in the plasma and on the surface of the cathode. Based on the results reported [88, 96], it is possible even now to conclude that emission in group spots on copper is predominantly thermionic. This can be attributed to the fact that the energy used to vaporize atoms (predominant process in spots of this type) may be about the same as the energy required to "vaporize" electrons, or may even be higher.

It follows from the solution of equations (5.5)-(5.11) that minimum current to the spot depends substantially on the cathode material erosion rate. Therefore, when analyzing a specific type of spot characterized by a certain current to the spot, it is necessary to use parameter G, measured under rigidly controlled conditions.

This conclusion also pertains to analysis of processes in cathode spots with relatively low current, $I \simeq 20$ A, and displacement rate, $\nu \simeq 10^2$ cm/s, formed in extraneous gas at atmospheric pressure, when the expansion of products of erosion of an electrode is somewhat slower [27, 46, 61]. Taking into account considerable variation of I_{min} with G (see Fig. 5.5), it can be assumed that the mechanism sustaining such spots also corresponds to the thermal model. In describing them, however, it is necessary to take into account the influence of the oxidation process (in air), preheating of the electrode by current (spots of this type appear 10^{-4} s after the discharge is initiated) and other factors characterizing conditions at which the spots appear.

Without detailed analysis that will surely require additional empirical data, it can be assumed that all spots can be subdivided into two large groups: thermal spots, when thermal vaporization of cathode material takes place from the whole surface of the spot and is caused by the heat released during bombardment of the spot by ions of the near-cathode gas, and local-thermal spots, when formation of a neutral medium occurs from individual sections of the spot. The second group may include spots such as those with explosive "vaporization" during heating of micro nonuniformities on the surface of the cathode as a result of the Nothingham effect and Joule heat dissipation [80, 82].

However, it should be pointed out that the explosive model is probably more applicable to the transition stage between breakdown and arcing, while description of the arc process in terms of this model encounters certain difficulties. One of the difficulties encountered is the low electric field strength in the near-cathode region of the arc ($E \simeq 10^8$ V/cm), required for explosive emission. Even though the presence of ionized gas in the inner-electrode gap leads to a small drop of the critical field, its level still remains sufficiently high.

The nonuniformities contribute to the arc discharge process. Plasma generated by explosion of micro tips expands at a rate of 10^6 cm/s during 10^{-8} s, in a hemisphere of 10^{-2} cm radius [82]. Ions of plasma from this volume incident to the cathode do not effectively heat the whole surface of the spot, but only the existing nonuniformities that characterize the local thermal vaporization, and maintain the concentration of particles at the required level. The experimental data show that concentration of charged particles in type II spots is adequately high (exceeding 5×10^{17} cm^{-3}) [30, 40].

Estimates using these data show that the thickness of the layer of spatial charge is approximately 10^{-6} cm, i.e., considerably smaller than the characteristic dimensions of micro-nonuniformities which are on the order of 10^{-4} cm.

Consequently, potential and electric field distributions around any protrusion on a cathode will be analogous to that near a smooth surface, and the incoming flux of energy from ions will occur from practically all sides of the protrusion.

This makes it possible to estimate the temperature of the protrusion during the first stage, on the assumption that it is completely isolated (thermally) from the solid body. Then, the temperature gradient will be

$$\Delta T \sim j_i u_{eff}(1+2\ell_1/r_1)/r_1 c\gamma ,$$

where ℓ_1 is the height, r_1 is the radius of the cross section of the protrusion. It can easily be seen that for copper at $\ell_1/r_1 \simeq 4$, $j_1 \gtrsim 10^4$ A/cm^2 and $t \simeq 10^{-6}$ s, when $\Delta T \gtrsim 10^3$ K.

Subject to these simplifications and the assumption that the vapor pressure in spots of type I is close to an equilibrium value calculated from the temperature of the protrusion (high concentration of plasma particles at the cathode), investigation of the (5.5)-(5.11) system of equations has shown that its solution exists when $I \simeq 10$ A, $j \simeq 10^4$ A/cm^2, $t \simeq 10^{-6}$ s and $G \simeq 10^{-4}$ g/s. The lifetime of such a spot may be estimated, assuming complete vaporization of protrusions on the surface of the cathode in the region of the spot. Assuming that, for copper, their mass $\pi r_1{}^2 \ell_1 \gamma \simeq 3 \times 10^{-11}$ g and the number of protrusions is 10-10^2, the erosion rate determined will result in complete vaporization during the time, 10^{-6}-10^{-5} s. This is in agreement with the experimental data [30, 80]. Thus, as a result of the small dimensions, high thermal resistance of protrusions may lead to high heating of individual parts of the surface of the cathode providing the necessary vapor pressure in the near-cathode region generated by vaporization of the cathode material.

Further improvement of the local-thermal model must continue along the path of investigating effects associated with nonuniformity of vaporization processes, electron emission, thermal conductivity, etc. Under conditions being considered, this may lead to an improvement of the form of equation for the total current.

In conclusion, refer to analysis of the influence of

the electric field near the cathode on the spot parameters. With this in mind, parameters of the near-cathode region of the arc for a copper cathode were calculated, assuming $E = 2E_M$ for various values of current to the spot, lifetime of the spot, work function, and cathode drop. When $I = 10$ A, the erosion rate varies across $0\text{–}10^{-4}$ g/s. Analysis of the results of this calculation (Table 5.4, Fig. 5.6) shows, primarily, a temperature change for the first solutions, i.e., the temperature drops by approximately 300 K. This process

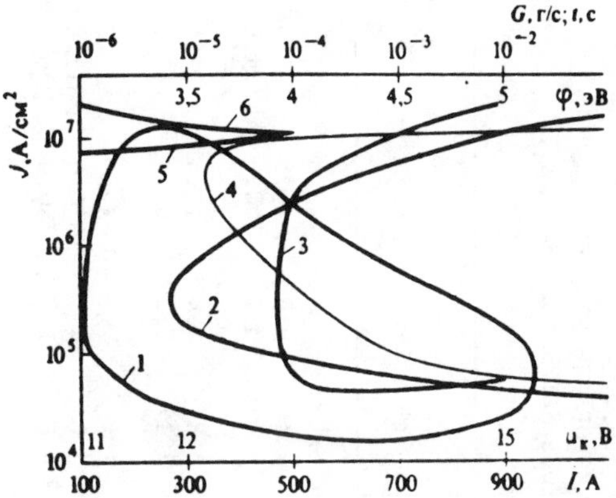

Fig. 5.6 Variation of current density in the cathode spot of a copper electrode with current (1), cathode potential drop (2), work function (3), lifetime (4,6), and electrode material erosion rate (5) ($E = 2E_M$). 1-3 and 4-6 - stationary and nonstationary modes, respectively; 5 - $t = 10^{-4}$ s.

intensifies with decreasing work function of the cathode. The concentration of particles of the near-cathode plasma decreases with decreasing temperature. As a result of increase in the degree of gas ionization, current density remains at almost the previous level. Minimum cathode drop and work function increase somewhat (see curves 2 and 3 in Fig. 5.6).

Second solutions differ considerably. First of all, this refers to the current density that drops sharply (by 2-3 orders) in comparison with calculations on the assumption $E = E_M$, especially in dependences of j on u_k and ϕ (curves 2 and 3). Such a result shows that, when second solutions of the system of equations correspond to specific numerical values, it is necessary to analyze parameters of the region of space charge.

Curve 5 illustrates the dependence of current density on electrode material erosion rate when $t = 10^{-4}$ s, and curve 6, on lifetime of the spot when $G = 10^{-4}$ g/s. Both

Table 5.4
CALCULATION OF PARAMETERS OF THE SPOT FOR A SOLID CATHODE MADE OF VARIOUS MATERIALS ($E = 2E_M$)

I,A	G,g/s	Solutions	α	s	T_K,K	w_e,eV	$E\cdot10^{-6}$ V/cm	$f.10^{-4}$ A/cm^2	$(n_a+n_i)\cdot10^{-19}$cm^{-3}	$n_i\cdot10^{-18}$ cm^{-3}	$\nu_o 10^{-3}$, cm/s
				Steady-state							
200	$2\cdot10^{-2}$	1	0.224	0.47	3380	1.42	9.8	4.5	2.46	5.5	1.9
		2	0.96	0.955	3940	4.67	43.1	1100.0	11.0	106.0	100.0
110	$1.4\cdot10^{-2}$	1	0.49	0.53	3390	1.8	1.47	11.4	2.5	12.4	5.6
		2	0.93	0.75	3530	3.17	2.5	64.4	3.9	36.0	20
				Lifetime $t = 10^{-4}$ s							
10	10^{-4}	1	0.97	0.95	3950	5.0	44	1070.0	11.0	110	9.2
	10^{-6}	1	0.97	0.94	3900	4.86	41.7	740.0	10	97.8	6.0

curves are constructed for relatively low current, $I = 10$ A. It can be seen that current density at $G < 10^{-4}$ g/s is almost constant. An estimate of the terms of heat balance of the cathode in calculations performed at low currents and material erosion rates has shown that the importance of reverse electrons may prove to be substantial, and sometimes even predominant.

In analyzing the mechanism of sustaining the spot, one of the principal dependences is that of j on I (see curve 1 in Fig. 5.6). Analysis shows not only the existence of a minimum current to the spot, but also of a maximum current. This can be explained as follows. Curve 1 was calculated using an exponential experimental dependence of erosion on current [27, 28]. Therefore, the analysis indicates that, when $I > 900$ A, energy losses associated with cathode material erosion exceed the flux of heat to the cathode by bombardment with ions. In this case, the (5.5)-(5.11) system of equations has no solution. Actually, the experimental dependence of G on I [27, 38] used in the calculations is applicable up to $I \simeq 7 \times 10^2$ A. In addition, when $I > 10^3$ A, cathode material erosion may occur in vapor and other phases.

However, the dependence of j on I shown in Fig. 5.6 can nevertheless form a basis for qualitative explanation of experimentally observed current restriction in individual spots and their tendency to fragment at relatively high discharge current. But detailed measurement of the parameter G under specific conditions is required for qualitative analysis of this effect.

5.3.3.2. Film Cathode

A distinguishing characteristic of an arc spot burning on thin metal film cathodes is their relatively high current density ($j \simeq 10^6$ A/cm^2) in comparison with spots on solid cathodes ($j \simeq 10^4$ A/cm^2). According to estimates, spots with such high current density, even at relatively high displacement rates of spots, are capable of strongly overheating material of the film at only a fraction of an ampere of current to the spot. In this connection, analysis of parameters of such spots in terms of the thermal model is quite promising. The results of such an analysis are especially interesting if one takes into account that experiments with film cathodes provide the data that form the basis of the experimental method of autographs used in determining current density in the spot.

A theoretical description of processes on film cathodes will be performed, primarily on the basis of the results obtained elsewhere [16, 88], and also the results of additional calculations that make it possible to compare parameters of spots for various materials, the influence of thermophysical constants (they are different from those for solid cathodes), etc.

Detailed experimental investigations of arc spots on film cathodes (metal film deposited on glass or metallic substratum) [25] were carried out in order to clarify the basic mechanism of the dynamics of cathode spot development. Most of the critical data were obtained as a result of investigations of traces of autographs left by a cathode spot.

A spot referred by the author as "cell" on not very thick film (less than 1 μm) leaves a clear trace of a completely removed metal. The width of the path, δ, of a single cell and its length, ℓ (path covered during the lifetime of the cell $\ell = \nu t_*$, where ν is the displacement rate being measured, t_* is the lifetime), make it possible to calculate total erosion and the specific erosion rate of cathode material.

$$G^* = \delta d \ell \gamma \, , \quad G = G^*/t_* = \delta d \nu \gamma \, . \tag{5.15}$$

Here, d is the thickness of the film, γ is the metal density. It was established that characteristics of a cathode spot (δ, G, ν, t_*, cell current I_d, etc.) depend considerably on the thickness of the film. However, certain experimental data (especially the fact that threshold and cell currents coincide) led to the conclusion that a spot on a thick film ($d \simeq 1.5$ μm) is identical to that on a solid cathode.

Theoretical analysis reported [25] led the authors to the conclusion that the characteristic size (r) of a cell (region in which current flows), i.e., cathode spot, is many times less than the cross section of an autograph, δ, and, consequently, the size of the autograph cannot be used as a characteristic size of a cathode spot. The value of δ was identified with the size of the fusion isotherm for a moving point ($r \ll \delta$) of a heat source. Omitting details, let us point out only certain internal contradictions of the theoretical model suggested [25].

In all of the models of propagation of spots described in Ref. 25, it is assumed that the spot is displaced after vaporization of metal beneath it. Even the concept of "cumulative" vaporization of the film is utilized. On the other hand, all theoretical calculations neglect energy used for vaporization.

Comparing energy Q_δ required to vaporize a mass of the metal measured experimentally from an autograph, with energy Q_1, reaching the cathode spot at the expense of ions has shown that $Q_\delta >> Q_1$. This led the author of Ref. 25 to the conclusion that the vaporization process cannot be significant, and that the principal part of energy supplied is spent on melting the metal within the volume of the autograph and on thermal conductivity. However, such a conclusion leaves unclear the mechanism of removing the metal from the surface of the cathode and its connection with "cumulative" vaporization.

Another contradiction arises in connection with the relation between the size of the cell, r, and that of the autograph, δ. Calculation of the temperature field in the solid and in the film for a concentrated source [115-117] shows that no sharp changes of temperature gradient occur in the region bound by the fusion isotherm. From this, it follows that, when $r << \delta$ in a relatively large region outside the cell, surface temperature must be sufficiently high ($T >> T_m$). However, it is not clear why these sections of the surface do not participate in emission and current transfer.

From our point of view, the contradictory results reported [25] are associated primarily with an inaccurate determination of the ion component of current and inadequate formulation of the thermal problem of the phenomenon under consideration [88]. In this connection, consider the problem associated with the near-cathode region on a film cathode that may be obtained from equations (5.5)-(5.11), modified for this case [87, 88].

This system of equations describes (in view of its basic assumptions) processes in long-lifetime, slowly displaced spots, when gasdynamic and electrodynamic processes in the near-cathode plasma can be assumed to be steady-state (or depending on time as on a parameter). A characteristic feature of such spots is that the surface of the metal in the region of the spot is heated to high temperature that determines the vaporization rate of metal (erosion depends primarily on vaporization) and the electron emission.

It is natural that such a spot is formed as a result of a certain nonstationary process described by a system of equations, (5.5)-(5.11). Displacement of the spot also corresponds to a certain transient process. Furthermore, if the displacement is sufficiently small (the lifetime of a spot at the same place is considerably greater than the time required to establish the gasdynamic processes in plasma), the spot may be investigated using stationary equations. In these equations, the temperature of the surface of the spot (and, therefore, all parameters of the near-cathode region) is a function of time determined by the solution of the nonstationary thermal problem within the electrode. Spots of this type, i.e., group spots, appear at the surface of a solid copper cathode and cathodes made of certain other materials at sufficiently high currents to the spot ($I \simeq 200$ A).

On the other hand, spots with different properties (rapidly moving type I spots) appear at low discharge currents ($I \lesssim 10$ A) and during the initial stage of the discharge. As already noted, physical processes in these spots may differ radically from processes in group spots.

Unfortunately, experiments [25] provide no answer as to the type (based on physical properties and displacement mechanism) of the spot on the film cathode. This could be answered by directly observing the spot during its motion (not its trace after the experiment). However, such observations are not available.

The experimentally determined displacement rate of the spot on a film cathode was sufficiently high ($\nu \simeq 10^3$-10^4 cm/s) and comparable to the displacement rate of spots of type I on a solid cathode. Nevertheless, there are reasons to suspect that, in terms of its properties, a spot on a film cathode may be close to a group thermal spot. This will now be discussed in more detail. Conditions corresponding to copper film on glass will be considered (results of certain calculations will also be given for a bismuth film on glass).

Under experimental conditions, the lifetime of a spot at the same site, specified as time during which the spot is displaced a distance equal to its diameter, depends on the thickness of the film, $d \simeq 10^{-6}$-10^{-5} cm, and is equal to $\tau \simeq 2r/\nu = 2 \times 10^{-8}$-$5 \times 10^{-7}$ s, provided the radius of the spot, r, is assumed equal to the autograph, δ [25]. The last assumption appears to be natural if the spot is of thermal nature and erosion is caused by vaporization. Even though this assumption contradicts the theoretical model suggested [25], it corresponds to the results of calculations given below.

According to an estimate of the distance over which the metal during a lifetime, τ, is heated to temperatures comparable with the temperature of the spot, $\ell_T \simeq (Q_T/T_k c\gamma\pi d)^{1/2}$, where Q_T is the heat flux to the film by thermal conductivity, c is the heat capacity, T_k is the temperature in the spot. Assuming that $Q_T \approx Iu_k$, $\ell_T \simeq 10^{-4}$-10^{-3} cm. On the other hand, the width of the autograph depends on the thickness of the film and varies between 10^{-4} and 10^{-3} cm. Consequently, during the lifetime of a spot at the same site, the metal is heated to a depth comparable to the diameter of the spot. The lifetime of a cell (spot) $t_* = 10^{-6}-10^{-5}$ s at high currents $(I \gg I_d)$ depends on the thickness of the film [25]. At low currents, when only a single cell exists, its lifetime increases by hundreds of times. A comparison with a lifetime of a cell at the same site, τ, shows that, during a lifetime, it covers a distance that exceeds its diameter many times.

Hence, it appears plausible that the spot on the film cathode is displaced after vaporization of metal under the spot. The part of the surface where the spot is displaced is heated by thermal conductivity to a high temperature comparable with temperature of the spot.

In order to be able to describe parameters of such a spot by a system of equations, (5.5)-(5.11), with a variable spot temperature, it is necessary that time, τ, exceed considerably the longest time of the transient hydrodynamic process that, in this case, is convection heating of the plasma layer above the spot. This time can be defined as the time during which a particle of vapor covers a characteristic length of the ionization region: $\tau_i \sim \ell_i/\nu_o$, where $\ell_i \simeq (n_a \sigma_i)^{-1}$, ν_o is the vapor flow rate.

Assuming that parameters of the plasma do not differ significantly from the same parameters for a group spot on a solid cathode (i.e., $n_a \simeq 10^{20}$ cm^{-3}, $\sigma_i \simeq 10^{-16}$ cm^2, $\ell_i \simeq 10^{-4}$ cm), condition $\tau_i \ll \tau$ leads to the condition $\nu_o \gg 5 \times 10^3-2 \times 10^2$ cm/s. Results of the calculation performed below indicate compatibility of estimates given above with the reported experimental conditions [25], thus verifying applicability of the thermal model [87, 88] to a calculation of parameters of the spot on a film cathode.

It should be emphasized that, even though the spot displacement mechanism considered and accepted in future calculations coincides with the model given in Ref. 25, it is actually an assumption. It appears to be quite interesting to attempt to reveal the physical nature of the arc spot on a film cathode by means of experimental investigations of these spots with high temporal and spatial resolution (as was done when investigating spots on a solid cathode [30, 40]).

Generally speaking, one cannot exclude the possibility that certain spots are displaced along the surface of the film in a way analogous to type I spots [30, 40], or in some other way. Finally, it is possible that, depending on the external conditions, spots of various types appear even on a film cathode. Thus, the presence of several types of autographs may be associated with the appearance of several types of spots [25].

In determining the temperature field in a film, the spot on a film cathode is considered a mobile heat source. Therefore, the balance equation for energy on the surface of the spot in the (5.5)-(5.11) system of equations must be modified by changing the term describing the heat flux within the electrode. For the model of the displacement source accepted, the vaporization rate of the front, $\nu_* \simeq d/\tau \simeq \nu d/\delta$, and the angle of inclination of the plane of vaporization with respect to the substratum is determined from the relation $\alpha_o \simeq \nu_*/\nu \simeq d/\delta$. For the films under consideration, $\alpha_o \simeq 10^{-2}$.

Thus, in dealing with the temperature distribution in a film, a spot may be modeled by a circular source displaced at a rate, ν, along the surface of the film. In future calculations, the intensity of the source will be assumed to be Gaussian with characteristic length r. We will utilize a solution obtained in one report [116], a solution that does not take into account Joule heat released in the film.

For conditions considered here (thin films and short diameter source), the intensity of Joule heat release in the film in the region of the source is quite high. However, as a result of the high displacement rate of the source, Joule heating of the metal at points along the trajectory of the source proves insignificant ($\Delta T \simeq$ 100 K). This makes it possible to utilize without any modifications solution given in Ref. 116.

In estimating Joule heating of the film, it was assumed that the current from the spot flows along the radius, and that the current distribution over the thickness of the film is uniform. The current density distribution along the radius at a specified total current, I, is given by the relation,

$$j = I/2\pi dr \; .$$

To estimate the maximum heating of the metal, assume that thermal conductivity is zero. Then, the variation of temperature with time for any point in the film will be given by the equation,

$$\gamma c_\nu \frac{dT}{dt} = \frac{j^2}{\sigma_e} \; , \qquad (5.16)$$

where c_ν and σ_e are the thermal capacity and electrical conductivity of the film, respectively. Obviously, if the spot is displaced along a line, maximum heating will take place at points located along its trajectory, i.e., as the source approaches the points, they fall within the region of the highest heat release density.

The temporal variation of current density initially at a point a distance, R, from the spot and lying along its trajectory is given by the following relation

$$j = I/2\pi d\,(R-\nu t)\ .$$

Substituting this expression into equation (5.16) and integrating, assuming ν = const. and $R \gg r$, gives a formula for the temperature variation at point R at the time of arrival of the spot at this point:

$$\Delta T = I^2/\sigma_e \gamma c_\nu r \nu (2\pi d)^2\ . \qquad (5.17)$$

Calculations performed with this relation show that, for the range of film thickness under consideration and at electrical conductivity values measured under normal conditions in pure metal, temperature variation at the points along the trajectory of the spot is not too great. Taking into account the effect of the drop of electrical conductivity in very thin films ($d \simeq 10^{-6}$ cm [118]) results in a much higher heating temperature. However, because calculations using (5.17) provide an upper limit, Joule heating of the metal can be neglected, even in this case.

In all calculations, except thermal, the spot is modeled by a circle with a homogeneous distribution of parameters (within the boundaries of the spot, all parameters are constant). The temperature at the center of the spot, determined from the solution of the thermal problem (the possible error using such an approach has been investigated [115]), is assumed to be the temperature of the surface of the spot.

Using the solution of the thermal problem [116] and assuming that all of the erosion occurs in the vapor phase and that heat losses to the sublayer can be neglected, the energy balance equation for a film cathode can be represented in the following form:

$$(1-s)\, u_{eff}\, j + 1/2en_i \nu_{eT} kT_e \exp(-eu_k/kT_e)$$

$$- (\lambda_\nu + 2kT_k/m)\, Gj/I =$$

$$= \frac{4\pi\lambda d(T_k-300)\exp\left(-\frac{\nu^2 t_o}{2a}\right)}{\int_{t_o}^{t_o+t} \frac{dt'}{t'} \exp\left(-\frac{\nu^2 t'}{4a} - \frac{\nu^2 t_o^2}{4at'}\right)}\,, \tag{5.18}$$

where $t_o = r^2/4a$, t is the time during which the spot is displaced along the cathode, G is the electrode material erosion rate (see expression (5.15).

Physical meaning of the terms on the left-hand side of equation (5.18) was described in detail above. Its right-hand side is an expression for heat flux along the film removed through thermal conductivity. In calculating parameters of the spot using the energy balance equation (5.18) and equations (5.5)-(5.11), it is necessary to use the experimental values of ν and G, current I to the spot and the cathode potential drop, u_k. In specific calculations, I was assumed to be equal to the cell current I_d [25] that depends substantially on the thickness of the film. The experiments being analyzed [25] unfortunately lack data on the cathode drop, u_k, and its dependence on the thickness of the film. Such data for copper film on a glass substratum are also unavailable from experiments especially conducted to acquire data on the cathode potential drop [31, 32]. Hence, u_k was varied throughout a broad range of values in the calculations.

Results of the calculations of the dependence of current density, j, on the cathode drop, u_k, for films of various thickness are shown in Fig. 5.7. Because experimental determination of erosion (erosion rate G) of thick films becomes unreliable [25], we considered films not exceeding thickness of 5×10^{-5} cm.

The solutions obtained correspond to a quasi-stationary mode of the thermal problem, i.e., to the time $t > t_{**}$, where t_{**} is the time required to reach this mode. However, according to the method reported [116], for the range of parameters under investigation, $t_{**} \lesssim 10^{-6}$ s. This is many times shorter than the lifetime of the spot. Obviously, diameter of the spot cannot exceed the cross section of the autograph. Points corresponding to the condition $2r = \delta$ are plotted on curves in Fig. 5.7. The minimum possible current density, j^{π}_{min}, and $u_{k\delta}$ corresponding to these points are given in Table 5.5. The curves in Fig. 5.7 show that a minimum cathode potential drop corresponds to each thickness of the film. At lower cathode drops, the spot of the type being considered cannot exist (the system of equations has no solution).

Points along sectors of the curve, lying above points $2r = \delta^*$ correspond to a real spot ($2r \lesssim \delta$).[3] The solution obtained also makes it possible to evaluate the maximum current density, j_{max} (minimum radius), of the spot. Obviously, the cross section of the autograph cannot exceed the cross section of the melting isotherm, δ_π. Therefore, condition $\delta = \delta_\pi$ places an upper limit on the possible current density values. The value δ_π is determined from the solution of the following problem:

$$T_\pi \left(\frac{\delta_\pi}{2}\right) = \frac{Q_{t\pi}}{d\pi\lambda d}\, e^{\frac{\nu^2 t_o}{2a}} \times \int_{t_o}^{t_o+t} \exp\left(-\frac{\nu^2 t_o + (\delta_\pi 2)^2}{4at'} - \frac{\nu^2 t'}{4a}\right) \frac{dt'}{t'}\,, \tag{5.19}$$

where Q_T is the heat flowing into the film by thermal conductivity that depends on the current density and is given by the solution of the problem of parameters of the spot, $Q_{T\pi}$ is the value of Q_T corresponding to j_{max}, T_π is the melting temperature.

The values of j_{max}, r_{min}, $u_{k\ max}$ determined on the assumption $\delta = \delta_\pi$ are given in Table 5.5. A comparison of r_{min} and r_{max} shows that they differ by approximately 100 %. In view of this, it follows that the condition $2r = \delta$ is actually fulfilled. Certain experimental data directly verify that the difference between the radius of the spot and the cross section of the autograph is less than the difference between r_{min} and r_{max}. For example, it is noted elsewhere [25] that the edges of the autograph are fused. The report's author assumes that the metal may be pulled to the edges of the autograph by surface tension forces.

3 Condition $2r \lesssim \delta$ eliminates the nonuniqueness of the solution of a system of equations in a natural way.

Thus, the calculated and the experimental data make it possible to conclude that the cross section of an autograph may serve as a measure of the characteristic dimension of the spot and, subsequently, current density within it.

Table 5.5 also shows the concentration of particles $n_a + n_i$; degree of ionization, α; electron component of current, s; electric field strength, E; energy of electrons, w_e; cathode temperature, T_k, and vapor removal rate, ν_o, calculated on the assumption of minimum current density ($2r = \delta$). It can be seen that the influence of the change of energy of plasma electrons on parameters of the spot is considerably less for film cathodes than for a solid cathode and, for all practical purposes, is insignificant (see also Fig. 5.7, dashed curve 1).

Results of the calculations show that the gas in the near-cathode region is highly ionized and the electron component of current is relatively small, $s \lesssim 0.75$. Comparison of the energy fluxes, Q_v, required to vaporize material of the film energy, $Q_{T\delta}$, scattered in the film through thermal conductivity, and the total energy to the spot (Table 5.6) shows that they are all comparable. These facts contradict conclusions reached elsewhere [25] ($s \simeq 0.95$, $Q_v = 0$), even though the physical model of the spot assumed here and in that reference are the same,

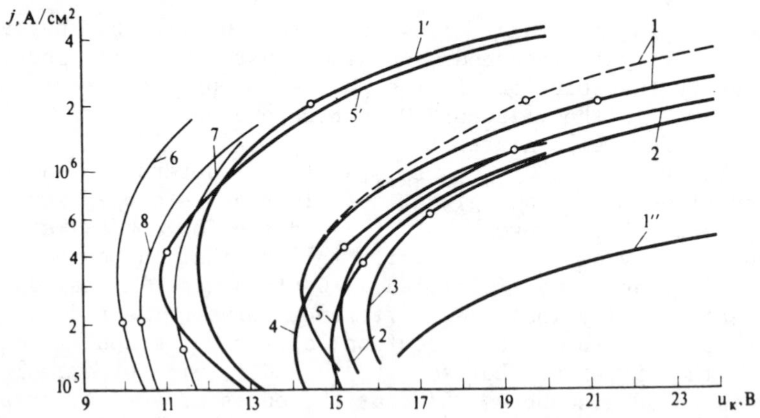

Fig. 5.7 Variation of the current density with the cathode potential drop for films of various thickness (μm).

1,1′ 1′′ - 0.017; 2 - 0.025; 3 - 0.06; 4,5 - 0.12; 5 - 0.14; 6,8 - 0.068; 7 - 0.11; 1-5, 1′, 1′′, 5′ - copper film on glass; 6-8 - bismuth film on glass; 1′, 5′ - calculations assuming heat of vaporization 10 times smaller than heat of vaporization of a solid body; 1′′ - calculation assuming E is equal to $2E_M$.

i.e., a spot continuously displaced because of vaporization of the metal of the spot.

Computations have shown that, as the energy of plasma electrons in the plasma energy balance equation increases, components of the energy balance for a film cathode do not change significantly (see Table 5.6). It is quite noticeable that heat losses through thermal conductivity of the film are considerably smaller (approximately 4 times) than analogous quantities for a solid cathode (see 5.3.3.1.). These losses also differ considerably for relatively thick films, indicating basic difference in the importance of thermal processes in film and in solid cathodes, and also make it possible to explain such small currents to the spot in films.

If the cathode potential drop was known from the experimental data, curves in Fig. 5.7 could be used to determine j and the size of the spot. Lack of data on this parameter makes it impossible to reach a final conclusion on whether the assumed model corresponds to a real spot on a film cathode.[4]

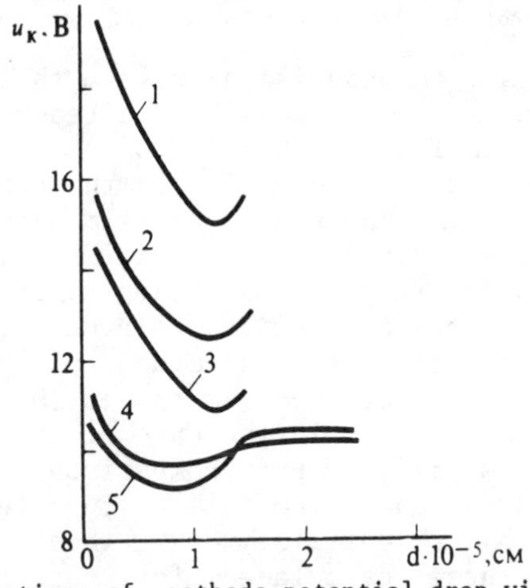

Fig. 5.8 Variation of cathode potential drop with thickness of the copper film, subject to the condition $2r = \delta$.

1,2 - calculated taking into account and neglecting heat losses, respectively, caused by thermal conductivity in metal film; 3 - calculation for heat of vaporization decreased 10 times in comparison with the heat of vaporization for a solid cathode; 4,5 - experimental data [31] for an indium film on tungsten and titanium substrata, respectively.

4 Hence, in this connection, acquisition of data on the dependence of u_k on the thickness of the film is very important.

Table 5.5 RESULTS OF THE CALCULATION

Variants	$d \cdot 10^4$, cm	I_d, A	$v \cdot 10^{-3}$, cm/s	$G \cdot 10^4$, g/s	$u_{K\delta}$, V	α	s	w_e, eV	T_K, K	$j^*_{min} \cdot 10^{-5}$ A/cm^2
							Copper			
1	0.017	0.1	13	0.46	21.25	0.83	0.761	3.2	4000	20.2
1*					19.5	0.74	0.748	2.96	4100	21.7
1**					19.5	0.77	0.745	2.94	4000	17.5
1°					20	0.82	0.706	2.43	3420	3.4
1^{oo}					20	0.92	0.812	2.05	3512	7.9
1_0					20	0.91	0.844	3.33	3342	16.4
2	0.025	0.15	10.1	0.86	19	0.58	0.696	2.39	3980	10.7
2*					18.5	0.57	0.698	2.43	4062	12.7
3	0.06	0.4	5.7	2.6	17	0.38	0.641	1.99	3965	5.8
3*					17	0.38	0.644	2.02	4025	6.7
4	0.12	0.9	3.4	5.6	15	0.28	0.616	1.80	3977	4.07
4*					15	0.278	0.617	1.82	4007	4.34
5	0.14	1.2	3.4	8.2	15.5	0.256	0.599	1.74	3945	3.3
5*					15.5	0.257	0.602	1.76	3981	3.6
							Bismuth			
6	0.068	0.06	3.2	1.3	10	0.029	0.568	1.29	3780	2.0
7	0.11	0.08	2.7	2.3	11.25	0.025	0.520	1.24	3686	1.33
8	0.068	0.08	2.7	2.3	10.5	0.025	0.550	1.25	3740	1.55

Comment. Variants 1-8 were calculated from a system of equations for a film cathode with constants corresponding to a solid metal cathode; 1*-5*--energy carried away by plasma electrons to the cathode is increased by a factor of $2u_k$; 1** - usual calculation, but

Nevertheless, the condition $2r = \delta$ that appears to be quite justified makes it possible to reach an interesting qualitative conclusion based on results of the calculations. The dependence of the cathode potential drop, u_k, on the thickness of the film when $2r = \delta$ (curve 1 in Fig. 5.8) displays a minimum, with the cathode potential drop for thick films being close to the cathode potential drop on a solid cathode (u_k = 15 V). Nonmonotonic variation of the cathode potential drop with decreasing thickness of the film and a sharp increase for very thin films is noted in the case of film cathodes on a metal substratum [31]. For comparison, Fig. 5.8 shows curves plotted for an indium film on tungsten and titanium (curves 4 and 5). Qualitatively, these curves have the same shape as those calculated for copper film on glass (curves 1 and 3). However, the experimentally measured cathode potential drops for all film cathodes with a metal substratum are below the cathode potential drop on a solid cathode [25, 31]. Calculated curves for a copper film on glass lie above the similar dependence for a solid cathode.

Such qualitative difference (even though not verified experimentally) may be associated with the difference in thermal conditions, i.e., in the case of a metal substratum, a certain amount of heat may be transferred to the substratum by thermal conductivity. A change of

OF PARAMETERS OF THE SPOT ON FILM CATHODES

$E \cdot 10^7$, V/cm	$\nu_o \cdot 10^{-4}$, S/s	$(n_a+n_i)\cdot 10^{20}$ cm^{-3}	$j_{max}\cdot 10^{-5}$ A/cm^2	u_{kmax}, V	$\delta=2r_{max}\cdot 10^4$ cm	$\delta_k=2r_{min}\cdot 10^4$ cm
2.4	7.2	1.2	60	35	2.5	1.4
2.25	6.3	1.5	66	35	2.5	1.4
2.27	6.3	1.2	-	-	2.5	1.4
2.1	5.4	0.28	-	-	2.5	1.4
2.6	9.6	0.37	-	-	2.5	1.4
3.4	11.0	0.6	-	-	2.5	1.4
1.9	4.9	1.2	45	32	4	2
1.9	4.9	1.4	50	32	4	2
1.5	3.1	1.16	30	27	9	4
1.5	3.2	1.3	33	27	9	4
1.25	2	1.2	20	22	16	7.4
1.2	2	1.3	21	22	16	7.4
1.16	2	1.1	15	21	20	10
1.16	2	1.2	15	21	20	10
1.1	0.1	11.1	3.5	10	6	4.7
1.0	0.1	10.3	2.5	11.25	8	6.4
1.0	0.1	11.0	-	-	-	-

with u_k corresponding to the variant 1*; 1-8, 1*-5*, 1** - $E = E_M$; 1°, 1^{oo}, 1_0 - $u_k = 20$ V, $E = 2E_M$; 1° - for heat of vaporization, λ_v is decreased 10 times compared with a solid; 1_0 - λ_v is increased 10 times and vapor pressure of the metal increased 3 times in comparison with the equilibrium value for a solid.

Table 5.6
CALCULATION OF ENERGY FLUXES ASSOCIATED WITH THERMAL CONDUCTIVITY AND VAPORIZATION OF THE MATERIAL OF THE FILM

Energy flux cal/s(V)	$d \cdot 10^4$ cm				
	0.017	0.025	0.06	0.12	0.14
Q	$\frac{0.1017}{0.102}$	$\frac{0.185}{0.183}$	$\frac{0.555}{0.548}$	$\frac{1.188}{1.182}$	$\frac{1.725}{1.706}$
Q_v	$\frac{0.0643}{0.0646}$	$\frac{0.118}{0.119}$	$\frac{0.363}{0.364}$	$\frac{0.782}{0.783}$	$\frac{1.143}{1.145}$
$Q_{T\delta}$	$\frac{0.0374}{0.0373}$	$\frac{0.067}{0.064}$	$\frac{0.192}{0.184}$	$\frac{0.406}{0.399}$	$\frac{0.582}{0.563}$
	(1.558)	(1.86)	(2.0)	(1.88)	(2.02)

Note: Numerator shows heat fluxes calculated subject to the condition that the energy to the cathode carried away by plasma electrons is increased by a factor of $2u_k$, parentheses show the values of $Q_{T\delta}$ in volts.

certain parameters of the spot on a film cathode has been observed when the glass substratum is replaced with a metal one. For example, it is known that, for bismuth films on glass, the electron transfer coefficient is 3-1/2 times greater than for a film deposited on molybdenum [25].

In view of the fact that in the experiments considered that the technology of depositing films was the same, their properties can be assumed to be identical. Consequently, qualitative difference in cathode material erosion rate, and other characteristics of the spot (cell current, displacement rate, etc.) may be associated with different thermal conditions, verifying thermal mechanism of maintaining the spot.

To get some idea on the heat distribution between the substratum and the film, estimate the terms of the thermal balance for a bismuth film deposited on a molybdenum plate. Consider a film with the following parameters [25]:

$$d = 4.8 \times 10^{-6} \text{ cm}, \qquad \nu = 2.5 \times 10^{3} \text{ cm/s} ,$$

$$I_d = 7 \times 10^{-2} \text{ A}, \qquad G = 5.9 \times 10^{-5} \text{ g/s} ,$$

$$\delta = 5 \times 10^{-4} \text{ cm} .$$

It is assumed that $u_k = 10$ V is close to the experimental values for bismuth on tungsten [31].

Under conditions being considered, the heat loss to vaporization when the incident heat flux $Q = 1.8 \times 10^{-2}$ cal/s (it was assumed that $s = 0.7$) is $Q_v = 10^{-2}$ cal/s. From this, it follows that $Q_T = 8 \times 10^{-3}$ cal/s, where Q_T is the heat that in a general case is transferred through thermal conductivity into the film and to the substratum. However, calculations using formula (5.19) show that the heat required to heat only the film to boiling temperature is approximately equal to Q_T and, consequently, only a small part of the incident heat flux reaches the substratum. It should be noted that calculation of the temperature of substratum, even for maximum conditions ($Q = Q_T$), leads to $\Delta T < 1$ K, i.e., the substratum almost always remains cold, even if it absorbs a certain amount of heat.

In general, relation between terms of the heat balance may change if one takes into account deviation of the coefficient of thermal conductivity of the film from the corresponding value for a solid cathode that was used in both in the estimate obtained here and in prior discussion.

Therefore, qualitative difference between the experimental data on the cathode potential drop (and other spot parameters) and the results of the calculations based on the methodology proposed may be associated with a difference between thermophysical properties of real films and a solid electrode made from the same metal. The nature of variation of thermophysical properties of material of the films depends on the technology of deposition and, possibly, other causes. Differences between properties of films and solids is indicated by the fact that the nature of the spot on a film cathode with a substratum of the same metal differs from that of a spot on a solid, i.e., spot displacement and erosion are analogous to a spot on a different stratum (only the film is eroded) [31]. The same cause may be responsible for the effect of an electrical conductivity drop in thin films [118].

To illustrate the influence of thermophysical constants of the film on the results of the calculations, Fig. 5.8 includes curve 2 calculated on the assumption $\lambda = 0$. It can be seen that a decrease of thermal conductivity and, consequently, losses through thermal conductivity, in this case (film on glass) lead to a considerable decrease of the cathode potential drop (other parameters of the spot, i.e., temperature, charged particle concentration, etc., do not change very much).

Calculations in which the heat of vaporization of a film cathode is decreased 10 times in comparison with that of a solid body lead to an even larger decrease of the cathode potential drop (curves 3 in Fig. 5.8 and 1′ and 5′ in Fig. 5.7). It can be seen that, for relatively thick films, $u_k \simeq 11.2$ V and is close to values measured under conditions when the film is sputtered on the metal substrata [32]. The same tendency of decreasing u_k is observed in calculations that take into account the feasibility of increasing film vapor tension in comparison with that of the solid at the same temperature of the surface of the cathode (see Table 5.5). Computations show that simultaneous measurement of the heat of vaporization and vapor pressure may exert a strong influence on u_k.

The cathode potential drop calculated may also depend strongly on the assumption made during computations, i.e., on the computation scheme chosen, which depends on the expression for the electric field strength. Computational dependence of j on u_k subject to the condition $E = 2E_M$ (see Fig. 5.7) is shifted sharply toward larger values of

u_k. Hence, in theoretical analysis of the dependence of the cathode potential drop on the thickness of the film, it is important to take into account all the factors listed above.

The influence of thermophysical constants of the film may also be considered, using results of the calculations for bismuth on glass (see curves 6-8 in Fig. 5.7 and Table 5.5), showing that $u_k \simeq 10$ V. This is considerably less than the analogous value for copper and close to the data for bismuth on a metal substratum ($u_k \simeq 8.3$ V). Other parameters (see Table 5.5) also differ from the results of the calculations for copper film.

These facts indicate that, in conducting experimental investigations using film cathodes, it is necessary to control the thermophysical properties of films. Unfortunately, such data are not available in published results [31, 32].

We shall now consider the feasibility of investigating physical processes in a spot on a solid cathode, using the results of experiments performed on film cathodes. Such analysis would be possible if, with increasing thickness of the film, the spot on a film cathode would transform into a spot on a solid cathode (as was actually assumed in Ref. 25).

However, if the spots on the film correspond to the model considered, i.e., if thermal and thermionic processes are the predominant ones and erosion occurs through vaporization, one can state with a certainty that the properties of spots must change drastically in any attempt to transform our analysis to a solid electrode.

Spots of the first type exist on a solid cathode at a low current characteristic of a film cathode. Because, at average heat fluxes, the temperature in the spot is too low to generate a conducting medium in the near-electrode region, such spots cannot be described within the framework of the thermal model. The feasibility of sustaining thermal spots on a film cathode and possibility of describing these spots by a system of equations considered, depend on the high thermal resistance of thin films that result in a high temperature in a spot at a relatively low heat flux (comparable with a flux in spots of the first type).

Nevertheless, in terms of physical characteristics, spots on metal films are closest to group, low-mobility spots on a solid cathode, which appear at high discharge current. However, comparison of the results of the calculations for film (see Table 5.5) and solid (see Table 5.3 and Table 5.4) cathodes made from the same material shows that parameters of the near-cathode regions of such spots may differ considerably. This is most clearly demonstrated by high values of current density (in agreement with the experimental data).

Hence, an arc spot on a film cathode is apparently an independent physical object, investigation of which is obviously of considerable interest and might serve to extend understanding of physical processes in cathode spots. However, it is unlikely that qualitative regularities determined during such an investigation can be transferred to arc spots burning on a pure solid cathode. Apparently, this can be done in the case of sputtered or oxidized solid cathodes. The results presented in this subsection may be used to analyze the mechanism of sustaining a constricted discharge under conditions existing in an MHD generator, where thin films are formed on metal electrodes in a channel as a result of deposition of seed from the gaseous phase of the products of combustion flow.

5.3.3.3. Method of Autographs

Foregoing analysis of the near-cathode processes of an arc discharge on solid and film electrodes makes possible a somewhat different consideration of the reliability of the results of determining current density in a cathode spot by means of autographs. In general, an answer to this question requires solution of the thermal problem within the electrode from the known specific heat flux in the spot. It is then necessary to compare characteristic size of the spot (current flow region), the autograph, and the melting zone of the metal, and to establish a correspondence between the luminescent region determined optically and the current-conducting zone determined by means of the method of autographs. In this analysis (see also References 79 and 119), the required specific heat flux will be determined by solving the (5.5)-(5.11) system of equations.

It should be recalled that the principal deficiency of the method of autographs is the uncertainty of the concept of the "path," i.e., regions with different types of damage (craters, melting, change of structure, etc.), the dimensions of which differ, may be located within the boundaries of the faulted sector. This and certain technical difficulties of recording the dimensions of the path, current to the spot, etc., explain the fact that the current density determined by various authors using the method of autographs differ by many orders of magnitude ($j = 10^4$-10^{11} A/cm^2).

These difficulties and certain theoretical estimates resulted in serious criticism of the limits of applicability of the method of autographs, and have led to the conclusion that this method cannot be used to estimate the size of the cathode spot and, consequently, the current within it [25, 27].

Overall, the results of the calculations of the thermal mode of a solid electrode [79, 117] conducted with the use of relations for a propagating, normally distributed heat source (displacement rate, power and its distribution were given parametrically) [116] show that the configuration of the temperature field depends considerably on the current density and the displacement rate of the spot. For heat sources displaced relatively slowly ($\nu \lesssim 10$ cm/s), the melting isotherm is almost a circle and, subsequently, is similar to the assumed shape of the spot.

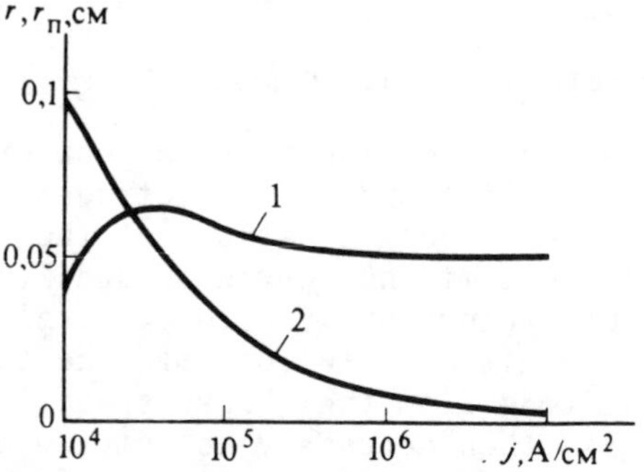

Fig. 5.9 Dependence of the radius of the melting isotherm r_π(1) and radius of the spot (2) on current density of the copper electrode. 1,2 - calculation using thermal conductivity equation for a solid and equation for the total current, respectively.

The dependence of the radius of the melting isotherm, r_Π, on the current density at I = 300 A for a copper electrode (curve 1 in Fig. 5.9) is not unique, generally speaking, i.e., two or more considerably different values of j correspond to the same value of r_π. A choice of specific values of r_π may be made when establishing a relation between the results of solution of the thermal problem and parameters of the spot. Therefore, Fig. 5.9 also shows the dependence of the radius of the spot, r, on j, acquired using the equation for the total current (curve 2). Both curves intersect at the same point ($j \simeq 5 \times 10^4$ A/cm^2), i.e., at low displacement rates, the dimensions of the spots between the melted zone and the current-conducting zone may be interrelated.

A more detailed justification of the validity of the method of autographs will be provided for a specific type of spots on copper, taking into account emission and plasma processes in the near-cathode region. First, however, we will present the results of solution of the thermal problem of a relatively high displacement rate of the heat source. When it exceeds 10^2 cm/s, the melting isotherm is strongly elongated along the axis of displacement. Furthermore, according to calculations, the length of the melting isotherm in the range of 10^2-10^3 cm/s displacement rates varies considerably (in comparison with its width) [79]. The dependence of the length of the melting isotherm on the displacement rate is characterized by a maximum and, consequently, is also a nonunique function. An important fact is that the elongated shape of the melted path may appear during displacement of the heat source which may have a noncircular shape. In identifying the area of the path with dimensions of the current-conducting zone of propagating spots, it is necessary to take this characteristic into account, e.g., using solution of the thermal problem in the solid.

The conclusion reached applies to continuously displaced spots. Autographs of discretely moving spots can be described in the low-displacement rate approximation.

As a rule, discussions of the validity of the method of autographs did not specify the mechanism responsible for destruction of the electrode by a cathode spot. However, it is known that, depending on the discharge conditions, different types of cathode spots may appear. The spots lead to different electrode erosion processes. In spots of the first type, i.e., rapidly displaced spots with low current to the spot (1-10 A), which result in a low degree of erosion, thermal processes are local phenomena. This can be attributed to the fact that, during the short lifetime of the spot at its characteristic energy flux levels, the entire surface of the metal underlying the spot is not heated to the high temperature necessary to melt or vaporize it. Sections of the surface with intense erosion in such spots occupy an area considerably smaller than the luminescent region of the spot [30, 79]. Obviously, if the size of the spot (current-conducting surface) in this case is associated with the size of the luminescent region, current density calculated from the erosion area (method of autographs) will exceed considerably the actual value determined from the luminescent area, i.e., for spots of the first type, the method of autographs is no longer valid as a technique for determining current density in the cathode spot.

It is interesting that a large quantity of published experimental data apparently refers to spots of the first type. Consequently, the critique of the method of autographs used in these papers is correct, although detailed investigation of the erosion mechanism, even in this case, will make it possible to establish a relationship between size of the path and the current-conducting region.

At the same time, it is known that, in a number of cases, the cathode spot is of thermal origin, i.e., erosion in the spot depends on the high surface temperature and is characterized primarily by cathode material vaporization. This type of spot includes group spots [27, 61], i.e., spots on film cathodes, when the film is not too thick [25] and cathode spots on metal electrodes of an MHD generator channel (see 5.4.).

Even though a large quantity of experimental data on spots on film cathodes is available, the thermal nature of such spots has not been established experimentally. However, the foregoing theoretical investigation shows that this is feasible and, for thermal spots on a film cathode, the cross section of the path left by the spot (autograph) differs very little from the size of the spot and, consequently, can be used to estimate current density in the cathode spot [16].

We shall show that, in the case of a group spot, good correspondence exists between the spot and the melting isotherm that represents the boundary of the path left by the spot on the surface of the metal [119]. Analysis of the temperature field has shown that such correspondence occurs at a current density on the order of $j \simeq 5 \times 10^4$ A/cm^2, characteristic of this type of spots.

A group spot is a stationary (or low-mobility), long-lifetime ($t \gtrsim 10^{-4}$ s) association of closely located spots. The total current to the group of spots is $I \simeq 200$ A. As a result of the spots in a group spot being close together, a group spot can be considered, thermally, a single formation. Investigation of the dynamics of the development of a group spot and the path it leaves [27, 39, 61] shows that, in this case, metal erosion occurs primarily by means of vaporization. Calculation of parameters of the spots (see Table 5.3) based on the concept of the thermal nature of a group spot is in good agreement with the experimental data on the current density in such spots, as determined from the size of the luminescent region [88].

The joining of individual spots into a group spot leads to formation of an intense, stationary heat source heating the surface of the metal in the area of the spot to a temperature on the order of 4000 K. This is sufficiently high for the emission current density and vapor density near the surface to reach the levels required to sustain a spot. Naturally, the metal outside the spot is also heated. In particular, it leads to expansion of the region of molten metal bounded by the melting isotherm. A path clearly marked by a zone of molten metal remains after the spot is extinguished and the metal is cooled.

A finer structure of the path is also possible. It is associated with individual spots in a group, and also occurs when erosion takes place not only in the vapor phase, but also in a form of drops, particles, etc. Erosion depends considerably on fabrication of the electrode, especially its degasing. Proper fabrication produces conditions such that erosion occurs only in the vapor phase [27, 39]. In order to compare the size of the melting isotherm with that of the spot, we shall deal with the simplified problem of the heating of the metal near the group spot, using the model of the near-cathode region already considered. Relations making it possible to determine the temperature field in the body of the electrode correspond to a quasi-stationary spot, i.e., to a state that appears a certain time after the occurrence of the spot. The initial state from which the spot is developed, as well as the physical processes in the near-cathode plasma that determine dynamics of its development, have hardly been investigated. In this connection, we shall use a model of spot development when dealing with the problem of metal heating.

Assume that a cloud of plasma with parameters close to the parameters in a developing spot is formed near the surface of the metal at the initial time, $t = 0$. At time $t > 0$, the metal is heated by the heat flux from the plasma to the surface of the electrode, which depends on the physical processes in plasma during development of the spot. Because thermal conductivity has a greater time lag than processes determining the state of the near-cathode plasma, the development time of the spot will depend on the time it takes to heat the surface of the metal in the region of the spot to a temperature at which vaporization and emission will occur.

Assume that the size of the spot and the cathode drop do not change substantially during the development process of a group spot (such variation was not reported in experiments [27, 61]). Furthermore, the energy flux lost by the spot through thermal conductivity in the initial stage, $Q_o \simeq I_i u_k$, where I_i is the ion current close to (in view of the assumption concerning the initial state of the plasma) the ion current in the developed spot. As the surface of the metal is heated, the energy flux lost by the spot through thermal conductivity will decrease because of evaporation, emission, etc., and will approach that in a developed spot, Q_∞.

Let us define the development time, t_o, as the time during which the temperature at the center of the spot reaches a value in a stationary spot (in a thermal model the distribution of parameters in a spot is assumed to be homogeneous, corresponding to the center of the spot). To estimate t_o, assume that at $t = 0$ heat flux equal to Q_∞ and having a Gaussian distribution is incident on a circle of radius r_∞ (where r_∞ is the radius of the spot being developed) on the surface of a semi-infinite electrode. Solution of the problem for center of the spot [116],

$$T = Q_\infty \arctan\,(4at/r_\infty^2)^{1/2}/\lambda\pi^{3/2} r_\infty\ ,$$

shows that conditions characteristic of a group spot on copper ($I = 200$ A and Q_∞ and r_∞ calculated from equations (5.5)-(5.11)) are attained at $t_o = 8 \times 10^{-4}$ s when $T/T_\infty = 0.7$, where T_∞ is the temperature of the developed spot (curve 1 in Fig. 5.10). Because, in view of the problem formulation, the value of t_o obtained is the upper limit of t (initial radius of the plasma cloud is smaller than r_∞), it can be assumed that, under conditions being considered,

$$t_o \lesssim 5 \times 10^{-4}\ \text{s}\ .$$

Heating of the metal continues even after the spot has been developed and its parameters have reached their stationary values (heat source continues to function and, in view of the formulation of the problem, its intensity does not change). Heating of the metal can be roughly estimated from the temporal variation of the radius of the melting isotherm, r_∞ (see curve 2 in Fig. 5.10). A comparison of curves in Fig. 5.10 shows that the temperature distribution in the electrode becomes stationary much more quickly than that of parameters of the near-cathode region.

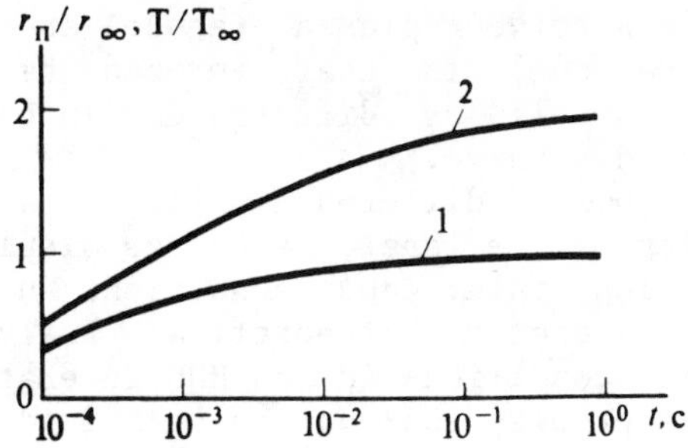

Fig. 5.10 Dependence of relative temperature of spot (1) and relative radius of molten zone (2) on time.

Nevertheless, even for spots characterized by a long lifetime ($t \simeq 1$ s), the size of the melting isotherm differs from the size of the spot by only 100 %, whereas this difference is zero for spots with a lifetime $t = 10^{-3}$ s. The lifetime of group spots recorded in experiments described in the literature [27, 61], 2×10^{-3} s, was limited by duration of the discharge. On the basis of general considerations, it appears that group spots may exist even longer.

It should be noted that, according to published reports [27, 39, 61], the characteristic path of a group spot (lifetime of a spot $t \simeq 2 \times 10^{-3}$ s) is on the order of 5×10^{-2} cm, the size of the spot determined optically is $(2\text{-}5) \times 10^{-2}$ cm and the size of the spot and the molted zone, calculated for the stationary case, are approximately 3.3×10^{-2} and 6.5×10^{-2} cm, respectively. Hence, the size of the melting isotherm may serve as a reliable estimate of the size of the group spot and, consequently, current density within it.

Therefore, the method of autographs based on the relationship between the measured size of the molten path with the size of the spot is quite acceptable for estimates of current density in thermal spots. It is likely that further investigation of the erosion mechanism will make it possible to apply this method to other types of spots.

5.4. Constricted Discharge on Metal Electrodes in an MHD Generator Channel

An interesting feature of operation of electrodes of an MHD facility is the flow of high-temperature, ionized gas, containing alkali metal seed, around the electrodes. As was already noted, in terms of the nature of burning in the near-cathode region, the arc discharge appearing under these conditions between plasma (anode) and cooled, metal electrodes (cathode) is characterized by a number of similarities, as well as differences in comparison with other types of discharge.

First, this difference is manifested in characteristics of the spot, such as displacement rate, lifetime, erosion rate, and conditions under which arc spots appear. Therefore, theoretical analysis of an arc discharge under conditions in an MHD generator represents an independent problem that has to be solved to be able to design mostly new types of electric power generators.

5.4.1. Theoretical Analysis. It should be noted that, even though the problem is important, the number of attempts to describe constricted discharge in an MHD generator is very limited, and the results of these investigations are incomplete and contradictory. There are no commonly accepted views concerning physical processes that occur under the conditions noted. Consequently, there is no single model describing parameters of an arc spot and erosion caused by it.

This is verified by the most typical studies [4, 11, 12, 120], which actually suggest two different mechanisms for the electric current flow in an arc discharge on a copper cathode. Because the results described in these references are extremely important to the understanding of investigations of the near-cathode region of an arc in an MHD generator described in the following text, we shall analyze them in detail.

One of the first published attempts to consider discharge processes in the near-electrode region in the combustion products flow of natural gas with potassium salt additive [11] analyzed parameters of processes occurring in the discharge column, the geometry of which is that of a segmented cone with an aperture angle $\beta_0 = 60^0$ and an axis perpendicular to the plane of the cathode. The top of the cone lies within the electrode. The diameter of the upper base of the cone is equal to the diameter of the cathode spot, d_k, which is assumed to be equal to 0.1 mm. The height of the segmented arc cone is assumed to be equal to the thickness of the boundary layer, $\delta_k = 1$ mm. The energy balance equations in gas and space charge for this layer are written in curvilinear coordinates. Space charge is assumed to be formed as a result of changes of the electrical field intensity caused by changes of electrical conductivity. Space charge was calculated by solving Poisson equation simultaneously with the continuity equation in the form div j = 0. The energy balance in gas was written on the assumption that Joule heat in an element of arc is equal to heat losses resulting from convection, radiation, and thermal conductivity.

This leads to a second order differential equation in terms of the temperature. Solution of the above-noted equations made it possible to determine the temperature of the gas and its gradient (heat flux) in it, the relation between the concentration of positive and negative particles η (a ratio of ion/electron current components), potential distribution and its gradient as a function of distance from the surface of the cathode.

The temperature, equal to the temperature of the flux, and the fact that $\eta = 1$ were chosen to be boundary conditions on the right-hand end of the arc cone. The energy balance on the cathode spot in the form similar to relation (5.3), but with the heat flux from the arc determined by the temperature gradient at the boundary of the column with the surface of the cathode, and plasma radiation taken into account, will probably be used on the left-hand end of the arc cone. The terms of the energy balance associated with cooling of the cathode by electrons emission and neutralization energy of ions on cathodes are small and were neglected. The heat flux to the cathode caused by kinetic energy of ions is simply not discussed. In calculations of parameters characterizing properties of gases (thermal conductivity, electrical conductivity, etc.) and radiation and convection heat losses, the numerical computations done were of thermodynamic processes for a mixture of gases containing approximately 20 components, the choice of which was determined by temperature conditions.

Generally speaking, solution of the system of equations being used [11] makes it possible to determine the following quantities at the boundary of the arc cone with the electrode surface (in the cathode spot): electric field strength, E; cathode temperature, T_k; electron current density, j_e; relative electron current, potential drop in the boundary layer, u_b, and also heat flux into the electrode, Q_T, reduced to the total current. These parameters may be estimated if the temperature gradient at the outer boundary of the boundary layer is known. Because it was not known, it was chosen so as to maintain energy balance at the cathode. The following parameters of the spot at current $I = 2$ A were obtained by solving the problem: $E \simeq 10^7$ V/cm, $T_k \simeq 1800$ K, $u_b \simeq 2000$ V, $\eta \simeq 10^7$, $j_e \simeq 10^4$ A/cm^2, $Q_T = 130$ W/A. Taking into account that arbitrary values of d_k and angle β_0 were used in numerical calculations, in general, calculated values cannot correspond to the experimental parameters of a cathode spot [11]: $Q_T \simeq 13$ W/A, $u_b \simeq 130$ V, and the electron transfer coefficient, $G \simeq 10^{-6}$ g/C. The value of G calculated from the vaporization rate of copper at the temperature of the cathode was 3×10^{-7} g/s or 1.5×10^{-7} g/C, i.e., also considerably below the experimental value. In view of this, the authors of the referenced report [11] suggest eliminating the difference between calculations and the experimental data by optimizing geometry of the arc discharge, taking into account thermionic emission equation that depends on the cathode temperature and the electric field near it and, also, by using the minimum potential drop condition on the boundary layer. However, the report [11] does not include the investigation suggested and its authors refer to the principally new method they suggest for analyzing current transfer equations on electrodes of an MHD generator as the principal result of their work.

In our view, the principal deficiency of the report [11] is that the formation mechanism of charge carriers in a narrow region directly at the cathode was not considered, and the incoming flux of the kinetic energy of ions was not taken into account. In addition, without being stated, it was assumed that the required ion current at the boundary of plasma with the cathode is always produced by potassium ions in the combustion products flow, the concentration of which was calculated to be 10^{15} cm^{-3} [11]. It can be easily seen that the current density corresponding to such a concentration of ions does not exceed 1 A/cm^2, in other words, when $j \simeq 10^4$-10^5 A/cm^2, actually all of the current must be transported by the emission electrons.

Because the work function of electrons was assumed [11] to be 4.5 eV, this emission current density can be attained only at high spot temperature (on the order of 4000 K). First of all, this fact contradicts the conclusion of low temperature in arc spots of an MHD generator [11]. Secondly, it requires that the metal atoms evaporated from a spot during current transfer, subject to the condition that high temperature can be sustained by the heat flux from the arc, be taken into account.

In addition, intense energy release in the near-cathode plasma must lead to separation of the electron temperature from the ion temperature. This, in its turn, must affect the rate of ionization processes. Even though the possibility of deviation from the ionization equilibrium was noted in the report [11], this fact was not taken into account. Also, the participation of seed material in the discharge that, according to the data in the same paper, settles on the surface of the metal electrode was also neglected.

In terms of assumptions made, Ref. 120 is close to later publications [4] and [12]. In Ref. 4, and in Ref. 11, it is assumed that ion current to the cathode is by means of seed ions in the combustion products flow. It is assumed that, in a certain region near the cathode, equal to the ionization length, all potassium atoms are completely ionized ($\alpha = 1$) and the relation for the ion current density is assumed to be

$$j_i = 1/4\ n_a \nu_T\ .$$

The following five equations are written for the above-noted ionization region:

1) The electron emission equation in the form of the Richardson-Dushman equation for thermionic emission. It differs from Ref. 11 in that the work function, $\phi = 2.2$ eV, depends on potassium precipitated on the surface of the electrode.

2) Energy balance equation is used in the same form as in referenced publication [11]. However, the kinetic energy of ions, $j_i u_k$, is taken into account.

3) Total current equation connecting the current of the spot with its area and current density.

4) Electron energy equation in the ionization region that takes into account the fact that the energy of emission electrons, $j_e u_k$, is spent on ionization and heat transfer of the external boundary of the characteristic region.

5) Relation for the temperature of electrons taken from another report [120]:

$$w_e = 3/2kT_e = 1/2(u_i - u_a^*)\ ,$$

where u_a^* is the potential of the first excited state of the seed atom.

This system of equations makes it possible to determine only five parameters of the spot: u_k, I, r, j, j_e. Other authors [4] assume an arbitrary value for the sixth parameter (T_k) and vary it throughout a broad range of temperatures (1200-2000 K).

Even if one neglects the fact that T_k is a free parameter, the principal deficiency of Refs. 4 and 11 is the assumption concerning the predominant importance of potassium atoms in the combustion products flow. Because it was taken into account that the cathode in the spot is heated primarily by ion current (energy flux $j_i u_k$) and the value of j_i is small because of the small values of n_a, solution of this system of equations led to unjustifiably large cathode potential drops. Thus, on cold (500-700 K) electrodes, u_k = 200-500 V [4].

Another major deficiency of Ref. 4 is the use of quite arbitrary expression of T_e, written out above [120], according to which the electron temperature is specified by atomic constants of seed and is a constant quantity independent of burning conditions of the discharge. In reality, T_e depends apparently on the ratio of the incoming flux and energy losses in the ionization region and on parameters of the cathode spot (j_e, u_k, etc.). Even though the energy balance of electrons is written out, it is used to determine other parameters of the spot.

Nevertheless, the approximate analysis performed [4] made it possible to establish certain qualitative agreement with the experimental data. Similar to the experiment, the results of which are summarized elsewhere [4], total current in the channel of the generator increases linearly with increasing temperature of the electrode. The spot in the experiment with copper electrodes leaves a molten trace. This agrees with the results of the calculations (T_k exceeds melting temperature). Calculated data for spot parameters [4] were used to process electrical characteristics of an MHD generator. Good agreement was obtained between these results and the experimental data.

The point of view pertaining to the mechanism sustaining a cathode spot on metal electrodes presented in one report [12] is completely opposite of that in two others [4, 11]. Analyzing possible sources of atoms and molecules in the ionization region, capable of providing ion current to the cathode (seed atoms, combustion products, electrode material), the authors reached the conclusion that, under conditions considered, the seed and combustion products in a gas flow do not provide the experimentally observed current density even at $\alpha = 1$. Therefore, it is concluded that the principal contribution to the ion current transfer is made by atoms of the vaporized cathode material. The system of equations remains the same as in Ref. 4, but emission is assumed to be thermal field emission (ϕ = 4.5 eV) and, therefore, the McKeown equation [77] is used to determine the electric field intensity at the cathode. The degree of ionization, α, is assumed to be another free parameter in addition to

T_k. The principal parameters to be determined are current to the spot and the near-cathode drop, which are also used to find the electrical characteristics of an MHD generator.

The assumed model of the cathode spot [12] is actually analogous to the theory of the near-electron region of an arc with a vaporizing cathode [78]. A closed system of equations for such a model is given above (see 5.3.2.) and in other references [87, 88]. According to the experimental data, the lifetime of cathode spots in a vacuum at a current of 5.5 A is low (on the order 10^{-6} s) [12]. Therefore, within the framework of the thermal model [87, 88], such a type of spot is described only at j $\gtrsim 10^7$ A/cm^2 (see 5.3.3.1.). This value of j exceeds considerably the values reported in experiments in an MHD generator (see Table 5.1). This is the principal deficiency of one approach used [12]. In addition, this work neglects completely the specific conditions of an MHD generator, namely, the presence of seed in its channel. Hence, in view of the free parameters and contradictory assumptions, not one of the models suggested uniquely describes processes in cathode spots appearing in an MHD generator.

5.4.2. Model and Solution of the System of Equations. From the analysis performed, it follows that, in constructing models of cathode processes of a constricted discharge in an MHD generator channel, the most important problem pertains to the source of the neutral medium responsible for ion production in the near-electrode region. Because, during heating of the cathode, the atoms and ions in the near-electrode plasma are partially scattered into the surrounding medium, the cathode erosion mechanism is closely associated with the neutrals generation process.

Analysis of discharge on metal electrodes [14, 15, 34] under conditions of an MHD facility will be conducted on an example of specific, published results of experimental investigations [6, 7, 13, 48, 66] (see 5.2.). The data from these references on the dynamics of development of an arc spot on MHD generator cathodes that protrude into the combustion products flow, obtained by high-speed photography, lead to the conclusion that two types of spots are predominant in regions under consideration. These are spots propagating at a relatively high displacement rate (type I spots) in the zone affected by the flux, and spots almost stationary (type II) in the separation zone of the flow, which burn at the boundary with a thick layer of seed.

Analysis of parts of the electrode surface indicates that, at the relatively low temperature, cathode erosion in the zone affected by the flow is insignificant (interactions of type I spots) and that clearly observed traces of destruction in the form of holes, craters, etc. (interactions of type II spots) appear in the separation zone of the flow. It should be recalled that cathode spots on copper in an ionized gas flow seeded with potassium are characterized by a displacement rate of $\nu \simeq 10\text{-}10^2$ cm/s; radius, $r \simeq 10^{-2}$ cm; current to the spot, $I = 5\text{-}15$ A; lifetime, $t \gtrsim 10^{-4}$ s, and that the heat flux into the electrode when current flows, $Q_T \simeq 20$ W/A.

Unfortunately, in almost all experiments, except those described in one report [13], a very important spot parameter, the amount of seed deposited on the surface of the cathode (G_π), which is vaporized by the heat of cathode spots, was not measured. In view of its importance, in future analysis, we shall estimate G_π, using indirect empirical data.

As has already been noted, not only electrode material, but also seed deposited on it from the gas flow (potassium compounds) can participate in burning of arc spots under conditions in an MHD facility [11, 39, 65]. When the current is switched on, the initial seed layer on the electrode (thickness of the layer, $d_\pi \simeq 0.1$ cm) is rapidly cleared by spots (in approximately one minute), and the spots then burn on the "pure" surface [7, 13, 67]. When the current is switched off, the seed layer is deposited during a period of several minutes. According to the experimental data (see 5.2.), erosion of copper is insignificant ($G \simeq 10^{-6}$ g/C). Therefore, spots primarily vaporize the film of previously deposited seed (potassium compounds).

In the first approximation, its erosion rate in spots, G_π, may be estimated assuming that the volume of seed, Σd_π, is vaporized during a time, t_π. From this, it follows that $G_\pi = \Sigma d_\pi \gamma_\pi / N t_\pi$, where Σ is the area of the surface of electrode on which spots are burning, γ_π is the seed density, N is the number of spots. Taking into account that the diameter of the electrode is 40 mm, $d_\pi \simeq 0.1$ cm, $t_\pi \simeq 10^2$ s, and $N \simeq 3\text{-}10$, we obtain $G_\pi \simeq 10^{-3}$ g/s.

A more accurate calculation of G_π may be performed using a method based on analogy between heat and mass transfer [121]. This analysis shows that the specific mass flux of potassium compounds is 5×10^{-4} g/(cm$^2 \cdot$s) [14]. Taking into account that the precipitated seed is completely removed by arc spots, and the area of the surface on which the spot is burning, $\Sigma = \pi D^2/8 \simeq 6$ cm^2, we get G_π = (3-10) x 10^{-4} g/s as a function of the number of spots.

This result verifies the approximate estimate of G_π performed above and indicates considerably higher degree of erosion in spots of potassium compounds in comparison with that on pure copper. It should be noted that the value of G_π, determined by weighing the amount of seed carried away by the spots, is approximately 10^{-4} g/s [13]. The nature of the path left by a cathode spot during its displacement on a relatively thick layer of seed may be judged from Fig. 5.11 [13, 67].

An estimate of the temperature of electrodes heated by the heat flux from an arc spot may also serve to verify the low erosion rate of copper. Assuming that the

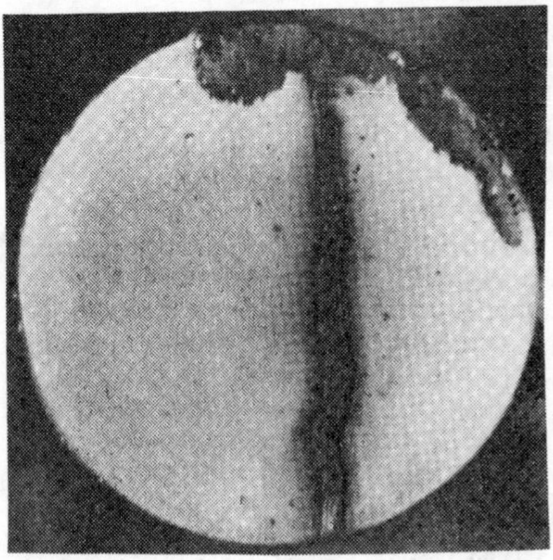

Fig. 5.11 Trace left by a cathode spot during its displacement on a copper electrode coated with a film of potassium compounds [13, 67].

electrode is a semi-infinite body, the temperature of the spot in a steady-state mode at heat fluxes, Q_T, measured in experiments and spot sizes observed is approximately 1000 K. Even though exact calculations, taking preheating into account, lead to a somewhat higher value of T_k, the copper vapor tension at this temperature is, nevertheless, low and, therefore, it is not surprising that erosion of copper is not too high and is considerably lower than the amount of deposited seed removed. Consequently, in certain cases, copper may not participate in the formation of charge carriers of the near-cathode plasma of an arc discharge (this requires that $T_k \simeq 3000$ K). See 5.3.3.1.).

As has already been noted, the potassium atom concentration in the flow is too low (on the order of 10^{16} cm^{-3}) and cannot be responsible for current transfer in spots. At the same time, as a result of vaporization of seed deposited on the electrode near the cathode, a layer of vapor may be formed that can influence processes in the spot.

In some reports [39, 65], it was suggested that one should take into account participation of seed film in arc burning. However, as a rule, this pertained to the relatively thick seed layer ($d_\pi \simeq 0.1$ cm). In the reports, the fact was established of the existence of a diffuse discharge on a copper electrode covered with molten potash at a current of 10 A and current density of 5-10 A/cm^2 [39]. When the thick layer of seed was removed, it was assumed that the discharge burns directly on the electrode and that the mechanism of sustaining the spots is similar to that in vacuum or in air. The suggestion that the small amount of seed remaining on the surface of the electrode after removal of the principal thick layer participates in burning of arc spots was first made in References 14, 34.

Consider the burning mechanism of cathode spots under conditions in an MHD generator [14, 15, 34], taking into account the seed on the surface of the electrode, utilizing the theory describing processes on solid and film cathodes in vacuum [16, 87, 88]. According to this model, a spot can exist on a film cathode, the displacement rate of which is determined by the vaporization rate of material under it. What's more, it was shown that, in this case, the radius of the spot is close to the width of its autograph.

Before proceeding to a description of the model of a spot burning under MHD generator conditions, let us obtain the necessary estimates. It is assumed that a thin film of seed, d_o, that determines the occurrence of spots, remains after removal of the initially deposited thick layer of seed, d_π. The thickness of this film of seed, determined from the formula $d_o = G_\pi/2r\nu\gamma$ given in published reports [16, 25] is on the order of 10^{-5}-10^{-4} cm. Here, we used parameter G_π, obtained for a thick film, even though for a thin film it may be different as a result of a change of its content. However, this is unimportant, because only an approximate estimate is needed. When applying the system of equations to a discharge on film cathodes, it is very important to include the correct term describing heat outflow through thermal conductivity in the energy balance equation at the cathode. The specific form of this term depends primarily on the thermophysical properties and the thickness of the film as well as the type of substratum [16]. For electrodes with a thermal conductivity coefficient comparable to the analogous coefficient for a sufficiently thick film, it is necessary to take into account the temperature distribution both inside the body and along the thickness of the film. Discharge on very thick films can apparently be considered as discharge on a solid electrode. It has been estimated that, for discharge conditions in an MHD generator, the film on the surface of the electrode is so thin ($d_f \simeq 1$ μm) that the temperature gradient, ΔT_k, along its thickness can be neglected in comparison with T_k. An approximate calculation may be performed using the formula,

$$\Delta T_k / T_k = q_T d_\pi / \lambda_\pi T_k ,$$

where the heat flux across a unit's surface is $q_T \simeq 10^4$ cal/(cm^2 x s). A result of this calculation will depend to a considerable degree on the coefficient of thermal conductivity of material of the film, λ_π. At present, no experimental data on this quantity are available for films in an MHD generator. The conclusion reached is valid at $\lambda_\pi \gtrsim 10^{-2}$ cal (cm x s x K), which apparently corresponds to conditions considered. In this case, the cathode energy balance equation may be taken in the form similar to expression (5.9) and (5.10), where the term taking into account heat transfer into the cathode depends on thermal conductivity of the solid substratum. The temperature of the film will be equal to the temperature of the electrode surface in the region of the spot.

When deriving a system of equations that takes into account the influence of atoms of the electrode material, it will be assumed, for generality, that the cathode surface is a source of potassium and copper atoms, the partial pressure of which is given by the equilibrium value at the temperature of the cathode in the spot. The basis of such an assumption is the fact that, during operation of the MHD generator, there is a constant flux to the cathode of potassium ions participating in the current transfer, which are neutralized on the surface of the cathode. Therefore, the film deposited on the cathode may be enriched considerably with potassium atoms.

Unfortunately, no experimental data are available at present on the composition of the seed deposited on the electrode in the presence of an electric current flow with arcs. Available experimental data acquired without current, when the tubes are cooled [122], indicate that seed deposits consist primarily of K_2CO_3.

However, calculations performed with the model suggested here, subject to the condition that the K_2CO_3 film is evaporated with subsequent decomposition, have shown that, in this case, cathode spots cannot exist, because of very low vapor tension of such a film [123]. Apparently, direct verification of the model accepted here has resulted from spectral analysis of charged particles directly in the spot. However, such experiments have not been described in scientific literature.

The assumption of equilibrium pressure of electrode material atoms requires experimental and theoretical justification that, at present, encounters certain difficulties (see 5.3.3.1.). Nevertheless, it allows us to perform a rather simple analysis of the importance of atoms vaporized from the electrode during near-cathode processes, provided the deviation of the actual pressure from the equilibrium pressure is small. Subject to the conditions imposed by the assumptions, a heat spot with high plasma particle pressure at a relatively high degree of ionization may occur in the near-cathode region of an MHD generator. In this case, the ion concentration in a thermal arc spot can be estimated from an expression for a random component of ion current: $n_i \simeq 3 \times 10^{14} j(1-s) \simeq 3 \times 10^{18}(1-s) cm^{-3}$.

Because the ion current component in thermal spots is relatively high (see 5.3.3.) and the degree of ionization of the plasma of the spot, $\alpha \simeq 10^{-1}-10^{-2}$, the heavy particle concentration turns out to be on the order of 10^{19} cm^{-3}. The high atom concentration makes it possible to use analysis of the model given above for an arc with a cathode that is being vaporized, taking into account the film of easily ionized material on the surface. It is necessary that the vapor flow rate be much lower than the thermal velocity of atoms, i.e., that the condition $\alpha \times 10^5/4A_i j_i << 1$, be fulfilled. In the case under consideration, this relation is on the order of 10^{-3} and, consequently, the condition that the vapor flow rate be low is obviously fulfilled.

Based on the above analysis and conclusions reached elsewhere [16, 87, 88], one can suggest the following model that is valid under the existing conditions. The cathode spot appearing on the electrodes under conditions existing in an MHD facility (in the presence of an ionized gas flow seeded with potassium) is sustained primarily by vaporization of a thin film of seed constantly deposited from the flow on the surface of the cathode. The near-cathode region consists of a layer of space charge in the ionization (or diffusion) zone. In general, the atoms of material vaporized from the electrode in a certain thermal regime may participate in diffusion processes of the near-cathode plasma. The importance of a metal electrode is reduced primarily to the transfer of heat and of electric charges. Spot displacement to a new site is determined by vaporization of the film down to the surface of the solid electrode and the regularities of the gas flow around it. The film and the cathode are heated by bombardment of ions that transfer their potential and kinetic energy. Cooling occurs primarily through thermal conductivity of the copper electrode, cathode material vaporization, and electron emission. The electrons emitted, passing through the region of the space charge formed by the ions near the cathode, heat the near-cathode plasma in which new ions are generated by means of thermal ionization and ionization by an electron beam emitted by the cathode.

Assume that overall changes of parameters of the mixture near the cathode as a result of expansion are small. If the connection between the cross section of the gas jet, F(x), and the coordinate, x, is of the form $F(x) = F_o(1 + (x \, \mathrm{tg} \, \psi/r))$ [106], a noticeable change in F(x) at $\psi \simeq 45°$ will occur at distances exceeding radius of the spot, r, i.e., on the order of 10^{-2} cm at a current to the spot $I \simeq 10$ A. This means that, during expansion of the gas, parameters such as j remain practically constant over this length and, for rough calculations, it can be assumed that $\rho\nu = \mathrm{const} = G$, where ν is the gas velocity, ρ is its density, G is the copper and potassium erosion rate. The length of the arc, x_a, where parameters j and u_π vary very little, can be determined in another way, i.e., using energy relations. Comparing the energy flux across a unit surface area of a cross section of a volume of a gas jet, q_1 and q_2, released in this volume per unit time, we obtain

$$x_a \simeq q_1/q_2 \simeq u_f \sigma_e / j \ ,$$

where σ_e is the electrical conductivity of gas in the jet. Substituting typical values for a copper arc ($u_f \simeq 5$ V, $j \simeq 10^4$ A/cm², $\sigma_e \simeq 20$ S/cm) we get $x_a \simeq 10^{-2}$ cm. This is comparable with the radius of the spot. The estimate of x_a turns out to be very necessary for analysis of gas flow parameters in a channel of an MHD facility that takes into account arc current flow in the boundary layers.

We shall also assume that total pressure of gas, p, along the jet as a result of the energy flux from the electromagnetic field is practically constant over a certain distance, ℓ. Estimates using equations describing the motion of gas [106, 107] show that, at particle concentrations given above, $\ell \gtrsim 10^{-3}$ cm and $\nu \simeq (x)^{1/2}$. Because $p \simeq G\nu$, variation of the total pressure is also slight. The fact that parameters at distances ℓ and x_a are practically constant can be attributed to the small variation of gas conductivity under conditions being considered.

Consider elementary processes within a certain zone, ℓ_p, characterized by a large number of collisions of particles of various types and ion generation. Under conditions considered, the gas in this zone consists primarily of five components: potassium and copper atoms and electrons (concentration

$$n_{a_k},\ n_{i_k},\ n_{a_M},\ n_{i_M} \text{ and } n_e ,$$

respectively). Collisions between these particles will determine ionization and recombination processes and will influence the motion of ions to the cathode. Because the concentration of particles is high, we can neglect radiation from the region under consideration [87].

Let us determine the characteristic interaction lengths, which depend on the relations between particle concentrations and effective cross sections of the elementary processes. The ratio of ionization potentials

$$u_{i_k} \text{ and } u_{i_M}$$

is such that, at concentrations

$$n_{i_k} \text{ and } n_{i_M}$$

being compared, the ratio of concentrations of neutral atoms of copper with potassium ($T_e \simeq 1$ eV) satisfies the condition

$$n_{a_M}/n_{a_k} \simeq \exp\left((u_{i_M} - u_{i_k})/kT_e\right) \simeq 10^2 .$$

Taking into account the estimates above,

$$n_{i_k} \simeq n_{a_k} .$$

In other words, under conditions being considered, the degree of ionization of potassium can be considerably higher than that of copper. To be specific, we shall assume

$$n_{a_M} \simeq 10^{20}\ \text{cm}^{-3}\ .$$

Interaction of particles at electron energy on the order of 1 eV may be subdivided into the following categories [22, 124-127]: Coulomb, characterized by the cross sections

$$\sigma_{ei_k} \sim \sigma_{ei_M} \sim \sigma_{ee} \sim \sigma_{i_k i_k} \sim \sigma_{i_M i_M}$$

$$\sim \sigma_{i_k i_M} \sim 10^{-13}\ \text{cm}^2\ ;$$

polarization (interaction of particles with potassium atoms) with cross sections

$$\sigma_{ea_k} \sim \sigma_{i_M a_k} \sim \sigma_{a_k a_k} \sim \sigma_{a_M a_k} \sim 10^{-14}\ \text{cm}^2\ ;$$

exchange, characterized by exchange cross sections

$$\sigma_{i_M a_M} \sim \sigma_{i_k a_k} \sim 10^{-14}\ \text{cm}^2\ ;$$

elastic with cross sections

$$\sigma_{ea_M} \sim \sigma_{i_k a_M} \sim \sigma_{a_M a_M} \sim 10^{-15}\ \text{cm}^{-2}\ .$$

Finally, let us separate collisions leading to ionization of atoms by electrons of the beam ($u_k \simeq 10$ eV) with cross sections

for potassium $\sigma_{i_k} \sim 10^{-14}\ \text{cm}^2$

for copper $\sigma_{i_M} \sim 10^{-16}\ \text{cm}^2$

and scattering of the beam of electrons by charged particles with cross sections

$$\sigma_{k\pi} \sim 10^{-15}\ \text{cm}^2 .$$

According to estimates, for the range of parameters given, relaxation of fast electrons of the beam as a result of collisions occurs at a distance that is approximately the same for potassium and copper atoms, i.e.,

$$\ell_p \sim \ell_{pk} \sim \ell_{pM} \sim 10^{-4}\ \text{cm} .$$

The mean free path of various types of particles

$$+ (D_{i_M a_M} - D^{i_M}) \frac{dn_a}{dx} + (D_{i_M a_k} - D_{i_M a_M}) \frac{dn_{a_k}}{dx}$$

$$+ \frac{eE_\pi}{kT} (n_e D_{i_{Me}} - n_{i_k} D_{i_M i_k}) ,$$

$$I_{a_k} = (\Theta D_{a_{ke}} - (1+\Theta) D^{a_k}) \frac{dn_e}{dx} + D_{a_k i_M} \frac{dn_{i_k}}{dx}$$

$$+ D_{a_k i_k} \frac{dn_{i_M}}{dx} + (D_{a_k a_M} - D^{a_k}) \frac{dn_a}{dx} - D_{a_k a_M} \frac{dn_{a_k}}{dx} ,$$

$$I_{aM} = (\Theta D_{a_{Me}} - (1+\Theta) D^{a_M}) \frac{dn_e}{dx} + D_{a_M i_M} \frac{dn_{i_M}}{dx}$$

$$+ D_{a_M i_k} \frac{dn_{i_k}}{dx} - D^{a_M} \frac{dn_a}{dx} + D_{a_M a_k} \frac{dn_{a_k}}{dx} . \qquad (5.20)$$

where

$$n_a = n_{a_k} + n_{a_M}; \quad n_e = n_{i_M} + n_{i_k};$$

$$\Theta = T_e/T; \quad c_\alpha = \rho_\alpha/\rho; \quad \rho = \sum_\alpha \rho_\alpha;$$

$$\rho_\alpha = n_\alpha m_\alpha \text{ when } \alpha, \beta = e, i_k, i_M, a_k, a_M, \alpha \neq \beta .$$

An expression for a flux of particles of each type, taking into account the convective term, has the form during collisions with specific types of other particles is within the limits of 10^{-6}-10^{-4} cm $\lesssim \ell_p$. The mean free path of all particles in gas is considerably less than ℓ_p. The length required to establish equilibrium temperature of electrons is on the order of 10^{-5}-10^{-6} cm.

At distances $x < \ell$, then, the near-cathode gas is a quasi-neutral plasma (Debye length is considerably shorter than ℓ_p) that can be investigated by means of hydrodynamic techniques. The plasma particle distribution function is assumed to be Maxwellian and the temperature of electrons may differ from the temperature of heavy particles, T, that, for simplicity, is assumed to be equal to the temperature of the cathode in the spot, T_k.

The following expressions for particle fluxes of each type were obtained from analysis of transfer equations for a five-component, partially-ionized mixture [128].

$$I_e = \Theta^{-1}\Big((D_{ei_M} - (1+\Theta)D^e)\frac{dn_e}{dx} + (D_{ea_M} - D^e)\frac{dn_a}{dx} + (D_{ea_k} + D_{ea_M})\frac{dn_{a_k}}{dx} - \frac{e\Theta}{kT_e}E_\pi(n_{i_k}D_{ei_k} + n_{i_M}D_{ei_M})\Big),$$

$$I_{i_k} = (\Theta D_{i_ke} - (1+\Theta)D^{i_k})\frac{dn_e}{dx} + D_{i_ki_M}\frac{dn_{i_M}}{dx} + (D_{i_ka_M} - D^{i_k})\frac{dn_a}{dx} + (D_{i_ka_k} - D_{i_ka_M})\frac{dn_{a_k}}{dx} + \frac{eE_\pi}{kT}(n_eD_{i_ke} - n_{i_M}D_{i_ki_M}),$$

$$I_{i_M} = (\Theta D_{i_M e} - (1+\Theta)D^{i_M}) \frac{dn_e}{dx} + D_{i_M i_k} \frac{dn_{i_K}}{dx}$$

$$\Gamma_\alpha = I_\alpha + n_\alpha \nu \ .$$

The diffusion coefficients may be represented in the following form

$$D^{\alpha} = \sum_{\beta} c_{\beta} D_{\alpha\beta} \ ,$$

$$D_{e\beta} = (0.3(1-0.6\Theta^{-1})\ \nu_{eT} K_{\beta}^{(1)})\ n_{i_k} \sigma_{ei_k} \ ,$$

$$D_{i_k\beta} = HK_{\beta}^{(2)}/(n_e \sigma_{i_k i_M} + n_{a_M} \sigma_{i_M a_M}) \ ,$$

$$D_{i_M\beta} = HK_{\beta}^{(3)}/(n_e \sigma_{i_k i_M} + n_{a_M} \sigma_{i_M a_M}) \ ,$$

$$D_{a_k\beta} = HK_{\beta}^{(4)}/(n_{i_k} \sigma_{a_k i_k} + n_{a_M} \sigma_{a_k a_M}) \ ,$$

$$D_{a_M\beta} = HK_{\beta}^{(5)}/(n_{a_M} + n_{i_M})\ \sigma_{i_M a_M} \ , \qquad (5.21)$$

where

$$H = 0.42\nu_{iT} \ .$$

In obtaining expressions for the diffusion coefficients, the terms on the order of m_e/m were neglected. It was taken into account that the mass $m_k \simeq m_M = m$. The coefficients $K_{\beta}^{(n)}$ are complex functions of arguments such as n_α/n_β and $n_\alpha \sigma_{\alpha\beta}/n_\beta \sigma_{\beta\alpha}$. These expressions are very unwieldy and will not be reproduced here. Numerical estimates show that

$$K_{\beta}^{(1)} \sim K_{\beta}^{(3)} \sim K_{\beta}^{(4)} \sim K_{i_M}^{(5)} \sim 0.1 \ ,$$

$$K_{\beta}^{(2)} \sim K_{\beta \neq i_M}^{(5)} \sim 1 \ . \qquad (5.22)$$

Let us note that comparison of the diffusion coefficients obtained with analogous coefficients for a three-component gas [87] shows that an increase in the number of components of the ionized mixture may lead to considerable changes of these coefficients. For example, $D_{i_M a_M}$ is ten times smaller than D_{ia} [87] (see also 5.3.2.).

Expressions (5.21) and (5.22) make it possible to establish the following relation between the coefficients of diffusion:

$$D_{ei_k} \simeq D_{ei_M} \simeq D_{ea_k} \simeq D_{ea_M} >> D_{a_M e}$$

$$\simeq D_{a_M i_k} \simeq D_{a_M a_k} \simeq D_{i_k e} \simeq D_{i_k i_M}$$

$$\simeq D_{i_k a_k} \simeq D_{i_k a_M} > D_{a_k e} \simeq D_{a_k i_k} \simeq D_{a_k a_M}$$

$$\simeq D_{i_M e} \simeq D_{i_M i_k} \simeq D_{i_M a_k} \simeq D_{i_M a_M}$$

$$\simeq D_{a_K i_M} \simeq D_{a_M i_M} . \qquad (5.23)$$

Equations of continuity for potassium and copper ions, respectively, have the form

$$\frac{d\Gamma_{i_k}}{dx} = \sigma_{i_k} n_{a_k} N_e(x)$$

$$+ \beta n_e^o n_{i_k}^o n_e (1 - n_e n_{i_k} / n_e^o n_{i_k}^o) ,$$

$$\frac{d\Gamma_{i_M}}{dx} = \sigma_{i_M} n_{a_M} N_e(x)$$

$$+ \beta n_e^o n_{i_M}^o n_e (1 - n_e n_{i_M} / n_e^o n_{i_M}^o) , \qquad (5.24)$$

where

$$N_e(x) = N_{ew} \times \exp\left(- {}_o\int^x (n_{a_M} \sigma_{i_M} + n_{a_k} \sigma_{i_k} + 2\sigma_{k\pi} n_e)\, dx\right) .$$

Here and below, quantities with "o" correspond to equilibrium conditions.

Equations comprising (5.24) take into account that ions of the near-cathode plasma are produced by ionization by an electron beam emitted from the cathode, N_{ew}, and as a result of thermal ionization of atoms by plasma elections. The ions disappear as a result of triple recombination with coefficient β.

In solving the problem of ion concentration, one may use conditions analogous to those used in one report [87] as boundary conditions. In that publication [87], it was assumed that the particle flux at the wall, which is a function of the ion concentration at the boundary, n_w, is specified, and that the concentration at the opposite boundary of the region is at equilibrium. In general, because radiation is low, deviation from the equilibrium value may occur only as a result of convection loss of particles. However, because $G_M \simeq 10^{-6}$ g/C and $G_k \simeq 10^{-4}$ g/C are also small, under conditions being considered, the convective term is also relatively small. Energy equations for electrons of a five-component ionized gas may be obtained in the following form [87]:

$$\frac{dq_e}{dx} = eE_\pi I_e + eu_k \frac{dN_e}{dx} + Q^{**} + Q^* , \tag{5.25}$$

where

$$q_e = \mu_{ee} I_e + 2.5\ kT_e I_e ;$$

$$Q^{**} = u_{i_k} \left(\frac{dI_{i_k}}{dx} + \frac{d(\nu n_{i_k})}{dx} \right) + u_{i_M} \left(\frac{dI_{i_M}}{dx} + \frac{d(\nu n_{i_M})}{dx} \right) ;$$

$$\mu_{ee} = 0.5\ R^+ ;$$

$$R^+ = (2 + R_1^+)/(1+1.3R_1^+/3) ;$$

$$R_1^+ = (n_{a_k} \sigma_{ea_k} + n_{a_M} \sigma_{ea_M})/n_e \sigma_{ee} ;$$

$$Q^* = 3n_e m_e m^{-1} k(T_e - T) \sum_\beta \tau_{e\beta}^{-1} .$$

Analysis of (5.25) shows that taking into account heavy particles of potassium and copper together leads to a change of certain terms: Q**, associated with inelastic processes, and $\mu_{ee}I_e$, characterized by interaction of plasma particles. In the range of parameters considered, $R^+ \simeq 2$-2.3 and, consequently, $\mu_{ee} \simeq 1$-1.15. For a three-component mixture of ionized copper vapors, $\mu_{ee} \simeq 0.5$ [87] (see also 5.3.2).

In deriving (5.25), the term associated with the heat flux caused by thermal conductivity was omitted. In general, evaluation of the importance of thermal conductivity requires determination of the temperature distribution in the relaxation layer, with proper boundary conditions (for example, second or third order) on the right-hand side. In a stationary case, this problem is reduced to analysis of the thermal conduction equation, in which heat sources and heat sinks depend on the coordinate.

For conditions under consideration, when the thermal conductivity coefficient is $\lambda \simeq 10^{-3}$ cal / (cm x s x K), an estimate of the maximum electron temperature gradient, even neglecting heat sinks, provides a value considerably smaller than T_e [87]. Also, note that, as a result, $\lambda/\rho \simeq 1$, the coefficient of thermal conductivity of the mixture, $a/\lambda c\rho$, is relatively large. This leads to a rapid equalization of electron temperature, even in the presence of fluctuations. When integrating equation (5.25), these considerations made it possible to assume T_e = const.

Consideration of the influence of various types of particles on the structure of the ion current at the cathode requires an estimate of the importance of diffusion terms in equations describing the concentration distribution of ions. In performing estimates a ratio of the coefficients of diffusion (5.23) will be used here. Excluding the electric field strength by means of the equation for the electron flux in equations for potassium and copper ion fluxes, (5.20), taking into account expression (5.24), gives the following equation in dimensionless form:

$$-\frac{d^2(\bar{k}f/\bar{k}_o)}{dy_k^2} + \Theta\frac{d}{dy_k}\left(\frac{\bar{k}}{\bar{k}_o}\frac{df}{dy_k}\right) = 1 - \frac{\bar{k}f^2}{\bar{k}_o f_o^2},$$

$$-\frac{d^2((1-\bar{k})f/(1-\bar{k}_o))}{dy_M^2} + \Theta\frac{d}{dy_M}$$

$$\times\left(\frac{1-\bar{k}}{1-\bar{k}_o}\frac{df}{dy_M}\right) = 1 - \frac{1-\bar{k}}{1-\bar{k}_o}\frac{f^2}{f_o^2}, \qquad (5.26)$$

where

$$f = n_e/n_e^o; \quad y_k^2 = \beta n_e^o x^2/D_{i_k a_k} \; ;$$

$$y_M^2 = \beta n_e^o x^2/D_{a_M i_M} \; ;$$

$$\bar{k} = n_{i_k}/n_e; \quad 1 - \bar{k} = n_{i_M}/n_e \; .$$

For simplicity, the terms associated with ionization by the beam electrons, convective particle transfer, and the total gas pressure gradient are neglected in (5.26). Analysis of the (5.26) system of equations leads to the conclusion that the nonequilibrium (or diffusion) region of length, ℓ_d, for potassium is different than that for copper, i.e., one is

$$(D_{a_m i_m}/D_{i_k a_k})^{1/2}$$

times the other. In both cases, however, the length of this region is on the order of the mean free path of the ion of the respective species in air. Estimates taking into account numerical values of parameters given above show that $\ell_{dM}/\ell_M \simeq 0.6$ and $\ell_{dk}/\ell_k \simeq 0.2$, where ℓ_M and ℓ_k are the mean free paths of the copper and potassium particles, in the same order. These relations were obtained for a coefficient of recombination corresponding to the electron energy on the order of 1 eV. When T_e increases to 2-3 eV, under conditions of the mixture considered, diffusion length may exceed the mean free path of particles.

Hence, in determining ion fluxes to the cathode, if $T_e \simeq 1$ eV, the diffusion terms may be neglected and the ion concentration may be assumed to be equilibrium and constant near the electrode. After integration of equation (5.25), this makes it possible to represent the expression for the electron energy balance in the following simplified form:

$$(eu_k + 2kT_k) = (2.5 + 0.5R^+)(1 + \alpha\zeta)kT_e \tag{5.27}$$

$$+ (\alpha\zeta + 1 - s)\,(u_{i_k} + (u_{i_M} - u_{i_k})s_{i_M}) + 2k(T_e + T_k)j_{eT}/j \; ,$$

where

$$\zeta = eG/Im; \quad G = G_M + G_k \ ;$$

$$\alpha = n_e/(n_e+n_i); \quad j_{i_M} = s_{i_M} j_i \ ;$$

$$j_i = e(\Gamma_{i_M} + \Gamma_{i_k}); \quad s = j_e/j \ ;$$

$$j_e = j_{eo} - j_{eT} \ .$$

Here, j_{eT} is the electron flux to the cathode from the plasma. The term characterizing Joule energy release is actually several times greater than the term describing energy transfer by electrons during elastic collisions with heavy particles. However, even this term is small ($E_\pi \simeq 10^3$ V/cm) in comparison with kT_e. These terms in the region ℓ_p may be considered to be small (on the order of 1 eV) additions (with opposite signs) to the cathode potential drop. Calculations (see 5.3.3.1.), taking into account variations of u_k show that small changes do not exert much influence on the solution of the problem under consideration. Therefore, the terms discussed were omitted in deriving expression (5.27).

The (5.20)-(5.27) system must include equations for the total emission current, electric field strength (see 5.3.2.), and, also, cathode energy balance, which have the following form under conditions being considered:

$$(1-s)u_{eff}j + 2kT_e j_{eT} = (Q_T I + (\lambda_{ev.m} + 2kT_k/m)G_m$$

$$+ (\lambda_{ev.m} + 2kT_k/m)\ G_k)\ j/I + \sigma T_k^4 \ ;$$

$$(1-s)\ u_{eff} = (1-s)\ (u_k + u_{i_k}$$

$$+ s_{i_M}(u_{i_M} - u_{i_k})) - \phi + (e^3E)^{1/2} \ ,$$

$$IQ_T = 2\lambda(I/j)^{1/2}(1/2\pi \arctan(4\pi ajt/I))^{-1}(T_k - T_o) \ . \tag{5.28}$$

Equation (5.28) takes into account the foregoing estimates, according to which thermal resistance of the film on the surface of a solid cathode may be neglected [14]. The meaning of individual terms of this equation was already discussed. An attempt to write a system of equations, taking into account the influence of the electrode atoms under conditions existing in an arc in the channel of an MHD generator, has been made [129]. However, in that report, diffusion equations for each of the components were written neglecting cross-terms and the diffusion coefficients for multi-component plasma and, in reality, reduced to those obtained in an earlier study [87].

The (5.20)-(5.28) system of equations with u_k and I determined from experimental data makes it possible to determine j, T_k, r, E, s, and also plasma parameters. No direct measurements of u_k are available for cathode spots burning on electrodes of an MHD generator. Instead of this quantity, the experimental data on the heat flux into the electrode, Q_T, will be used here. The relation between Q_T and u_k is given by equation (5.28).

More generally, under conditions existing in an MHD generator, the displacement rate of a spot must be taken into account in the energy balance on the cathode for spots of type I. Because it is relatively low (on the order 10^2 cm/s) and undergoes relatively little displacement during the lifetime of a spot, the fact noted will be considered during analysis of the temporal variation of parameters of the spot.

Analysis of the (5.20)-(5.28) system of equations shows that the degree of participation of copper and potassium atoms in processes on the cathode depends strongly on the thermionic emission properties of the surface. At low work function values, the cathode temperature proves to be low (approximately 1000 K) and the importance of copper atoms becomes negligibly small. In this case, the complex (5.20)-(5.28) system of equations is reduced to simple equations for a three-component mixture, (5.5)-(5.11), in which ions are formed by ionization of potassium atoms.

Let's first consider a case where the work function is close to its value for pure potassium. Then, it will be shown that such conditions may be close to those existing in spots of type I on electrodes of an MHD generator. Next, an analysis will be conducted under conditions in which the work function of the electrode exceeds the work function of potassium (T_k > 1000 K).

A system of equations for the near-cathode region of the arc was investigated at various characteristic currents to the spot (5-15 A) and at heat flux into the electrode as a function of the mean temperature of its surface, T_o, and the lifetime of the spot at the same site (Fig. 5.12, Table 5.7). Some of the calculations were also made using the following relation [14, 95]:

$$h = \ell_i , \qquad (5.29)$$

where h is the width of the region of positive space charge near the cathode, ℓ_i is the mean-free path of an

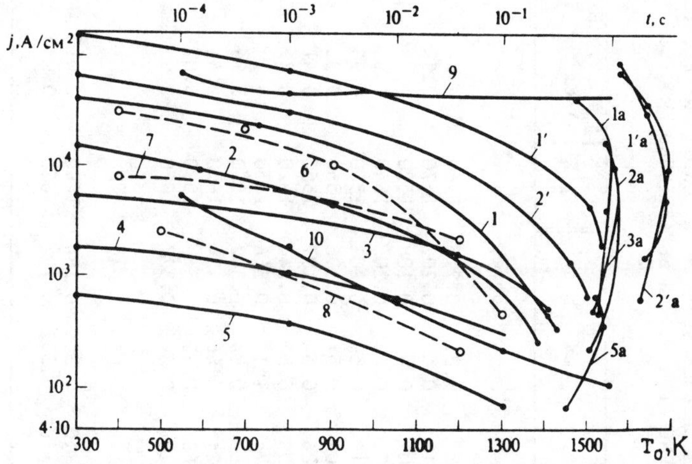

Fig. 5.12 Variation of current density in the spot with mean electrode temperature T_o (1-8, 1', 2'), temperature in the spot T_k (1a-3a, 5a, 1'a, 2'a), and with lifetime of spot T_o = 300 K [9, 10].

1-5, 9, 10, 1a-3a, 5a, 1', 2', 1'a, 2'a - calculations using experimental value of heat flux into the electrode of 20 W/A and 10 W/A, respectively; 6 [7, 66], 7 [6], 8 [6] - experimental data; 1, 2, 6, 7, 9, 1', 2', 1a, 2a, 1'a, 2'a - copper; 3-5, 8, 10, 3a, 5a - steel; 1, 3-5, 9, 1', 1a, 3a, 5a, 1'a - I = 5 A; 10 - I = 10 A; 2, 2', 2a, 2'a - I = 15 A; 1, 2, 1', 2', 1a, 2a, 1'a, 2'a - established process; 3-5, 3a, 5a - t = 10^{-4}, 10^{-3}, and 10^{-2}, respectively.

ion that depends on the collision cross section between ion and potassium atoms. A relation connecting parameters u_k, n_a, and j can be obtained from condition (5.29) and the (5.20)-(5.28) system of equations [14, 95]. This indicates that, if agreement exists between the data in

Table 5.7
PARAMETERS OF THE CATHODE SPOT UNDER CONDITIONS EXISTING IN AN MHD GENERATOR FOR AN ESTABLISHED PROCESS

Q_T, W/A	I,A	T_o,K	u_k,V	α	w_e,eV	s	T_K,K	$j \cdot 10^{-4}$, A/cm^2	$E \cdot 10^{-6} \cdot$ V/cm	$(n_a+n_i) \cdot 10^{-20}$, cm^{-3}	$n_i \cdot 10^{-18}$, cm^{-3}
					Copper						
20	5	300	23	0.08	0.81	0.216	1460	4.1	5.6	1.1	8.8
		1000	22.8	0.013	0.6	0.202	1570	0.98	2.8	1.65	2.1
		1400	22.7	0.001	0.4	0.188	1520	0.044	0.5	1.36	0.1
	15	300	22.8	0.023	0.65	0.206	1540	1.55	3.5	1.5	3.3
		600	22.8	0.012	0.59	0.201	1570	0.96	2.7	1.65	2
		1000	22.8	0.004	0.51	0.196	1560	0.32	1.6	1.6	0.7
15	60	300	22.8	0.006	0.52	0.197	1550	0.41	1.8	1.54	0.88
		300	17.8	0.008	0.56	0.244	1620	0.79	2.2	1.9	1.6
	5	300	13.3	0.27	1.1	0.366	1450	16	8.6	1.04	28
		800	12.9	0.08	0.83	0.342	1580	7.26	5.9	1.7	13
		1200	12.8	0.02	0.66	0.424	1650	2.41	3.4	2.13	4.24
10	15	300	13	0.068	0.82	0.34	1580	6.6	5.6	1.7	11.6
		800	12.8	0.024	0.68	0.327	1650	2.9	3.8	2.1	5.1
					Steel - 1x18H10 T*						
20	10	300	22.7	$2.7 \cdot 10^{-1}$	0.356	0.184	1460	0.013	0.32	1.05	0.03
10	10	300	12.7	$6.1 \cdot 10^{-4}$	0.392	0.296	1620	0.053	0.52	1.93	0.1

*Russian classification of a very good stainless steel

Tables 5.7 and 5.8 and the data in Ref. 14, it is possible to replace some experimental parameters (e.g., Q_T) with this relation. Such a comparison shows that the spot parameters obtained here (excluding j) differ considerably from analogous data in that reference [14], where s ≃ 0.01, T_k ≃ 1300-1100 K, $(n_a + n_i) \simeq 10^{19}$ cm^{-3}. Calculations performed with equations (5.20-5.28) provide results that differ significantly with respect to all parameters ($j \simeq 10^6$ A/cm^2, $Q_T \simeq 2$ W/A, s ≃ 0.7, $u_k \simeq 8$ V) from those given in Tables 5.7 and 5.8 and the experimental values of Q_T and j [6, 7, 11]. The use of equation (5.29) with the (5.20)-(5.28) system of equations makes its solution critical to small variation of parameters of the near-cathode plasma. Therefore, the results of the calculations are not in agreement with the experimental data. When relation (5.29) is used to evaluate the size of the spot (current density) without the use of equations describing the near-cathode gas, it should be remembered that other parameters (s, T_k, α) may differ significantly from the actual parameters satisfying conservation equations of plasma.

Let us proceed to a more detailed discussion of the results of the solution of the system of equations for the discharge under conditions existing in an MHD generator. First, it is necessary to clarify the influence of the mean electrode temperature, T_o, heated by a high-temperature gas flow on the spot parameters. Analysis of the results of the calculations (see Fig. 5.12, Tables 5.7 and 5.8) shows that, if current density (curves 1-3, 5, 1′, 2′) and, consequently, the area of the spot, πr^2 = I/j, may vary significantly with increasing T_o, then temperature of the spot (curves 1a-3a, 5a, 1′a, 2′a) in the range investigated depends weakly on the electrode temperature. An interesting fact is that a solution of this system of equations exists up to the values of T_o close to T_k (see Fig. 5.12). Numerical analysis at $T_o \simeq T_k$ shows that there is no solution of the system of equations and, for the spot radius, r → ∞. Physically, this means that the degree of discharge constriction decreases considerably only at electrode temperature approximately equal to the temperature of the cathode spot. In other words, conditions for transition into a distributed discharge mode on cooled metal electrodes of an MHD generator may be attained only when $T_o \simeq T_k$. Apparently, this may explain the reported presence of arc discharges in an MHD generator throughout a wide range of temperatures [6, 7].

Table 5.8
PARAMETERS OF A CATHODE SPOT UNDER CONDITIONS EXISTING IN AN MHD GENERATOR FOR AN ESTABLISHED PROCESS WHEN Q_T = 20 W/A

t, s	I, A	T_0, K	u_k, V	α	w_e, eV	s	T_K, K	$j \cdot 10^{-4}$, A/cm^2	$E \cdot 10^{-6}$, V/cm	$(n_a+n_i) \cdot 10^{-20}$, cm^{-3}	$n_i \cdot 10^{-18} \cdot$ cm^{-3}
					Copper						
10^{-3}	5	300	22	0.09	0.83	0.218	1460	4.5	5.9	1.08	9.8
		1000	22.8	0.016	0.615	0.203	1560	1.16	3	1.56	2.5
					Steel, 1x18H10 T*						
10^{-4}	5	300	22.8	0.007	0.544	0.198	1570	0.55	2.1	1.7	1.2
		1200	22.8	0.002	0.457	0.192	1550	0.15	1.1	1.5	0.32
10^{-2}	5	300	22.7	0.001	0.42	0.189	1520	0.067	0.72	1.3	0.15
		1300	22.7	0.0001	0.33	0.183	1440	0.006	0.21	0.97	0.014
10^{-4}	10	300	22.8	0.007	0.54	0.198	1580	0.54	2	1.7	1.16
10^{-3}	5	300	22.7	0.0025	0.47	0.193	1560	0.19	1.2	1.6	0.4
		1300	22.7	0.0005	0.383	0.187	1530	0.031	0.5	1.4	0.07

*Russian classification of a very good stainless steel

Overall, it should be noted that the principal consistencies in the variation of parameters are in agreement with the experimental data for spots of type I (see Fig. 5.12, Tables 5.7 and 5.8). In particular, similarly to the experiment [6, 7], the area of the spot decreases considerably (j decreases) with increasing mean electrode temperature. Furthermore, in the case of steel electrodes, as well as copper electrodes. this effect is manifested significantly at $T_o \simeq 900$ K. Numerical analysis of the influence of the lifetime of the spot on its parameters, assuming Q_T = const. (see curves 9 for copper and 10 for steel in Fig. 5.12 and Tables 5.7 and 5.8) shows that a discharge on steel differs from a discharge on copper by being considerably nonstationary. Qualitative dependence on time also is in agreement with the conclusion that spots that last longer have greater area [6, 7].

The results of the calculations also demonstrate the influence of the heat flux, Q_T, the experimental data on which are characterized by scattering (see 5.2.), on the parameters of the spot, which undergo changes when Q_T decreases by approximately 100 %. It is interesting that all other parameters (I, t, electrode material) in the model under consideration exert absolutely no influence on u_k. Therefore, the accuracy of the solution of the system of equations with respect to u_k depends on the accuracy of measurement of the experimental value of Q_T.

But, it should be kept in mind that these conclusions and numerical results at high electrode temperature (T_o > 1200 K) are valid, provided assumptions made in the calculations (a semi-infinite body model, preservation of a layer of seed on the surface of the cathode) remain valid. According to the experimental data, in the absence of current, the seed deposited is melted at $T_o \simeq 1100$ K and no seed deposits are present on walls at $T_o \gtrsim 1200$ K (in some cases, as a result of seed deposition, this temperature may be somewhat higher) [122]. In the latter case, the mechanism of electric discharge on pure electrodes must be principally different from the one suggested above. For example, it is necessary to take into account that surface ionization may occur during collision of potassium atoms of the products of combustion with a metal electrode, the electron work function of which is $\phi \gtrsim 4.5$ eV (Cu, Fe, Ni, etc.). According to the experimental and theoretical data [127], it follows that, for the materials listed, the efficiency of surface ionization of potassium in the range 1000-2000 K, $n_i/n_a \simeq$ 0.86-0.8, i.e., almost all atoms are ionized. At potassium concentration in combustion products characteristic of conditions in an MHD generator ($n_a \simeq 10^{16}$ cm^{-3}) and their complete ionization near the electrode, the ion current density is on the order of 10^2 A/cm^2, i.e., close to the values of j on electrodes at high temperature [66]. Consequently, surface ionization of potassium may be the determining process in an arc discharge. For the method of sustaining the discharge noted above (a detailed analysis of which is beyond the scope of this work), thermophysical properties of the electrodes at a fixed T_o will exert very little influence on parameters of the discharge (ions are generated by the products of combustion), and erosion is determined by chemical processes and the rate of vaporization of the heated metal. The intensity of such erosion can be controlled, using chemically stable and relatively refractory materials.

Where the seed on the electrode participates in sustaining the discharge, analysis of erosion processes may be conducted, using the model considered.

5.4.3. Erosion Processes on the Cathode. As a rule, erosion of metal electrodes is associated with either electrochemical or chemical processes in the near-electrode regions, or with thermal influence of arc spots. In addition, mechanical effects of solid particles of combustion products flow has been noted. Experimental investigation of erosion processes of metal electrodes has been conducted and reported [1, 13, 48, 49]. Some data are also provided elsewhere [4, 11].

Hence, at the present time, there is no single point of view with respect to the mechanism of erosion of metal electrodes. This applies especially to conditions at low electrode temperature (T_o < 900 K). Naturally, at relatively high temperature electrodes are corroded in an oxidizing medium. Even in that case, however, the importance of corrosion in destroying the cathode in the presence of arc spots is not clear. Consider the wear of electrodes, using copper electrodes operating in a constricted discharge mode, as an example.

Such an investigation is also necessary in connection with estimates available in the literature [1, 11, 48] of the dependence of the rates of vaporization of cathode material from the spot on temperature, on the basis of which it was concluded that the importance of this process and erosion of the cathode is insignificant. In general, the results of such estimates are approximate, i.e., the temperature of the spot (parameters exerting a strong influence on the rate of vaporization) is, as a rule, determined using equations describing individual processes, or only from the thermal conductivity equation, and are useful only for obtaining qualitative estimates. In addition, visual observation of traces left by the cathode spots on electrodes support the assumption that, under certain conditions (see 5.2.), thermal interaction of the constricted discharge is the predominant factor in destruction of electrodes [3-5, 13, 48] and clearly illustrate the high temperature in the arc spot. Therefore, we shall discuss the feasibility of heating the surface of the electrode in the spot up to the temperature at which erosion depends on the cathode material vaporization, based on analysis of the solution of the system of equations describing the near-cathode region of an arc discharge in an MHD generator.

Let us use the results of the calculations for fast spots of type I, given in foregoing sections, and evaluate the cathode material removal by these spots. Because the average temperature in the spot is considerably higher than the melting temperature, T_m, it appears possible that the molten metal will be set in motion by the gas flow around it, or will be removed by a force on the spot resulting from pressure of the near-cathode gas.

Taking into account that the melting radius, r_π, is approximately equal to the radius of the spot (see 5.3.3.) [119], it is possible to obtain the lower estimate of the mass of material being removed from the volume of the melted crater. In this case, the material removal rate of a metal cathode is expressed as $G_m = Id'\gamma/jt$, where d' is the depth of the melting isotherm to be determined by solving the heat problem in the body, γ is the electrode material density. For copper at $I \simeq 10$ A, $j \simeq 10^4$ A/cm^2, $d' \simeq 10^{-3}$ cm, $t \simeq 10^{-4}$-10^{-3} s, we obtain $G_m \gtrsim 10^{-3}$-10^{-2} g/s. This is considerably above the experimentally measured values on cooled electrodes ($T_o \simeq 500$ K); on the order of 10^{-7} g/C [6, 11], 10^{-7}-10^{-6} g/C [1, 13, 48, 49]. At $I \simeq 10$ A, the erosion rate lies within the limits 10^{-5}-10^{-6} g/s. Although the fact that the calculated value is considerably higher than the experimental value does not preclude some splattering, it nevertheless makes possible an assumption that the principal part of the molten metal under the spot is not removed.

In order to estimate the importance of vaporization in cathode erosion, it is necessary to determine the flux of metal atoms leaving the surface and returning to it. The flux of atoms leaving the region of the spot can be found from solution of the diffusion problem, taking into account the influence of the electric field of the near-cathode plasma (under conditions being considered, the degree of ionization of copper atoms is high). The metal atom can be considered to have left the cathode if it traverses the region with high particle pressure ($n_a \simeq 10^{20}$ cm^{-3}) and enters the chemically active medium.

In view of the absence of a solution to such a problem, it is possible to use a formula describing vaporization in a vacuum [108] to estimate the maximum cathode material erosion rate under conditions being considered. Such a computation at various T_o shows that the specific erosion rate varies little, and is approximately 10^{-4} g/(cm^2 x s). This can be attributed to the small variation of the temperature of spots of type I. At the same time, total erosion rate varies between 10^{-8}-10^{-6} g/s when the temperature varies between 300 and 1200 K. An increase in T_o leads to an increase in the area of the spot and, consequently, to erosion by this spot.

At low temperature (500-800 K), the rate (10^{-8}-10^{-7} g/s) calculated from the vaporization rate [108] proved to be considerably smaller than the experimental value (see 5.2.). This difference makes it possible to reach the following conclusion concerning erosion caused by different types of arc spots burning on cooled metal electrodes protruding into the flow in an MHD generator.

According to the foregoing assumption, a type I arc spot in an MHD generator, moving along the surface of the electrode, burns on a thin seed of film, the thermophysical and thermionic emission properties of which are close to those of potassium. When the spot moves to a new site, the vaporized film is restored by a mass transfer process. The calculated low temperature of the spot and its erosion are attributable to these properties of the film.

Therefore, if the properties of a real film of seed are similar to those forming the basis of the calculation, it is not surprising that the calculated cathode material erosion rate is considerably lower than the experimental value. This difference indicates that the importance of fast spots of type I in cathode erosion is insignificant. This conclusion does not contradict the experimental data (see Fig. 5.1), according to which most of the destruction is observed in this zone, where the slow spots of type II are concentrated.

As already noted, spots of type II differ from spots of type I primarily by the small variation of their sizes and displacement rates, i.e., their considerably longer lifetime at the same site. It appears that, if the burning conditions of type II and type I spots are the same, they must be described within the framework of the system of equations suggested (see 5.4.2.), with their peculiarities taken into account.

However, calculations have shown that taking into account not only the lifetime of spots of type II but, also, parameters of the nonmoving, stationary spot (subject to the condition that the surface parameters remain the same) still does not make it possible to explain the experimentally observed erosion. The assumption of concentration of high currents and possible overheating of the cathode in the region of the site of type II spots leads to an analogous conclusion. This is verified by calculations for the current to the spot, I = 60 A, and the electrode temperature, T_o = 1400 K (see Tables 5.7 and 5.8).

The investigations noted permit the assumption that the conditions for burning of discharge in type II spots differ from conditions in type I spots. This can be attributed to the type II spots being hardly displaced and, hence, burning on the sputtered surface of the cathode, the properties of which can be changed considerably during the time of the discharge. For example, emissivity decreases as a result of vaporization or changes in properties of an alkali film. In the limiting case, thermionic emission of electrons may occur from pure metal.

Certain data verifying the influence of the degree of deposition of alkali seed as a function of the electrode temperature and concentration of seed atoms in gas on the thermionic characteristics of a metal cathode are given in the literature [130, 131]. One cannot exclude the possibility that the influence of the parameters of seed may also change somewhat the work function of the cathode within the zone affected by type I spots. However, the absence of clear traces of destruction (channels) during direct observation of electrodes (see Fig. 5.1a) in this zone makes it possible to assume that the temperature in the spots is low. A more accurate analysis may be performed when more qualitative data become available about cathode material erosion by individual types of spots.

The calculated temperature of the spot depends on the thermionic emission constants of the cathode material (see 5.3.3.). Analysis of processes in the near-cathode region of a vacuum arc [88] also shows that the parameter determining the temperature of the spot and, consequently, erosion is the work function of the cathode. At the same time, an exceedingly low cathode material erosion away from the cathode, compared with that in vacuum, as well as the existing differences in parameters of spots, indicate effective participation of seed in the burning process of discharge under MHD generator channel conditions. Type II spots in an MHD generator can burn on previously melted metal mixed with seed.

Estimates show that, as a result of the high tension of potassium vapors (provided deviation from equilibrium value is small), in this case, the number of their atoms turns out to be predominant. For such a surface, the work function, ϕ, differs from that of a pure film by its capacity to vary within a range between ϕ_k (its value for potassium) and ϕ_m (its value for copper). Therefore, when investigating parameters of type II spots, we will use the above given scheme of calculating a fast spot, assuming that ion generation occurs as a result of potassium ionization and that electron emission occurs from the surface, the work function of which is specified parametrically and varies between ϕ_k and ϕ_m.

Note that the vapor tension of potassium under conditions being considered also depends on the composition and structure of the surface of the cathode. In type II spots, where the temperature is relatively high, potassium may also be vaporized from a decomposed layer of seed at the boundary with which the spots have appeared.

We will now analyze the results of the calculations for $I = 2\text{-}25$ A and $T_o = 300\text{-}1200$ K. Parameters of the spot depend equally on the work function of the cathode material and the electric field strength (both quantities enter the exponential factor describing emissivity of an electrode [88]). As an illustration of this influence, we have calculated the cathode material erosion rate as a function of the electrode thermionic emission properties at various values of the mean temperature, T_o, and current to the spot, I, for two cases: $E = E_m$ (Fig. 5.13a) and $E = 2E_m$ (Fig. 5.13b). To clarify the influence of the heat flux diverted by the electrode on erosion, calculations were performed at two of its values, 20 W/A (continuous curves) and 10 W/A (dashed curves).

As was assumed, curves in Fig. 5.13 show that erosion of the cathode depends substantially on the work function of the emitting surface in the spot. The spot current also influences the rate of erosion. Even when the work function, ϕ, is one-half in comparison with that for potassium, the erosion rate of cathode material caused by vaporization increases by three orders of magnitude, and may become comparable with that observed in experiments.

When the current increases threefold, from 5 A (curve

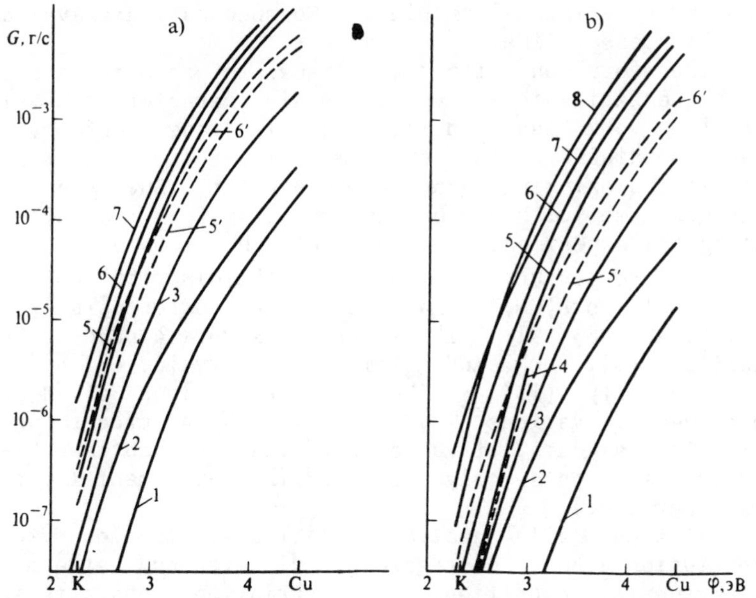

Fig. 5.13 The dependence of erosion rate of cathode on work function for $E = E_M$(a) and $E = 2E_M$(b).

1-8 - Q_T = 20 W/A; 5′, 6′ - Q_T = 10 W/A; 1-2 - I = 2 A; 3 - I = 5 A; 4-7, 5a, 6a - I = 15 A; 8 - I = 25 A; 4 - T_o = 300 K; 1, 5, 5′ - T_o = 600 K; 3, 6, 8, 6′ - T_o = 900 K; 2, 7 - T_o = 1200 K.

3) to 15 A (curve 6) (T_o = 900 K), the rate of erosion may increase by an order of magnitude. However, the dependence on current may be more complex, because of the fact that both the heat flux and the electrode temperature, which are a function of current, vary at the same time. At a constant current and ϕ, the influence of Q_T and T_o is still less than the influence of I. The erosion rate of cathode material calculated for $E = 2E_M$, assuming that all other conditions are equal, is considerably smaller than for $E = E_M$. Furthermore, in the second case, the influence of the remaining parameters (I, T_o, Q_T) is manifested to a much lower degree than in the first case.

Consequently, if one knows the actual values of ϕ and E under conditions being considered, it is possible to explain the high electrode erosion qualitatively, and to obtain a quantitative estimate as well. However, if it is possible to improve the value of E from the solution of a correctly formulated problem for a layer of space charge near the cathode, determination of the work function, ϕ, of the cathode in an MHD generator channel is a very complex experimental problem. No such data are available at the present time.

Therefore, an attempt was made to process the result of the experiments so as to obtain characteristic values of the work function of the electrode surface, ϕ_e, corresponding to the experimental erosion rates under conditions existing in an MHD generator. Such processing was performed with use of the results of the calculations using the experimental values of the erosion rate, G_e, taking into account the data on the distribution of the number of spots, N, as a function of current (Table 5.9) [1, 6, 13, 48, 49]. The experiments were conducted on EE facility [1], U-02, and "Temp" facility [6, 48, 49], and K-1 facility [13]. The required values of ϕ_e were obtained at various T_o, Q_T, E. When processing the experimental data, it was assumed that erosion is achieved by type II spots only (G_{eo} is the experimental erosion rate per spot).

First of all, analysis of Table 5.9 leads to the conclusion that, disregarding differences between the experimental conditions, the variational characteristic work function of the surface of electrodes occupied by type II spots under conditions existing in an MHD generator is almost insignificant. Some increase of ϕ_e is observed at high electrode temperature (on the order of 1200 K). Apparently, this can be attributed to variation of the properties of seed and conditions under which it is deposited. The following average values of the work function, ϕ_{ea}, in eV were calculated from ϕ_e in Table 5.9, assuming $E = E_M$ and $E = 2E_M$:

$E = E_M$	$E = 2E_M$	
2.65	3.0	[13]
2.8	3.15	[48]
2.91	3.37	[1]
3.17	3.7	[49]

Overall, for all the experiments $\phi_{ea} \simeq 2.8$ eV when it is assumed that $E = E_M$, and $\phi_{ea} \simeq 3.1$ eV, when it is assumed that $E = 2E_M$. The results of Ref. 49 stand somewhat apart. This can probably be attributed to the higher electrode temperature in these experiments. At low electrode temperature ($T_o \lesssim 900$ K), according to results in that report [49], erosion exceeds several times analogous data of other experiments (see Table 5.9). This can be attributed to the difference between experimental conditions and to the dynamics of behavior of cathode spots that depends on total current. Unfortunately, Ref. 49 does not provide data on the dependence of erosion directly on current. In processing the experimental

Table 5.9
WORK FUNCTION OF ELECTRONS CORRESPONDING TO THE EXPERIMENTAL DATA ON THE WROSION RATE OF ELECTRODES IN AN ARC DISCHARGE UNDER CONDITIONS EXISTING IN AN MHD GENERATOR

I_0, A	$G_e \cdot 10^6$, g/s	I, A	$G_{eo} \cdot 10^6$, g/s	q_e, eV		Ref.
				$E=E_M$	$E=2E_M$	
5	0.5	5	0.5	2.66	3.05	[13]
5*	5	5	5	2.88	3.3	
135	27	15	3	2.56	2.86	
100	15	15	2	2.5	2.8	
200	92	15	7	2.66	2.98	
60*	80	15	20	2.68	3.0	[48]
45	35	15	12	2.72	3.06	
6	3	6	3	2.92	3.36	
4.5*	12.7	4.5	12.7	3.04	3.46	
15	7.2	15	7.2	2.66	3.0	
15	7.2	5	2.4	2.88	3.3	
12.5	1.88	4	0.69	2.8	3.2	[1]
12.5	1.88	2	0.31	3.0	3.6	
12.5	8.75	4	2.92	2.96	3.4	
20	8.0	5	2.0	2.9	3.3	
15	15	15	15	2.76	3.1	[49]
15*	150	15	150	3.04	3.4	
15**	600	15	600	3.24	3.56	
5**	200	5	200	3.5	4.1	

Note: Calculations were performed for Q_T = 20 W/A and T_o = 600 K. (* - calculations for T_o = 900 K; ** - calculations for T_o = 1200 K).

results, it was assumed that the current to the spot is equal to 5 A and 15 A [7, 66].

Analysis of the results of the calculations shows that a 50 % decrease of Q_T results in a 1-5 % increase of ϕ_e. It is of interest to consider the influence of taking into account the variation of the emission properties of the surface of the cathode on all of the parameters of the spot, including the electron energy, w_e; atom concentration, n_a; and ion concentration, n_i, in the near-cathode plasma, as well as on the degree of its ionization, α. Table 5.10 lists the characteristic values of the spot parameters calculated for a steady-state regime for I = 15 A. It can be seen that the spot temperature and, to a greater degree the total concentration of particles and the cathode material removal rate undergo change.

Thus, the numerical analysis of the parameters of the near-cathode region of the arc makes it possible to assume that the erosion of metal electrodes can be attributed to vaporization of cathode material and to a considerable

Table 5.10
PARAMETERS OF THE SPOT AND EROSION RATE OF THE CATHODE UNDER CONDITIONS IN AN MHD GENERATOR FOR VARIOUS COMPUTATIONAL SCHEMES

μg/s	ϕ, eV	α	s	w_e, eV	$(n_a+n_i)\cdot10^{-20}$ cm^{-3}	$j\cdot10^{-4}$, A/cm^2	$E\cdot10^{-6}$, V/cm	T_K, K	G, μg/s
$2E_M$	2.26	0.019	0.203	0.595	0.7	0.6	4.3	1370	0.01
	3.0	0.008	0.199	0.594	4.0	1.6	7.0	1875	7.9
	3.5	0.006	0.196	0.599	8.2	2.7	9.0	2240	130.0
E_M	2.26	0.012	0.202	0.592	1.65	0.95	2.7	1570	0.03
	3.0	0.006	0.197	0.598	7.18	2.45	4.4	2160	7.6
	3.5	0.005	0.190	0.606	12.9	3.9	5.6	2570	785.0
$2E_M$**	2.26	0.064	0.336	0.737	0.69	2.3	6.7	1360	0.003
	3.0	0.027	0.326	0.735	4.0	6.6	11.4	1880	2.0
	3.5	0.020	0.320	0.744	8.3	11.0	15	2250	34.0
$2E_M^*$	2.26	0.002	0.0191	0.426	1	0.065	1.4	1450	0.49
	3.0	0.003	0.192	0.50	4.8	0.58	4.3	1954	51.2
	3.5	0.003	0.191	0.526	9.1	1.23	6.3	2300	450.0
E_M^*	2.26	0.002	0.192	0.45	1.6	0.13	1.0	1560	1.5
	3.0	0.002	0.191	0.517	7.21	0.94	2.72	2160	200
	3.5	0.002	0.190	0.541	12.9	1.88	3.9	2564	1600

Comments. Calculations were conducted for Q_T = 20 W/A and T_o = 600 K; * - calculations with T_o = 1200 K; ** - calculations for Q_T = 10 W/A.

degree depends on deposition of seed influencing emission properties of the surface (effective work function).

Under conditions being considered, on the average, ϕ_e is approximately 3 eV and is the characteristic of emission properties of a copper electrode operating under conditions existing in a plasma flow with seed consisting of potassium compounds. This value of ϕ_{ea} may be used to estimate the erosion rate of the cathode, G, using the model developed (see 5.4.2.) under analogous conditions, when designing electrode walls of high power-output MHD generators. However, it is expedient to conduct direct measurements of the work function of the cathode and the temperature of electrons as a function of current for final verification of the mechanism of erosion, analysis of the numerical results, etc.

Depending on the parameters of the discharge, the calculated values of w_e (see Tables 5.7, 5.8, 5.10) fall within the range 0.4-1 eV. The electron temperature determined as a result of measurements conducted in the near-cathode region of the arc of a model facility, in the presence of ash, is 0.55 eV. The measurements were conducted in a very limited range of discharge current (8-10 A) [39].

Therefore, even though theoretically calculated electron temperature is in agreement with the experimental values of T_e, a detailed experimental investigation of this parameter as a function of various conditions may provide more complete data on the validity of assumptions on which calculations are based. In addition, such experiments are required in view of a lack of reliable data on the rate of vaporization and vapor tension of deposited seed, on the temperature, content, and structure of the seed solution in the surface layer of the metal and their influence on the concentration of atoms at the cathode.

To estimate the importance of copper atoms in the discharge, one can assume that the concentration depends on the saturated vapor pressure of the pure element, and that the potassium concentration is lower than that determined from equilibrium pressure. Under these assumptions, calculations show that, because the importance of potassium atoms becomes considerable only at concentrations comparable with

$$n_{i_M},$$

practically all parameters of the spot prove to be close to the values for a spot in a vacuum. An exception is T_e, which, in view of the energy equation for electrons (5.27), is somewhat lower, and the electron component of current, which also changes somewhat in comparison with the results of analogous calculations for pure potassium and copper. However, in a vacuum arc, i.e., in an arc with a vaporized cathode, a spot with low current ($I \lesssim 10$ A) can be described only at values of j considerably (two orders of magnitude) higher than the analogous quantity for a discharge under conditions existing in an MHD generator (see 5.3.3.1.). Lower j, values comparable to the experimental values (see 5.2.) may be attained at an unjustifiably high cathode potential drop ($u_k \gtrsim 40$ V), as was actually done in one study [129]. It should also be noted that, because the electrode temperature in these calculations is high (on the order of 4000 K), erosion of copper is comparable to erosion under conditions of a vacuum, the rate of which exceeds by two orders of magnitude the erosion rate of electrode material of an MHD generator.

The latter fact provides justification for considering the conclusion that the processes in the near-cathode plasma of an arc spot of an MHD generator as a function of primarily potassium vapor pressure. An important aspect of the problem is the differentiation between the pressure of potassium vapor from solution and the pressure for the pure substance. This can be achieved, utilizing thermodynamic analysis of the solution process of potassium in copper under complex conditions existing in an MHD generator [132].

Some change of the dependence of potassium vapor pressure on temperature in comparison with the equilibrium pressure for pure potassium does not alter the principal concept of this phenomenon. Parameters of spots may be described within the framework of the model considered, but at a different cathode temperature within the spot. Therefore, investigation of the thermodynamics of the solution is required not so much to understand the mechanism sustaining the discharge, as for qualitative analysis of erosion processes of electrodes of an MHD generator channel.

Overall, it should be noted that the theoretical analysis of the near-cathode arc spot processes under conditions existing in an MHD generator made it possible to explain certain important experimental data, and to develop necessary concepts of interpretation techniques useful in designing facilities.

References

[1]Motulevich, V. P. (ed.), "Open Cycle MHD Generators," collection of articles translated from English, 1972, 836 p.

[2]Petrick, M., and B. Ya. Shumyatskiy (eds.), Open-Cycle Magnetohydrodynamic Electric Power Generation, a joint USA/USSR publication, 1978.

[3]Kirillin, V. A., A. E. Sheyndlin, V. V. Kirillov, et al., "Certain Results of Investigations on the U-02 MHD Modeling Facility," Fifth International Conference on Magnetohydrodynamic Electric Power Generation, Munich, April 1971, Vienna, IAEA, 1971, Vol. 1, 1971, p. 353-370.

[4]Dicks, J. B., Y.C.L. Wu, L. W. Crawford, et al., "Characteristics of a Family of Open-Cycle Series MHD Generators," translation of an unspecified English paper.

[5]Dicks, J. B., D. Denzel, S. Witkowski, et al., "Temporal and Spatial Distribution of Electrical Parameters in an MHD Generator Channel," translation of an unspecified English paper.

[6]Zelikson, Yu. M., V. V. Kirillov, E. P. Reshetov, and B. D. Flid, "Certain Regularities of Operation of Metal Electrodes of an MHD Generator," Teplofizika vysokikh temperatur, Vol. 8, No. 11, 1970, p. 193-202.

[7]Zalkwind, V. I., V. V. Kirillov, A. P. Markina, et al., "Experimental Investigation of Cathode Spots on Metal Electrodes Protruding into the Flow," Zhurnal prikladnoy mekhaniki i tekhnicheskoy fiziki, No. 2, 1974, p. 17-23.

[8]Baranov, V. Yu., D. D. Malyuta, and F. P. Ulinich, "Current Flow Across Boundary Layers," Teplofizika vysokikh temperatur, Vol. 11, No. 3, 1973, p. 457-464.

[9]Zauderer, B., "Electrical Characteristics of the Linear Hall and Faraday Generators at Small Hall Parameters," Journal of the American Institute of Aeronautics and Astronautics, Vol. 5, No. 3, 1967, p. 378-581.

[10]Dicks, J. B., Y.C.L. Wu, D. Denzel, et al., "Certain Results of Investigation of a Generator with Diagonally Conducting Walls," translation of an unspecified English paper.

[11]Adams, R. C., and E. Robinson, "Electrode Processes in MHD Generators," Proceedings IEEE, Vol. 56, No. 9, 1968, p. 1519-1535.

[12]Muehlhauser, J. W., and J. B. Dicks, "Arc Spots and Voltage Losses in Hall Generator," Proceedings of the 14th Symposium on Engineering Aspects of Magnetohydrodynamics, Tullahoma, Tennessee, April 1974, University of Tennessee Space Institute, 1974, p. VIII-2.1-VIII.2.8.

[13]Zykova, N. M., and T. S. Kurakina, "Investigation of Discharge on Cooled Copper Electrodes in Combustion Products Flow,"

Teplofizika vysokikh temperatur, Vol. 14, No. 5, 1976, p. 1088-1090.

[14]Beylis, I. I., V. I. Zalkind, and A. S. Tikhotskiy, "Cathode Spots on Metal Electrodes under Conditions of an MHD Generator Channel," Teplofizika vysokikh temperatur, Vol. 15, No. 1, 1977, p. 158-163.

[15]Beylis, I. I., "Near-Cathode Region of a Constricted Discharge on Metal Electrodes of an MHD Generator," Teplofizika vysokikh temperatur, Vol. 15, No. 6, 1977, p. 1269-1275.

[16]Beylis, I. I., and G. A. Lyubimov, "Theoretical Description of an Arc Spot on a Film Cathode," Zhurnal tekhnicheskoy fiziki, Vol. 46, No. 6, 1976, p. 1231-1239.

[17]Engel′, A., and M. Shteenbeck, "Physics and Technique of an Electric Discharge in Gas," translation from German, Vol. 1, 1935; Vol. 2, 1936.

[18]Finkel′nburg, B., and G. Meker, Electric Arcs and Thermal Plasma, translation from German, 1961, 369 p.

[19]Holm, P., "Electrical Contacts," translation from English.

[20]Loob, L., "Basic Processes of Electrical Discharges in Gases," translation from English.

[21]Rozhanskiy, D. A., Physics of Gas Discharge (Fizika gazovogo razryada), Moscow-Leningrad, Publishing House ONTI NKTP SSSR, 1937.

[22]Granovskiy, V. L., Electric Current in Gases, General Electrodynamic Problems (Obshchiye voproxy elektrodinamiki gasov), Moscow-Leningrad, Publishing House Gostekhteorizdat, Vol. 1, 1952.,

[23]Granovskiy, V. L., Electric Current in Gases, Steady State Current, (Elektricheskiy tok v gazakh, ustanovivshiysya tok), Moscow, Publishing House Nauka, Vol. 2, 1971.

[24]Ecker, G., "Electrode Components of the Arc Discharge," Ergibn, exakt. Naturwissenschaften, Bd. 33, 1961, p. 1-104.

[25]Kesayev, I. G., Cathode Processes of an Electric Arc (Katodnyye protesessy elektricheskoy dugi), Moscow, Publishing House Nauka, 1968.

[26]Kesayev, I. G., Cathode Processes of a Mercury Arc and its Stability (Katodniye protsessy rtutnoy dugi i voprosy ee ustoychivosty) Moscow-Leningrad, Publishing House Gosenergoizdat, 1961.

[27]Rakhovskiy, V. I., Physical Foundation of Commutation of Electric Current in Vacuum (Fizicheskiye osnovy kommutatsii toka v vakuume), Moscow, Publishing House Nauka, 1970.

[28]Lyubimov, G. A., and V. I. Rakhovskiy, "Cathode Spot of a Vacuum Arc," Uspekhi fizicheskikh nauk, Vol. 125, No. 4, 1978, p. 665-706.

[29]Rohatgi, V. K., and S. A. Aisenberg, "Impedance and Fraction of Ion Current at the Cathode of a Plasma Accelerator," Journal of the American Institute of Aeronautics and Astronautics, Vol. 7, No. 3, 1969, p. 144-149.

[30]Kantsel', V. V., "Experimental Investigation of Near-Cathode Regions of an Electric Arc Discharge," dissertation for a candidate of physical mathematical sciences, Institute of High Temperatures, AS SSSR, 1973.

[31]Grakov, V. E., "Cathode Potential Drop of Vacuum Arcs on Pure Metals," dissertation for a candidate of physical-mathematical sciences, Minsk, Belorussian State University (BGU), 1967.

[32]Grakov, V. E., "Cathode Drop of Arc Discharge on Pure Metals," Zhurnal tekhnicheskoy fiziki, Vol. 34, No. 8, 1964, p. 1482-1493.

[33]Kesayev, I. G., "Regularities of Cathode Drop and Threshold Currents of Arc Discharge on Pure Metals," Zhurnal teknicheskoy fiziki, Vol. 34, No. 8, 1964, p. 396-404.

[34]Baranov, N. N., V. J. Molotkov, L. D. Poberezhesky, et al., "Investigation of Physical Processes in the Near-Electrode Region of MHD Generators," Proceedings of the 15th Symposium on Engineering Aspects of Magnetohydrodynamics, Philadelphia, Pa., May 1976, University of Pennsylvania, 1976, p. 1.4.1.-1.4.8.

[35]Ogiwara, H., "La chute de potential cathodique d'electrode froide du generateur MHD," Japanese Journal of Aplied Physics, Vol. 3, No. 3, 1969, p. 388-392.

[36]Rubin, E. S., and R. H. Eustis, "Effect of Electrode Size on the Performance of a Combustion Driven MHD Generator," translation of an unspecified English paper.

[37]Isakayev, M. E., and Yu. S. Arbuzov, "Heat Fluxes and Voltage-Current Characteristics of Discharge between Electrodes in Transverse Flow of Seeded Combustion Products Plasma," Teplofizika vysokikh temperatur, Vol. 12, No. 3, 1974, p. 657-659.

[38]Kantsel', V. V., T. S. Kurakina, V. S. Potokin, et al., "On the Influence of Thermophysical Parameters of Material on Erosion of Electrodes in a Concentrated Vacuum Discharge," Zhurnal tekhnicheskoy fiziki, Vol. 38, No. 6, 1968, p. 1074-1078.

[39]Rakhovskiy, V. I., "Electrode Erosion in a Constricted Discharge," Sibirskiy otdel AN SSSR, Izvestiya, Seriya tekhnicheskikh nauk, Vol. 3, No. 1, 1975, p. 11-27.

[40]Rachovskii, V. I., "Experimental Study of the Dynamics of Cathode Spots Development," IEEE Transactions on Plasma Science, Vol. PS-4, No. 2, 1976, p. 81-82.

[41]Kutzner, I., and Z. Zalucki, "Electrode Erosion in the Vacuum Arcs," International Conference on Gas Discharges, September 1970, London, Institute of Electrical Engineers, 1970, p. 87-94.

[42]Plyutto, A. A., V. N. Ryzhkov, and A. G. Kapin, "High Flow Rate Plasma Fluxes of Vacuum Arcs," Zhurnal eksperimental'noy i teoreticheskoy fiziki, Vol. 47, No. 8, 1964, p. 494-507.

[43]Lyubimov, G. A., "Working Group on Electrode Phenomena," Teplofizika vysokikh temperatur, Vol. 17, No. 3, 1979, p. 662-666.

[44]Daalder, J. E., "Erosion and Origin of Charged and Neutral Species in Vacuum Arcs," Journal of Physics D, Aplied Physics, Vol. 8, 1975, p. 1647-1659.

[45]Pravoverov, I. L., and A. I. Struchkov, "Erosion of Pure Metals in an Electric Arc," Elektrotekhnika, No. 1, 1976, p. 100-101.

[46]Kimblin, S. U., "Electrode Erosion and Ionization Processes in Near-Electrode Regions of Vacuum Arcs and at Atmospheric Pressure," Experimental Investigations of Plasmotrons (Eksperimental'nyye issledovaniya plazmotronov), M. F. Zhukov (ed.), Novoskibirsk, Publishing House Nauka, 1977, p. 226-253.

[47]Isaenkov, Yu. I., A. D. Iserov, V. V. Kirillov, et al., "Investigation of an MHD Generator Channel of the U-25 Facility," Teplofizika vysokikh temperatur, Vol. 12, No. 2, 1974, p. 399-412.

[48]Zelikson, Yu. M., "Investigation of Parameters of Electrodes of an Open-Cycle MHD Generator," dissertation for a candidate of technical sciences, Institute of High Temperatures, Moscow, 1974.

[49]Zalkind, V. I., V. I. Kondrat'yev, V. V. Kirillov, et al., "Investigation of Erosion of Metal Electrodes in Open-Cycle MHD Generator Channel," Preprint IVTAN, No. 3-015, Moscow, 1977, 15 p.

[50]Cobine, J. D., and C. J. Gallagher, "Current Density of the Arc Cathode Spot," Physical Review, Vol. 75, No. 10, 1948, p. 1524-1530.

[51]Dunkerley, H. S., and D. L. Schaefer, "Observation of Cathode Arc Tracks," Journal of Aplied Physics, Vol. 26, No. 11, 1955, p. 1384-1385.

[52]Sommerville, J. M., and W. R. Blevin, "Current Densities in Cathode Spots of Transient Arcs," Physical Review, Vol. 76, No. 7, 1949, p. 982-986.

[53]Sommerville, J. M., W. R. Blevin, and N. H. Fletcher, "Electrode Phenomena in Transient Arcs," Proceedings of the Physical Society of London, Vol. 65, 1952, p. 963-965.

[54]Vroe, H., "Vacuum Arcs on Tungsten Cathodes," Nature, Vol. 182, 1958, p. 338-340.

[55]Sanger, C. C., and R. E. Secker, "Arc Cathode Current Density Measurement," Journal of Physics D, Aplied Physics, Vol. 4, 1971, p. 1940-1945.

[56]Basharov, R., E. N. Gavrilovskaya, O. A. Malkin, et al., "Investigations of Cathode Spots of an Impulsive Discharge between Parallel Electrodes," Zhurnal tekhnicheskoy fiziki, Vol. 35, No. 10, 1965, p. 1855-1859.

[57]Froome, K. D., "The Behaviour of the Cathode Spot on an Undisturbed Mercury Surface," Proceedings of the Physical Society of London, Vol. B62, 1949, p. 805-812.

[58]Froome, K. D., "The Behaviour of the Cathode Spot on an Undisturbed Liquid Surface of Low Function," Proceedings of the Physical Society of London, Vol. B63, 1950, p. 377-382.

[59]Froome, K. D., "Current Densities of Free Moving Cathode Spots on Mercury," British Journal of Aplied Physics, Vol. 4, 1953, p. 91-93.

[60]Grakov, V., and V. Hermoch, "Formation of Cathode Spots on Electrodes of Short-Time High Intensity Electric Discharge," Czechoskovak Journal of Physics, Vol. B13, No. 7, 1963, p. 509-517.

[61]Zykova, N. M., "Investigation of Dynamics of Development of Cathode and Anode Spots of an Electric Arc," dissertation for a candidate of physical-mathematical sciences, Institute of Physics of the Siberian branch of the Soviet Academy of Sciences, Krasnoyarsk, 1968.

[62]Maycock, J., and J. Raite, "Calculated Characteristics of an MHD Generator Channel," translation of an unspecified English paper.

[63]Baryshev, Yu. V., V. O. German, Yu. P. Kukota, et al., "Recording Arc Discharges on a Cathode of a Model of an MHD Facility," Zhurnal prikladnoy mekhaniki i tekhnicheskoy fiziki, No. 2, 1968, p. 54-55.

[64]Rosa, R. J., Magnetohydrodynamic Energy Conversion, New York, McGraw Hill Book Company, 1968.

[65]Zalkind, V. I., V. V. Kirillov, Yu. A. Larionov, et al., "Microarc Operating Mode of Electrodes of an MHD Generator," Zhurnal prikladnoy mekhaniki i tekhnicheskoy fiziki, No. 1, 1970, p. 130-134.

[66]Zalkind, V. I., and A. P. Markina, "Experimental Investigation of Cathode Spots on Metal Electrodes Coated with a Film of Alkali Metal Compounds Exposed to a Low-Temperature Plasma Flow," Preprint of Institute of High Temperatures (IVTAN), No. 3-014, Moscow, 1977, 10 p.

[67]Kurakma, T. S., "Investigation of Processes in Near-Electrode Regions of a Discharge and Their Influence on Electrode Destruction in Combustion Products Flow," dissertation of a candidate of technical sciences, Institute of High Temperatures (IVTAN), 1978.

[68]German, V. O., M. P. Zektser, G. A. Lyubimov, et al., "Investigation of Discharge between Cold Insulation Wall Modules,"

Teplofizika vysokikh temperatur, Vol. 10, No. 4, 1972, p. 874-881.

[69]German, V. O., M. P. Zektser, G. A. Lyubimov, et al., "Experimental Investigation of a Discharge between Modules of a Cold Insulating Wall," Fifth International Conference on Magnetohydrodynamic Electrical Power Generation, Munich, April 1971, Vienna, IAEA, 1979, Vol. 1, 1971, p. 307-320.

[70]Lyubimov, G. A., "Breakdown Condition of a Near-Electrode Layer in a Flow of Ionized Gas," Zhurnal prikladnoy mekhaniki i tekhnicheskoy fiziki, Vol. 11, No. 3, 1973, p. 16-23.

[71]Oliver, D. A., and C. D. Maxwell, "Interaction of Magnetohydrodynamic Plasma with Boundaries," AIAA Paper, No. 108, 1977, p. 13-25.

[72]Oliver, D. A., "Interelectrode Breakdown on Electrode Walls Parallel and Inclined to the Magnetic Field," Proceedings of the Sixth International Conference on the 6th International Conference on MHD Electrical Power Generation, 1975, Washington, D.C., Vol. 1, 1975, p. 329-343.

[73]Burenkov, D. K., Yu. L. Dolinsky, and V. I. Zalkind, "Study of the Maximum Hall Voltage and Interelectrode Breakdown in the Channel of an Open-Cycle MHD Generator," Proceedings of the 16th Symposium on Engineering Aspects of Magnetohydrodynamics, Pittsburgh, Pa., May 1977, University of Pittsburgh, 1977, p. vi.6.19-vl.6.34.

[74]Burenkov, D. K., V. I. Zalkind, V. V. Kirillov, et al., "Investigation of Electrical Characteristics of the Boundary Layer on Metal Surfaces in Open-Cycle MHD Generator Channels," Teplofizika vysokikh temperatur, Vol. 14, No. 2, 1976, p. 359-364.

[75]Kovbasyuk, V. I., N. N. Baranov, A. D. Iserov, et al., "Apearance of Interelectrode Arcs on Electrical Fluctuations in an MHD Channel," Teplifizika vysokikh temperatur, Vol. 15, No. 6, 1977, p. 1294-1302.

[76]Langmuir, I., "The Interaction of Electron and Positive Ion Space Charge in Cathode Sheet," Physical Review, Vol. 33, No. 6, 1929, p. 954-962.

77McKeown, S. S., "The Cathode Drop in an Electric Arc," Physical Review, Vol. 34, 1929, p. 611-614.

[78]Lee, T. H., and A. Greenwood, "Theory for the Cathode Mechanism in Metal Vapor Arc," Journal of Aplied Physics, Vol. 32, No. 5, 1961, p. 916-923.

[79]Beylis, I. L., "Theoretical Investigation of Cathode Processes in a Vacuum Arc Discharge," dissertation for a candidate of physical-mathematical sciences, Institute of High Temperatures (IVTAN), Moscow, 1973.

[80]Beylis, I. I., V. V. Kantsel′, and V. I. Rakhovskiy, "Explosive Model of a Rapidly Displaced Cathode Spot," Electrical Contacts

(Elektricheskiye kontakty), Moscow, Publishing House Nauka, 1975, p. 14-16.

[81]Mitterauer, J., "Dynamische Fieldemission (DF - Emission): Eine neue Modelvorstellung des Kathodenflecks an kalten Kathoden. - Acta Phys. Austr., 1973, Bd. 37, p 175-192.

[82]Mesyats, G. A., "Explosive Processes on a Cathode in a Gas Discharge," Zhurnal tekhnicheskoy fiziki, pis'ma, Vol. 1, No. 19, 1975, p. 885-888.

[83]Ecker, G., "Theory of the Vacuum Arc, The Stationary Cathode Spot," General Electric Report, No. 71-C-195, July 1971, 42 p.

[84]Ecker, G., "Present-Day Development of the Theory of Near-Electrode Regions of an Electric Arc," Teplifizika vysokikh temperatur, Vol. 11, No. 4, 1973, p. 865-870.

[85]Ecker, G., "Theory of Cathode Phenomena, Experimental Investigation of Plasmotrons" (Eksperimental'nyye issledovaniya plasmotronov), Zhukovskiy, M. F. (ed.), Novosibirsk, Publishing House Nauka, 1977, p. 155-207.

[86]Beylis, I. I., and V. I. Rakhovskiy, "Theory of Cathode Mechanism of an Arc Discharge," Teplifizika vysokikh temperatur, Vol. 7, No. 4, 1969, p. 620-625.

[87]Beylis, I. I., G. A. Lyubimov, and V. I. Rakhovskiy, "Diffusion Model of the Near-Cathode Region of a Concentrated Arc Discharge," AN SSSR, Doklady, Vol. 203, No. 1, 1972, p. 71-74.

[88]Beylis, I. I., "Theoretical Investigation of Parameters of a Cathode Spot of a Vacuum Arc Discharge," Zhurnal tekhnicheskoy fiziki, Vol. 44, No. 2, 1974, p. 400-419.

[89]Osadin, B. A., "Theory of Cathode Spot of a Concentrated Vacuum Arc," Zhurnal tekhnicheskoy fiziki, Vol. 37, No. 11, 1967, p. 2061-2966.

[90]Goloveyko, A. G., "Elementary and Thermophysical Processes on a Cathode During a Strong Impulsive Discharge," Inzhenerniy fizicheskiy zhurnal, Vol. 14, No. 3, 1968, p. 478-487.

[91]Kulyapin, V. M., "Elements of Qualitative Theory of Arc Discharge Cathode Processes," Zhurnal tekhnicheskoy fiziki, Vol. 41, No. 2, 1971, p. 381-386.

[92]Kozlov, N. P., and V. I. Khvesyuk, "Theory and Calculation of Near-Cathode Regions of Electric Arcs," Correspondence 1, Vol. 31, No. 10, 1971, p. 2135-2141; Correspondence 2, p. 2141-2150.

[93]Kozlov, N. P., and V. I. Khvesyuk, "Theory and Calculation of the Near-Cathode Regions of Electric Arcs," Problems of Low-Temperature Physics (Voprosy fiziki nizkotemperaturnoy plazmy), Minsk, Publishing House Nauka i tekhnika, 1970, p. 503-507.

[94]Hall, A. W., "Cathode Spot," Physical Review, Vol. 126, No. 5, 1962, p. 1603-1610.

[95]Lyubimov, G. A., "Models Describing the Near-Cathode Region of a Concentrated Arc Discharge," Zhurnal tekhnicheskoy fiziki, Vol. 43, No. 4, 1973, p. 888-893.

[96]Beylis, I. I., and G. A. Lyubimov, "Parameters of the Near-Cathode Region of an Arc Discharge," Teplofizika vysokikh temperatur, Vol. 13, No. 6, 1975, p. 1137-1145.

[97]Ecker, G., "Theoretical Investigation of a Cathode Spot in a Vacuum," Teplofizika vysokikh temperatur, Vol. 16, No. 6, 1978, p. 1297-1304.

[98]Rukhovskiy, V. I., "On the Problem Concerning Aplication of the Method of the Existence Diagram to Estimating Parameters of a Cathode Spot," Zhurnal tekhnicheskoy fiziki, Vol. 16, No. 6, 1978, p. 1305-1306.

[99]Zektser, M. P., "Influence of Joule Heat Release in a Metal Cathode on the Temperature of a Cathode Arc Spot," Zhurnal tekhnicheskoy fiziki, Vol. 15, No. 1, 1977, p. 218-221.

[100]Kurakina, T. S., V. S. Potoskin, and V. I. Rakhovskiy, "Investigation of the Bridge Stage During Current Commutation in Vacuum," Zhurnal tekhnicheskoy fiziki, Vol. 37, No. 2, 1967, p. 330-334.

[101]Potokin, V. S., and V. I. Rakhovskiy, "Investigation of the Bridge Stage During Current Commutation in Vacuum," Zhurnal tecknicheskoy fiziki, Vol. 37, No. 2, 196, p. 330-334.

[102]Beylis, I. I., "On the Problem Concerning an Analytical Aproximation of the Nordheim Function," Teplofizika vysokikh temperatur, Vol. 9, No. 1, 1971, p. 184-186.

[103]Shcherbinin, P. P., "Electric Field in the Near-Electrode Plasma Layer," Zhurnal tekhnicheskoy fiziki, Vol. 42, No. 12, 1972, p. 2490-2500.

[104]Ecker, G., "Unified Analysis of the Metal Vapor Arc," Zeitschrift fur Naturforsh., Bd. 28a, 1973, p. 417-428.

[105]Baksht, F. G., B. Ya. Moyzhes, and V. A. Nemchinskiy, "Calculation of the Near-Electrode Layer in a Low-Temperature Plasma," Teplofizika vysokikh temperatur, Vol. 47, No. 2, 1977, p. 297-303.

[106]Lyubimov, G. A., "On Dynamics of Cathode Jets," Teplofizika vysokikh temperatur, Vol. 47, No. 2, 1977, p. 297-303.

[107]Zektser, M. P., and G. A. Lyubimov, "Theoretical Investigation of High Velocity Plasma Fluxes from a Cathode Spot of a Vacuum Arc," Zhurnal tekhnicheskoy fiziki, Vol. 49, No. 1, 1979, p. 3-11.

[108]Deshman, S. D., "Scientific Foundation of Vacuum Technology," translated from English.

[109]Kokoin, I. K., "Tables of Physical Constants" (Tablitsy

fizicheskikh velichin), Moscow, Publishing House "Atomizdat," 1976, 1006 p.

[110]Selikatova, S. M., and I. A. Lukatskaya, "Initial Stage of a Vacuum Arc Switch," Zhurnal tekhnicheskoy fiziki, Vol. 42, No. 7, 1972, p. 1508-1515.

[111]Belkin, G. S., and M. E. Danilov, "Measurement of Energy into Electrodes during Burning of an Arc in a Vacuum," Zhurnal tekhnicheskoy fiziki, Vol. 11, No. 3, 1973, p. 598-601.

[112]Daalder, J. E., "Energy Dissipation in the Cathode of Vacuum Arc," Journal of Physics D, Aplied Physics, Vol. 12, 1979, p. 761-763.

[113]Zektser, M. P., and G. A. Lyubimov, "Electrode Heating by the Cathode Spot of a Vacuum Arc," Journal of Physics D, Aplied Physics, Vol. 12, 1979, p. 761-763.

[114]Lapshin, V. A., "Investigation of Cathode Potential Drop in Vacuum Arcs with Cathodes of Alloys and Materials with Heterogeneous Microstructure," Theoretical Physics, Plasma Physics (Teoreticheskaya fizika, Fizika plasmy), Minsk, Nauka i tekhnika, 1975, p. 59-60.

[115]Beylis, I. I., and V. I. Rakhovskiy, "Taking into Account Heat Losses During Heating of an Electrode by a Propagating Arc Discharge," Inzhenerniy fizicheskiy zhurnal, Vol. 19, No. 4, 1970, p. 678-681.

[116]Rykalin, N. N., <u>Calculation of Thermal Processes During Welding (Raschesty teplovykh protessov pri svarke)</u>, Moscow, Publishing House Mashgiz, 1951.

[117]Beylis, I. I., G. V. Levchenko, V. S. Potokin, et al., "Heating of a Body Being Affected by a Moving, High-Intensity, Concentrated Heat Source," Fizika i khimiya obrabotki materialov, No. 3, 1967, p. 19-24.

[118]Chopra, K. L., "Electric Phenomena in Thin Films," translated from English.

[119]Beylis, I. I., and G. A. Lyubimov, "Determination of Current Density in a Cathode Spot of an Arc Discharge by Means of the Method of Autographs," Zhurnal tekhnicheskoy fiziki, Vol. 46, No. 10, 1976, p. 2181-2184.

[120]Nicols, L. D., and M. A. Mantenieks, "Analytical and Experimental Studies of MHD Generator Cathodes Emitting in Spot Mode," paper of the American Society of Mechanical Engineers, NWA/HT-51, 1969, p. 1-12.

[121]Laricheva, M. A., and I. L. Mostinskiy, "Investigation of Potassium Hydroxide Vapor Condensation from Seeded Combustion Products Flow," Teplofizika vysokikh temperatur, Vol. 8, No. 3, 1972, p. 660-663.

[122]Mostinskiy, I. L., "Ionizing Seed in Open-Cycle MHD Generators Operating on Natural Gas," dissertation for doctor of technical sciences, Institute of High Temperatures (IVTAN), 1972.

[123]Zakharko, Yu. A., I. L. Mostinskiy, and V. D. Cherkass, "Importance of Decomposition of Potassium Carbonate During Evaporation of Its Drops in a High-Temperature Gas Flow," Teplofizika vysokikh temperatur, Vol. 13, No. 2, 1975, p. 386-391.

[124]Massey, H.S.W., and E.H.S. Burhop, "Electronic and Ionic Impact Phenomena," Oxford University Press, N.J., 1952.

[125]Sena, L. A., Collision of Electrons and Ions with Gas Atoms (Stolknoveniye elektronov i ionov s atomami gasa), Moscow-Leningrad, Publishing House Gostekhteorizdat, 1948.

[126]Smirnov, B. M., Introduction to Plasma Physics (Vvedeniye v fiziku plasmy), Moscow, Publishing House Nauka, 1975.

[127]McDaniel, I., "Collision Processes in Ionized Gases," translated from English.

[128]Polyanskiy, V. A., "Diffusion and Conductivity in Partially Ionized Multitemperature Gas Mixture," Zhurnal prikladnoy mekhaniki i tekhnicheskoy fiziki, No. 5, 1964, p. 11-17.

[129]Vasyutkin, A. M., "Influence of Easily Ionized Seed on Parameters of a Near-Cathode Region of an Arc Discharge," Zhurnal tekhnicheskoy fiziki, Vol. 48, No. 3, 1978, p. 494-501.

[130]Koester, J. K., M. Sajben, and E. E. Zukoski, "Investigation of Thermionic Emission from Electrodes into a Seeded Dense Plasma," Direct Thermal Energy Conversion into Electric Energy and Fuel Elements (Pryamoye preobrazovaniye teplovoy energii v elektricheskuyu i toplivnyeye elementy), Review of Non-Soviet Literature, Informatsionnyy byulleten', Moscow, Publishing House VINITI, No. 10(99), 1970, p. 14-21.

[131]Psarouthakis, J., "Variation of the Work Function of Tungsten in Barium and Cesium or Strontium and Cesium Vapors," Direct Thermal Energy Conversion into Electric Energy and Fuel Elements (Pryamoye preobrazovaniye teplovoy energii v elektricheskuyu i toplivnyeye elementy), Review of Non-Soviet Literature, Informatsionnyy byulleten', Moscow, Publishing House VINITI, No. 9(98), 1970, p. 101-111.

[132]Kirillin, V. A., A. E. Sheyndlin, and E. E. Shpil'rayn, Thermodynamics of Solutions (Termodinamika rastvorov), Moscow, Publishing House Energiya, 1980.

PROGRESS IN ASTRONAUTICS AND AERONAUTICS SERIES VOLUMES

VOLUME TITLE/EDITORS

***1. Solid Propellant Rocket Research** (1960)
Martin Summerfield
Princeton University

***2. Liquid Rockets and Propellants** (1960)
Loren E. Bollinger
The Ohio State University
Martin Goldsmith
The Rand Corporation
Alexis W. Lemmon Jr.
Battelle Memorial Institute

***3. Energy Conversion for Space Power** (1961)
Nathan W. Snyder
Institute for Defense Analyses

***4. Space Power Systems** (1961)
Nathan W. Snyder
Institute for Defense Analyses

***5. Electrostatic Propulsion** (1961)
David B. Langmuir
Space Technology Laboratories, Inc.
Ernst Stuhlinger
NASA George C. Marshall Space Flight Center
J.M. Sellen Jr.
Space Technology Laboratories, Inc.

***6. Detonation and Two-Phase Flow** (1962)
S.S. Penner
California Institute of Technology
F.A. Williams
Harvard University

***7. Hypersonic Flow Research** (1962)
Frederick R. Riddell
AVCO Corporation

***8. Guidance and Control** (1962)
Robert E. Roberson
Consultant
James S. Farrior
Lockheed Missiles and Space Company

***9. Electric Propulsion Development** (1963)
Ernst Stuhlinger
NASA George C. Marshall Space Flight Center

***10. Technology of Lunar Exploration** (1963)
Clifford I. Cummings and Harold R. Lawrence
Jet Propulsion Laboratory

***11. Power Systems for Space Flight** (1963)
Morris A. Zipkin and Russell N. Edwards
General Electric Company

***12. Ionization in High-Temperature Gases** (1963)
Kurt E. Shuler, Editor
National Bureau of Standards
John B. Fenn, Associate Editor
Princeton University

***13. Guidance and Control—II** (1964)
Robert C. Langford
General Precision Inc.
Charles J. Mundo
Institute of Naval Studies

***14. Celestial Mechanics and Astrodynamics** (1964)
Victor G. Szebehely
Yale University Observatory

***15. Heterogeneous Combustion** (1964)
Hans G. Wolfhard
Institute for Defense Analyses
Irvin Glassman
Princeton University
Leon Green Jr.
Air Force Systems Command

***16. Space Power Systems Engineering** (1966)
George C. Szego
Institute for Defense Analyses
J. Edward Taylor
TRW Inc.

***17. Methods in Astrodynamics and Celestial Mechanics** (1966)
Raynor L. Duncombe
U.S. Naval Observatory
Victor G. Szebehely
Yale University Observatory

***18. Thermophysics and Temperature Control of Spacecraft and Entry Vehicles** (1966)
Gerhard B. Heller
NASA George C. Marshall Space Flight Center

***19. Communication Satellite Systems Technology** (1966)
Richard B. Marsten
Radio Corporation of America

*Out of print.

***20. Thermophysics of Spacecraft and Planetary Bodies: Radiation Properties of Solids and the Electromagnetic Radiation Environment in Space** (1967)
Gerhard B. Heller
NASA George C. Marshall Space Flight Center

***21. Thermal Design Principles of Spacecraft and Entry Bodies** (1969)
Jerry T. Bevans
TRW Systems

***22. Stratospheric Circulation** (1969)
Willis L. Webb
Atmospheric Sciences Laboratory, White Sands, and University of Texas at El Paso

***23. Thermophysics: Applications to Thermal Design of Spacecraft** (1970)
Jerry T. Bevans
TRW Systems

24. Heat Transfer and Spacecraft Thermal Control (1971)
John W. Lucas
Jet Propulsion Laboratory

25. Communication Satellites for the 70's: Technology (1971)
Nathaniel E. Feldman
The Rand Corporation
Charles M. Kelly
The Aerospace Corporation

26. Communication Satellites for the 70's: Systems (1971)
Nathaniel E. Feldman
The Rand Corporation
Charles M. Kelly
The Aerospace Corporation

27. Thermospheric Circulation (1972)
Willis L. Webb
Atmospheric Sciences Laboratory, White Sands, and University of Texas at El Paso

28. Thermal Characteristics of the Moon (1972)
John W. Lucas
Jet Propulsion Laboratory

29. Fundamentals of Spacecraft Thermal Design (1972)
John W. Lucas
Jet Propulsion Laboratory

30. Solar Activity Observations and Predictions (1972)
Patrick S. McIntosh and Murray Dryer
Environmental Research Laboratories, National Oceanic and Atmospheric Administration

31. Thermal Control and Radiation (1973)
Chang-Lin Tien
University of California at Berkeley

32. Communications Satellite Systems (1974)
P.L. Bargellini
COMSAT Laboratories

33. Communications Satellite Technology (1974)
P.L. Bargellini
COMSAT Laboratories

34. Instrumentation for Airbreathing Propulsion (1974)
Allen E. Fuhs
Naval Postgraduate School
Marshall Kingery
Arnold Engineering Development Center

35. Thermophysics and Spacecraft Thermal Control (1974)
Robert G. Hering
University of Iowa

36. Thermal Pollution Analysis (1975)
Joseph A. Schetz
Virginia Polytechnic Institute

37. Aeroacoustics: Jet and Combustion Noise; Duct Acoustics (1975)
Henry T. Nagamatsu, Editor
General Electric Research and Development Center
Jack V. O'Keefe, Associate Editor
The Boeing Company
Ira R. Schwartz, Associate Editor
NASA Ames Research Center

38. Aeroacoustics: Fan, STOL, and Boundary Layer Noise; Sonic Boom; Aeroacoustic Instrumentation (1975)
Henry T. Nagamatsu, Editor
General Electric Research and Development Center
Jack V. O'Keefe, Associate Editor
The Boeing Company
Ira R. Schwartz, Associate Editor
NASA Ames Research Center

39. Heat Transfer with Thermal Control Applications (1975)
M. Michael Yovanovich
University of Waterloo

40. Aerodynamics of Base Combustion (1976)
S.N.B. Murthy, Editor
Purdue University
J.R. Osborn, Associate Editor
Purdue University
A.W. Barrows and J.R. Ward, Associate Editors
Ballistics Research Laboratories

41. Communications Satellite Developments: Systems (1976)
Gilbert E. LaVean
Defense Communications Agency
William G. Schmidt
CML Satellite Corporation

42. Communications Satellite Developments: Technology (1976)
William G. Schmidt
CML Satellite Corporation
Gilbert E. LaVean
Defense Communications Agency

43. Aeroacoustics: Jet Noise, Combustion and Core Engine Noise (1976)
Ira R. Schwartz, Editor
NASA Ames Research Center
Henry T. Nagamatsu, Associate Editor
General Electric Research and Development Center
Warren C. Strahle, Associate Editor
Georgia Institute of Technology

44. Aeroacoustics: Fan Noise and Control; Duct Acoustics; Rotor Noise (1976)
Ira R. Schwartz, Editor
NASA Ames Research Center
Henry T. Nagamatsu, Associate Editor
General Electric Research and Development Center
Warren C. Strahle, Associate Editor
Georgia Institute of Technology

45. Aeroacoustics: STOL Noise; Airframe and Airfoil Noise (1976)
Ira R. Schwartz, Editor
NASA Ames Research Center
Henry T. Nagamatsu, Associate Editor
General Electric Research and Development Center
Warren C. Strahle, Associate Editor
Georgia Institute of Technology

46. Aeroacoustics: Acoustic Wave Propagation; Aircraft Noise Prediction; Aeroacoustic Instrumentation (1976)
Ira R. Schwartz, Editor
NASA Ames Research Center
Henry T. Nagamatsu, Associate Editor
General Electric Research and Development Center
Warren C. Strahle, Associate Editor
Georgia Institute of Technology

47. Spacecraft Charging by Magnetospheric Plasmas (1976)
Alan Rosen
TRW Inc.

48. Scientific Investigations on the Skylab Satellite (1976)
Marion I. Kent and Ernst Stuhlinger
NASA George C. Marshall Space Flight Center
Shi-Tsan Wu
The University of Alabama

49. Radiative Transfer and Thermal Control (1976)
Allie M. Smith
ARO Inc.

50. Exploration of the Outer Solar System (1976)
Eugene W. Greenstadt
TRW Inc.
Murray Dryer
National Oceanic and Atmospheric Administration
Devrie S. Intriligator
University of Southern California

51. Rarefied Gas Dynamics, Parts I and II (two volumes) (1977)
J. Leith Potter
ARO Inc.

52. Materials Sciences in Space with Application to Space Processing (1977)
Leo Steg
General Electric Company

53. Experimental Diagnostics in Gas Phase Combustion Systems (1977)
Ben T. Zinn, Editor
Georgia Institute of Technology
Craig T. Bowman, Associate Editor
Stanford University
Daniel L. Hartley, Associate Editor
Sandia Laboratories
Edward W. Price, Associate Editor
Georgia Institute of Technology
James G. Skifstad, Associate Editor
Purdue University

54. Satellite Communications: Future Systems (1977)
David Jarett
TRW Inc.

55. Satellite Communications: Advanced Technologies (1977)
David Jarett
TRW Inc.

56. Thermophysics of Spacecraft and Outer Planet Entry Probes (1977)
Allie M. Smith
ARO Inc.

57. Space-Based Manufacturing from Nonterrestrial Materials (1977)
Gerard K. O'Neill, Editor
Princeton University
Brian O'Leary, Assistant Editor
Princeton University

58. Turbulent Combustion (1978)
Lawrence A. Kennedy
State University of New York at Buffalo

59. Aerodynamic Heating and Thermal Protection Systems (1978)
Leroy S. Fletcher
University of Virginia

60. Heat Transfer and Thermal Control Systems (1978)
Leroy S. Fletcher
University of Virginia

61. Radiation Energy Conversion in Space (1978)
Kenneth W. Billman
NASA Ames Research Center

62. Alternative Hydrocarbon Fuels: Combustion and Chemical Kinetics (1978)
Craig T. Bowman
Stanford University
Jorgen Birkeland
Department of Energy

63. Experimental Diagnostics in Combustion of Solids (1978)
Thomas L. Boggs
Naval Weapons Center
Ben T. Zinn
Georgia Institute of Technology

64. Outer Planet Entry Heating and Thermal Protection (1979)
Raymond Viskanta
Purdue University

65. Thermophysics and Thermal Control (1979)
Raymond Viskanta
Purdue University

66. Interior Ballistics of Guns (1979)
Herman Krier
University of Illinois at Urbana-Champaign
Martin Summerfield
New York University

67. Remote Sensing of Earth from Space: Role of "Smart Sensors" (1979)
Roger A. Breckenridge
NASA Langley Research Center

68. Injection and Mixing in Turbulent Flow (1980)
Joseph A. Schetz
Virginia Polytechnic Institute and State University

69. Entry Heating and Thermal Protection (1980)
Walter B. Olstad
NASA Headquarters

70. Heat Transfer, Thermal Control, and Heat Pipes (1980)
Walter B. Olstad
NASA Headquarters

71. Space Systems and Their Interactions with Earth's Space Environment (1980)
Henry B. Garrett and Charles P. Pike
Hanscom Air Force Base

72. Viscous Flow Drag Reduction (1980)
Gary R. Hough
Vought Advanced Technology Center

73. Combustion Experiments in a Zero-Gravity Laboratory (1981)
Thomas H. Cochran
NASA Lewis Research Center

74. Rarefied Gas Dynamics, Parts I and II (two volumes) (1981)
Sam S. Fisher
University of Virginia at Charlottesville

75. Gasdynamics of Detonations and Explosions (1981)
J.R. Bowen
University of Wisconsin at Madison
N. Manson
Université de Poitiers
A.K. Oppenheim
University of California at Berkeley
R.I. Soloukhin
Institute of Heat and Mass Transfer, BSSR Academy of Sciences

76. Combustion in Reactive Systems (1981)
J.R. Bowen
University of Wisconsin at Madison
N. Manson
Université de Poitiers
A.K. Oppenheim
University of California at Berkeley
R.I. Soloukhin
Institute of Heat and Mass Transfer, BSSR Academy of Sciences

77. Aerothermodynamics and Planetary Entry (1981)
A.L. Crosbie
University of Missouri-Rolla

78. Heat Transfer and Thermal Control (1981)
A.L. Crosbie
University of Missouri-Rolla

79. Electric Propulsion and Its Applications to Space Missions (1981)
Robert C. Finke
NASA Lewis Research Center

80. Aero-Optical Phenomena (1982)
Keith G. Gilbert and Leonard J. Otten
Air Force Weapons Laboratory

81. Transonic Aerodynamics (1982)
David Nixon
Nielsen Engineering & Research, Inc.

82. Thermophysics of Atmospheric Entry (1982)
T.E. Horton
The University of Mississippi

83. Spacecraft Radiative Transfer and Temperature Control (1982)
T.E. Horton
The University of Mississippi

84. Liquid-Metal Flows and Magnetohydrodynamics (1983)
H. Branover
Ben-Gurion University of the Negev
P.S. Lykoudis
Purdue University
A. Yakhot
Ben-Gurion University of the Negev

85. Entry Vehicle Heating and Thermal Protection Systems: Space Shuttle, Solar Starprobe, Jupiter Galileo Probe (1983)
Paul E. Bauer
McDonnell Douglas Astronautics Company
Howard E. Collicott
The Boeing Company

86. Spacecraft Thermal Control, Design, and Operation (1983)
Howard E. Collicott
The Boeing Company
Paul E. Bauer
McDonnell Douglas Astronautics Company

87. Shock Waves, Explosions, and Detonations (1983)
J.R. Bowen
University of Washington
N. Manson
Université de Poitiers
A.K. Oppenheim
University of California at Berkeley
R.I. Soloukhin
Institute of Heat and Mass Transfer, BSSR Academy of Sciences

88. Flames, Lasers, and Reactive Systems (1983)
J.R. Bowen
University of Washington
N. Manson
Université de Poitiers
A.K. Oppenheim
University of California at Berkeley
R.I. Soloukhin
Institute of Heat and Mass Transfer, BSSR Academy of Sciences

89. Orbit-Raising and Maneuvering Propulsion: Research Status and Needs (1984)
Leonard H. Caveny
Air Force Office of Scientific Research

90. Fundamentals of Solid-Propellant Combustion (1984)
Kenneth K. Kuo
The Pennsylvania State University
Martin Summerfield
Princeton Combustion Research Laboratories, Inc.

91. Spacecraft Contamination: Sources and Prevention (1984)
J.A. Roux
The University of Mississippi
T.D. McCay
NASA Marshall Space Flight Center

92. Combustion Diagnostics by Nonintrusive Methods (1984)
T.D. McCay
NASA Marshall Space Flight Center
J.A. Roux
The University of Mississippi

93. The INTELSAT Global Satellite System (1984)
Joel Alper
COMSAT Corporation
Joseph Pelton
INTELSAT

94. Dynamics of Shock Waves, Explosions, and Detonations (1984)
J.R. Bowen
University of Washington
N. Manson
Universite de Poitiers
A.K. Oppenheim
University of California
R.I. Soloukhin
Institute of Heat and Mass Transfer, BSSR Academy of Sciences

95. Dynamics of Flames and Reactive Systems (1984)
J.R. Bowen
University of Washington
N. Manson
Universite de Poitiers
A.K. Oppenheim
University of California
R.I. Soloukhin
Institute of Heat and Mass Transfer, BSSR Academy of Sciences

96. Thermal Design of Aeroassisted Orbital Transfer Vehicles (1985)
H.F. Nelson
University of Missouri-Rolla

97. Monitoring Earth's Ocean, Land, and Atmosphere from Space – Sensors, Systems, and Applications (1985)
Abraham Schnapf
Aerospace Systems Engineering

98. Thrust and Drag: Its Prediction and Verification (1985)
Eugene E. Covert
Massachusetts Institute of Technology
C.R. James
Vought Corporation (Ret.)
William F. Kimzey
Sverdrup Technology AEDC Group
George K. Richey
U.S. Air Force
Eugene C. Rooney
U.S. Navy Department of Defense

99. Space Stations and Space Platforms — Concepts, Design, Infrastructure, and Uses (1985)
Ivan Bekey
Daniel Herman
NASA Headquarters

100. Single- and Multi-Phase Flows in an Electromagnetic Field
Energy, Metallurgical, and Solar Applications (1985)
Herman Branover
Ben-Gurion University of the Negev
Paul S. Lykoudis
Purdue University
Michael Mond
Ben-Gurion University of the Negev

101. MHD Energy Conversion: Physiotechnical Problems (1986)
V.A. Kirillin
A.E. Sheyndlin
Soviet Academy of Sciences

102. Numerical Methods for Engine-Airframe Integration (1986)
S.N.B. Murthy
Purdue University
Gerald C. Paynter
Boeing Airplane Company

103. Thermophysics Aspects of Re-Entry Flows (1986)
James N. Moss
NASA Langley Research Center
Carl D. Scott
NASA Johnson Space Center

(Other Volumes are planned.)